IN QUEST OF
The Stars and Galaxies

Jones and Bartlett Titles in Physical Science

Astronomy Activity and Laboratory Manual
Alan W. Hirshfeld

Climatology
Robert V. Rohli & Anthony J. Vega

Environmental Oceanography: Topics and Analysis
Daniel C. Abel & Robert L. McConnell

Environmental Science, Eighth Edition
Daniel D. Chiras

Environmental Science: Systems and Solutions, Fourth Edition
Michael L. McKinney & Robert M. Schoch

Essentials of Geochemistry, Second Edition
John V. Walther

Igneous Petrology, Third Edition
Alexander McBirney

In Quest of the Solar System
Theo Koupelis

In Quest of the Universe, Sixth Edition
Theo Koupelis

Invitation to Oceanography, Fifth Edition
Paul R. Pinet

Invitation to Organic Chemistry
A. William Johnson

Organic Chemistry, Third Edition
Marye Anne Fox & James K. Whitesell

Outlooks: Readings for Environmental Literacy, Second Edition
Michael L. McKinney

Principles of Atmospheric Science
John E. Frederick

Principles of Environmental Chemistry, Second Edition
James E. Girard

IN QUEST OF
The Stars and Galaxies

Theo Koupelis
Edison State College

JONES AND BARTLETT PUBLISHERS
Sudbury, Massachusetts
BOSTON   TORONTO   LONDON   SINGAPORE

World Headquarters
Jones and Bartlett Publishers
40 Tall Pine Drive
Sudbury, MA 01776
978-443-5000
info@jbpub.com
www.jbpub.com

Jones and Bartlett Publishers Canada
6339 Ormindale Way
Mississauga, Ontario L5V 1J2
Canada

Jones and Bartlett Publishers International
Barb House, Barb Mews
London W6 7PA
United Kingdom

Jones and Bartlett's books and products are available through most bookstores and online booksellers. To contact Jones and Bartlett Publishers directly, call 800-832-0034, fax 978-443-8000, or visit our website, www.jbpub.com.

Substantial discounts on bulk quantities of Jones and Bartlett's publications are available to corporations, professional associations, and other qualified organizations. For details and specific discount information, contact the special sales department at Jones and Bartlett via the above contact information or send an email to specialsales@jbpub.com.

Production Credits
Chief Executive Officer: Clayton Jones
Chief Operating Officer: Don W. Jones, Jr.
President, Higher Education and Professional Publishing: Robert W. Holland, Jr.
V.P., Sales: William J. Kane
V.P., Design and Production: Anne Spencer
V.P., Manufacturing and Inventory Control: Therese Connell
Publisher, Higher Education: Cathleen Sether
Acquisitions Editor: Molly Steinbach
Senior Editorial Assistant: Jessica S. Acox
Editorial Assistant: Caroline Perry
Production Manager: Louis C. Bruno, Jr.
Senior Marketing Manager: Andrea DeFronzo
Text Design: Anne Spencer
Cover Design: Scott Moden
Illustrations: Precision Graphics, Carlisle Communications, Elizabeth Morales, and Graphic World, Inc.
Senior Photo Researcher: Christine Myaskovsky
Composition: Graphic World, Inc.
Cover and Title Page Image: Courtesy of NASA and The Hubble Heritage Team (AURA/JTJI)
Printing and Binding: Courier Kendallville
Cover Printing: Courier Kendallville

Library of Congress Cataloging-in-Publication Data
Koupelis, Theo.
  In quest of the stars and galaxies / Theo Koupelis. — 1st ed.
    p. cm.
  ISBN 978-0-7637-6630-6 (alk. paper)
  1. Stars—Textbooks. 2. Galaxies—Textbooks. I. Title.
QB801.7.K68 2010
523.8—dc22   2009027215
6048

Printed in the United States of America
14 13 12 11 10   10 9 8 7 6 5 4 3 2 1

About the cover: A "light echo" illuminates dust around supergiant star V838 Monocerotis.

To Alex and Billy, for shining so bright.

BRIEF CONTENTS

CONTENTS

CHAPTER 13

Interstellar Matter and Star Formation 367

CHAPTER 14

The Lives and Deaths of Low-Mass Stars 389

Appendixes

BOXED FEATURES

ADVANCING THE MODEL

HISTORICAL NOTE

TOOLS OF ASTRONOMY

ACTIVITIES

I recognize that many students using this text are not planning to become astronomers or perhaps even scientists in general. Yet, I strongly believe in the importance of every student (and every person, for that matter) understanding how science operates and how we know what we know about our world. Without an understanding of the basic concepts of science, we cannot understand the scientific and technological progress around us. We cannot contribute informed and useful opinions on how this progress is used. Above all, we cannot distinguish science from pseudoscience. A habit of skeptical thought and a reliance upon observation and testing to verify an idea are valuable not just for scientists but for everyone. Towards that end, this book was written as an introduction to astronomy, as well as an introduction to the general methods of science.

The book begins with the historical development of the science and technology of astronomy. I frequently stop our tour of the universe to discuss how new scientific theories gradually replaced older ones, building upon established information by adding new observations and new insights into how the universe works. We will begin our journey through the stars and galaxies in our universe by discussing the most familiar of stars—the Sun. We then describe the properties of stars and how to measure them. Our discussion of interstellar matter and the formation of stars leads into a narrative of the lives and deaths of low-mass and massive stars. As we continue to move out through the universe, we investigate first our own Milky Way Galaxy, followed by an account of the diversity of galaxies out there in the universe. Our journey through the universe concludes with a discussion of cosmology, and the search for intelligent extraterrestrial life.

Up-to-Date Content

In Quest of the Stars and Galaxies is a one-semester text drawn from the new, sixth edition of *In Quest of the Universe*. Even though the basic goals of the book have remained the same as in the fifth edition, this volume represents a thorough rewrite of the book that includes a rethinking of every argument and an update of all the information presented in all its forms (text, numbers, tables, and figures). New information can be found throughout the book, but the following are some of the major changes and updates:

- An expanded discussion of non-optical telescopes is included in Chapter 5, where space telescopes are now discussed in a *Tools of Astronomy* box.

- In Chapter 10, the discussion on the Kuiper belt and Oort cloud has been expanded, based on the discoveries of new trans-Neptunian objects. Also, the section on the "Importance of the Solar System Debris" includes an in-depth discussion of the connections between meteoritic material and life in our solar system.

- Based on observations by *TRACE, SOHO,* and *Hinode*, new insights into how the Sun's magnetic field controls the release of mass and energy into the solar atmosphere are presented in Chapter 11.

- The discussion on various methods of distance measurements has been expanded in Chapters 14 and 17.

- Based on observations by *HST, XMM-Newton,* and *Chandra*, new insights on the distribution of dark matter are described in Chapter 17, along with new information on early star formation based on *Spitzer* data.

- A clear and in-depth description of all the latest findings on cosmology is included in Chapter 18. These new results have signaled the dawn of the era of precision cosmology.

- A great number of new images are included from a variety of sources that cover the entire electromagnetic spectrum and are the result of many international collaborative efforts. I have tried to select pictures that are useful as well as exciting, illustrating many of the awesome events that take place throughout the cosmos.

Pedagogical Features

- *Data Pages* on individual planets, the Moon, and the Sun should help students in reviewing the material and focusing on important concepts; these data pages can be found in Appendix C. A data page for planetary nebulae is included in Chapter 14.

- *Marginal definitions* and *marginal notes* are located throughout the text. Definitions appear next to key terms highlighted in the text; these definitions are collected in the Glossary at the end of the book. Marginal notes often refer to relevant material in other chapters or provide additional facts about the topic under discussion.

- *Quotations* appear in the margin to give students an idea of what astronomers and physicists themselves thought about science or about their work. Other quotations show what people in other disciplines or other cultures think about the issues involved in astronomy.

- *Student misconceptions* are addressed by placing in the margin typical questions often asked by students. Each such question appears in the chapter section that provides the answers to the question. Students can use these questions as guidelines for review, making sure they understand the correct explanation for each question.

- *Historical Note* boxes discuss historical developments and biographies of astronomers.

- *Advancing the Model* boxes discuss concepts covered in the text in more depth and/or cove additional material.

- *Tools of Astronomy* boxes discuss mathematical and observational tools astronomers use to describe the universe.

- *Worked Examples* illustrate how astronomers determine some of the data we now accept as accurate descriptions of celestial objects.

- *Try One Yourself* problems involve a more mathematical understanding of astronomy.

- A built-in *Study Guide* at the end of each chapter provides three sets of exercises. New exercises have been added and some old ones have been revised. Answers to even-numbered exercises are included at the end of the book.

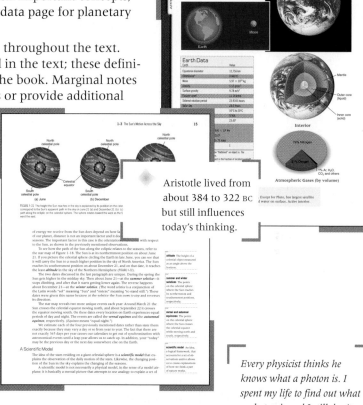

Aristotle lived from about 384 to 322 BC but still influences today's thinking.

Every physicist thinks he knows what a photon is. I spent my life to find out what a photon is and I still don't know it.
Albert Einstein

Is an arcminute or an arcsecond a unit for measuring time?

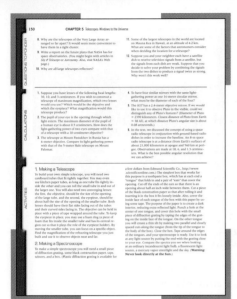

- *Recall Questions* are mostly multiple choice but also include other review questions.

- *Questions to Ponder* are generally more complex than the Recall Questions and sometimes ask students to relate material in the chapter to ideas from earlier chapters. Some questions ask students for personal evaluations of issues or to research topics in other resources.

- *Calculations* pose numerical problems for students to solve, based on examples in the chapter.

- *Activities* are also included in the Study Guides, featuring observational astronomy and drawing of scale models, among other things.

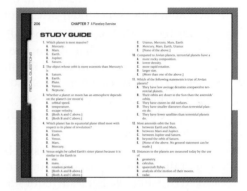

Ancillary Materials

To assist the instructor with teaching this course and provide students with the best in teaching aids, Jones and Bartlett Publishers has prepared a complete ancillary package. All the electronic materials are compatible with Windows and Macintosh systems. Additional information and review copies of any of the following items are available through your Jones and Bartlett publisher's representative.

For the Instructor

WebAssign® is the premier online independent teaching and learning environment, guiding over three million students through their academic careers since 1997. It was developed by teachers, for teachers, allowing you to:

- Create and distribute assignments using questions specific to this textbook
- Grade, record, and analyze student responses and performances instantly
- Offer practice exercises, quizzes, and homework
- Upload resources to share and communicate with your students seamlessly

For more detailed information and to sign up for free faculty access, please visit www.webassign.net. To find out how students can purchase access to WebAssign bundles with this textbook, please contact your Jones and Bartlett publisher's representative.

The *Instructor's Media CD-ROM* provides instructors who have adopted the text with the following traditional ancillaries:

- The **PowerPoint**® **Image Bank** provides a library of all the illustrations, photographs, and tables, to which Jones and Bartlett Publishers holds the copyright or has permission to reprint digitally, inserted into PowerPoint slides. Instructors can easily copy individual images into existing PowerPoint slides or add more images to the PowerPoint Lecture Outline Slides.

- The **PowerPoint Lecture Outline Slides**, prepared by Gregory J. Black of the University of Virginia, combine text and images into a presentation that is both educational and engaging. Using the Microsoft® PowerPoint program, instructors can modify and edit the slide show to fit individual presentation needs.

- Seventy **animations of key illustrations** in the text have been created, expanding the reach of the figures beyond the static page. Illustrations in the text accompanied by the icon ◯ indicate that a corresponding animation is on the Instructor's Media CD-ROM and on the Starlinks student companion Web site. Because so many concepts in astronomy involve motion, these animated illustrations are invaluable tools for helping students understand the concepts.

Online Instructor's Resources

- The **Instructor's Manual**, prepared by David J. Ennis of Ohio State University, contains chapter summaries and outlines, learning objectives, teaching suggestions and classroom activities, questions to test student understanding, and answers to all of the end-of-chapter questions that appear in the book.

- The **Test Bank**, prepared by George Spagna of Randolph-Macon College, features nearly 3000 questions in plain text files that can be easily integrated with your existing course management software.

For the Student

Increase your in-the-field observing success with the ***Starry Night® Pro version 6 DVD***, provided free with this text. Use the Events Finder to choose targets, or make your own Observing List for a specific night or for specific objects of interest. Print out three-view star-hopping charts customized to your equipment that effortlessly guide you to challenging objects. With extensive data sets, advanced telescope control, and comprehensive observational tools, Starry Night Pro is a powerful program designed for the serious astronomer. Transform your computer into a sophisticated virtual observatory!

Starlinks is the Jones and Bartlett Web site that accompanies *In Quest of the Stars and Galaxies*. This site, http:// physicalscience.jbpub.com/starlinks, hosts a wide array of additional resources and activities for independent study, review, and research.

- A **Study Guide** includes learning objectives and summaries for each chapter of the book, which you can download and use as note-taking guides.

- A set of original **Animations** that correspond to images in the book will help you to visualize the key concepts of motion in astronomy.

- An **Interactive Glossary**, on-screen **Flashcards**, and electronic **Crosswords** offer fun and easy ways for you to review the key terms.

- To help with reading comprehension and exam preparation, **Study Quizzes and Short Answer Quizzes** for each chapter provide you with short Web-based tests, the results of which can be submitted directly to your instructor.

- Additional **NetQuestions** are Web-based exercises that encourage independent research.

- **Web Explorations** recommends external Web sites that provide further information about topics covered in the text.

> **Quest Ahead to Starlinks:**
> **http://physicalscience.jbpub.com/starlinks**
>
> **Starlinks** is this book's online learning center. It features **eLearning**, which contains chapter quizzes and other tools designed to help you study for your class. You also can find **online exercises**, view numerous relevant **animations**, follow a guide to **useful astronomy sites** on the Internet, or even check the latest **astronomy news** updates.

Acknowledgments

Good teachers learn from their students. Attention to students' questions helps instructors to discover which concepts are most difficult and require special explanation. The same is true for textbook writers, and I thank my students over the years for their questions.

Although it is the author's name that appears on the cover of a textbook, all textbook writers know that "their" book is far from theirs alone. The sweat of many brows moistens the pages of every book. My thanks to the Jones and Bartlett team for all their excellent work and especially Louis Bruno for his infinite patience and dedication to this project.

Special thanks to Karl Kuhn for giving me the opportunity to work with him and learn from him. I hope future editions of this text will do justice to his original vision for it.

I owe a debt of gratitude to my teachers through the years, great teachers and friends—examples that have been very hard to follow. I thank them for everything they taught me and for providing me with a never-ending challenge.

My support and guidance during the entire process of revising/updating this text came from the two stars in my sky, Alex and Billy. Thank you for being there, for everything you are teaching me every day, and for making the world such a beautiful place.

Reviewers play a major role in the success of a book. I would like to thank all those reviewers who have spent time commenting on the revisions and page proofs for *In Quest of the Stars and Galaxies*.

Jill Bechtold, University of Arizona

Peter A. Becker, George Mason University

John H. Bieging, University of Arizona

Gregory J. Black, University of Virginia

Juan Cabanela, Minnesota State University, Moorhead

Michael W. Castelaz, University of North Carolina, Asheville and Pisgah Astronomical Research Institute

Peter Deutsch, Pennsylvania State University, Beaver Campus

David B. Green, Pepperdine University

David Goldwater, Community College of Southern Nevada

Bruce Gronich, University of Texas, El Paso

David Heiden, Fairfield High School

Michael Inglis, SUNY Suffolk

Douglas R. Ingram, Texas Christian University

Jonathan Keohane, Hampden Sydney College

Lauren Likkel, University of Wisconsin, Eau Claire

Joe Mast, Eastern Mennonite University

Mark McConnell, University of New Hampshire

Jim Mihal, Milwaukee Area Technical College

Scott Miller, University of Louisville

Slawomir Piatek, New Jersey Institute of Technology

Pamela Peterson, Anoka-Ramsey Community College, Cambridge Campus

Cameron Reed, Alma College

Sheldon Schafer, Bradley University

George Spagna, Randolph-Macon College

Walther Spjeldvik, Weber State University

Martin Steinbaum, Iona College

Rosemary Walling, University of Alaska Southeast

Marguerite Zarrillo, University of Massachusetts, Dartmouth

I invite instructors using *In Quest of the Stars and Galaxies* to contact me directly with comments and suggestions. It is you who know best what should be changed in future editions. For any questions and suggestions about the ancillary materials, please contact Jones and Bartlett Publishers at info@jbpub.com

Theo Koupelis
Edison State College
tkoupelis@edison.edu

The Quest Ahead

ON THE NIGHT OF MARCH 23, 1993, AMATEUR ASTRONOMER David Levy photographed part of the sky near the planet Jupiter. His friends, fellow astronomers Carolyn Shoemaker and her husband Eugene Shoemaker, spotted something unusual in the picture: a comet that had broken up into about 20 pieces. The comet became known officially as Comet Shoemaker-Levy 9 (the ninth comet discovered by these three sky-watchers) and unofficially as the "string of pearls" comet.

When astronomers announced the news of the comet on June 1, 1993, they had traced its path closely enough to tell that it had come under the influence of Jupiter's powerful gravitational field. They had deduced that the comet was pulled apart by Jupiter's gravity in July of 1992. They predicted that the comet would crash into Jupiter on or about July 25, 1994.

By October 1993, astronomers had refined their predictions. The comet was then stretched out over a distance of 3 million kilometers (1.8 million miles) in orbit around Jupiter. The first piece was predicted to hit Jupiter at 4:01 PM (Eastern daylight savings time in the United States) on July 16, and the last piece was predicted to hit at 4:00 AM on July 22, 1994; however, nobody knew whether the impacts would be big or small or even observable by telescopes on Earth.

Comet Shoemaker-Levy 9 was broken by Jupiter's gravitational forces into more than 20 fragments (bottom photo), all following the same path toward a collision with Jupiter. This photo of Jupiter (top photo) shows the impact spots left by some of the fragments.

All cross references to chapters, sections, figures, and tables pertain to the main text, *In Quest of the Universe, Sixth Edition. In Quest of the Solar System* contains Chapters 1–11 and 19 of the main text. *In Quest of the Stars and Galaxies* contains Chapters 1–5 and 11–19 of the main text.

By the time of the predicted impacts, the entire world was watching, linked together by the Internet and the television. The *Hubble Space Telescope* was trained on Jupiter, as was the *Galileo* space probe then approaching the Jupiter system, as well as most major observatories around the world and untold numbers of amateur telescopes. It has been said that more telescopes were aimed at the same spot—Jupiter—than ever before or since, and the viewers were not disappointed.

Fragment A of the comet hit Jupiter at 4:11 PM Eastern daylight time, only 10 minutes off the prediction made 10 months earlier. Traveling at 60 kilometers/second (130,000 miles/hour), each fragment created a fireball in Jupiter's upper atmosphere reaching up as high as 3000 kilometers (1900 miles) and leaving darkened areas in the atmosphere as wide across as the diameter of the Earth. Scientists estimated that some of the larger impacts, caused by fragments about 3 kilometers across, released energy equivalent to 6 million megatons of TNT. (The largest nuclear bomb ever detonated produced a blast equivalent to about 5 megatons.) The event made front-page headlines around the world, featuring photographs of fireballs and dark impact spots across the giant planet.

> Six million megatons is about 600 times the estimated arsenal of the world.

How were astronomers able to predict the time of impact with such accuracy? What did they learn from observing the aftereffects of this stupendous collision? Many scientists believe that a similar collision with an asteroid occurred on Earth 65 million years ago, heralding the end of the age of dinosaurs. Since seeing the visible proof of such collisions on Jupiter, many people worry about the chances of further catastrophes on our own planet. Astronomers say that the chances of such an event happening here are roughly once in 300,000 years, but how can they make such a prediction?

You will learn the answers to some of these questions in this book. You will find that although collisions of comets and planets are very rare in our immediate neighborhood of space, the vast regions of the universe show tremendous activity. Stars are born in dense clouds of gas while other stars explode, spewing forth gas back into space. Galaxies collide over periods of millions of years, hurtling stars around at fantastic speeds. The universe is a wild and fabulous showcase, and we will see some of its highlights, looking out from our quiet corner called Earth.

Astronomy is the oldest of the sciences, but it is still changing rapidly today. It is a study rich in interesting personalities and exciting discoveries. Its long history and recent advances make it an excellent example of the progressive nature of science.

> *Nothing is rich but the inexhaustible wealth of nature. She shows us only surfaces, but she is a million fathoms deep.*
> Ralph Waldo Emerson

> **star** A self-luminous celestial object.

Science is fundamentally a quest for understanding, and in the case of astronomy, the subject of the quest is the entire universe. The study of astronomy covers the full range of matter from smallest to largest. On the small end, it includes atoms and their internal workings, for astronomers need to know about the nuclear reactions that make *stars* shine; this knowledge one day may lead to peaceful applications of thermonuclear reactions. Astronomers also study Earth's neighbors, the planets, and in doing so help us solve problems right here on Earth. For example, a study of weather patterns on Venus and Jupiter has advanced our understanding of weather on Earth. In addition, studying Mars and Venus helps us understand the past and future of our own planet. Information we collected using unmanned spacecraft and Earth-based telescopes has revolutionized our understanding of how our solar system formed billions of years ago and has evolved since then. In particular, our neighborhood was and, to a lesser extent, still is a shooting gallery; we can see the scars of collisions on the surfaces of Mercury, the Moon, and other planetary satellites. Understanding the composition of the objects in our vicinity may one day prove extremely valuable in our efforts to find additional sources of materials for our own survival. The apparent weightlessness and very low density of space may one day lead to the manufacture of exceptional substances. In some cases, astronomy's results are more trivial but nevertheless practical: materials developed for the *Apollo* missions to the Moon are now used in tennis rackets.

> Planets shine by reflected sunlight, but stars have an internal source of energy for their light. In Chapter 11 we discuss this stellar energy source.

> **nebula** (plural **nebulae**) An interstellar region of dust and/or gas.

Astronomers also study the largest groupings of billions of stars, and they ask big questions, the answers to which tell us something about ourselves, both as individuals and as a species. Our studies of stars and *nebulae* help us better understand the natural cycle of the formation of a star, its evolution, its death, and the recycling of some of its material into a new star.

Our studies of *galaxies* help us better understand the creation and evolution of our universe. Studying astronomy causes us to change our basic outlook as we better appreciate where humans fit into the magnificent universe. Our species and our planet seem to be dwarfed by the immensity of the cosmos. In a sense, astronomy teaches humility.

At the same time, one marvels that humans have been able to learn so much about the universe in which we live and that such tiny beings are able to comprehend the huge, complex universe. From this we may decide that size is not so important after all—at least not when compared with the intelligence and spirit of our species. We should take pride in our accomplishments but also continue pushing the frontiers of knowledge.

galaxy A group of gravitationally bound stars, ranging from a billion to perhaps a thousand billion stars.

1-1 The View from Earth

In our modern world, few of us can escape from city lights to sit under a clear sky and enjoy its splendors. Fortunately, our ancestors had the time and inclination to do so. The sky on a clear, dark night is a wondrous sight, and before electric lights and television, watching the sky must have been a very popular activity.

The Sun is described in Chapter 11.

As your eyes adjust to the darkness, perhaps the most impressive spectacle that you see is the myriad of twinkling stars in the sky, ranging from dim to bright and from lone stars to clusters of many stars (**FIGURE 1-1**). What is truly astounding is that each of these stars is simply another sun, shining by the same processes that take place in our Sun.

The nature of the Sun and the tremendous energy it produces have been mysteries through the centuries, but now we know that the Sun is a sphere made up almost entirely of hydrogen and helium and that its energy comes from nuclear fusion reactions in its core. As you sit under the stars some night, try to think of each of them as a sun, and imagine how far away they must be to look as dim as they are.

We discuss the stars in Chapters 12 through 15.

FIGURE 1-1 The night sky illuminates a fir tree in this photo. Brighter stars appear larger in a photograph because they overexpose the film. Their apparent size is not related to their actual size. Return to this figure after you gain some experience in recognizing stars and constellations on the night sky, and try to find the Southern Cross, Carina, and alpha and beta Centauri; these are some of the most significant southern stars and constellations.

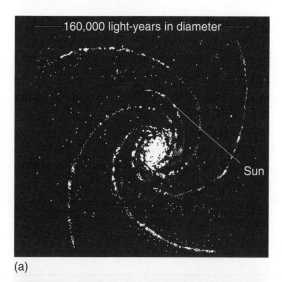

160,000 light-years in diameter

Sun

(a)

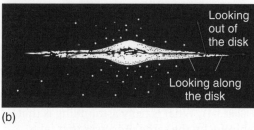

Looking out of the disk

Looking along the disk

(b)

FIGURE 1-2 (a) Our Sun is one of a few hundred billion stars that form a spiral disk somewhat like this simplified, face-on drawing of our Galaxy. (b) This simplified edge-on drawing of the Milky Way shows that when we see it in the sky, we are looking along its disk, and distant stars are not distinct but instead appear as a bright haze. When we look in other directions, we see right out of the Galaxy. (c) This wide-angle photo shows the Milky Way [in the constellations Sagittarius (left) and Scorpius (right)].

(c)

Milky Way Historically, the diffuse band of light that stretches across the sky. Today the term refers to the **Milky Way Galaxy**, the group of a few hundred billion stars of which our Sun is one. It is this group of stars that causes the diffuse band of light.

Nebulae are involved in both the birth and the death of stars.

How many stars are there? Only about 5000 can be seen with the naked eye from Earth, but telescopes reveal hundreds of billions of stars clustered in a giant disk that we call the Galaxy (**FIGURE 1-2a**). The wondrous *Milky Way*, a hazy white area that stretches across the sky, is caused by our view along the disk of the Galaxy (Figure 1-2b and c).

Besides the Milky Way you can see a few smaller patches of dim light in the sky. We call them nebulae (**FIGURE 1-3**) because of their intrinsic, nebulous appearance. They are giant clouds of gas, illuminated by light from stars within them. Though we compare them with clouds, we know today that the gas within them is extremely sparse, sparser than a laboratory vacuum here on Earth. A nebula is visible to us only because it is very large; our line of sight passes through so much material that although any given part of it provides only slight illumination, the nebula as a whole is visible.

Other objects that were formerly called "nebulae" are separate galaxies from ours, groups of millions to hundred of billions of stars (**FIGURE 1-4**). They appear to the eye as tiny, hazy splotches only because they are so far from us.

The patient observer is likely to see a meteor, a quick flash of light across the sky. Once called "falling stars," meteors are caused by (mostly) rocky visitors from beyond the Earth that glow as they burn in our atmosphere. A few times in your life, you will be able to see a comet, which will probably appear as the largest object in the sky.

If you watch for a few hours, you will see that the stars are not stationary at all, but move across

FIGURE 1-3 This is a photo (not a painting!) of a small area of the sky. The Eagle Nebula is a very luminous cluster of stars surrounded by dust and gas. The three pillars at the center of the image were made famous in an image by the *Hubble Space Telescope*.

the sky, most of them rising in the east and setting in the west (**FIGURE 1-5**). The entire sky seems to be on a bowl rotating around us. You will make another observation as you watch the sky through the seasons: different stars are visible at different times of the year because the entire sky shifts to the west very slightly from night to night. Why are we privileged to be at the center of this majestic spectacle?

As you watch the sky from night to night and begin to remember patterns of the positions of stars, you will see that a few "stars" do not stay in the same place relative to the others. These are the *planets*—wanderers on an apparently irregular path across the sky. If the Moon is visible, it represents a magnificent view, varying from night to night but always interesting.

What we know about these celestial objects would have astounded our forefathers. We know that the Moon is a solid planet-like object about one fourth the diameter of the Earth. The planets range in size from tiny Mercury, with a diameter about 38% of Earth's, to Jupiter, a giant more than 11 times wider than Earth. As for the Sun, the celestial object most important to us, its diameter is about 10 times Jupiter's and 109 times Earth's.

If you have the opportunity to use binoculars or a small telescope to view the heavens, a myriad of sights will be available. As we proceed with our quest for understanding in this text, we will point out many of the objects that are accessible to the backyard astronomer.

FIGURE 1-4 This distant spiral galaxy (named NGC-2997) is in the constellation Antlia. Although this galaxy is not visible to the naked eye, a few other galaxies are. The bright stars in the photo are in our Galaxy and are much closer to us than NGC2997.

Galaxies are the subject of Chapters 16 and 17.

planet Any of the eight (Mercury to Neptune) large objects that revolve around the Sun, or similar objects that revolve around other stars.

The Moon is discussed in Chapter 6 and the planets in Chapters 7 through 10.

Pluto's diameter is about 19% of Earth's. Pluto's designation as a planet was changed in 2006 (see Chapter 10).

FIGURE 1-5 Night sky at ESO-Paranal in Chile. This time exposure shows the motion of the stars. The lines are actually parts of circles, whose center is not in the picture. You can make photographs like this with any camera capable of taking time exposures. Choose a dark, starry night, and point the camera toward the part of the sky that you want to photograph. Fix the camera so that it won't move, and open the shutter for an hour or two. (See margin note by Figure 1-8 on page 7.)

1-2   The Celestial Sphere

celestial sphere The imaginary sphere of heavenly objects that seems to center on the observer.

celestial pole The point on the celestial sphere directly above a geographic pole of the Earth.

As we watch the sky during the night from any mid-latitude location on Earth, we see some stars rising in the east and setting in the west. Stars above the poles of the Earth move in concentric circles, centered on a spot in the sky above each pole (**FIGURE 1-6**).

Another observation that can be made by even a casual observer is that the stars stay in the same patterns night after night. The Big Dipper seems to retain its shape through the ages as it moves around the North Star; in reality, its shape is very slowly changing, as we describe in the next section. It is easy to see why the ancients concluded that the stars act as if they were glued on a huge sphere that surrounds and rotates around the Earth. **FIGURE 1-7** illustrates this *celestial sphere*.

Its axis of rotation passes through the sphere at the center of the circles of Figure 1-6, and at the center of similar circles for the stars in the Southern Hemisphere. The photograph in Figure 1-6 was taken from the Northern Hemisphere, and the point at the center of the circles, the *north celestial pole*, is exactly above the North Pole of the Earth.

FIGURE 1-6 The stars in this time exposure of the northern sky seem to move in a counterclockwise direction around one point. Polaris, the North Star, formed the short, bright streak near the center of the circles. Polaris also moves around this center because our North Star is not located exactly above the Earth's rotation axis. Color differences between the streaks were caused by variations in brightness, not by the stars' actual colors.

Above the Earth's South Pole, we find a corresponding point on the celestial sphere called the **south celestial pole**. The motions of the stars make it appear that the Earth is sitting still at the center of the celestial sphere as it rotates around us. To picture the motion of the celestial sphere, you might think of it as spinning on rods that extend straight out from the Earth's North and South Poles. These rods would connect to the celestial sphere at the north and south celestial poles.

Constellations

It is natural, when we look at the sky, to look for order—for a pattern. This desire for order is basic to science. Indeed, we can see (or imagine) patterns in the stars. The ancients saw similar patterns and identified them by associating them with beings in their particular mythology. **FIGURE 1-8a** is a photograph of the **constellation** Orion, and Figure 1-8b shows a drawing of the hunter Orion in Greek mythology. He is fighting off Taurus, the bull at the upper right. From the Northern Hemisphere, if you look at the evening sky in December, January, February, or March, you will see the stars of Orion. You may find it easy to imagine the three closely spaced stars as the belt of the hunter. And if you have a good dark sky, you can see the stars that form his sword's sheath hanging from the belt as well as the stars forming his upraised arm.

The mythological creatures of most other constellations are much more difficult to imagine. Ursa Major—Latin for Great Bear—is a prominent constellation of the northern sky (**FIGURE 1-9**). This group of stars will likely not remind you of a bear, but you might recognize the pattern of stars in the bear's rear and tail: the Big Dipper. The group of stars that form Ursa Major was identified as a bear by ancient civilizations

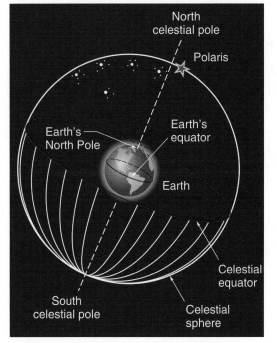

FIGURE 1-7 Because of their daily motion, objects in the sky appear to be on a sphere surrounding the Earth. (The celestial sphere is an imaginary sphere whose size is really much larger compared with Earth than it appears in this diagram.)

The word *constellation* comes from Latin, meaning "stars together."

constellation An area of the sky containing a pattern of stars named for a particular object, animal, or person.

Trails of stars that are on the celestial equator are straight lines. The rise and set points of such stars mark true east and west, and the angle that the trail of such a star makes with the horizon reveals the geographic latitude from which the picture was taken. A star that is almost exactly on the celestial equator is Mintaka (delta Orionis), the rightmost of the three stars in Orion's belt, as shown in Figure 1-8. (See Figure 1-5.)

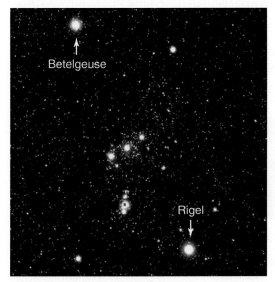

(a)

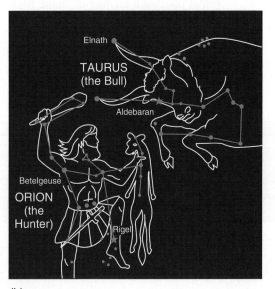

(b)

FIGURE 1-8 (a) From the Northern Hemisphere, the stars of the constellation Orion are visible high in the winter sky. (b) Ancient Greeks pictured Orion as a hunter warding off Taurus the Bull.

FIGURE 1-9 This hand-colored engraving of Ursa Major by Johann Bayer appeared in *Uranometria* in 1603. This was the first attempt at a celestial atlas and began the practice of naming stars after the constellation in which they appear.

Do the shapes of constellations change through time?

in North America, Europe, Asia, and Egypt. Historian Owen Gingerich has suggested that the labeling of these stars as a bear may have originated in Asia or Europe as far back as the Ice Age and then gradually spread to other cultures. The Greeks associated it with a bear before the time of Homer, and the constellation is mentioned in the *Odyssey*. According to the Greek story of the origin of the constellation, the bear once had been a nymph who attracted the attention of Zeus, the father of the other gods. This caused Hera (the wife of Zeus) to be jealous and to change the nymph into a bear. Finally, to protect the bear from hunters, Zeus grabbed it by the tail and flung it into the sky. This is why Ursa Major has a longer tail than earthly bears. One legend in Native American culture held that hunters had chased the big bear onto a mountain from which it leaped into the sky. The bowl of our Big Dipper is now the bear, and the handle is made up of the hunters who followed.

We know that as early as 2000 BC the Sumerians had defined several constellations, including a bull and a lion. Today we realize that the constellations have no real identity. They are accidental patterns of stars, much the same as patterns you might have seen in the clouds when you were a young dreamer watching them pass overhead. The stars don't change their patterns as quickly as the clouds, however; so perhaps it was natural for ancient peoples to attribute real meaning to them. Besides, the stars were in the "heavens," and their association with the gods seemed natural.

Why do we say that the patterns are accidental? First, the various stars are located at different distances from Earth. This means that if the Earth were in a greatly different position, we would see different patterns. For example, the stars of Orion are not all close to one another. Consider the two stars Rigel and Betelgeuse (Figure 1-8b). Rigel is nearly two times as far from Earth as is Betelgeuse. In addition, stars gradually move relative to one another along different directions and at different speeds. Given enough time, the shape of any constellation will change, but the distances from Earth of the stars in a constellation are so large that thousands of years must pass for a change in shape to be easily recognized. FIGURE 1-10 shows how the Big Dipper is changing.

Despite their artificial nature, constellations are used by astronomers today to identify parts of the sky. For example, Halley's Comet was in the constellation Aquarius on Christmas Day in 1985. The ancient Greeks had no constellations in areas of the sky that did not have bright stars nor in areas they could not see from their location

50,000 years ago   Today   50,000 years from now

FIGURE 1-10 The center drawing shows the Big Dipper as it is today; the arrows indicate the direction of motion of its stars as seen on the sky. From these motions, we can conclude that it once had the shape shown in the left drawing and will some day have the shape shown in the right one.

in the Northern Hemisphere; as a result, others have had to be added to their list. In addition, the constellations of ancient cultures had poorly defined boundaries, and thus, modern astronomers have had to establish definite ones. By international agreement of astronomers, the entire celestial sphere has been divided into 88 regions ("constellations") with specific boundaries. About half of these regions correspond to constellations identified by the ancient Greeks, and the names we know them by are Latin translations of the original names. FIGURE 1-11 shows the constellation Cygnus and its boundaries.

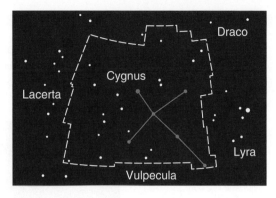

FIGURE 1-11 The constellation Cygnus was seen as a swan, but we often call it the Northern Cross, as outlined here. The official boundaries of Cygnus are shown as white lines. Draco, Lyra, Vulpecula, and Lacerta are neighboring constellations. (From the Northern Hemisphere, you can find Cygnus in the evening sky in late summer and early fall.)

Measuring the Positions of Celestial Objects

When people speak of objects in the sky, they sometimes talk about how far apart they are. What they are probably referring to is their **angular separation**, or the angle between the objects as seen from here on Earth. FIGURE 1-12 illustrates this angle. We say, for example, that the angular separation of the two stars at the ends of the arms of the Northern Cross (Figure 1-11) is about 16 degrees. As FIGURE 1-13 indicates, this says nothing about the actual distance between the two stars.

It is often necessary in astronomy to discuss angles much smaller than one degree. For this purpose, each degree is divided into 60 **minutes of arc** (or 60 **arcminutes**, or 60'), and each minute is divided into 60 **seconds of arc** (or 60 **arcseconds**, or 60"). Although these units are very similar in definition to units of time, they are not units of time, but units of angle. As an example of the use of these smaller units, a good human eye can detect that two stars that appear close together are indeed two stars if they are separated by about 1 arcminute or more. In Chapter 5 we explain how the use of a telescope allows us to detect such double stars if they are separated by as little as one arcsecond—3600 times smaller than one degree.

There is an easy way to estimate angles in the sky. Make a fist and hold it at arm's length. The angle you see between the opposite sides of your fist is about 10 degrees (FIGURE 1-14a). For estimating smaller angles, the angle made by the end of your little finger held at arm's length is about one degree (Figure 1-14b). You can use these rules to estimate the angular separations of stars.

angular separation The angle between lines originating from the eye of the observer toward two objects.

minute of arc One sixtieth of a degree of arc.

second of arc One sixtieth of a minute of arc.

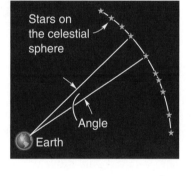

FIGURE 1-12 The two stars, when viewed from Earth, have an angular separation as shown.

Is an arcminute or an arcsecond a unit for measuring time?

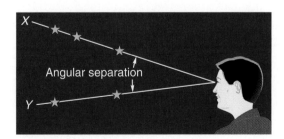

FIGURE 1-13 The angular separation of stars says nothing about their distances apart. All of the stars that lie along line X have the same angular separation from the stars that lie along line Y.

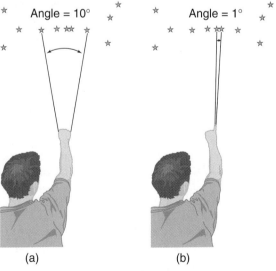

FIGURE 1-14 (a) Your fist held at arm's length yields an angle of about 10°. (b) Your little finger held at arm's length cuts off an angle of about 1°. Both the Sun and the Moon have a diameter of about 0.5°.

EXAMPLE

Mizar and Alcor are two stars in the Big Dipper that can be distinguished by the naked eye (**FIGURE 1-15**). They are separated by 12 arcminutes. Mizar, the brighter of the two, reveals itself in a telescope to be two stars, separated by 14 arcseconds. Express each of these angles in degrees.

FIGURE 1-15 The Big Dipper is part of the constellation Ursa Major. (Refer to Figure 1-9 to see the full drawing of Ursa Major, the Big Bear.)

Mizar and Alcor

Alkaid

SOLUTION   Because 60 arcminutes equals one degree, we can create the conversion factors

$$\frac{1°}{60'} = 1 \qquad \text{and} \qquad \frac{60'}{1°} = 1.$$

We choose the form of the conversion factor that causes the given unit to cancel, giving us the unit we want. In this case, multiply the 12 arcminutes by the first ratio, thus converting its value to degrees:

$$12' \times \frac{1°}{60'} = 0.20°.$$

The second part of the problem is done in like manner, but another conversion factor is needed to convert arcseconds to arcminutes:

$$14'' \times \frac{1'}{60''} \times \frac{1°}{60'} = 0.0039°.$$

TRY ONE YOURSELF
The top star in the head of Orion (Figure 1-8b) is a double star with a separation of 4.5 arcseconds. Express this angle in degrees.

The answers to each Try One Yourself is in Appendix H.

Celestial Coordinates

To describe the location of an object on Earth, we use the coordinate system of longitude and latitude whose center is the center of the Earth. The use of these two numbers uniquely defines the position of the object on Earth for anyone. Similarly, astronomers specify locations of objects in the sky by using a coordinate system

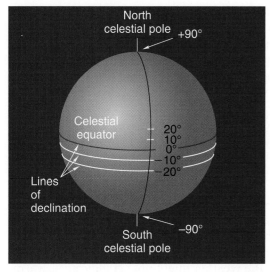

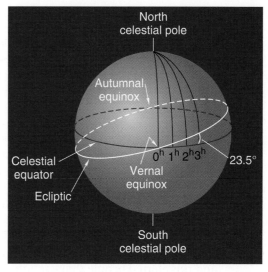

(a)   (b)

FIGURE 1-16 (a) Declination measures the angle of a star north or south of the celestial equator. Angles north are positive, and angles south are negative. (b) Right ascension measures the angle around the celestial equator eastward from the vernal equinox; as we see in the next section, this point is where the Sun crosses the equator moving northward. Angles are expressed in hours and parts of an hour, with 24 hours encompassing the entire circle. The ecliptic is the apparent path of the Sun on the celestial sphere; the angle between the planes of the ecliptic and the celestial equator is about 23.5 degrees.

inscribed on the celestial sphere, the imaginary sphere centered on the Earth. The equatorial coordinate system describes the location of objects by the use of two coordinates, *declination* and *right ascension*.

The declination of an object on the celestial sphere is its angle north or south of the **celestial equator** (**FIGURE 1-16a** and Figure 1-7). Angles north of the equator are designated positive, and those south are negative. Thus, the scale ranges from +90 degrees at the North Pole to −90 degrees at the South Pole. For example, Sirius (the second brightest star in our sky after the Sun) has a declination of −16°43′.

The right ascension of an object states its angle around the celestial sphere, measuring eastward from the vernal equinox; as we see in the next section, this is the location on the celestial equator where the Sun crosses it moving north. Instead of expressing the angle in degrees, however, it is stated in hours, minutes, and seconds (Figure 1-16b). These units are similar to units of time, with 24 hours around the entire circle. Sirius has a right ascension of 6h 45m 9s, or—as it is usually written—$6^h45^m.15$, as 9 seconds is 0.15 minutes.

declination A star's angle north or south of the celestial equator.

right ascension A star's angle around the celestial equator measuring eastward from the vernal equinox.

celestial equator A line on the celestial sphere directly above the Earth's equator.

1-3   The Sun's Motion Across the Sky

Like the stars, the Sun and Moon also seem to move around the Earth as the hours pass, rising in the east and setting in the west. Watching the Sun's motion over a few days might lead us to conclude that it is moving at the same rate as the stars, but if we carefully observe the stars in the sky immediately after sunset, we will see that they change as the weeks and months go by. We can show this by mapping the stars that surround the Earth, and locating the position of the Sun on this map by observing the stars that appear just after the Sun sets and again just before the Sun rises. The Sun is between those two groups of stars. If we do this again 2 weeks later, we would find that the Sun has moved and is now farther toward the east on our map. Thus, it appears that the Sun moves around the Earth, but not as fast as the sphere of stars. Instead, it seems to move constantly eastward among the stars.

You may want to see Figure 2-14, which shows how today's model explains the apparent motion of the Sun among the stars.

◆ See the Advancing the Model box on page 16 for more details on leap years.

How long does the Sun take to get all the way around the sphere of stars? You could determine the approximate time by drawing the pattern of stars you see in the western sky after sunset tonight and then waiting until you see that same pattern again. If you did this over a number of cycles, perhaps you could determine that the time is about 365 days. Being an alert observer, you would undoubtedly notice that this cycle coincides with the cycle of the seasons on Earth. Finally, you—or perhaps your descendants to whom you hand down your data—would decide that the time for the Sun's cycle through the stars exactly fits the cycle of the seasons and that 365 days is the length of the year. (The time the Sun takes to return to the same place among the stars is actually not a whole number of days. It is about 365.25 days. Every 4 years we add 1 day to our year to make up for this difference. This is our leap year.)

Early in history, people noticed that certain star patterns appear in the sky at the same time every year. Even before the development of the calendar, the arrival of these patterns was used as an indication of a coming change of seasons. For example, the arrival of the constellation Leo in the evening sky meant that spring was coming, and this alerted people that even though the weather might not yet indicate it, warm days were on the way.

The Ecliptic

For an observer on Earth, as the Sun moves among the stars, it traces the same path year after year.

FIGURE 1-17a is a map of the stars in a band above the Earth's equator extending 30 degrees on either side of the equator. Figure 1-17b shows the relationship of these stars to the Earth. In fact, the dashed line drawn straight across the map is directly above the Earth's equator and is the celestial equator. The other line on the map corresponds to the Sun's apparent path among the stars and shows that it is sometimes north of the celestial equator and sometimes south. The apparent path that the Sun takes among the stars is called the *ecliptic*, and it is not the same as the celestial equator. (The name comes from the fact that an eclipse can occur only when the Moon is on or very near this line. This is discussed further in Section 1-5.)

The constellations through which the Sun passes as it moves along the ecliptic are called constellations of the *zodiac*.

FIGURE 1-18 shows the 12 major constellations of the zodiac. (Actually, the Sun spends almost 3 weeks in a 13th constellation, Ophiuchus, which is a much longer time than it spends in Scorpius.)

ecliptic The apparent path of the Sun on the celestial sphere.

zodiac The band that lies 9° on either side of the ecliptic on the celestial sphere.

FIGURE 1-17 (a) A map of the stars within 30 degrees of the equator. Picture this map wrapped around the Earth as shown in (b). Compare (b) with Figure 1-16b.

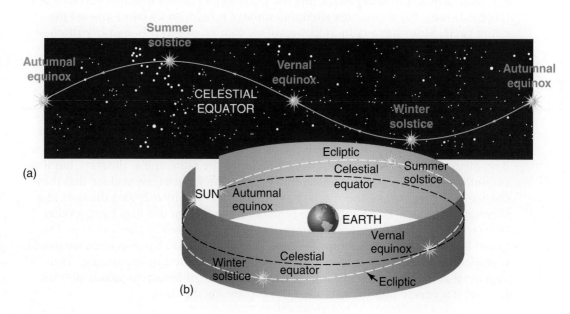

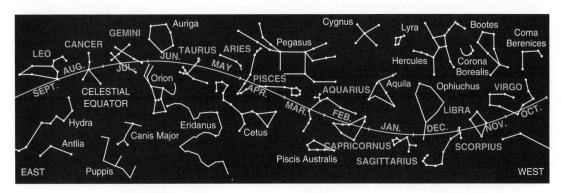

FIGURE 1-18 The constellations of the zodiac lie along the ecliptic. In addition to the 12 zodiacal constellations that most people are familiar with, there is a 13th—Ophiuchus—that the Sun passes through. You may want to look at Figures 2-14 and 2-15 for an explanation of the Sun's apparent motion among the stars.

The months indicated on the ecliptic show the Sun's locations at various times. On March 21, the Sun is in the constellation Pisces and is crossing the equator on its way north. The changing position of the Sun on the celestial sphere is related to the changing of the seasons, as we now explain.

The Sun and the Seasons

There are three easily observed differences between the behavior of the Sun in winter and summer:

Why do we have seasons?

- For an observer in the Northern Hemisphere, the Sun rises and sets farther north in the summer than in the winter. **FIGURE 1-19** shows the Sun's rising and setting points at various times of the year as well as the Sun's path across the sky in each case. Although we may say that the Sun rises in the east, it actually rises exactly east only when it is on the celestial equator—at a particular time in March and again in September.

- The Sun is in the sky longer each day in summer than in winter. Figure 1-19 shows why this happens, for the Sun's path above the horizon is much longer in June than in December. This effect is one reason for the seasons. Because the summer Sun is in the sky longer, we collect more solar energy during the daytime, and less time is available at night for our surroundings to lose the energy they have gained.

- The third difference in the Sun's observed behavior results in two more reasons for our seasonal differences. As Figure 1-19 illustrates, the Sun reaches a point higher in the sky in summer than in winter, and **FIGURE 1-20** emphasizes this by showing the Sun's path and its location at midday in late June and in late December. When the Sun is higher in the sky, its light hits the surface at an angle that gets closer to 90 degrees. Consider shining a flashlight directly down onto

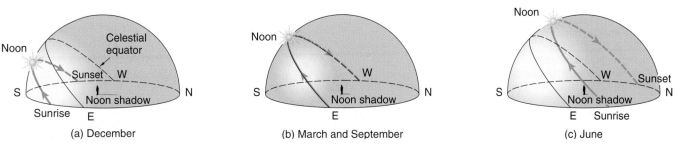

FIGURE 1-19 The Sun's apparent path across the sky of the Northern Hemisphere in (a) December, (b) March or September, and (c) June. The Sun rises and sets at different places on each date, and its noontime altitude differs in each case.

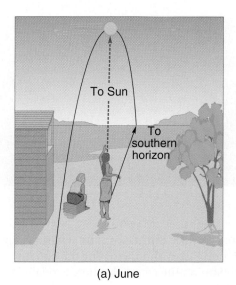

(a) June   (b) December

FIGURE 1-20 The Sun's apparent path across the sky in the Northern Hemisphere in (a) summer and (b) winter. In each case we have drawn a line from south to north straight over the person's head (the **meridian**). The Sun moves across the sky along the yellow line and reaches a much higher position in the sky in summer than it does in winter.

You may want to see Figure 2-15, which shows that according to today's model, the tilting of the Earth's axis with respect to its orbital plane explains the seasons.

Is the change of our distance from the Sun an important factor that causes seasons?

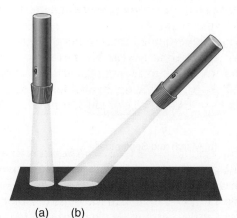

FIGURE 1-21 (a) The light from the flashlight shines perpendicularly onto the surface, whereas in (b) it strikes the surface at an oblique angle. The fact that the same amount of light illuminates more surface in (b) means that each little part of the surface is less illuminated in that case. Relate this to the noontime Sun's position in the sky in summer and winter.

(a)   (b)

a surface in one case and shining it at an angle to the surface in another (**FIGURE 1-21**). In the second case, the same amount of light is spread over more surface area, and thus, each portion of the surface receives less light. The same is true for sunlight; thus, a given area, in a given amount of time, receives more energy from the Sun in June than in December. In addition, because the Sun is never high in the sky in winter, its light must pass through more atmosphere in winter than in summer. As a result, each portion of the lit surface receives less direct light, as more of it is scattered and absorbed in the atmosphere. (See Figure 8-2 for this effect.)

If you live in the Southern Hemisphere, you know that this explanation is backward for your part of the Earth. In the Southern Hemisphere, the Sun gets higher in the sky in December than in June, and the seasons are reversed from those in the Northern Hemisphere. While people are enjoying summer in Canada, Australians are feeling the chill of winter.

It is logical for someone to believe that the difference in seasons is directly related to the change in distance of the Earth from the Sun. That is, you may believe that when we are closer to the Sun we have summer, and when we are farther from the Sun we have winter. After all, you do feel warmer when getting closer to a fireplace; however, if this were the only reason for the seasons, then how could we have two different seasons, for the two hemispheres, at the same time? In reality, the distance of the Earth from the Sun does not vary too much during a year, at least as a percentage of its average distance. (Also, the Earth is closer to the Sun in January and farther from the Sun in July.)

By analogy, suppose you were standing at a distance of 1 meter from a fireplace and then walked toward or further from it by less than 2 centimeters. It obviously would not make too much of a difference in the amount of energy you received from the fireplace. Thus, even though in principle the amount

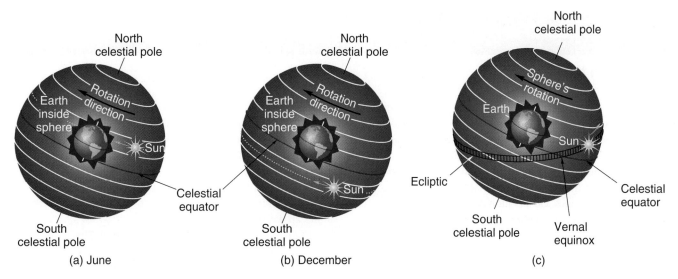

FIGURE 1-22 The height the Sun reaches in the sky is explained by its position on the celestial sphere. The dashed lines correspond to the Sun's apparent path in the sky on June 21 (a) and December 21 (b). (c) The Sun acts as if it follows a path along the ecliptic on the celestial sphere. The sphere rotates toward the west as the Sun gradually moves along it toward the east.

of energy we receive from the Sun does depend on how far we are from it, in the case of our planet, distance is not an important factor and it does not really influence the seasons. The important factor in this case is the orientation of the Earth with respect to the Sun, as shown in the previously mentioned observations.

To see how the path of the Sun along the ecliptic relates to the seasons, refer to the star map of Figure 1-18. The Sun is at its northernmost position on about June 21. If you picture the celestial sphere circling the Earth in late June, you can see that it will carry the Sun to a much higher position in the sky of North America. The Sun reaches its southernmost position on about December 21, and on that date, it reaches the least *altitude* in the sky of the Northern Hemisphere (**FIGURE 1-22**).

The two dates discussed in the last paragraph are unique. During the spring the Sun gets higher in the midday sky. Then about June 21—at the **summer solstice**—it stops climbing, and after that it starts getting lower again. The reverse happens about December 21—at the **winter solstice**. (The word *solstice* is a conjunction of the Latin words "sol" meaning "Sun" and "sistere" meaning "to stand still.") These dates were given this name because at the solstice the Sun *seems to stop* and reverses its direction.

The star map reveals two more unique events each year: Around March 21 the Sun crosses the celestial equator moving north, and about September 22 it crosses the equator moving south. On these dates every location on Earth experiences equal periods of day and night. The events are called the **vernal equinox** and the **autumnal equinox**, respectively. (*Equinox* means "equal night.")

We estimate each of the four previously mentioned dates rather than state them exactly because they may vary a day or so from year to year. The fact that there are not exactly 365 days per year causes our calendars to get out of synchronization with astronomical events until a leap year allows us to catch up. In addition, your "today" may be the previous day or the next day somewhere else on the Earth.

A Scientific Model

The idea of the stars residing on a giant celestial sphere is a **scientific model** that explains the observation of the daily motion of the stars. Likewise, the changing position of the Sun in the sky explains the changing of the seasons.

A scientific model is not necessarily a physical model, in the sense of a model airplane; it is basically a mental picture that attempts to use analogy to explain a set of

altitude The height of a celestial object measured as an angle above the horizon.

summer and winter solstices The points on the celestial sphere where the Sun reaches its northernmost and southernmost positions, respectively.

vernal and autumnal equinoxes The points on the celestial sphere where the Sun crosses the celestial equator while moving north and south, respectively.

scientific model An idea, a logical framework, that accounts for a set of observations and/or allows us to create explanations of how we think a part of nature works.

ADVANCING THE MODEL

Leap Year and the Calendar

People in early times began keeping records of celestial events because they realized that certain cycles observed in the sky corresponded to familiar events in their surroundings, such as the arrival of migratory birds and the ripening of fruits. The agricultural revolution that began around 10,000 BC led to the need to better understand the celestial cycles and to predict the seasonal changes. Observational evidence collected through the years showed that there is a correlation between the seasons and the apparent motion of the Sun among the stars. As a result, the first calendars were created by the end of this period. The Indians of the American Southwest followed a lunar calendar (of 29.5 days average per lunar cycle), resulting in a $29.5 \times 12 = 354$-day year, which obviously needed an annual adjustment. A solar calendar (of 360 days) was used in Mesopotamia and Egypt. The 360-day year follows a Babylonian practice of using the sexagesimal (60-based) system, instead of our current decimal (10-based) system. (Our division of the circle into 360 degrees is probably a result of this Babylonian approximation for the year and the fact that 360 is divisible by the numbers 2, 3, 4, 5, and 6.)

Observations of solstices and equinoxes offered a way for early civilizations to reset the calendar every year. Archaeological evidence suggests that numerous astronomical traditions of keeping a "calendar" sprang up independently around the world—for example, at Stonehenge, England around 2800–2200 BC, and by the Plains Indians of Wyoming around 1500–1760 AD. Such evidence includes special alignments or designs of temples that mark solstices or other events, legends, and historical records of observation ceremonies carried out at the sites. Buildings designed with astronomical orientations were also built by the Egyptians (the great pyramids around 3000 BC), the Aztecs, Mayans, and Incas.

Our present calendar comes primarily from the Roman calendar. This calendar started its year in March; the Latin words for the numbers 7 through 10 are "septem," "octo," "novem," and "decem." By the time of Christ, January and February had been added, giving us the 12 months we have now. The months of the calendar were based on the Moon's period of revolution around the Earth and alternated in length between 29 and 30 days to match the average of 29½ days between full Moons. Twelve of these lunar months totaled only 354 days, which was defined as the length of the standard year. To make up for the fact that this calendar quickly got out of synchronization with the seasons, an entire month was inserted about every 3 years—a sort of "leap month."

No single authority controlled the calendar, and by the time of Julius Caesar (100–44 BC), the date assigned to a specific day differed widely between different communities. In some countries, nearby communities did not even agree on what year it was. Caesar reformed the calendar in 46 BC, making the months alternate between 31 and 30 days, except for February, which had 29. Thus, the *Julian calendar* had 365 days, and (almost) like ours, it added 1 day at the end of February every 4 years to make it correspond more closely to the time the Sun takes to return to the vernal equinox.* This latter time determines the seasons and is called the *tropical year*. Thus, the Julian year had an average of 365.25 days.

The tropical year is 365.242190 days long. This means that the seasons on Earth repeat after that period of time, and the difference between the tropical year and the average 365.25 days of the Julian calendar caused the calendar to gradually get out of synchronization with the seasons. The difference between the Julian and tropical years is about 0.0078 days per year, which corresponds to about 3 days for every 400 years. By the year 1582, the vernal equinox occurred on March 11 rather than March 25, as it had when Julius Caesar instituted the calendar. It was time for more reform. Pope Gregory XIII declared that 10 days would be dropped from the month of October so that October 15 followed October 4 that year. That restored the vernal equinox to March 21, which corresponded to what the Catholic church wanted for establishing the date of Easter each year. To keep the calendar from having to be adjusted this way again, Pope Gregory instituted a new leap year rule: every year whose number was divisible by four would be a leap year—as in the Julian calendar—unless that year was a century year not divisible by 400 (such as 1800 or 1900). Thus, 1700, 1800, and 1900 were not leap years, but the year 2000 was a leap year. This rule takes care of the 3 days per 400 years discrepancy mentioned previously here. Thus, the Gregorian year has an average of 365.2425 days.

Roman Catholic countries accepted the *Gregorian calendar*, but most other countries chose to stick with the old calendar. England changed in 1752, at which time it was necessary to omit 11 days. Russia did not adopt the Gregorian calendar until 1918.

The Gregorian calendar reform, however, is also an approximation, albeit a better one than the Julian calendar reform. Because the difference between the Gregorian and tropical years is about 0.0003 days per year, there is a difference of about 1 day for every 3300 years. As a result, a proposal has been made to include an exception to the Gregorian calendar: the years 4000, 8000, 12,000, and so on will not be leap years, as they would have been according to the original Gregorian calendar. Such a calendar will have an average of 365.24225 days and will be accurate enough that it will not have to be revised for about 20,000 years.

* The vernal equinox, defined in this chapter, determines the moment when spring begins.

observations in nature. We thus are able to say that the stars appear as if they are on a sphere rotating around the Earth. Later in the book, we encounter some scientific models that cannot be represented well by a physical construction.

We can combine our explanation for the seasons with our model of the celestial sphere. Imagine a path *along the ecliptic* (Figure 1-22c). As the celestial sphere rotates around the Earth, the Sun moves along this path following the sphere's general motion on a daily basis, as shown in Figures 1-22a and b. Gradually, however the Sun creeps back eastward along the track. As the months pass, the Sun changes its position among the stars. It moves along the path with such speed that in about 365 days it is back to where it started.

The Sun in our model moves from north to south of the equator and back again. This corresponds to the observed motion of the real Sun. We are not saying that there is actually a real track on which the Sun moves, but our model provides this picture that helps us describe the motion of the Sun. That is one of the functions of a scientific model: to allow us to make sense of a set of observations. Our model does not actually explain the Sun's motion in the sense of telling us why it occurs, but it provides a mechanism that allows us to say, "OK, that makes sense now."

The model we have constructed is a *geocentric model*—an Earth-centered model. In Chapter 2 we examine reasons why this model has been replaced (but not entirely abandoned).

geocentric model A model of the universe with the Earth at its center.

1-4   The Moon's Phases

Like the stars and the Sun, the Moon also *seems* to orbit the Earth, but it does so in such a way that its same face points toward Earth at all times. At first thought, you might be tempted to say that the Moon does not *rotate*, but this is not so.

Consider FIGURE 1-23. The Moon is shown with a spot on its surface. In (a), you see that if the Moon did not rotate, this spot would always face the same way in space. In (b), that spot continues to face the Earth as the Moon goes around its orbit, but this means that the Moon must rotate once for every *revolution* around the Earth.

The fact that the rotation period and revolution period of the Moon are exactly equal seems remarkable and cannot be attributed to mere coincidence. In fact, this is another phenomenon that can be explained by the law of universal gravitation, as will be described in Chapter 3.

The photographs in FIGURE 1-24 show the *phases* of the Moon at various times during a period of about a month. The cause for the Moon's phases has been known since antiquity; to explain them we need only consider three objects: the Earth, the Sun, and the Moon.

The Moon circles the Earth, completing one orbit in slightly less than a month. This causes its position in the sky, relative to the Sun, to change with time. Figure 1-24a shows various positions of the Moon in its orbit around the Earth. The Sun is out of the picture, far to the left. The drawings of the Earth and the Moon are dark on the side away from the Sun because sunlight does not reach that side of them. Consider the Moon in position *A*. Most of the side that faces the Earth is dark, and only a small portion of that side is lit by the Sun. Figure 1-24b(A) shows how the Moon appears from Earth when it is in position *A*. We call such a Moon a *waxing crescent* Moon.

When the Moon is at position *B* on its monthly

rotation The spinning of an object about an axis that passes through it.

revolution The orbiting of one object around another.

phases (of the Moon) The changing appearance of the Moon during its revolution around the Earth, caused by the relative positions of the Earth, Moon, and Sun.

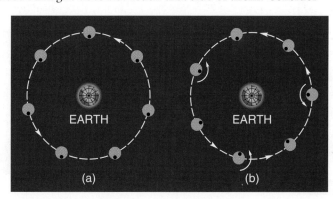

FIGURE 1-23 The Moon rotates as it revolves. If it did not rotate, a dot on its surface would always point in the same direction in space, as in (a). Instead, the Moon always keeps the same face pointed toward Earth (b). (The Moon-Earth system is not drawn to scale.)

FIGURE 1-24 The Moon in various phases. (a) Seen from above the Earth's North Pole, no light reaches the half of the Moon shown as black. The half of the Moon's surface facing the Sun is always lit; the portion shown in white can be seen by an observer on Earth, whereas the lined portion cannot be seen. The elongation of the Moon, stated as an angle either east or west of the Sun, is indicated for the waxing gibbous phase and for the waning crescent phase. We are viewing the system from above the north, and thus, east is counterclockwise. (The Moon–Earth system is not drawn to scale.) (b) These photos of the Moon in phases correspond to the positions shown in (a). The Earth-Moon distance is not the same for each photo.

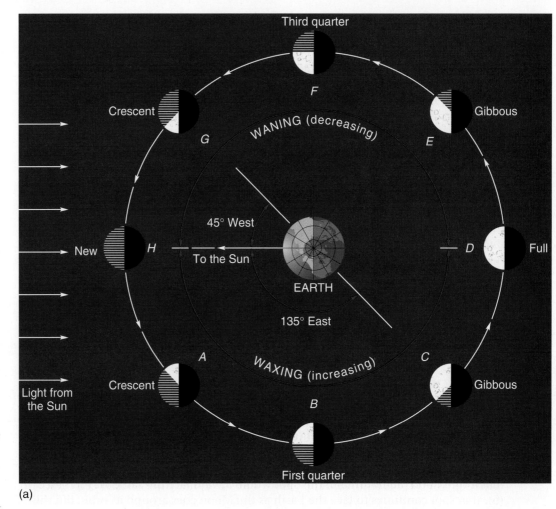

(a)

◆ Terms related to the phases of the Moon are defined based on their elongation as shown in **TABLE 1-1**.

elongation The angle of the Moon (or a planet) from the Sun in the sky. The angle of elongation is illustrated in Figure 1-24a.

TABLE 1-1

Terms Relating to Moon Phases

Phase	Elongation (in degrees)
Waxing	0–180° east
Waning	0–180° west
Crescent	0–90° east or west
Gibbous	90°–180° east or west
New	0
First quarter	90° east
Full	180°
Third quarter	90° west

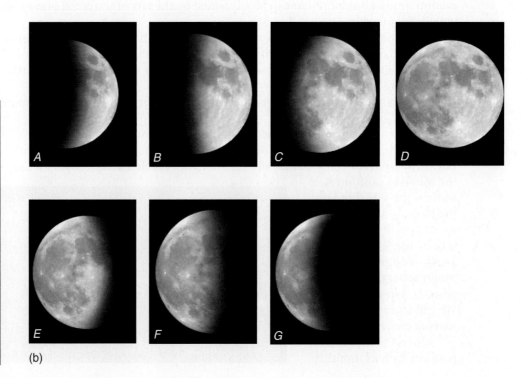

(b)

trip around the Earth, we call it a ***first-quarter*** Moon, for if we start the cycle when the Moon is at position *H*, it is now one quarter of the way around.

From Earth, a ***waxing gibbous*** Moon will appear as in the photograph of Figure 1-24b(C). The word "waxing" is derived from an old German term that means "growing." Between points *H* and *D*, the Moon is waxing; the visible portion is growing nightly from a thin crescent near *H* toward the ***full Moon*** (when the Moon is at position *D*). When we see the Moon in a gibbous phase, most of its sunlit side is facing the Earth.

Observe the photograph of the Moon when it is in position *E*. It is again in a gibbous phase, but here the gibbous phase is called ***waning gibbous*** because from night to night the lit portion that we observe is decreasing (waning) in size.

At position *F* the Moon is again in a quarter phase, the ***third quarter***, or ***last quarter***. Then around position *G* we have the ***waning crescent*** phase, and, finally, the Moon is back to where we start the cycle, at position *H*. Because we (arbitrarily) begin the cycle here, we call this a ***new Moon***. In this position, the Moon is not visible in our sky because no sunlight strikes the side facing Earth. Only at this phase can an eclipse of the Sun occur. Because it takes about a month for the Moon to revolve around the Earth, you may expect to see a solar eclipse approximately once a month. This does not occur, as you may know from personal experience. In the next section we see why such an eclipse does not occur every time there is a new Moon; before you read it, try to think of a reason why this may be so. (Also, check the activities at the end of this chapter to get a better understanding of the Moon's phases and of eclipses.)

Because the Earth moves in its orbit while the Moon revolves around it, we must distinguish between two revolution periods of the Moon. Refer to **FIGURE 1-25**. At position *A*, the Moon is full, and an imaginary line connecting the Earth and the Moon is pointing to the right of the figure toward the distant stars. About 27.3 days later, the Earth reaches point *B*, whereas the Moon, as it revolves around the Earth, is at such a position that an imaginary line connecting the Earth and the Moon is pointing again to the right of the figure toward the distant stars. The time interval between these

quarter (phase) The phase of a celestial object when half of its sunlit hemisphere is visible.

gibbous (phase) The phase of a celestial object when between half and all of its sunlit hemisphere is visible.

full (phase) The phase of a celestial object when the entire sunlit hemisphere is visible.

crescent (phase) The phase of a celestial object when less than half of its sunlit hemisphere is visible.

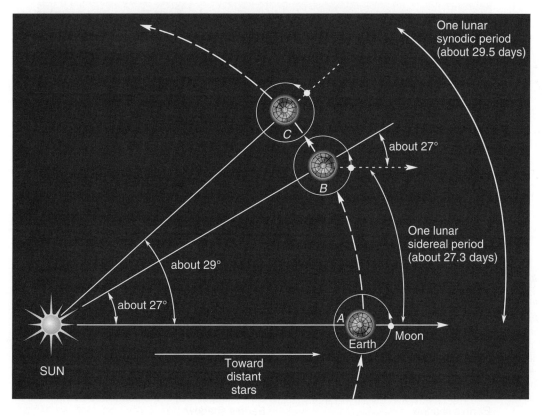

FIGURE 1-25 This drawing shows the difference between the Moon's sidereal period and its synodic period. When the Earth is at point *A*, the Moon is full. At point *B*, the Moon has completed one sidereal period, but about 2 more days are required for it to reach full phase again (which occurs when the Earth is at point *C*), at which time one synodic period will have been completed.

sidereal period The amount of time required for one revolution (or rotation) of a celestial object with respect to the distant stars.

synodic period The time interval between successive similar alignments of a celestial object with respect to the Sun.

lunar month The Moon's synodic period, or the time between successive similar phases.

The lunar month is 29^d 12^h 44^m 2^s.9.

lunar eclipse An eclipse in which the Moon passes into the shadow of the Earth.

umbra The portion of a shadow that receives no direct light from the light source.

two successive similar alignments between the Earth, Moon, and distant stars is the *sidereal period* of the Moon (about 27.3 days).

Notice, however, that the Moon is not full at this time. A certain amount of time must pass for the Earth and Moon to reach an alignment with the Sun (at position *C*) so that the Moon is again full. The time interval between two successive similar alignments between the Earth, Moon, and Sun is the **synodic period** of the Moon (about 29.5 days, a *lunar month*).

A calculation at the end of the chapter shows you how to find the difference of about 2 days between the sidereal and synodic periods. (Keep in mind that the Earth covers 360° around the Sun in about 365 days, whereas the Moon covers 360° around the Earth in about a month. That is, the Earth moves about 1° per day and the Moon moves about 12° per day on the sky.)

1-5 Lunar Eclipses

During its orbit of the Earth, the Moon sometimes enters the Earth's shadow; when this occurs, sunlight is blocked from reaching the Moon. This phenomenon is known as a *lunar eclipse* (FIGURE 1-26). You might wonder why this doesn't happen at the time of every full Moon. There are several reasons. First, as we have seen, the Moon and Earth are very small compared with their distance apart: the Moon is 30 Earth diameters away. Thus, it is unlikely that they will align so accurately that one eclipses the other. Think of trying to align a grapefruit, a Ping-Pong ball 12 feet away, and a distant object.

Second, the Earth's full shadow—called the *umbra*—tapers down to a point in a direction opposite to the Sun. Along this direction, the Earth's umbra is smaller than the Earth. This is because the source of light—the Sun—is so much larger than the Earth. At the distance of the Moon (at point *A*, for example), the width of the umbra is only three fourths of the diameter of the Earth. The Moon thus is less likely to pass

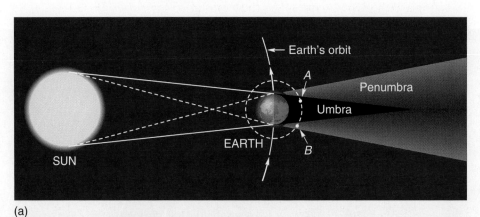

(a)

(b) This observer is looking at the Earth and Sun from point *A*, in the umbra of the Earth's shadow.

(c) This observer is looking at the Earth and Sun from point *B*, in the penumbra of the Earth's shadow.

FIGURE 1-26 (a) Point *A* is in the umbra of the Earth's shadow. Point *B* is in the penumbra, where light from part of the Sun hits it. Distances are not to scale in the drawing. (b) View from point *A*. This observer is looking at the Earth and Sun from point *A*, in the umbra of the Earth's shadow. (c) View from point *B*. This observer is looking at the Earth and Sun from point *B*, in the penumbra of the Earth's shadow.

through the shadow than if the shadow were the size of the Earth. If the Moon is at point *B* of Figure 1-26, on the other hand, it is only in partial shadow, for light from the left part of the Sun (as seen from the Moon) is hitting it. When the Moon is here, in the **penumbra**, it will not receive the full light from the Sun and will appear dim to Moon watchers on Earth. The penumbral shadow increases in size at greater distances from Earth, but it is not equally dark across its width. Right next to the umbra, the shadow is very dark, but it gets brighter and brighter out toward its edge. When the Moon passes through the outer penumbra, we don't even notice the darkening.

The third and most important factor in explaining why a lunar eclipse does not occur at each full Moon is that the Moon's plane of revolution is tilted relative to the Earth's plane of revolution around the Sun. Consider **FIGURE 1-27**, which shows both the Earth's and the Moon's orbits. The tilt of the Moon's orbit with respect to the Earth's is actually only 5 degrees, but we have exaggerated it in the drawing. When the Earth is at positions *B* and *D*, its shadow cannot hit the Moon. In fact, only when the Earth is near points *A* and *C* can the Moon pass through its shadow. These points represent the two **eclipse seasons** that occur each year.

Thus, in most cases of a full Moon, the Moon will not be in the Earth's shadow but will be either north or south of it. The plane of the Moon's orbit changes relatively little as the year progresses, and so eclipses can only occur about twice a year.

Types of Lunar Eclipses

At the Moon's average distance, the umbra has a diameter of about 9200 kilometers. Because the diameter of the Moon is less than 3500 kilometers, the Moon can easily be covered by the umbra; however, the Moon might not pass right through it. **FIGURE 1-28** shows three possible paths of the Moon through the shadow of the Earth. If the Moon moves along path *A*, it only passes through the penumbra, producing a **penumbral lunar eclipse**. In this case, it darkens slightly as it does so, but such an effect is not obvious from Earth and is only noticeable if the Moon passes into the darkest part of the penumbra, near the umbra.

If the Moon follows path *B*, it slowly darkens as it moves toward the umbra. **FIGURE 1-29** is a triple exposure of the Moon during an eclipse. The Moon is moving along a path such as path *B*, and the exposure at the right side of the photo shows the Moon while only part of it is in the Earth's umbral shadow. As the Moon continues to move into the umbra, the shadow slowly moves across its surface until the Moon appears as it does in the center exposure, where we see a **total lunar eclipse**.

Depending on the Moon's distance from Earth, it may take an hour from the time of first contact with the umbra until the eclipse reaches totality. The Moon can stay in the shadow for up to about 1.5 hours. On the left side of Figure 1-29, we see the Moon leaving the Earth's umbra again.

Why don't we have a lunar eclipse every month?

penumbra The portion of a shadow that receives direct light from only part of the light source.

eclipse season A time of the year during which a solar or lunar eclipse is possible.

penumbral lunar eclipse An eclipse of the Moon in which the Moon passes through the Earth's penumbra but not through its umbra.

total lunar eclipse An eclipse of the Moon in which the Moon is completely in the umbra of the Earth's shadow.

Eclipse seasons occur when the Earth is at points *A* and *C* in Figure 1-27. Eclipse seasons are slightly less than 6 months apart because the orientation of the Moon's orbit changes slightly as time passes.

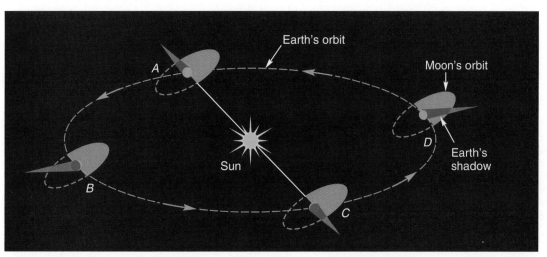

FIGURE 1-27 When the Earth is very near either *A* or *C*, lunar eclipses can occur, but when it is at other points in its orbit, the Moon does not pass through its shadow. If the Earth's orbital plane around the Sun were this page, then the dashed half of the Moon's orbital plane around the Earth would be under this page and the other half would be above this page.

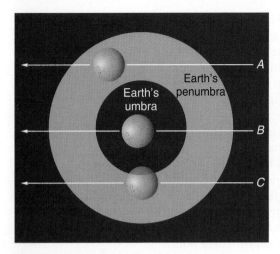

FIGURE 1-28 Three of the possible paths of the Moon through the Earth's shadow. Path *A* produces only a penumbral eclipse. Path *B* produces a total eclipse, and *C* produces a partial eclipse.

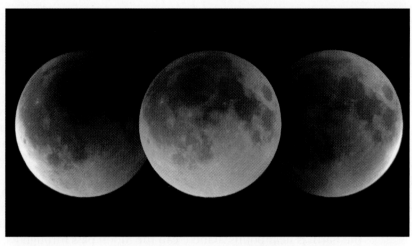

FIGURE 1-29 This is a triple-exposure photo of the Moon taken before (left), during (middle), and after (right) the total eclipse of August 28, 2007. This composite image is centered on the same sky background to show how much the Moon moved during the eclipse.

partial lunar eclipse An eclipse of the Moon in which only part of the Moon passes through the umbra of the Earth's shadow.

If the Moon follows path *C* of Figure 1-28, it is never entirely covered by the umbra, and we see only a ***partial lunar eclipse***. The dark shadow creeps across the Moon, covering (in the case shown) only the top part of the Moon.

An eclipse of the Moon, especially a total eclipse, is a beautiful sight. The totally eclipsed Moon is not completely dark, however. Even when the Moon is completely in the umbra, some sunlight strikes the Moon. This light has been refracted (and scattered) by the Earth's atmosphere, as shown in **FIGURE 1-30**. As the light passes through the atmosphere, however, the blue end of the spectrum is scattered away more than the red is, and thus, the light that makes it to the Moon is more reddish. After reflection from the Moon's surface, this light must again pass through the Earth's atmosphere; thus, the light seen by an observer on the ground is mostly red. For this reason the eclipsed Moon appears a dark red color (as in the middle image in Figure 1-29).

TABLE 1-2 shows the dates of coming lunar eclipses. You cannot be sure that you will be able to see any particular lunar eclipse, however, for two reasons. First, to be able to see it, you must be on the dark side of the Earth when the eclipse occurs. This means that, on the average, only half of the people on Earth have a chance to see a given lunar eclipse. For this reason, the last column of the table indicates where each eclipse will be visible. Second, there is the weather factor. A cloudy night can ruin a long-planned eclipse-viewing party.

FIGURE 1-30 Though the Moon is in total eclipse, some light is refracted toward the Moon by the Earth's atmosphere. As the light passes through the atmosphere, however, the blue end of the spectrum is scattered away more than the red is, so the light that makes it to the Moon is more reddish. After reflection from the Moon's surface, this light must again pass through the Earth's atmosphere and thus the light seen by an observer on the ground is mostly red. (This selective scattering also explains why the Sun looks red at sunsets and sunrises and why the sky is blue.)

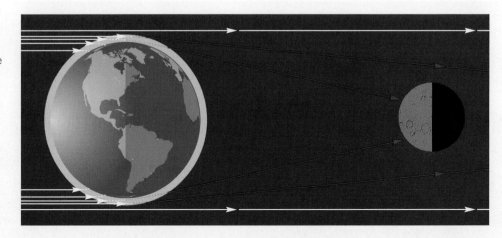

TABLE 1-2

Dates of Total/Partial Lunar Eclipses, 2009-2012

Date	Type	Visible from the Following Regions
Dec. 31, 2009	Partial	Europe, Africa, Asia, Australia
June 26, 2010	Partial	East Asia, Australia, Pacific, West Americas
Dec. 21, 2010	Total	East Asia, Australia, Pacific, Americas, Europe
June 15, 2011	Total	S. America, Europe, Africa, Asia, Australia
Dec. 10, 2011	Total	Europe, E. Africa, Asia, Australia, Pacific, N. America
June 4, 2012	Partial	Asia, Australia, Pacific, Americas

1-6 Solar Eclipses

We have seen that a lunar eclipse occurs when the shadow of the Earth falls on the full Moon. An eclipse of the Sun—a **solar eclipse**—occurs when the Moon, in its new phase, passes directly between the Sun and the Earth so that the Moon's shadow falls on the Earth. There is a major difference between these events, however. The Earth's size is such that its umbral shadow reaches back into space nearly a million miles, and at the distance of the Moon, it is easily large enough to cover the entire Moon. The umbral shadow of the Moon, however, reaches only about 377,000 kilometers (234,000 miles). Compare this with the average distance from the Earth to the Moon of about 384,000 kilometers (239,000 miles). If the Moon stayed at this average distance, its dark shadow (its umbra) would never fall on the Earth.

The Moon, however, follows an eccentric orbit, coming as close as 363,300 kilometers (218,000 miles) to Earth. So it does get close enough that its umbra can reach the Earth. When this occurs, we can experience one of the most spectacular of natural phenomena, a **total solar eclipse**. FIGURE 1-31 shows two cases of the Sun, Moon, and Earth being aligned when the Moon is close enough for its umbra to reach the Earth. Even when the Moon is at its closest, the width of its umbral shadow at the Earth's distance is only about 130 kilometers (80 miles). The width of the shadow depends on where it hits the Earth's surface, but it seldom exceeds 400 kilometers (250 miles).

This explains why relatively few people ever experience a total solar eclipse. As the Moon moves along, its shadow swings in an arc across the surface of the Earth (Figure 1-31). This strip may be many thousands of kilometers long, and you have to be within this strip at the exact moment of totality, in clear weather, to see the total eclipse of the Sun. At totality, the sky is dark enough that planets and the brightest stars can be seen in the sky. The appearance of the Sun is shown in FIGURE 1-32. What you see around the dark disk where the Moon blocks out the Sun is the glowing

solar eclipse (or **eclipse of the Sun**) An eclipse in which light from the Sun is blocked by the Moon.

total solar eclipse An eclipse in which light from the normally visible portion of the Sun (the photosphere) is completely blocked by the Moon.

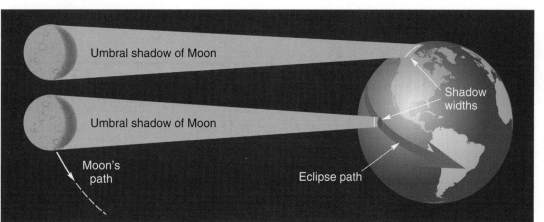

FIGURE 1-31 If the Moon's umbra strikes the Earth at an angle, a wider area on Earth will experience a total eclipse. The Moon's motion causes the path of a total solar eclipse to sweep across the Earth. The lower eclipse shown moves primarily across water and, therefore, would be seen by few people.

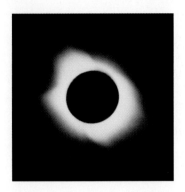

FIGURE 1-32 During a total solar eclipse, the glowing light of the Sun's outer atmosphere—the corona—is visible. This photo of the June 30, 1992 eclipse was taken from the window of a DC-10 30,000 feet above the ground.

corona The outer atmosphere of the Sun. (We discuss the Sun and its corona in Chapter 11.)

partial solar eclipse An eclipse in which only part of the Sun's disk is covered by the Moon.

TABLE 1-3

Dates of Total/Annular Solar Eclipses, 2009-2012

Date	Type of Eclipse	Visible from the Following Regions
July 22, 2009	Total	India, Nepal, China, Central Pacific
January 15, 2010	Annular	Central Africa, India, Myanmar, China
July 11, 2010	Total	South Pacific, Easter Island, Chile, Argentina
May 20, 2012	Annular	Asia, Pacific, N. America
November 13, 2012	Total	Australia, New Zealand, S. Pacific, S. America

outer atmosphere of the Sun, called the ***corona***. This is a layer of gas that extends for millions of miles above what normally appears to be the surface of the Sun. The gas glows because of its high temperature, but the glow is so much dimmer than the light we receive from the main body of the Sun that it is observed only during an eclipse. The opportunity to observe the corona is one of the scientific values of an eclipse, although today we are able to block out the Sun by artificial means to observe its outer layers.

A total solar eclipse is truly an awesome experience. If you get a chance to travel to the path of totality of a solar eclipse (**TABLE 1-3**), don't pass it up. It is one of nature's grandest spectacles. The next total solar eclipse to cross the continental United States will occur on August 21, 2017 (and the next after that will be on April 8, 2024).

The Partial Solar Eclipse

FIGURE 1-33 shows not only the Moon's umbra, but also its penumbra. Within the penumbra one sees a ***partial solar eclipse***. The penumbra covers a much greater portion

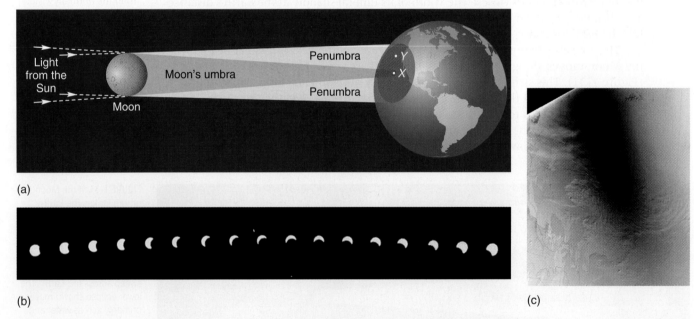

(a)

(b)

(c)

FIGURE 1-33 (a) A person at point *X* sees a total solar eclipse, whereas a person at *Y* sees a partial solar eclipse, with only the southern part of the Sun blocked by the Moon. (Compare to Figure 1-26a for a *lunar* eclipse.) (b) This series of photos shows the progression of a partial solar eclipse as would be seen by a person standing at point *Y*. (c) The umbra and penumbra in the Moon's shadow can clearly be seen in this true-color image of the November 13, 2003 total solar eclipse over Antarctica. The tip of the roughly 500-km long shadow is pointing toward Africa. The Sun was about 15° above the horizon, just rising over Antarctica, when this image was taken from space.

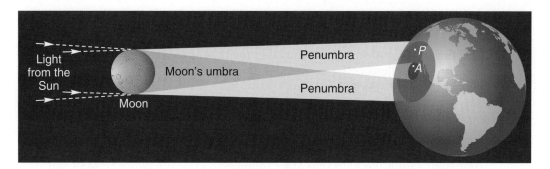

FIGURE 1-34 When the Moon is far away during a solar eclipse, the eclipse will be annular. The person at point *A* sees the annular eclipse, whereas the person at *P* sees a partial eclipse. (Compare to Figure 1-26a for a *lunar* eclipse.)

of the Earth's surface, stretching about 3000 kilometers (2000 miles) from the central path of totality, and thus, most of us have the opportunity to see a few partial solar eclipses during our lifetimes.

In a partial solar eclipse, the dark disk of the Moon moves across the Sun, but its path is not perfectly aligned with the Sun and it does not move across the center of the Sun. The closer you are to the path of totality, the more of the Sun is blocked out by the Moon.

The Annular Eclipse

A total solar eclipse can occur only when the Moon is directly between the Sun and the Earth, and the Moon is close enough to the Earth that its umbral shadow reaches the Earth. When the Moon is at its average distance from the Earth, it is a little too far away to cause a total eclipse on Earth. Therefore, somewhat fewer than half the solar eclipses that occur are total. **FIGURE 1-34** illustrates what happens when the Moon is too far from Earth to allow a total eclipse: its disk is not large enough to cover the Sun, even when it is directly centered on the Sun. As a result, an observer at point *A* would see at eclipse maximum something similar to that shown by the photograph in **FIGURE 1-35**. The Latin word *annulus* means ring, and from the figure you can see why such an eclipse is called an ***annular eclipse***. Note the spelling; this is not an annual eclipse. Slightly over half of the solar eclipses are annular.

annular eclipse An eclipse in which the Moon is too far from Earth for its disk to cover that of the Sun completely, and thus, the outer edge of the Sun is seen as a ring.

There are about twice as many total or annular solar eclipses as total lunar eclipses, but you are much less likely to see a solar eclipse because of the narrow path of the shadow.

(a)

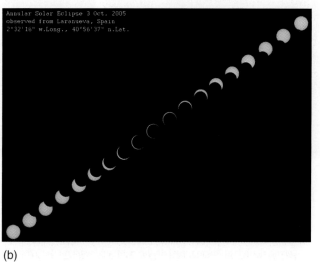

(b)

FIGURE 1-35 (a) During an annular eclipse, we can see the entire ring—annulus—of the Sun around the Moon. This photo shows the annular eclipse of May 30, 1984. The irregularity of the ring is due to mountains and valleys on the Moon's surface. (b) This series of photos shows the progression of the Moon's motion across the Sun during an annular eclipse.

1-7   Observations of Planetary Motion

Thus far we have barely mentioned the other major class of objects in the sky: the planets. Without a telescope we can see five planets: Mercury, Venus, Mars, Saturn, and Jupiter. "Planet" comes from a Greek word meaning "wanderer," and wander they do. Like the Sun and Moon, the planets move around among the stars on the celestial sphere. The Sun and Moon always move eastward among the stars, but the planets sometimes stop their eastward motion and move westward for a while. They lack the simple, uniform motion of the Sun and Moon.

Why is East on the left on sky photographs?

FIGURE 1-36a shows the path that Mars followed on the sky between May and November 2003. Mars moved eastward (to the left) and southward (downward) between May and late July 2003 but then started moving backward, heading west! It moved westward until late September 2003 and then resumed its eastward motion for about 2 years, at which time (September 2005) it began another of its backward "loops" (Figure 1-36b). We call this backward motion ***retrograde motion***. Retrograde motion is characteristic of planets, including those discovered in modern times.

retrograde motion The east-to-west motion of a planet against the background of stars.

Although the planets move among the stars in a seemingly irregular manner, there are limits to where they move, for they never get more than a few degrees from the ecliptic. Mercury and Venus have an additional limitation on their motion. These two never appear very far from the position of the Sun in the sky. We only see them either in the western sky shortly after the Sun has set or in the eastern sky shortly before sunrise. Mercury appears so close to the Sun that it is hard to find, even if we know where to look. This is because even when Mercury is at its maximum ***elongation***, we have to look for it in the semibright sky during dawn or twilight.

elongation The angle in the sky from an object to the Sun.

You never see Mercury or Venus high in the sky at night. Any model of the planets must explain not only their observed retrograde motion, but why they stay near the ecliptic as well as the peculiar behavior of Mercury and Venus.

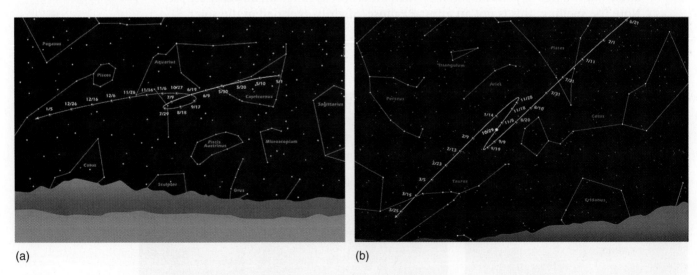

(a)   (b)

FIGURE 1-36 (a) Mars' retrograde motion in 2003. If you charted Mars' position on the night sky you will find that, in general, Mars moves eastward from one night to the next; however, every 2 years or so, Mars changes direction and moves westward for about 2 months before resuming its eastward motion again. Courtesy NASA/JPL-Caltech. (b) Mars' retrograde motion in 2005. To understand why maps and photographs of the sky show east and west reversed from Earth maps (on which north is at the top, east on the right, and west on the left), think of yourself lying face down on the ground with your head toward the north. Your right arm points toward the east. Suppose you now turn over to look up to the sky; your right arm is now toward the west! Thus the difference occurs because when we look at a map of the Earth we are looking down, but when we view a map of the sky we are looking up. Courtesy NASA/JPL-Caltech.

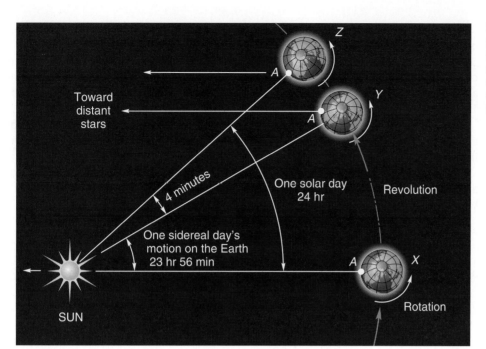

1-8　Rotations

As a planet revolves around the Sun, it also rotates around its axis. In discussing the rotation periods of the planets, we must distinguish between a solar day and a sidereal day. The ***solar day*** is defined as the time between successive passages of the Sun across the ***meridian*** so that the length of a solar day on Earth is 24 hours; however, this is not the same amount of time that the Earth takes to complete one rotation. With respect to the stars, the Earth rotates once in 23 hours, 56 minutes.

Refer to **FIGURE 1-37** to see why there is a difference between the solar day and the *sidereal day*, which is the amount of time required for an object to complete one rotation with respect to the stars. In that figure, we have designated a particular location on the Earth as A. When the Earth is at location X in the figure, point A faces the Sun directly. When the Earth has rotated once with respect to the stars, moving from X to location Y, a *sidereal day* would have passed. At this location, point A does not face the Sun directly; for this to occur again, the Earth must rotate an additional angle of about 360/365 degrees. This is because the Earth moves about 360/365 degrees per day around the Sun as it rotates; we have exaggerated this motion in the figure so that you can see the effect. Because it takes about 24 hours to rotate through 360 degrees, this additional rotation requires 24/365 hours or 4 minutes. The Earth's sidereal period is 23 hours and 56 minutes, but the Earth has to rotate for another 4 minutes to complete a *solar day* (location Z).

From the point of view of an observer on Earth, it is the Sun and the stars that move around the Earth. As such, a star will rise 4 minutes earlier from one night to the next.

solar day The amount of time that elapses between successive passages of the Sun across the meridian.

meridian An imaginary line that runs from north to south, passing through the observer's zenith.

sidereal day The amount of time that passes between successive passages of a given star across the meridian.

1-9　Units of Distance in Astronomy

In everyday life we use different units of distance for different types of measurements. We might use centimeters and meters (or inches and feet in the United States) to measure distances around the house. Although it would be possible to continue to

1 mile is about 1.6 kilometers. 1 yard is about 0.9 meters.

astronomical unit A distance equal to the average distance between the Earth and the Sun. This is about 150 million kilometers or 93 million miles.

light-year The distance light travels in a year.

Is a light-year a unit of time or distance?

use these units when we describe distances across the country or across the Earth, it is much more convenient to use kilometers (or miles) in these cases.

To measure distances in the solar system, kilometers and miles are too small to be convenient. In this case we use the ***astronomical unit (AU)***, which is defined as the average distance between the Earth and the Sun. We thus are able to say that Mars is 1.5 AU from the Sun and that Venus gets as close as 0.3 AU to the Earth.

When we go from considering distances within the solar system to distances between stars, the astronomical unit is too small to be very useful. A unit that is handy in this case is the ***light-year***, defined as the distance light travels in 1 year.

The speed of light is tremendous—a bit less than 300,000 kilometers/second or about 186,000 miles/second—so a light-year is indeed a great distance. Because a year has approximately

$$365 \, \text{days} \times \frac{24 \, \text{hours}}{\text{day}} \times \frac{60 \, \text{minutes}}{\text{hour}} \times \frac{60 \, \text{seconds}}{\text{minutes}} = 3.2 \times 10^{7} \, \text{seconds},$$

a light-year is about

$$300,000 \, \frac{\text{km}}{\text{second}} \times 3.2 \times 10^{7} \, \text{seconds} = 9.6 \times 10^{12} \, \text{km},$$

or about 9.6 trillion kilometers or about 63,000 AU. This is about 6 million million miles, or 6 trillion miles. (See the accompanying Tools of Astronomy box for an explanation of this notation. You see here the reason it is needed.)

1-10   The Scale of the Universe

It is difficult for us to appreciate the sizes of objects in the universe and the distances between them (**FIGURE 1-38**). To try to get a sense of the objects that are the subject of astronomy, let us construct an imaginary scale model of part of the universe.

Suppose in our model we represent the Sun by a basketball, as in **FIGURE 1-39**. (A "standard" basketball is 9.4 inches in diameter.) On this scale, the Earth is the head

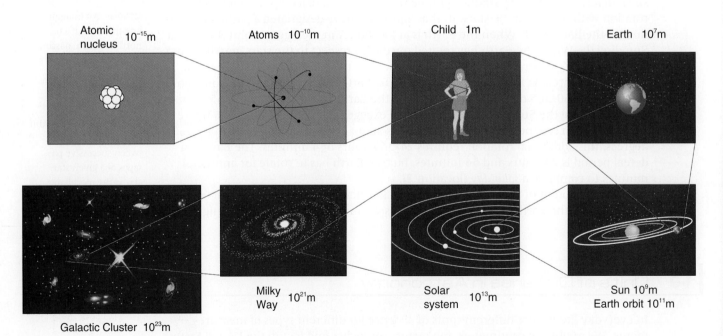

Atomic nucleus 10^{-15}m   Atoms 10^{-10}m   Child 1m   Earth 10^{7}m

Galactic Cluster 10^{23}m   Milky Way 10^{21}m   Solar system 10^{13}m   Sun 10^{9}m   Earth orbit 10^{11}m

FIGURE 1-38 The use of powers-of-ten notation (see the Tools of Astronomy box) makes it much easier to describe the sizes of astronomical objects, but the notation should not obscure the tremendous differences in size.

of a pin, about 84 feet away and invisible in the photo. The Moon is a dot the size of a period on this page and is about 2.6 inches from Earth. Jupiter, the largest planet, is the size of a grape about 146 yards away—almost one and a half football fields away. Pluto is a grain of sand about six tenths of a mile away. Thus, the entire solar system inside Pluto's orbit occupies an area of about 1.2 miles in diameter with a basketball-size Sun at the center.

Where is the nearest star (other than the Sun)? About 4460 miles away! If our model solar system is located in Detroit, Michigan, the nearest star might be in Honolulu, Hawaii (FIGURE 1-40). This is about the average distance between stars in our Galaxy; thus, we must imagine basketballs spread around randomly with approximately 4460 miles between adjacent ones. The diameter of the Galaxy on this scale would be about 164,000,000 miles.

By developing such models, we begin to appreciate the distances involved in the real universe. Such imaginary scale models will be constructed throughout the text. If you try to imagine the distances involved each time, the repetition will not be boring, but instead will be mind expanding.

FIGURE 1-39 If the Sun is a basketball, the Earth is the head of a pin 84 feet away.

Simplicity and the Unity of Nature

It is our human nature to try to simplify things. In the sky we see a tremendous variety of objects and phenomena. Astronomy provides a method of seeing order in the apparent confusion. The more we learn about the objects that make up the universe, the more patterns we see, and the more order we find. As our knowledge of the universe expands, we become more and more aware of the unity of the cosmos.

Each of the natural sciences studies a different aspect of the universe, but because the unity exists, the various sciences overlap in many areas. Astronomy is particularly close to physics, with the two fields becoming indistinguishable at times. Likewise, astronomy and geology combine as we attempt to understand the formation, evolution, and surface features of the planets. Chemistry and biology become areas that astronomers are concerned about as they study the compositions of astronomical objects and the possibility of extraterrestrial life, and meteorology aids the astronomer in studying weather patterns on other planets.

The application of physics in astronomy is now so common that *astrophysics* is often used synonymously with "astronomy."

astrophysics Physics applied to extraterrestrial objects.

1-11   Astronomy Today

We live in exciting times. During six Apollo spaceflights (1969–1972) humans walked and drove a vehicle on the Moon. In 1997, a robot vehicle sent us pictures from the surface of Mars. In 2004, two rovers at two different sites on the surface of Mars started searching for and collecting information on a range of rocks and soils that hold clues to the planet's past water activity. In 2008, another rover started digging down to the ice of a Martian arctic plain, sampling soil and ice for evidence of past

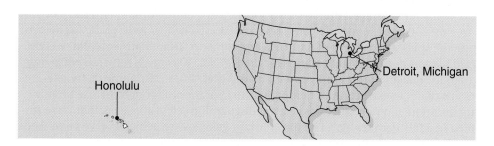

FIGURE 1-40 If our solar system with its basketball Sun is located in Detroit, the nearest star is at the distance of Honolulu.

A Model Universe with Sun as Basketball	
Earth	Head of pin, 26 m from Sun
Moon	Dot, 7 cm from Earth
Jupiter	Grape, 134 m from Sun
Pluto	Grain of sand, 1 km from Sun
Nearest star	7180 km from Sun
Our Galaxy	260,000,000 km diameter

TOOLS OF ASTRONOMY

Powers of Ten

Numbers in science—and particularly in astronomy—are sometimes extremely small or extremely large. For example, the diameter of a typical atom is about 0.0000000002 meters, and the diameter of the Galaxy is about 1,000,000,000,000,000,000,000 meters. Both are rather awkward numbers to use. We can avoid such clumsy numbers by using a variety of units, such as the astronomical unit within the solar system and the light-year for distances between stars, but sometimes it is inconvenient to switch units simply to avoid large and small numbers. Instead, scientists use *powers-of-ten* notation, also called *scientific* notation, or *exponential* notation. (The exponent is the power to which a number is raised.)

Scientific notation is simple because when the number 10 is raised to a positive power, the exponent is the number of zeros. For example:

$$10^1 = 10,$$
$$10^2 = 100,$$
$$10^3 = 1000.$$

Written in meters, the diameter of the Galaxy contains 21 zeros, thus, it is written as 10^{21} meters.

If the number to be expressed in this notation is not a simple power of 10, it is written as follows:

$$2100 = 2.1 \times 1000 = 2.1 \times 10^3,$$
$$305,000 = 3.05 \times 100,000 = 3.05 \times 10^5.$$

To change a number from scientific notation to regular notation, you simply move the decimal point to the right by a number of places equal to that indicated by the exponent, filling in zeros if necessary.

Rather than explain negative exponents, we will just give some examples and let you see the pattern:

$$10^0 = 1,$$
$$10^{-1} = \frac{1}{10^1} = 0.1,$$
$$10^{-2} = \frac{1}{10^2} = 0.01,$$
$$10^{-3} = \frac{1}{10^3} = 0.001,$$
$$2 \times 10^{-8} = \frac{2}{10^8} = 0.00000002,$$
$$4.67 \times 10^{-5} = \frac{4.67}{10^5} = 0.0000467.$$

In this case, the same rule is followed, moving the decimal point by a number of places equal to the number indicated by the exponent, but this time moving it to the left and supplying any needed zeros.

When two numbers written in scientific notation are multiplied, to get the final answer you multiply the coefficients of the two powers of ten and add their exponents. For example:

$$(3.2 \times 10^5) \times (2 \times 10^2) = (3.2 \times 2) \times 10^{(5+2)} = 6.4 \times 10^7$$

and

$$(4.2 \times 10^5) \times (3 \times 10^{-2}) = (4.2 \times 3) \times 10^{[5+(-2)]}$$
$$= 12.6 \times 10^3 = 1.26 \times 10^4.$$

In the case of a division, you divide the coefficients of the two powers of ten and subtract their exponents. For example:

$$\frac{3.2 \times 10^5}{2 \times 10^2} = \frac{3.2}{2} \times 10^{5-2} = 1.6 \times 10^3$$

and

$$\frac{4.2 \times 10^5}{3 \times 10^{-2}} = \frac{4.2}{3} \times 10^{[5-(-2)]} = 1.4 \times 10^7.$$

life on the planet's surface. After a 7-year journey, the *Cassini* orbiter entered Saturn's orbit in July 2004 and, 6 months later, released its attached *Huygens* probe for descent through the thick atmosphere of Saturn's largest moon, Titan. Farther afield, we've detected planets around distant stars. The *Hubble Space Telescope* is returning amazing images to us, images of things never seen before (**FIGURE 1-41**). Numerous new telescopes are under construction or being put into operation. Through the use of new instrumentation and new methods, we are discovering celestial objects that were undreamed of a few decades ago. The most distant galaxies are being observed and studied, and answers are being sought to questions as basic as the origin and the fate of the universe. Discoveries can be made only once, and the generations of people now alive are witnessing some of the most exciting discoveries ever.

Of what value is the science of astronomy? Will it help us advance toward worldwide prosperity? Probably not. Technological advances (sometimes beneficial and sometimes not) have followed almost every scientific advance. This, however, is not astronomy's purpose. Astronomy is a pure science rather than an applied science, and astronomers seek knowledge because knowledge is a reward in itself. Asking and answering questions is one of the things that make humans different from the other animals with whom we share our Earth. We are curious because we are human; we study the heavens because we are curious.

Conclusion

We know today that the universe is more wondrous than the most imaginative dreamer of old could have envisioned. To fully appreciate the wonder, however, one must understand the questions asked, the methods used, and the results produced by modern astronomy.

In our quest to understand the universe, we will journey through the solar system, to the stars, and then to distant galaxies, and we will come to appreciate that although our Earth is but a tiny rock circling an ordinary star, we humans have significance in the universe. The significance lies not in our size, but—at least in part—in the fact that our tiny brains have the ability to comprehend the spacious universe.

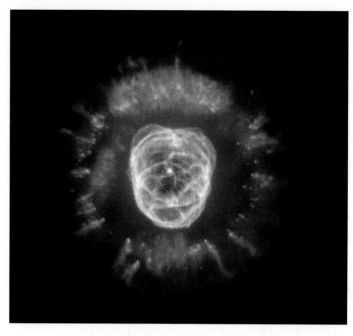

FIGURE 1-41 The "Eskimo" nebula, first spotted by William Herschel in 1787, looks like a face in a fur parka when seen through ground-based telescopes, but the *Hubble Space Telescope* shows it to be an inner bubble of gaseous material being blown into space by a dying star, surrounded by a ring of comet-shaped objects. The nebula is about 5000 light-years from Earth. Glowing gases in the nebula produce the colors in this image: nitrogen (red), hydrogen (green), oxygen (blue), and helium (violet).

STUDY GUIDE

RECALL QUESTIONS

1. The Milky Way that we observe with the naked eye is
   A. the path planets take across the sky.
   B. debris left by the motions of planets.
   C. debris left by comets.
   D. the asteroid belt.
   E. stars in the galaxy of which the Sun is a part.
   F. sunlight scattered from water vapor in the atmosphere.

2. The smallest planet is _____ and the largest is _____.
   A. Earth . . . Saturn
   B. Mercury . . . Earth
   C. Pluto . . . Earth
   D. Mercury . . . Jupiter
   E. Earth . . . Jupiter

3. In astronomical units, how far is the Earth from the Sun?
   A. 0.5
   B. 1.0
   C. 1.5
   D. 3.0
   E. 93,000,000

4. Which choice lists the objects in correct size from smallest to largest?
   A. Earth, Sun, solar system, Galaxy
   B. Sun, Earth, solar system, Galaxy
   C. Earth, Sun, Galaxy, solar system
   D. Earth, Galaxy, Sun, solar system
   E. [None of the above.]

5. Polaris is unique because it
   A. moves in a different direction than any other bright star.
   B. is the brightest star in the sky.
   C. twinkles more than any other bright star.
   D. is fairly bright and shows very little motion when viewed from Earth.
   E. [The premise is false. Polaris is not unique at all.]

6. In ancient times, people distinguished the planets from the stars because
   A. planets appear much brighter than any star.
   B. features on planets' surfaces could be seen whereas no star's features could be seen.
   C. planets move relative to the stars.
   D. planets differ in color from the stars.
   E. planets can be seen during the day.

7. Which of the following choices can be seen from Indiana?
   A. Stars near the north celestial pole.
   B. Stars near the ecliptic.
   C. Stars near the celestial equator.
   D. [All of the above can be seen from Indiana.]
   E. [None of the above can be seen from Indiana.]

8. On the first day of spring, the Sun rises
   A. north of east.
   B. directly east.
   C. south of east.
   D. [Any of the above, depending upon your location on Earth.]

9. What causes summer to be hotter than winter?
   A. The Earth is closer to the Sun in summer.
   B. The daylight period is longer in summer.
   C. The Sun gets higher in the sky in summer.
   D. [Both B and C above.]
   E. [All of the above.]

10. The Sun's apparent path among the stars
    A. is south of the celestial equator.
    B. is right on the celestial equator.
    C. is north of the celestial equator.
    D. is south of the celestial equator part of the time and north of it part of the time.
    E. crosses the north celestial pole once each year.

11. The Sun crosses the celestial equator on the first day of
    A. winter.
    B. spring.
    C. summer.
    D. fall.
    E. [Two of the above.]

12. Which of the following planets appear(s) to move through the background of stars?
    A. Venus
    B. Mars
    C. Jupiter
    D. [Both A and B above.]
    E. [All of the above.]

13. The ecliptic and celestial equators intersect at two points called the
    A. equinoxes.
    B. solstices.
    C. tropics.
    D. sidereal points.
    E. poles.

14. A minute of arc is
    A. a measure of how far the Sun moves during 1 minute of time.
    B. one-sixtieth of a degree.
    C. how far the Earth turns on its axis in 1 minute.
    D. 60 degrees.
    E. the angular diameter of the Sun.

15. Thirty arcminutes is about _____ degrees.
    A. 0.008
    B. 0.5
    C. 180
    D. 1800
    E. [None of the above.]

16. If the Moon was new last Saturday, what phase will it be this Saturday?
    A. Waning crescent.
    B. Waxing gibbous.
    C. At or very near first quarter.
    D. At or very near full.
    E. [Any of the above, depending upon other factors.]

17. If you observe the Moon rising in the east as the Sun is setting in the west, then you know that the phase of the Moon must be
    A. new.
    B. first quarter.
    C. full.
    D. third quarter.
    E. [Any of the above, depending upon other factors.]

18. Suppose that astronauts land somewhere on the Moon when it is new for Earth-bound observers. Which of the following statements would be true?
    A. Earth would appear full to the astronauts (assuming that they could see it).
    B. Around the landing site there might be bright sunlight.
    C. The landing site might be dark.
    D. [All of the above.]
    E. [None of the above.]

19. A lunar eclipse can occur
    A. only around sunset.
    B. only near midnight.
    C. only near sunrise.
    D. at any time of day or night.

20. The Sun is _____ during an annular eclipse as/than during a total solar eclipse.
    A. farther from Earth
    B. closer to Earth
    C. the same distance from Earth
    D. [No general statement can be made.]

21. You are likely to see more of which type of eclipse during your lifetime?
    A. Solar eclipse.
    B. Lunar eclipse.
    C. [No general statement can be made.]

22. Name at least seven types of celestial objects that are visible to the naked eye.

23. What causes the Milky Way we see in the sky?

24. What is a galaxy?

25. What causes a meteor?

26. How do stars near Polaris appear to move as we watch them through the night?

27. What is the ecliptic?

28. What are the approximate dates of the summer and winter solstices and what is their significance?

29. What is the origin of the word planet?

30. In what direction across the background of stars do the planets normally move?

31. Define *retrograde motion* of a planet.

32. Name the planets that are never seen far from the Sun in the sky.

33. In what part of the sky must you look to find the planets Mercury and Venus?

34. What is meant by the *elongation* of a planet?

35. Name eight different phases of the Moon in the order in which they occur.

36. Define umbra and penumbra.

37. Total (and annular) solar eclipses occur more frequently than do total lunar eclipses. Why, then, will you probably observe many more lunar eclipses than solar eclipses during your lifetime?

38. Why are penumbral lunar eclipses not easily detectable?

39. Why are some solar eclipses annular rather than total?

---

1. The text states that astronomy is "still changing rapidly" today. Report on a new discovery or new hypothesis in astronomy that you find in the news during the next 2 weeks. Suggested sources include national television news, newspapers, and news magazines.

2. Some people argue that astronomy does not produce anything useful for our lives and, therefore, it does not deserve public funding. How do astronomers answer this? What do you think?

3. Figure out a method not mentioned in this chapter that would allow you to measure the length of the year by astronomical observations.

4. Specify the two coordinates by which we describe locations in the sky. List the units of each and the maximum and minimum value of each.

5. How does our calendar adjust for the fact that the year is about 365.25 days long rather than an even 365 days?

6. The star in Orion's left leg (Rigel) is much farther away than the other bright stars in the constellation (see Figure 1-8). If you look at Orion in the sky, and then move far into space in the direction to your left, how will that star shift relative to the others?

7. Identify the position of the vernal equinox, autumnal equinox, and the two solstices on Figure 1-18.

8. Why do people who live near the equator not experience major seasonal changes? (The answer, "It is always hot there" is not sufficient.)

9. Find the Arctic Circle on an Earth globe. What is the astronomical significance of this line? Where is the Tropic of Capricorn and what is its significance?

10. If you look at the Moon tonight and then again tomorrow night at the same time, will it be farther east or farther west on the second night? If you observe it at 8 PM tonight and then again at 9 PM, will it appear farther east or farther west at the later time? Explain the discrepancy.

11. At about what time does the first quarter Moon rise? Set?

12. Will the Moon appear crescent or gibbous when it is at position X in **FIGURE 1-42**? At position Y? At about what time will the Moon rise if it is at position Y?

FIGURE 1-42 (Question 12)   **FIGURE 1-43** (Question 17)

13. If you see the Moon high overhead shortly before sunrise, about what phase is it in?

14. At the same time that people in Chicago see a total solar eclipse, what type of eclipse is seen by people in Evansville, Indiana? (Evansville is about 250 miles south of Chicago.) Answer the same question for a total lunar eclipse.

15. Even at the midpoint of most total lunar eclipses, the Moon is not uniformly illuminated. Instead, one edge of the Moon is usually much brighter than the rest. Why is this? What would be necessary in order for the red color to appear uniform across the Moon?

16. Discuss the danger involved in viewing a solar eclipse and describe four ways to view such an eclipse safely. (Hint: See the Activity "Observing a Solar Eclipse.")

17. **FIGURE 1-43** is a multiple exposure of the Moon and Venus as they set one night over Tulsa. Explain why the Moon's distance from Venus changes as time passes.

18. The expression "Once in a Blue Moon" is commonly used to describe a rare phenomenon. It was thought

that the origin of this expression had to do with the rare occurrence of two full Moons in the same month, the second one being called "Blue Moon." However, this is not the case. A literature search suggests that the term "Blue Moon" was assigned to the third full Moon in a season that had four full Moons. During which month(s) can we not have a "Blue Moon" if we adopt the first

(erroneous) explanation? During which month(s) do you expect to get "Blue Moons" if we adopt the second explanation? (Hint: You may want to consult *Sky & Telescope*; the magazine was actually the source of the erroneous modern convention for the meaning of the expression.)

CALCULATIONS

1. Using powers of ten, evaluate the following expressions:

$$(2 \times 10^{-2}) \times (3 \times 10^3)$$

and

$$\left(\frac{4 \times 10^{-4}}{2 \times 10^{-3}} \right)^2 .$$

2. Verify that one light-year is about 63,000 AU.

3. The speed of light is about 300,000 kilometers/second and the average distance between Earth and Sun (1 AU) is about 150 million kilometers. How long does it take light to travel from the Sun to Earth? (Hint: An object moving at a speed v for a time t covers a distance d given by $d = v \times t$).

4. Using scientific notation and data found in the appendices of the book, answer the following questions: (a) How many times greater is the Sun's radius than the Earth's? (b) The stars Rigel and Betelgeuse are both in the constellation Orion. Are they at the same distance from us? What does your answer suggest about the constellations? (c) How long does it take light to travel from Rigel to Earth? What does that tell you about the *current* state of this star?

5. The Crab Nebula is the remnant of a supernova explosion that was first seen by the Chinese in 1054 AD. The distance of this nebula from Earth is about 6000 light-

years. In what Earth year did the explosion actually occur?

6. Express an angle of 15 arcseconds in degrees.

7. Four degrees of angle is how many arcminutes?

8. Suppose one star is 3 arcminutes from another. What is this angle in degrees?

9. Given that the angle between the celestial equator and the ecliptic is 23.5 degrees, what is the declination and right ascension of the vernal equinox? Of the autumnal equinox? Of the winter solstice? Of the summer solstice?

10. Compare the distances reached by the umbral shadows of the Moon and the Earth. Compare the diameters of the two bodies. Discuss any relationship you see.

11. To show that the Moon's synodic period is about 2 days larger than its sidereal period, refer to Figure 1-25 and consider the following steps. (a) During one sidereal period (27.322 days), how many degrees has the Earth covered on its orbit around the Sun, from point *A* to point *B*? (Hint: The Earth covers 360° in about 365 days.) (b) Let us assume that we need an additional *x* days for a synodic period to be completed. During this period, both the Earth and the Moon move on their respective orbits. How many degrees does the Earth cover on its orbit around the Sun in this period? (c) To find *x*, set up a ratio by taking into account that the Moon covers 360° in a sidereal period.

ACTIVITIES

1. Building a Replica of the Solar System

Let us assume that we can build a replica of the solar system in which the Earth has a diameter of 1 mm. (Use relevant data from the appendices in your text.) In this replica, find the following: (a) the Sun's diameter (in cm), (b) the Earth-Moon distance (in cm), (c) the distances between the Sun and the planets (in m), (d) the distance to the nearest star other than the Sun (in km). Consult a map of your area (for example, from a phone book), and plot on it the positions of the planets (and the Moon), assuming the Sun is located at your school.

2. Star Charts and Observations

Obtain one of the star charts published monthly in astronomy magazines such as *Astronomy* and *Sky & Telescope*. On a clear, cloud-free night, use the star chart to locate as many constellations of the zodiac as you can. Can you locate the celestial equator? The ecliptic? Spend some time outdoors and see whether you could clearly distinguish the motion of the stars toward the west. Has Polaris moved? What is the overall pattern that the stars follow?

3. The Rotating Earth

On a clear dark night, find a good observing location, and draw a quick sketch of some stars high over your head. Look for a pattern so that you can remember these stars later. If you can, mark your map so that it shows which stars are toward the north, east, south, and west.

Take about 2 hours off; then go back to your observing location and find the stars you drew. How has their position changed? You probably know what this answer should be from having read the text. Your real job, however, is to stand under the stars and imagine that their motion is due to the rotation of the Earth. Try to "feel" the Earth turning under the stars. Which is easier to imagine, the stars on a sphere rotating around the Earth or the Earth spinning under the stars?

Finally, spend a half hour watching either a sunrise or a sunset. Better yet, watch the Moon rise or set, particularly when it is full or nearly full. Now picture this phenomenon as explained by the heliocentric system. Try to think of the Earth turning instead of the Sun or Moon moving. Imagine yourself on a little ball that is turning so that the place where the Sun's light hits the ball changes.

4. Do-It-Yourself Phases

This is an important activity that is worth the trouble if you want to understand the phases of the Moon. We will simulate the Sun/Earth/Moon system as it is shown in Figure 1-24. For the Moon you will need something like an orange, a grapefruit, or a softball. The softball would be best because it is rounder than the other choices. For the Sun, you can use a bright light across an otherwise dark room. (You could go outside and use the real Sun, but you must be careful not to look directly at it. This method would work best when the Sun is low in the sky.) Your head will be the Earth, and one of your eyes will be you.

Hold the ball (your Moon) out at arm's length so that it is nearly between you and your Sun. Now observe it as you move it around to the left until it is at 90 degrees to the Sun, the first-quarter position, as shown in FIGURE 1-44. Did you see its growing crescent as you were moving it? Continue to move it around your head and observe it as it changes phase. (When it gets directly behind you from the Sun, you will eclipse it.)

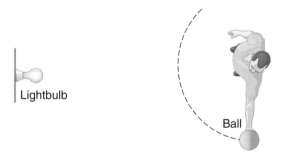

FIGURE 1-44 The person is holding a ball that represents the Moon at first quarter. When doing this, you should put the lightbulb farther away than indicated here, perhaps 10 to 15 feet away.

Now fix the Moon at the first-quarter position by laying it down on something in that position. Turn your head (and body) slowly around and around toward the left to simulate the Earth rotating. When the Sun is directly in front of you, it is noontime. As you lose sight of the Sun, it is sunset, about 6:00 PM. It is midnight when the back of your head is toward the Sun. When the Moon is at first quarter and it is sunset on Earth, where do you see the Moon?

Put the moon in various other positions and observe it as you rotate your head to simulate different times of day. Answer the following questions:

a. If the Moon is at third quarter, at about what time will it rise? At about what time will it set?

b. At about what time will a full Moon appear highest overhead?

c. If you see the Moon in the sky in mid-afternoon, about what phase will it be?

5. Observing the Moon's Phases

This exercise will take several nights. Find a calendar or daily newspaper that lists Moon phases, or ask your instructor for this information so that you can begin your observations 3 or 4 nights after the new Moon. Observations should start at sunset or shortly thereafter. At least four observations should be made, continuing for at least 3 more clear nights during the 2 weeks following your first night.

It is important that the observations be made from exactly the same place and at exactly the same time of night. Stand at the same place, not just in the same parking lot, for example. You might even go so far as to mark your location with chalk.

a. On each night of your observations, after you have arrived at your observing location, use a full sheet of paper to make a sketch of the position of the Moon relative to buildings, trees, and the like on the horizon. This sketch should show how things look to you and should not be a map. Label the buildings, such as "Student Union." Also include prominent stars you see near the Moon. Draw the Moon in the shape it appears and with the correct apparent size relative to objects on the horizon. Finally, write the date and time on your paper.

b. On one of the nights, repeat the observation after waiting about an hour. Use the same sketch and show the new position of the Moon.

c. When you have finished your four (or more) observations, make a general statement about how the Moon changed position and phase from one night to another. Look for a pattern in the Moon's behavior and explain that pattern based on the Moon's motion around the Earth.

6. Observing a Solar Eclipse

There is a misconception that the Sun emits especially harmful rays of some kind during a solar eclipse. This is a particularly anthropomorphic idea, for it would mean that somehow the Sun knows when Earth's Moon is about to block sunlight from the Earth. Naturally, the radiation emitted by the Sun during an eclipse is no different from that emitted at any other time.

Like most misconceptions, however, there is an element of truth in this idea. In fact, the Sun continuously emits radiation that is harmful to our eyes: infrared radiation.* If you were to look at the Sun anytime, this radiation would harm your eyes, but normally you are not able to look at the Sun. If you attempt it, your eye will quickly close because of the intense light. When the Sun is nearly totally eclipsed, however, its light is dim enough that you are able to look at it. Thus, during an eclipse it would be possible for a person to stare directly at the Sun for some time, all the while unknowingly absorbing the harmful infrared rays in his or her eyes.

There are a few safe methods of observing the Sun during an eclipse. First, we might use a telescope with a solar filter attached. This is a filter that blocks out some 99.99% of the Sun's light, allowing just enough through for us to see the Sun.

A more convenient way to use a telescope to observe an eclipse is illustrated in FIGURE 1-45, which shows the use of a telescope to project the partially eclipsed Sun onto a screen. In this case, a reflecting device was mounted on the telescope to cause the image to appear off to the side. This photo was taken when the Moon had progressed much of the way across the solar disk. It was near noon at the time, but the sky had still not darkened noticeably. Even a little of the Sun's light is enough to give us a bright day on Earth.

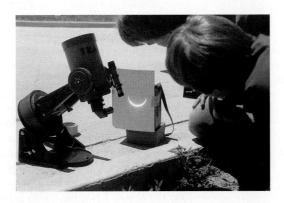

FIGURE 1-45 The telescope has a mirror (called a *star diagonal*) attached to it so that it projects the image of the partially eclipsed Sun onto the screen at the side.

A third method of safely observing a solar eclipse—one requiring little equipment—is by pinhole projection. To use this method, all that you need is a piece of cardboard with a hole in it and a piece of paper to use as a screen. **FIGURE 1-46** illustrates the method. Try holes of different sizes, from 1 millimeter to one made with a paper punch. (A hole made by a pin will probably be too small.) To see the image better, you might use large pieces of cardboard to shield your screen from reflected light or work in a dark room that has an opening facing the Sun. Block all of the opening except for your pinhole.

A fourth method of observing an eclipse is to obtain a safe filter through which you can view the Sun directly.

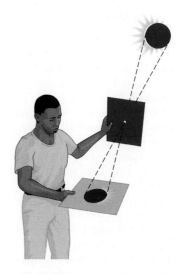

FIGURE 1-46 A pinhole projector can be used to view a solar eclipse, but you must shield your screen better from scattered light than is done here.

Extreme care must be exercised here, however. If your filter does not block the Sun's infrared radiation, it may damage your eyes. (Smoked glass is definitely not recommended.) To avoid the possibility of injury, it is recommended that you be very sure of your filter or that you use one of the other methods.

* The Sun also emits ultraviolet radiation that can harm the eyes; however, the primary danger during solar eclipses is infrared radiation.

1. The book *The Measure of the Universe*, by I. Asimov (Harper and Row, 1983), will take you for a trip through the universe using powers of ten.

2. The book *Powers of Ten* by Philip and Phylis Morrison (W. H. Freeman, 1982) also will take you for a tour through the universe using powers of ten. There is also a video based on the book (Pyramid Film and Video, 1989).

3. Read the article, "Solar Eclipses That Changed the World," by B. E. Schaeffer, in *Sky & Telescope* (May, 1994).

4. The Web sites for the magazines *Sky & Telescope* and *Astronomy* (and the magazines themselves), which cover the latest developments in astronomy for a general audience, have information about upcoming eclipses. An authoritative site for eclipses is maintained at NASA's Goddard Space Flight Center. At the same Center you also can find information about past, current, and future NASA missions (including *Lunar Prospector*) and pages devoted to our planet and the Moon.

5. *Spinoff* is a NASA publication that features successful transfers of NASA technology to the private sector, resulting in the development of numerous commercial products and services. Since the 1950s, space technology has been applied to tens of thousands of commercial products.

Quest Ahead to Starlinks:
http://physicalscience.jbpub.com/starlinks

Starlinks is this book's online learning center. It features **eLearning**, which contains chapter quizzes and other tools designed to help you study for your class. You also can find **online exercises**, view numerous relevant **animations**, follow a guide to **useful astronomy sites** on the Internet, or even check the latest **astronomy news** updates.

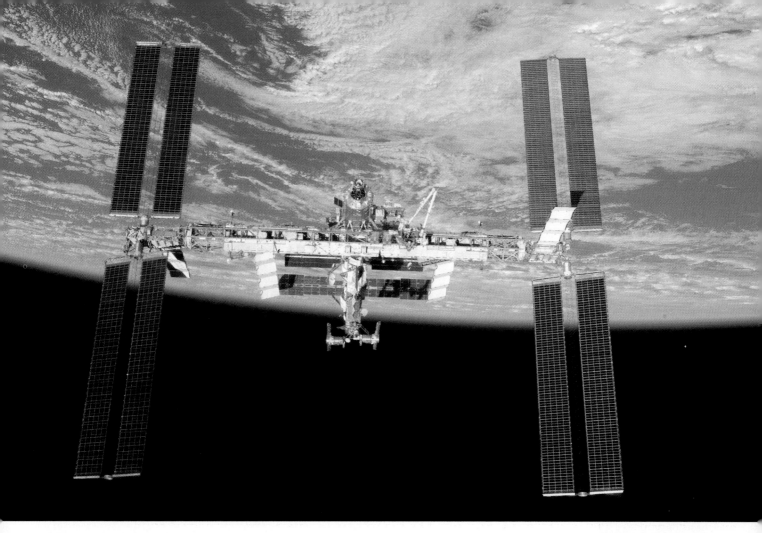

From an Earth-Centered to a Sun-Centered System

ANCIENT OBSERVERS WONDERED ABOUT THE OBJECTS AND PHENOMENA we described in Chapter 1. Through the methods of astronomy, we are answering questions asked by those ancient observers as well as questions far beyond their imagining. Our search toward understanding the universe fills us with awe and provides us with an endless stream of new questions. In this text, we try to describe not only the answers to some of the questions but also the methods used in the search.

Studying the methods that astronomers use actually means studying scientific methods in general, for although each of the sciences uses different instruments, the basic methods of inquiry are similar for all of the natural sciences. We thus explore the very nature of the scientific endeavor, showing how astronomers gather data, how those data are developed into theories, how various theories compete against one another, and how and why some theories are retained and others abandoned. Data and theories are the working materials of all of the sciences, and examples of the interplay between data and theory will be an essential part of our study.

All cross references to chapters, sections, figures, and tables pertain to the main text, *In Quest of the Universe, Sixth Edition. In Quest of the Solar System* contains Chapters 1–11 and 19 of the main text. *In Quest of the Stars and Galaxies* contains Chapters 1–5 and 11–19 of the main text.

The *International Space Station* moves away from the space shuttle *Atlantis*; in the backdrop, Earth's horizon and the blackness of space.

2-1 Science and Its Ways of Knowing

scientific method A never-ending cycle of hypothesis, prediction, data gathering, and verification.

hypothesis An educated guess made in describing the results of an experiment or observation. A hypothesis can be widely speculative but must be testable.

theory A synthesis of a large body of information that encompasses well-tested (by repeatable experiments) and verified hypotheses about certain aspects of the natural world. No theory can be proved true, but data can prove a theory to be false.

fact Generally, a close agreement by competent observers of a series of observations of the same phenomenon.

law or principle When a hypothesis has been repeatedly tested and has not been contradicted, it may become known as a law or principle.

scientific model An idea, a logical framework, that accounts for a set of observations and/or allows us to create explanations of how we think a part of nature works.

Science is a very human endeavor, and the science that a particular culture produces is in part a reflection of the characteristics of that culture. In astronomy, the human aspects of science are very visible.

Science is about discovering the orderliness of nature and the rules that govern this order. It is also the body of knowledge about nature that represents the collective efforts of people throughout history. Science progresses through the interaction of observations and experiments with theory. With our theories we make predictions, while our observations and experiments support, disprove, or place constraints on these theories. This information provides a means of determining the best description of nature. The characteristics of good science include the continuous search for new knowledge, the ability to make predictions and test them against observations, and the ability to repeat and verify any experimental result. Learning from observations and experiments and, as a result, refining our theories are not weaknesses in science—they are strengths.

It is said that the **scientific method** is the best approach for doing good science. This method is used to gain, organize, and apply new knowledge and was first introduced in the 16th century. Essentially, it involves the following series of steps: (1) recognize a problem; (2) make an educated guess, called a **hypothesis**, about the problem's solution; (3) predict the consequences of the hypothesis; (4) perform experiments to test the predictions; and (5) find the simplest general rule that organizes the hypothesis, prediction, and experimental outcome into a **theory**.

The scientific method is a never-ending cycle of hypothesis, prediction, data gathering, and verification. In reality, this method is not always used. For example, some of the most important discoveries in science were accidental, such as Becquerel's discovery of radioactivity by noting that a chunk of uranium ore caused a photographic plate to turn cloudy. The important ingredient is the continuous comparison between observations and theory.

It is important to emphasize again that a hypothesis is nothing but an educated guess made in explaining the results of an experiment or observation. A hypothesis must be testable and, at least in principle, it must be susceptible to being shown wrong. A useful hypothesis is one that makes predictions about nature that can be confirmed or refuted by observations. It is only considered to be a **fact** after it has been demonstrated by experiments.

In everyday life, the word *fact* implies something that is absolute. In science, however, it simply implies a generally close agreement among most competent scientists about a series of observations of the same phenomenon. Something that in science was considered a fact a few decades ago, thus, may turn out to be wrong. For example, we used to consider it a fact that the universe is static, whereas our current observations clearly show that we live in an expanding universe.

When a hypothesis has been tested over and over again and has not been contradicted, it may become known as a **law** or **principle**. In science we describe reality using **models**, which are developed sets of ideas used to describe some aspect of nature and are composed of hypotheses that are supported by observations and experiments. A model gives us a description of a phenomenon and allows prediction of future events, but it is not necessarily the truth or reality.

Finally, a theory is a synthesis of a large body of information that includes related hypotheses (which have been repeatedly tested and verified), giving us a self-consistent description of a certain aspect of the natural world. A theory can never be proven to be true, but data can prove a theory to be false. In everyday usage the word *theory* implies something that is nothing but a wild guess that has little to do with reality. In science, however, a *theory* represents a coherent framework describing an aspect of nature very well. For example, you might hear someone refer to Einstein's theory of relativity and say, "Well, it is *only* a theory," meaning that you must not put too much faith in it. Einstein's theory of relativity is indeed a theory but one that is well founded and has been experimentally verified many times.

Criteria for Scientific Models

Three criteria are applied to scientific models today.

- The first criterion is that the model must fit the data. It must fit what is observed.

- The second criterion is that the model must make predictions that allow it to be tested and that it must be possible to disprove the model—that is, to show that it needs to be modified to fit new data or perhaps needs to be discarded entirely. In other words, a scientific model must contain the potential seeds of its own destruction. If a model's prediction turns out to be correct, this serves as further evidence that the model is good, but it does not "prove" it. On the other hand, further evidence can always prove a model wrong. We can disprove a model, but we can never absolutely prove one. "Prediction" here does not necessarily refer to the future. It means that the theory itself indicates an observation that will either support it or disprove at least part of it. Sometimes the observation has already been made and needs only to be checked against the theory.

  Not all models make predictions that allow them to be tested and possibly be proven wrong. For example, suppose someone presents a model that says the planets appear to move the way they do because of a divine plan. Using this model, whatever we later find out about a planet can be explained by saying that it was a part of this divine plan. There would be no way to prove this model wrong. This model cannot be accepted as a good scientific model because it does not contain within itself predictions that allow it to be disproved.

- The third and final feature of a good model is that it should be aesthetically pleasing. This concept is difficult to define. It generally means that the model should be simple, neat, and beautiful. Today's idea of beauty in a scientific model is more along the lines of symmetry and simplicity. A model should be as simple as possible; that is, it should contain the fewest arbitrary assumptions. The philosophical choice that among two or more competing theories (which explain all known observations equally well) the best theory is the one that requires the fewest assumptions is known as ***Occam's razor***. It is named for a 14th-century Franciscan monk–philosopher who stressed that in constructing an argument one should not go beyond what is logically required. Using Occam's razor, one "cuts out" extraneous suppositions.

Occam's razor The principle that the best explanation is the one that requires the fewest unverifiable assumptions.

2-2   From an Earth-Centered to a Sun-Centered System

We currently have good explanations for what we see in the sky, and we understand the Earth's relationship to other astronomical objects. The search for this understanding began long before telescopes were invented. Even though many civilizations contributed in different ways to our understanding of the observed motions of stars and planets, our starting place will be two major theories that explain these motions, one of which is nearly 2000 years old. We examine these theories for two different purposes: first, to answer the question of where the Earth fits into the scheme of things, and second, to see how well they match the criteria for a good scientific theory. Finally, we show how and why one of these theories won out over the other and how the understanding of nature provided by the successful theory eventually allowed humans to travel beyond the Earth and leave footprints on the Moon.

Why does an astronomy text begin by looking back into history instead of plunging into today's astronomy? Furthermore, why start with an outmoded theory? There is good reason. To understand how astronomy—and science in general—works, we must look at how it progresses with time. We must see what led up to today's ideas. Science is a dynamic enterprise and the continual refinement of theories is one of its strengths. Every advance in science is necessarily incomplete and may well be partly inaccurate; this is a natural result of the fact that science is a human activity and thus includes people's biases. Looking back and understanding how a certain idea evolved allow us to better appreciate all of the specific influences on that idea from people of different times and nationalities. We can also better understand the basis of the idea and see how the scientific "knowledge filter" works, at least ideally, by recognizing and removing weaknesses such as personal biases during the different evolutionary stages of the idea.

2-3   The Greek Geocentric Model

To trace the evolution of our present model of the system of the heavens, we begin with the advanced Greek culture that lasted from about 600 BC until about 200 AD. Before looking directly at the Greek model, however, it is necessary to begin with a discussion of the world as seen by some of the ancient Greek philosophers, especially Aristotle.

There is a fundamental difference between the contributions to astronomy made by the ancient Greeks and those made by the other ancient civilizations. The Babylonians studied the heavens mostly because they were interested in interpreting daily events, in the same way as some people do today with astrology. The ancient Egyptians studied the heavens because they were interested in making predictions for agricultural purposes. The ancient Chinese believed in a kind of astrology in which events in the heavens influenced events on Earth; thus, they were very good at recording celestial events. The ancient Greeks, however, were interested in astronomy because of a pure philosophical desire to understand how the universe works. They believed in, and looked for, a sense of symmetry, order, and unity in the cosmos. They took the first steps in creating a unified model of the universe.

Had we never seen the stars, the sun, and the heavens, none of the words we have spoken about the universe would have been uttered. But now, the sight of day and night and the revolutions of the years have created number and given us a concept of time as well as the power of inquiring about the nature of the universe, and from this source we have derived philosophy. No greater good ever was or will be given by the gods to mortal men.
Plato, in Timaeus

Thales of Miletus (about 600 BC) believed that rational thought can lead to understanding of the universe. He also speculated that the Sun and stars were not gods, as was then usually thought, but balls of fire. Pythagoras (about 530 BC), one of the first experimental scientists, believed that nature can be described by numbers (that is, mathematics) and was the first to propose that Earth is spherical. His students, called the Pythagoreans, proposed around 450 BC that the universe is spherical, with a central "fire" containing a force that controls all motion. Around this fire, in order outward from the center, move Earth, the Moon, the Sun, the five planets, and the stars. This system predates by more than 2000 years Nicolaus Copernicus' revolutionary model of the planets moving around the Sun. The Pythagoreans were also the first to use the observed roundness of the Earth's shadow on the Moon as a supporting argument for the Earth being a spherical body. Plato (about 380 BC) taught the ideas of Thales and believed that the planets are spheres moving in circular orbits. He reasoned that astronomy contributed to the civilization of humanity.

Aristotle rejected the idea of the Pythagoreans that the Earth moves around a central "fire" and placed it back at the center of the solar system. Even though to us this choice seems to be obviously wrong, we must keep in mind that Aristotle's reasoning was supported by the observational evidence of his time. He argued that if Earth were moving, we ought to be able to see changes in the relative positions of the various stars in the sky, just as, if you drive down a highway, you see changes in the relative positions of nearby and distant trees. Such a shift in position due to motion is called *parallax*.

Hold your thumb in front of your face while you close one eye and look at the wall across from you (**FIGURE 2-1**). Now, without moving your thumb, view the wall with the other eye. Wink with one eye and then the other. You will see that your thumb seems to move from one spot on the wall to another. Also, the closer your thumb is to your face, the greater the shift is in its position with respect to the background. The explanation, of course, is that you are looking at your thumb from a different location each time you change eyes. In general, if our observing location changes, nearby objects will appear to move with respect to distant objects. This observation is given the general name ***parallax***.

parallax The apparent shifting of nearby objects with respect to distant ones as the position of the observer changes.

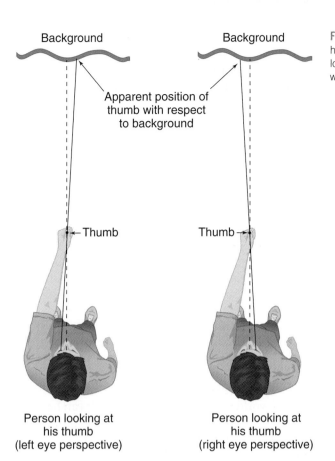

Background   Background

Apparent position of
thumb with respect
to background

←Thumb   Thumb→

Person looking at
his thumb
(left eye perspective)

Person looking at
his thumb
(right eye perspective)

FIGURE 2-1 To observe parallax, hold one thumb in front of you and look at it first with one eye and then with the other.

FIGURE 2-2 shows the Earth at two positions on opposite sides of its orbit. Just as your thumb shifted its apparent position when you alternately winked your eyes, if stars differ in their distances from Earth, a nearby star would be expected to shift its position relative to very distant stars. Such ***stellar parallax*** was first observed in 1838. All of the stars are so far away, however, that the greatest annual shift observed is only 1.5 arcseconds (corresponding to Proxima Centauri, the nearest star to the Sun), and this is why stellar parallax was not observed earlier.

A visual survey of the stars and constellations over time showed no evidence of such shift, and thus, Aristotle concluded that Earth must not move. This is a great example of how a perfectly correct logical argument can lead to wrong conclusions if based on incomplete data. One can hardly blame Aristotle for not being able to observe parallax, as the stars are so much farther away from Earth than anyone of his time could imagine.

Aristotle thought the Moon is spherical, probably by noticing that the line separating the lit side of the Moon from the unlit side changes its curvature as the phases progress. He believed that the Sun is farther away than the Moon because the Moon's

stellar parallax The apparent annual shifting of nearby stars with respect to background stars, measured as the angle of shift.

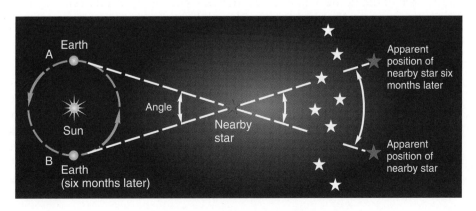

A Earth

Sun

B Earth
(six months later)

Angle

Nearby
star

Apparent
position of
nearby star six
months later

Apparent
position of
nearby star

FIGURE 2-2 As the Earth goes around the Sun, we see parallax of a nearby star as it shifts its position against background stars. The amount of parallax (half of the angle shown) is very much exaggerated here. For the nearest star, Proxima Centauri, the angle shown is only about 1.5 arcseconds (0.0004°). Accurate stellar parallaxes can currently be measured for stars up to about 1600 light-years away (corresponding to angles of about 4×10^{-3} arcseconds).

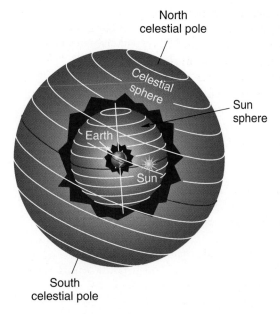

North
celestial pole

Celestial
sphere

Sun
sphere

Earth

Sun

South
celestial pole

FIGURE 2-3 To account for the Sun's path around the Earth (and among the stars), the Greek model located the Sun on a sphere that moves around the stationary Earth inside the celestial sphere of stars. The axis of the Sun's sphere is tilted with respect to the axis of the celestial sphere to account for the apparent path of the Sun in the sky.

Ptolemaic model The theory of the heavens devised by Ptolemy.

FIGURE 2-4 Ptolemy.

crescent phase shows that it passes between the Earth and Sun and that the Sun appears to move more slowly in the sky than the Moon. Aristotle used three observations to argue that the Earth is spherical, saying that only on a sphere do all falling bodies seek the center; that as a traveler goes north, more of the northern sky is exposed while the southern stars sink below the horizon; and that during lunar eclipses, the edge of the Earth's shadow is always circular.

Aristotle's system of the world made a great distinction between earthly things and the things of the heavens. He observed that matter on Earth and matter in the sky seem to behave fundamentally differently. For example, objects on Earth have an undeniable tendency to fall to the ground; they fall *down*. In fact, we might define the word "down" as the direction things fall. Objects in the sky don't do that; instead, they seem to move in circles around the Earth. Aristotle said that it is "natural" for earthly objects to move downward, that they seem *naturally* to seek the downward direction. On the other hand, it seems reasonable that heavenly objects move in circles. Something in the very nature of the objects seems to make them behave differently.

Aristotle saw another basic difference between the two types of objects: Although earthly objects always seem to come to a stop, heavenly objects just keep going. Thus, the "natural" motion of the two classes of objects appears to be different. This idea from Aristotle was a basic part of the world view of the Greeks, and it can be seen in their celestial models. They believed that there were two different sets of rules: one for earthly objects and one for celestial objects.

A celestial model developed by Greek thinkers before the time of Aristotle placed the stars on a sphere, similar to the model we described in Chapter 1. Instead of having the Sun move on a path on that sphere, however, the Sun was placed on another sphere that rotates around the Earth inside the sphere of the stars, as shown in **FIGURE 2-3**. To account for the fact that the Sun is sometimes seen north of the celestial equator and sometimes south, the Sun's sphere was thought to turn on a different axis from the one on which the stellar (celestial) sphere rotates. The difference between the tilts of the two spheres is shown in Figure 2-3.

There was good reason for the Greeks to use a sphere to carry the Sun, and the reason is linked to their admiration of geometry, which is traceable to the time of Pythagoras, the discoverer of the Pythagorean theorem. They sought geometric explanations for all natural phenomena. The Greeks then carried the idea of spheres much further and developed a model of the Earth, Sun, Moon, and planets using many spheres rotating around one another. For example, in Aristotle's model of the universe, every celestial object was attached to one of 55 crystalline spheres, each sphere having the Earth at the center and moving at different (but constant) speed around it. In order of distance from the Earth, Aristotle positioned the spheres carrying the Moon, Mercury, Venus, Sun, Mars, Jupiter, Saturn, the fixed stars, and finally the sphere of the Prime Mover. The Prime Mover caused the outermost sphere to rotate at constant speed, which in turn caused the next sphere to move, and so on. This model, however, could not explain the observations of retrograde motion and varying planetary brightness.

The Greek model we discuss most thoroughly is that of Claudius Ptolemy (**FIGURE 2-4**), who lived around 150 AD. In his book (which we call the *Almagest*), he presented a comprehensive model that lasted for almost 1400 years after his death. Today we know this geocentric model as the **Ptolemaic model**. Ptolemy abandoned the spheres of earlier Greek models, for he saw no need to imagine actual physical objects carrying the Sun, Moon, and planets. He still spoke of the stars, however, as being on the celestial sphere.

In accordance with the thinking of Aristotle, people of Ptolemy's time thought that things in the sky (the "heavens") must be perfect. It seemed reasonable that the heavens would feature the circle, a symmetrical shape with no beginning and no end. Indeed, the stars seem to move in circles around the Earth. It seemed natural that they would lie in a spherical arrangement and that the sphere would move around us at a constant speed. As we discussed in Chapter 1, the Moon changes its position among the stars from day to day just as the Sun does (but at a different speed).

The Moon fits the Ptolemaic scheme perfectly. All that is necessary is that it moves around the Earth as the closest heavenly body. In fact, the Moon fit the overall Greek philosophical approach. Being the closest to the imperfect Earth, the Moon might be expected to have imperfections on it. Indeed, it has dark and light areas. It is somewhere between the imperfect Earth and the perfect heavens.

The observed eastward motion of the planets also fits the Ptolemaic model. (In Section 1-7, we described the eastward and southward motion of Mars. The model can also account for the southward motion of Mars if the planet's orbital plane is placed at an angle with the celestial equator. Mars moves northward again later in the year. This is similar to how the model explains the Sun's motion.)

Any model of the planets must explain not only their observed retrograde motion, but also why they stay near the ecliptic and also the peculiar behavior of Mercury and Venus, which can never be seen high in the night sky. How did the Greeks account for the planets' unique motions? Were the heavens perhaps imperfect, with the planets not moving uniformly in a smooth circle?

A Model of Planetary Motion: Epicycles

Ptolemy was indeed able to use circles to make a model that fit the planets' motions. It simply took more than one circle for each planet. (Though this idea was originally used by Hipparchus, who lived about 300 years before Ptolemy, he did elaborate on it, and he is generally given credit for it.) FIGURE 2-5 shows Mars moving uniformly on a small circle (the *epicycle*) whose center moves uniformly around the Earth. By a correct choice of rotation speeds, an observed motion for Mars such as that shown in Figure 2-5 can be produced.

A person on Earth viewing Mars' motion among the stars sees the planet moving eastward most of the time. This corresponds to the eastward motion of the epicycle's

The Greeks listed the Sun and Moon as planets. This made seven planets, and this is the origin of our week having 7 days.

Having set ourselves the task to prove that the apparent irregularities of the five planets, the sun and the moon can all be represented by means of uniform circular motions, because only such motions are appropriate to their divine nature. We are entitled to regard the accomplishment of this task as the ultimate aim of mathematical science based on philosophy.
Ptolemy

epicycle The circular orbit of a planet in the Ptolemaic model, the center of which revolves around the Earth in another circle.

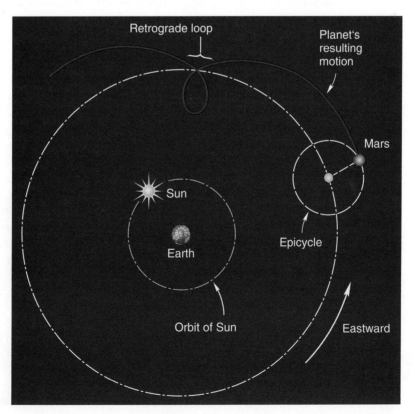

FIGURE 2-5 Mars' motion on its epicycle results in a looping path, which—when seen from Earth—appears as retrograde motion.

center around the Earth. Occasionally, however, the planet retrogrades. This occurs when the planet is inside the circular path of the epicycle's center and thus closer to the Earth. In this way, Ptolemy was able to preserve the idea of perfect circles and perfect uniform speeds for the heavenly objects. He could also explain why planets seem brightest when they are at the midpoint of their retrograde motion.

How, then, does the model explain why the planets never move far from the ecliptic? The answer is that the plane of their circular motion lies very close to the plane of the Sun's motion. The greater the angle between these two planes, the farther the planet will move from the ecliptic as it moves among the stars.

The different behavior of Mercury and Venus was explained by having the centers of their epicycles remain along a line between the Earth and the Sun (**FIGURE 2-6**). Although the center of other planets' epicycles could be anywhere on their circle around the Earth without regard to where the Sun is, Mercury and Venus are special cases.

The Ptolemaic model had many features that we look for in a scientific model. Even though the criteria for scientific models we discussed earlier in this chapter were not the standards of Ptolemy's times, it is instructive to apply them to the Ptolemaic model.

- The Ptolemaic model fits the data pretty well in that it gives an explanation for the motions of the heavenly objects. The stars' motions were explained by the rotation of the celestial sphere on which they reside. The Sun and the Moon were theorized to move in circles around the Earth. Explanation of the planets' motions required the use of epicycles and special treatment for Mercury and Venus, but when these features were included, the planets' motions were accounted for.

- The Ptolemaic model includes many testable predictions. The model made predictions for the locations of the planets at times in the future. One could use the model to predict that Jupiter would be at a particular place in the sky at a particular time the next year. Also, the Ptolemaic model holds that the Earth is stationary. It would predict that as knowledge advances and new methods are found to measure the motion of the Earth—either rotational motion or motion through space—no motion would be found. If further experimentation did not confirm this, the model would have to be adjusted drastically or abandoned.

- In the Ptolemaic model we find a form of symmetry in the sense that all objects travel in circles, and they basically obey the same rules as they move from east to

Actually, to make the model fit the data more accurately, Ptolemy either had to make the planets vary their speeds along their paths, or he had to move the Earth off-center among the planets' paths. He chose the latter method. After this and a few similar adjustments, the model fit pretty well but became even less aesthetically pleasing.

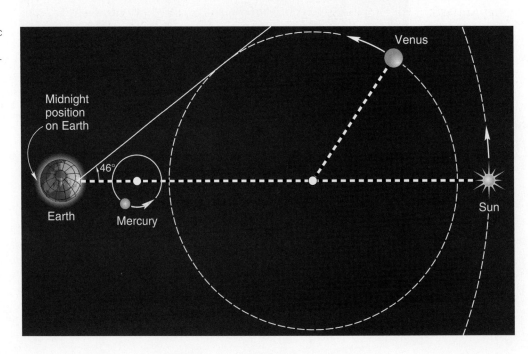

FIGURE 2-6 In the Ptolemaic model, the centers of Mercury's and Venus' epicycles stay between the Earth and the Sun. This accounts for the fact that the planets are never seen far from the Sun. Venus reaches a maximum of 46-degrees elongation. The drawing is not to scale.

west around the Earth. The model required the use of epicycles to explain retrograde motion, however, and included different rules for Mercury and Venus. The need for different rules for certain planets means that the model lacks an aspect of simplicity.

Remember, however, that the primary rule for a good model is that it fits the observations. An aesthetically pleasing model that did not come close to the real world would be almost useless. The Ptolemaic model did fit the data of the time, but to do so, it had to be continuously adjusted during the 1400 years of its reign.

2-4   Aristarchus' Heliocentric Model

About 400 years before Ptolemy, around 280 BC, the Greek philosopher Aristarchus had proposed a moving-Earth solution to explain the motions of the heavens. According to this model, the reason the sky seems to move westward is that the spherical Earth is spinning eastward. Although Ptolemy was aware of Aristarchus' model, he argued that if the Earth turned on an axis, it would be moving through the air around it, and, therefore, a tremendous wind should be observed in the opposite direction. Ptolemy referred to the moving-Earth model as being a "simpler conjecture." He saw that it was more aesthetically pleasing, but he believed it was not a good model because it presented obvious contradictions with the observation that the air stays where it is, along with everything loose on Earth. Ptolemy was unable to see that the Earth might be carrying the air and everything on it along as it rotates.

Aristarchus' model was a *heliocentric* model, constructed about 1800 years before Copernicus proposed the one that we use today (in a modified form). Aristarchus visualized the Moon in orbit around a spherical Earth and the Earth in orbit around the Sun. One could argue that Aristotle's argument, based on the lack of observable parallax for the stars, clearly shows that Aristarchus' heliocentric model is wrong. Why, then, did Aristarchus propose such a model? As shown below, the answer lies in the results he obtained about the relative distances and sizes of the Earth, Moon, and Sun.

Relative Distance to the Sun   When the Moon is exactly half lit, Aristarchus correctly stated that the angle between the Earth-Moon-Sun is 90 degrees (FIGURE 2-7). However, for the angle Moon-Earth-Sun, he used 87 degrees, instead of the correct value of 89.85 degrees. (A calculation at the end of the chapter offers a possible method that Aristarchus could have used to measure this angle.) Knowing the angles in the right triangle Earth-Moon-Sun, he was then able to obtain a relationship between

The gods did not reveal, form the beginning, all things to us; but in the course of time, through seeking, men find that which is better. But as for certain truth, no man has known it, nor will he know it; neither of the gods, nor yet all things of which I speak. And even if by chance he were to utter the final Truth, he would himself not know it; for all is but a woven web of guesses.

Xenophanes, Greek Historian, 6th century BC

heliocentric Centered on the Sun.

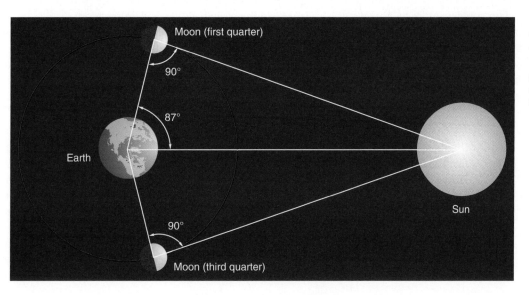

FIGURE 2-7 The positions of the Earth, Moon, and Sun when the Moon is exactly half lit. Aristarchus used this geometry to find the relative distances of the Moon and Sun from Earth.

ADVANCING THE MODEL

Astrology and Science

Astrology is the belief that the relative positions of the Sun, planets, and Moon affect the destiny of humans. It is often used to determine the nature of a person's personality or of certain circumstances in the future. To discover the astrological effect on an individual, an astrologer designs, or "casts," a horoscope.

Casting a horoscope involves determining the positions of significant heavenly objects at the moment of the person's birth. To do so, the zodiac is divided into 12 equal-sized *signs*, which have the same names as the 12 principal constellations of the zodiac, but differ from them in two ways. First, the constellations differ from one another in size, whereas all signs are the same size. The constellation Cancer, for example, is much smaller than Virgo or Pisces, but their signs are the same size. Second, the signs are at different locations in the sky from the constellations. At one time the positions corresponded fairly well, but because of *precession* (to be discussed in Chapter 3), the signs and constellations are now "off" by about one position. The sign of Aries, for example, is located mostly in the constellation Pisces. This means that a person born on March 22 is said by astrologers to be an Aries, while the Sun actually appeared to be in the constellation Pisces at the person's birth. (Some astrologers, called sidereal astrologers, use the constellations rather than the signs.)

Astrologers divide the sky above us into 12 *houses*, starting with the eastern horizon and proceeding around the sky. The houses do not rotate with the stars, but remain in the same position relative to Earth. Thus, the first house stays above the eastern horizon. The houses are arbitrarily associated with various areas of our lives, such as children, health, and enemies. Likewise, the planets are associated with various good or evil influences. Mars, for example, is associated with war.

When the previous information is determined for the moment of a person's birth, we have what is called the *natal chart* for that person. This chart could be made by anyone who knows the sky and has tables of solar, lunar, and planetary positions. It might be determined, for example, that the planet Mars was just above the eastern horizon at the time of your birth. This is where the astrologer's interpretation comes in, telling you about your personality and life pattern.

Scientific Criteria Applied to Astrology

Having briefly described the nature of astrology, we will now apply to it the criteria for a good scientific theory.

- **Does astrology fit the observations?** It claims to be able to account (at least in part) for our personalities and our characteristics. Can this be true? This question has been tested many times. Invariably, the tests show that astrologers are unable to identify the personality traits of people or to predict future events. (For example, read the report in the December 5, 1985, issue of *Nature* concerning the research done by Shawn Carlson at the University of California.) After numerous tests, we must conclude that astrology fails our first criterion.

- **Does it make verifiable predictions that could possibly prove the theory wrong?** In Carlson's study, astrologers claimed that if enough cases were considered, they could make verifiable predictions. Astrologers would say that it is not fair to test the accuracy of a horoscope of an individual person because it only predicts tendencies, and people do not necessarily follow their tendencies. In other words, you might be very different from what is predicted by your horoscope and still not contradict astrological predictions. If this is the case, there is no way that astrology can be proved incorrect. It therefore fails the second criterion.

- **Simplicity and aesthetics.** Astrology does not pass this test because it is not a single, unified theory at all, but instead is a group of arbitrary rules as to the effect of heavenly objects on people. One could hardly call it aesthetically pleasing.

Because astrology is a faith system, it cannot be disproved to those who believe. People who believe in it do not care whether scientists can explain the belief.

Scientists do not accept astrology as a valid theory. Some scientists strongly object to the publication of astrological columns in newspapers, but others claim that astrology does no harm if people use it only as a pastime and do not let it actually determine what they do with their lives.

---

the distances Earth-Moon and Earth-Sun. He found that the Sun is about 20 times farther from the Earth than the Moon is, instead of the correct value of about 390. However, this large difference of a factor of about 20 is due to the difference of only about 3 degrees in the choice of the angle Moon-Earth-Sun. What is important here is not the quantitative error, but the power and simplicity of the argument he used to find, 2000 years before anybody else, the relative distance to the Sun.

Relative Sizes of the Moon and Earth From the time it takes the Moon to move through the Earth's shadow during a total lunar eclipse, Aristarchus concluded that the Earth's diameter is about three times larger than the Moon's diameter (the correct ratio is about 3.7). The geometry by which he estimated the relative sizes of the Moon and Earth is shown in **FIGURE 2-8**. The fact that total solar eclipses just barely occur leads to the conclusion that the angular diameters of the Moon and Sun are about the same as seen from Earth. Aristarchus measured this angular diameter to be 0.5°, and thus he represented Earth's shadow by lines leading away from Earth and converging at an angle of 0.5°. These criteria specified the position and size of the Moon in the Earth's shadow.

Relative Sizes of the Moon and Sun Because the angular diameter of the Moon and the Sun are the same (0.5°) as seen from Earth, Aristarchus correctly concluded that

$$\frac{\text{radius of Moon}}{\text{radius of Sun}} = \frac{\text{distance from Earth to Moon}}{\text{distance from Earth to Sun}}.$$

Using the calculations mentioned here, he found that the Sun is about 20 times larger than the Moon, instead of the correct ratio of about 390; this error is the result of the difference of only 3 degrees in the choice of the angle Moon-Earth-Sun shown in Figure 2-7.

These calculations clearly showed the Sun being much bigger than Earth. As a result, Aristarchus concluded that the Sun, not Earth, must be the central body in the solar system, thus proposing the first heliocentric system. For this, an outraged critic declared he should be charged with impiety.

Aristarchus had made a map of the solar system but did not have the scale for it. If he knew the radius of the Earth and the distance to the Moon, then he could accurately find how big the Moon and Sun were and their exact relative distances.

Measuring the Size of the Earth

The first person to understand clearly the shape and approximate size of the Earth, 1700 years before Columbus, was Eratosthenes (276–195 BC). His calculation of the radius of the Earth is very simple (**FIGURE 2-9**).

Eratosthenes knew that at noon, during the summer solstice, the Sun shone directly down a well near Syene (the present city of Aswan, in Egypt), casting no shadow. This implied that the Sun was directly overhead, or at the *zenith* at that time.

In his city of Alexandria, on the same day, he noted that the Sun's direction was off the vertical by about 7 degrees. Because he was aware of Aristarchus' results, Eratosthenes could safely assume that the Sun is far away from the Earth, and thus, its light travels in almost parallel rays toward the Earth. He realized that the difference of 7 degrees had to be due to the curvature of the Earth. Because 7 degrees is about 1/50 of a full circle (7/360), Eratosthenes concluded that the Earth's circumference was about 360/7 or 50 times the distance from Alexandria to the site of the well. This distance was about 5000 stadia, where a stadium was a unit of distance equivalent to about 0.15 to 0.2 kilometers. If we assume that one stadium was 1/6 km, then the distance between the two cities is about 830 kilometers. Eratosthenes thus calculated the Earth's circumference to be 50 × 830 = 41,500 kilometers and thus the Earth's radius to be 41,500/2π or about 6600 km. This is amazingly close to the correct value of 6378 kilometers!

Combining the calculations of Aristarchus and Eratosthenes, the ancient Greeks had, for the first time,

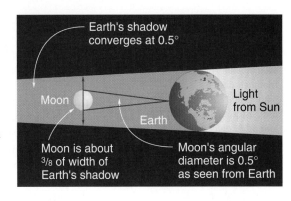

FIGURE 2-8 Aristarchus used the time it takes for the Moon to go through the Earth's shadow to find the relative size of the Earth and Moon.

zenith The point in the sky located directly overhead.

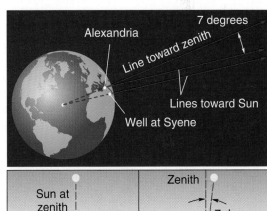

FIGURE 2-9 Eratosthenes calculated the size of the Earth by comparing the Sun's direction off the vertical at two locations at the same time. When the Sun was overhead at Syene, it was 7 degrees from the zenith at Alexandria.

measurements of the radii of the Earth, Moon, and Sun and the relative distances between them. Even though later astronomers obtained more accurate measurements, we would have to wait until 1769 AD to obtain the actual value of the astronomical unit and thus the true dimensions of the *solar system*.

solar system The system that includes our Sun, planets and their satellites, asteroids, comets, and other objects that orbit our Sun.

What is important here is not the accuracy of the measurements. It is the fact that simple logical arguments allowed the ancient Greeks to have a very good sense of the solar system more than 2000 years ago. Sometimes an idea is so revolutionary that it takes a long time for others to become convinced. In this case, the lack of observable parallax for the stars allowed the Ptolemaic model (instead of Aristarchus') to be the accepted model for the heavens for almost 1400 years, until a little after the time of Copernicus. Parallax of stars was successfully measured around 1838 AD. As a result, Aristotle's argument convinced one of the greatest observers of all times, Tycho Brahe (1546–1601), to believe in a geocentric universe decades after the Copernican revolution.

There is a tendency to think that Greek science was bad science because it is not today's science. This is simply not true. It is true that the Ptolemaic model is more primitive than today's. The Greeks did not have the accurate observations and extensive data we have today. Their model, however, did fit the data they had. In fact, it fits the casual observations of most people today. Can you think of any direct evidence obtainable without a large telescope that contradicts Ptolemy's model? What observations can you personally make, without relying on reference materials, which will show the 2000-year-old Ptolemaic model to be a poor model?

2-5   The Marriage of Aristotle and Christianity

During the 13th century, Saint Thomas Aquinas (1225–1274 AD), one of the greatest theologians and philosophers of the Christian church, incorporated the works of Aristotle and Ptolemy into Christian thinking. Aquinas insisted that there must be no conflict between faith and reason, and he blended the natural philosophy of Aristotle with Christian revelation. For the Aristotelian and Ptolemaic ideas we have discussed, the blend was an easy one. The idea of an Earth-centered universe fit comfortably with literal biblical interpretation, for it placed humans at the center of God's creation—the ultimate expression of the divine will (**FIGURE 2-10**).

The idea of a central, unmoving Earth was natural for early humans. Through the work of Aquinas, this easily accepted idea was shown to fit perfectly with Christian beliefs. So Aristotle's science—and with it the Ptolemaic model—became even more entrenched in Western culture. It was no longer just a natural, normal way of thinking about the world, but was part of Christian thinking and religious dogma.

Why have we presented material that seems to be more closely related to Church history than to science? The reason is that science does not exist apart from human culture. Science and scientists are part of the society in which they live, and it is necessary to know something of the flavor of an age to appreciate the work of the thinkers of that age—including those thinkers who today might be labeled as scientists. It, therefore, is important to emphasize that after the 13th century, the teachings of Aristotle and the Ptolemaic model were an ingrained part of Western thinking.

A very important characteristic of the Middle Ages was a great reliance on authority, particularly on authorities of the past. Today most of us have a much greater tendency to rely on our own thoughts, observations, experiences, and feelings than did people of the times we are discussing. Aquinas relied

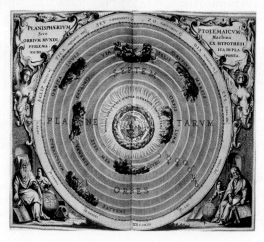

FIGURE 2-10 A medieval illustration from a world atlas of 1661, showing a model of the universe with the Earth at its center. This was the model approved by the Catholic Church, which had the power to punish anyone who argued against this model. Each planet is shown being carried by a chariot drawn by winged horses.

on the authority of the Bible, the authority of earlier churchmen, the authority of his superiors in the Church, and the authority of Aristotle. Arguments were often settled by reference to authorities rather than by personal experience or independent experimentation. Into this world came a man who was to cause a revolution in the way people think.

2-6   Nicolaus Copernicus and the Heliocentric Model

At the time when Columbus was journeying to America, a brilliant Polish student was studying astronomy, mathematics, medicine, economics, law, and theology. As was common for scholars of that time, Nicolaus Copernicus did not limit himself to studies in a particular discipline as we do today (perhaps necessarily). He is known today, however, for initiating the revolution that finally resulted in replacing the geocentric system of Ptolemy with a heliocentric—Sun-centered—system.

Copernicus worked on his heliocentric system for nearly 40 years. Even after all this work, he was slow to publish it. It was eventually published with the title *On the Revolutions of the Heavenly Spheres*, but it is often called *De Revolutionibus*, a shortened form of the original Latin title (**FIGURE 2-11**).

Copernicus apparently decided to develop and publish his model for three primary reasons. First, as the centuries had passed, it was found that Ptolemy's predicted positions for celestial objects were not in agreement with the carefully observed positions. For example, Ptolemy's model could be used to predict the position of Saturn on some night in the future. The prediction of its position for a night a few years later was accurate enough, but when the prediction was made for a few centuries in the future, the predicted position might differ from the carefully observed position by as much as 2 degrees. This difference corresponds to about four times the angular diameter of the Moon as observed from Earth. Ptolemy's model had been updated several times through the years so that it would be fairly accurate for awhile. These corrections were not refinements in the way it worked, but were "resettings" of each planet's position that were made to fit the latest observations. As we discussed in Section 2-1, a good model would not require such updating. Copernicus recognized this and sought a more accurate model.

A second reason Copernicus sought another model was that he became convinced after studying the ancient writings that a Sun-centered system would not only provide better data for use by the Church and by astrologers, but that it would also be a more aesthetically pleasing system. He linked astronomy very closely with his religious faith, arguing that placing the Sun at the center of everything is completely logical because the Sun is the source of light and life. The Creator, he said, would naturally place it at the center.

Finally, it has long been known that planets change their brightnesses as they shift positions among the stars. The Ptolemaic system accounted for this with its epicycles, for they would cause a planet to change its distance from the Sun and from Earth. Copernicus noticed, however, that Mars seemed to change in brightness even more than would be predicted by that model. The epicycles of Ptolemy were simply not large enough to explain the great changes in Mars' brightness (Figure 2-5). Thus, Ptolemy's model explained the basic phenomenon, but it did not explain it very precisely when

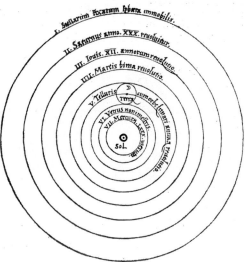

FIGURE 2-11 The Sun-centered model of the Universe that appeared in Copernicus' *De Revolutionibus*. It is said that a copy of the printed book was rushed to Copernicus so that he could see it as he lay on his death bed.

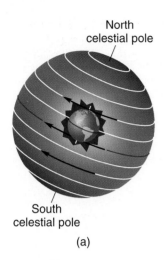 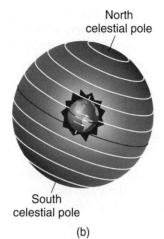

North celestial pole

North celestial pole

South celestial pole

South celestial pole

(a)

(b)

FIGURE 2-12 (a) The celestial sphere rotating toward the west around a stationary Earth produces the same observed motion of the stars as (b) the Earth rotating toward the east under a stationary celestial sphere.

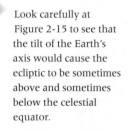 Try the first Activity at the end of this chapter to appreciate the change in outlook as one switches from a geocentric to a heliocentric system.

Look carefully at Figure 2-15 to see that the tilt of the Earth's axis would cause the ecliptic to be sometimes above and sometimes below the celestial equator.

accurate data were used. It was this observation that first prompted Copernicus to reconsider the heliocentric system of Aristarchus, which had lain hidden in obscurity for 2000 years.

We return to a discussion of the criteria for a good model after we have described the model developed by Copernicus.

The Copernican System

Copernicus' system revived many of the ideas of Aristarchus (**FIGURE 2-12**). In answering Ptolemy's argument that a rotating Earth would produce a great wind, Copernicus stated as an assumption that the air around the Earth simply follows the Earth around. How high above the Earth does this air extend? Copernicus did not answer this question, but his theory requires that the air does not extend all the way to the stars, or even to the Moon.

In Copernicus' heliocentric system, the Earth assumes the role of just another one of the planets, all of them revolving around the Sun. **FIGURE 2-13** shows the order of the planets that were known in the 16th century. If we consider the view of Figure 2-13 to be that of someone far out in space above the Northern Hemisphere of the Earth, the planets all move around in a counterclockwise direction, as shown by the arrows in this figure.

FIGURE 2-14 shows the Earth in a circular orbit around the Sun. Its path and the band of stars around the Sun do not appear as circles in the drawing because we are seeing them at an angle. (Remember, we are describing Copernicus' model, not today's model.) In the drawing, when viewed from Earth, the Sun can be thought of as located among certain stars. As the Earth moves around the Sun, the position of the Sun seems to be moving across the background of the stars, which is what we observe. Ptolemy explained it by having the Sun revolve around the Earth independent of the celestial sphere. Copernicus explained it by having the Earth move around a stationary Sun.

Recall from Figure 1-18 that the Sun appears to move from among the stars south of the equator to those north of the equator and back again. To fit this observation, Copernicus' model had the plane of the Earth's equator tilted with respect to the plane of its orbit around the Sun (**FIGURE 2-15**). (In Chapter 1 we discuss how and why this tilt causes the seasons on Earth.)

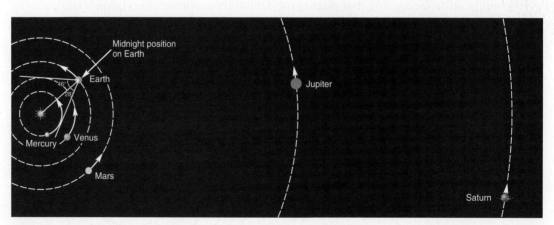

FIGURE 2-13 The planets that can be seen by naked-eye observations, shown in order from the Sun. Their relative orbital speeds are indicated by the lengths of the arrows. Planetary sizes and relative distances from the Sun are not to scale. This figure illustrates the heliocentric model's explanation for the fact that Mercury and Venus never get far from the Sun and can never be seen in the sky late at night. Compare it with Figure 2-6.

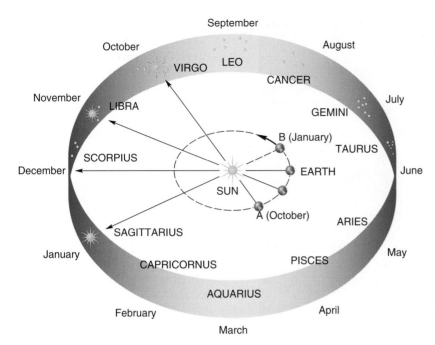

FIGURE 2-14 As the Earth moves around the Sun during the year, the Sun appears to move among the background stars. For example, when the Earth is in position A on its orbit (on October 1), the Sun appears to be in Virgo. (Of course, we cannot see the stars in Virgo or any stars for that matter during daytime.) On this date, however, an observer on Earth's midnight position sees Pisces all night long. Similarly, when the Earth is in position B (on January 1), the Sun appears to be in Sagittarius; however, an observer on Earth's midnight position sees Gemini all night long.

The Moon appears to complete about one revolution around the Earth each month. Copernicus accounted for this in his model by having the Moon on a sphere that revolves around the Earth in that time. For Copernicus, everything else circles the Sun, but the Moon revolves around the Earth.

We stated in Chapter 1 that someone watching the planets from Earth would see that these wanderers spend most of their time moving eastward against the background of the stars, but occasionally, according to a regular schedule that is different for each planet, they move in the opposite direction. Ptolemy explained this motion using epicycles. How does Copernicus' model handle retrograde motion? **FIGURE 2-16** shows several equally spaced positions of the Earth in its orbit, along with the corresponding positions of Mars. This is because Copernicus—like Ptolemy—assumed that the planets move at

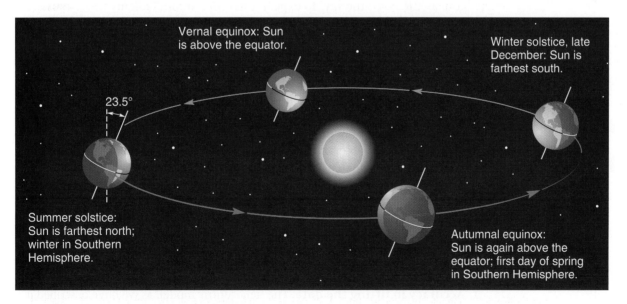

Vernal equinox: Sun is above the equator.

Winter solstice, late December: Sun is farthest south.

23.5°

Summer solstice: Sun is farthest north; winter in Southern Hemisphere.

Autumnal equinox: Sun is again above the equator; first day of spring in Southern Hemisphere.

FIGURE 2-15 The Copernican model explains the Sun's apparent motion north and south of the equator by having the Earth's equator tilted with respect to the planet's orbit around the sun.

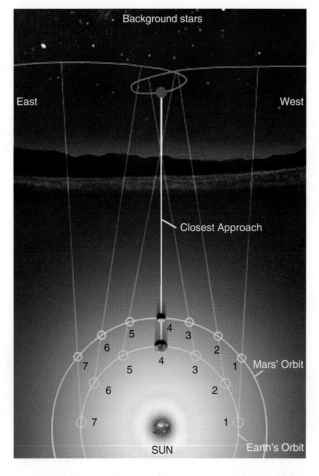

FIGURE 2-16 This drawing corresponds to Mars' 2003 retrograde motion. While Earth moves from position 1 to position 3, Mars appears to be moving eastward among the background stars. Then as Earth passes Mars, Mars appears to move backward before resuming its eastward motion. This is how the heliocentric model explains retrograde motion. At position 4, Mars is in the middle of its retrograde motion and closer to Earth; as a result, Mars looks brighter.

To picture how this works, think of running circular laps at constant speeds with a friend; you are running on the inside track and faster than your friend who is running on an outer track. Watch your friend against a distant background.

constant speeds. The lines from Earth to Mars illustrate the direction in which we on Earth see Mars. As both planets move on their orbits, Mars seems to move eastward (for the first three positions) among the background stars. Between locations 3 and 5, however, Mars follows a retrograde motion, after which it resumes its eastward motion. Mars reaches its closest distance to Earth at position 4, which explains why the planet looks brighter in the middle of its retrograde motion.

Copernicus was able to calculate the relative distances of the planets from the Sun. Two activities at the end of this chapter show the calculations for Mars and Venus. By such methods, Copernicus predicted that Mars is 1.5 times farther from the Sun than the Earth is. Similar calculations for the other planets yielded for Copernicus the values shown in the center column of **TABLE 2-1**. The table also shows today's measurements. Copernicus had a map of the solar system but without a scale; he did not know the value of the astronomical unit. How do we know the actual distances today? One way to determine distance is to bounce a radar beam from a planet and measure the time it takes for the beam to return. Knowing the speed of the radar signal, we can calculate the distance to the planet and from this the value of the astronomical unit.

Mercury and Venus never get very far from the Sun in the sky. To explain this observation, Ptolemy treated these planets as special cases by requiring that the centers of their epicycles remain along a line between the Earth and the Sun (Figure 2-6). As shown in Figure 2-13, in the Copernican system neither planet can get very far from the Sun as seen from Earth simply because they orbit the Sun at a distance less than the Earth's distance. No matter where Venus is in its orbit, it can never appear farther than 46 degrees away from the Sun as viewed from Earth. There is no way for an observer located at the midnight position on Earth to see Mercury or Venus at midnight.

TABLE 2-1		
Planetary Distances: Copernicus' Values versus Today's Values		
	Sun-to-Planet Distances	
Planet	Copernicus' Values (AU)	Today's Values (AU)
Mercury	0.38	0.387
Venus	0.72	0.723
Earth	1.00	1.000
Mars	1.52	1.524
Jupiter	5.2	5.204
Saturn	9.2	9.582

2-7   Comparing the Two Models

In Section 2-1 we discussed the scientific criteria for deciding the value of a proposed model. Let's use these criteria to compare the Copernican and Ptolemaic models.

1. Accuracy in fitting the data. The heliocentric model, as we have described it, is able to account qualitatively for the observed motions of the stars, the Moon, and the planets; however, there were differences between the predicted and observed planetary positions. Copernicus sought to increase the accuracy of his model by inserting small

HISTORICAL NOTE

Copernicus and His Times

Mikolaj Koppernigk (or Nicolaus Copernicus by his Latinized name; **FIGURE B2-1**) was born in Torun, Poland and was educated in Poland and Italy. During most of his life he worked for the Church, serving as canon of the cathedral of Frauenberg, Poland. He lived during an exciting and disturbing time, as a list of some of his contemporaries suggests: Henry VIII, Martin Luther, Michelangelo, Leonardo da Vinci, Christopher Columbus, and Raphael.

Copernicus entered the University of Bologna (Italy) in 1496 and was liberally educated in several subjects. His great interest in astronomy can be attributed to the close link between astronomy and religion at the time. There were two reasons for this link: First, it was important to the Church that its feast days be celebrated at the right time, and this required a correct calendar—a matter for astronomers. (The dates of Easter and the feasts that follow it depend even today upon accurate determination of the time of the vernal equinox.) Second, at the time of Copernicus, astrology was considered a more important study than astronomy, but astrology cannot function without accurate astronomical data.

FIGURE B2-1 Nicolaus Copernicus (1473–1543).

Copernicus lived about 100 years before the invention of the telescope. Measurements of the heavens had to be made with the naked eye, and yet accurate measurements were necessary. One important measurement was the exact time that a celestial object crossed the *meridian*. Copernicus made this measurement by having his house constructed with a narrow slot in one of its walls. By placing himself at the correct location and looking through the slot, he could determine accurately when a given object was crossing (or *transiting*) the meridian.

epicycles for the celestial objects to move on. This was necessary because he assumed—like Ptolemy—that the planets move at a constant speed. In fact, the observed motions did not correspond exactly to planets moving at constant speed. Copernicus' epicycles were smaller than Ptolemy's, and the motion of the planets around them was much slower so that each planet moved in a somewhat elongated circle.

Why did Copernicus not abandon the idea of circles altogether? The reason lies partly in the fact that the circle had such a long tradition and partly in Copernicus' belief that because the motions in the heavens are repetitive, "a circle alone can thus restore the place of a body as it was."

Even with his epicycles, Copernicus' model resulted in errors in predicting planetary positions of about 2 degrees over a few centuries of planetary motion. This is about the same error as in Ptolemy's model. No definitive observation could be made in the 1500s that would decide which of the two competing models fit the data more accurately. For that reason, using the evidence available in the 1500s, we must call our comparison of the models a draw, based on the criterion of data fitting. (Observing stellar parallax, for example, would have clearly shown that the Earth moves and thus supported the heliocentric model. Such observations would also have provided evidence that stars are not all at the same distance from Earth, something that both models assumed; however, stellar parallax was not observed until 1838, and the question had been settled by that time as a result of the contributions of Kepler and Galileo, which we discuss shortly.)

2. Predictive power. Both models made testable predictions. The Ptolemaic model predicted that stellar parallax should not exist, whereas the Copernican model predicted that it would; thus, both pass this test. (A prediction that is eventually disproved by new data is as acceptable as a prediction that is eventually supported by new data.)

The calculations of planetary distances based on the heliocentric model serve as an example of testable predictions made by the model. If later measurements had shown the distances to be wrong, the theory would have suffered a setback.

In the center of everything rules the sun; for who in this most beautiful temple could place this luminary at another or better place whence it can light up the whole at once?...In this arrangement we thus find an admirable harmony of the world, and a constant harmonious connection between the motion and the size of the orbits as could not be found otherwise. Copernicus

As the example of parallax shows, predictions are inherent in models. They need not be stated by the model's designer.

3. Simplicity: Mercury and Venus. Copernicus, and other thinkers of his time and the years following his death, preferred the heliocentric model to the Ptolemaic model even though it had no particular advantage as far as the criteria we have discussed so far. The main reason for this preference was due to our third criterion for a good model: sim-plicity. To make the comparison, it is instructive to see how the two models treated the motions of Mercury and Venus and the observed retrograde planetary motions.

Both models are able to explain why the observed motions of Mercury and Venus differ from those of the other planets, and both models are able to explain why Mercury and Venus are never seen high in the night sky.

The Ptolemaic model, however, required a special rule to bring this about (Figure 2-6). In the Copernican model, this was a natural consequence of the fact that Mercury's and Venus' orbits are inside the Earth's orbit (Figure 2-13).

Ptolemy used epicycles to explain retrograde motion. Each planet was assigned an epicycle of a certain size, a speed for its motion around the epicycle, and another speed for the center of the epicycle moving around the Earth. The speeds that were assigned did not form a pattern, and as we have seen, Mercury and Venus were special cases. Copernicus, on the other hand, saw a pattern in planetary speeds. According to his model, the farther from the Sun a planet is, the slower it moves in its orbit. This rule of motion was all that was needed to account for retrograde motion. (Copernicus found epicycles necessary to make his model more accurate but did not use them to explain retrograde motion.)

Overall, the descriptions of planetary motions offered by the heliocentric model were simpler, and Copernicus (and others) felt that his was a better model for ex-plaining the heavens. But when one considers both models in their full detail, the final verdict in favor of the heliocentric model seems less than overwhelming.

Why were people of Copernicus' time concerned with how simple and orderly a theory is? This concern arose from a religious belief that a Creator would not have created a disorderly, overly complicated universe. Scientists today would express their desire for simple models in different terms, but the desire is still present. One reason scientists search for a simple model is that our experience since the time of Copernicus tells us that when there are two competing models, the simpler one is more likely to fit new data—data from observations not yet made (Occam's razor). The belief that this pattern will continue with new models is a faith different from that of Copernicus and his supporters. It is a faith in the unity and beauty of nature, based on past experience.

Copernicus died in 1543, just as *De Revolutionibus* was being released. This single book started such an upheaval in people's thinking that the word "revolution" took on a second meaning—an upheaval, as in a social or political revolution. Copernicus began the revolution, but it was up to others to carry it forward to completion.

2-8   Tycho Brahe: The Importance of Accurate Observations

Three years after Copernicus died, Tycho Brahe (pronounced "bra" or "bra-uh") was born in Denmark (**FIGURE 2-17**). When he was a teenager studying law, he developed an interest in astronomy and learned that both the Ptolemaic and Copernican mod-els were based on recorded planetary data that were inaccurate. That is, he found discrepancies between different tables of observed planetary positions. He was con-vinced that before a decision could be made about which model was better or before a completely new and better model could be devised, there was a need for more accurate observations of the positions of the planets as they move across the back-ground of stars.

At this time, the telescope had not yet been invented, and thus, all of Tycho's observations were made with the naked eye. In an observatory built for him by King Frederick II of Denmark, Tycho built the largest observing instruments yet constructed. The mural in **FIGURE 2-18** shows some of his equipment. Because of the large size of this equipment, Tycho was able to measure angles to an accuracy of better than 0.1°, far more accurate than any measurements made before his time and close to the limit the human eye can observe.

In addition to making such accurate measurements, Tycho was careful to determine the accuracy of each individual measurement. For example, he would not only state the position of a particular planet, but would also give a value indicating the amount of uncertainty in his measurement. Because of this statement of uncertainty, when such data are later compared with the predictions of a particular model to test the model for accuracy, we can tell how close the predictions of the model should be expected to come to the reported measurements. For example, if Tycho stated that a particular measurement was accurate to 0.1° of angle, then we should expect a model's predictions to come within 0.1° of that measurement. On the other hand, if the model did not make the prediction within 0.1° of the measurement, it would have to be considered not accurate enough. The inclusion of a statement concerning the accuracy of measurements is now a common practice in science, but it was not in Tycho's time.

One tenth of a degree is the angle intercepted by a quarter when viewed face-on from a distance of 50 feet (15 meters).

Tycho's Model

You might expect that the quality of Tycho's observations would convince him of the correctness of the Copernican model. Exactly the opposite was true, for a reason having to do with parallax. Tycho argued that if the Earth were in motion around the Sun, as the Copernican model suggested, then nearby stars should show parallax as the Earth revolved around the Sun. Because he failed to observe any such parallax, he concluded that the Earth must be at the center of the universe and that the Copernican model was wrong. Tycho became famous for his observations of two very important phenomena early on in his career. In 1572 he observed a bright "star," what we now call a supernova explosion, which suddenly appeared in the sky and was observable for about 18 months, and in 1577 he observed a bright comet. What is important about these observations is that Tycho tried to measure the parallax of each of these two objects, but failed. As a result, he concluded that these objects must be very far from the Earth, showing that change occurs in the heavens, a view contrary to thousands of years of traditional belief.

Tycho's model of the heavens was a mix between the Ptolemaic and Copernican models (**FIGURE 2-19**). He positioned the Earth at the center with the Sun revolving around it, but he argued that the other planets were revolving around the Sun. Tycho's conclusion was wrong, but his argument for placing the Earth at the center was a very good one. It was the

FIGURE 2-18 This mural depicts Tycho's quadrant, which was designed to measure the time and the elevation angle of an object crossing the meridian. The quadrant itself consists of the slot at upper left and the large quarter circle. Tycho is shown as the observer at the edge of the mural at right center. In addition, he had himself painted on the quadrant (the large figure on the wall near center) to remind his assistants that he was watching. Tycho's quadrant had a radius of over 2 meters. The large quarter circle in the foreground is drawn to the scale of the people in front. The person shown partially on the drawing at right center is looking past a pointer on the degree markings and through the slot in the wall at upper left.

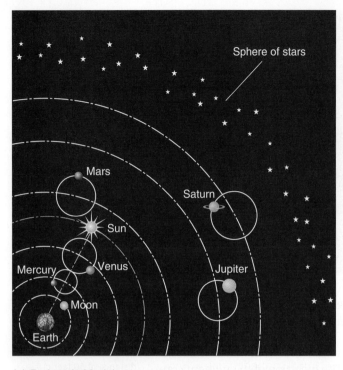
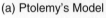

(a) Ptolemy's Model

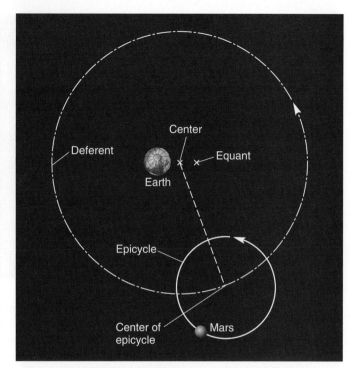

(b) Ptolemy's Model (modified)

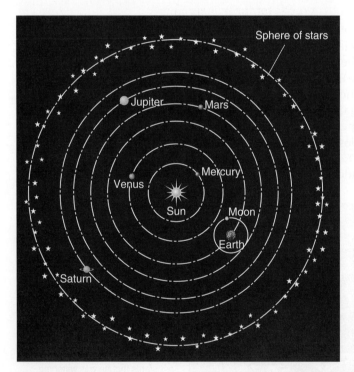

(c) Copernicus' Model

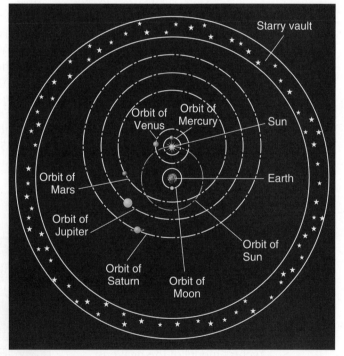

(d) Tycho's Model

FIGURE 2-19 The models of Ptolemy, Copernicus, and Tycho. The drawings are not to scale. (a) Ptolemy's Earth-centered epicycle theory (around 100 AD) of the layout of the universe, according to which the five visible planets move on epicycles around the Earth. The epicycles of the two innermost planets, Mercury and Venus, are centered on the line joining Earth to the Sun. (b) In this modification of his model, Ptolemy has the Earth displaced slightly from the center of the *deferent* (the circle on which the center of a planet's epicycle moves). The center of the epicycle moves at a constant speed as seen from the equant but not as seen from the Earth or the center. (c) Copernicus' Sun-centered theory of the layout of the universe, 1543 AD. The diagram is simplified; the planets all move on epicycles, similar to those in Ptolemy's theory, but here there are far fewer epicycles. Furthermore, Copernicus used no equants. (d) Tycho Brahe's Earth-centered model of the universe.

same argument used by the ancient Greeks, based on the lack of observable parallax for the stars; as we now know, even the nearby stars are so far away from us that their parallax is very difficult to measure. However, this clearly shows the importance of having many independent observations of the same phenomenon and the difficulty in understanding the universe based on data that will always be incomplete.

Night after night for 20 years, Tycho recorded data concerning the positions of the planets, with particular emphasis on Mars. This data will prove invaluable to Johannes Kepler, an assistant Tycho hired in 1600, the year before he died.

2-9 Johannes Kepler and the Laws of Planetary Motion

During his studies of theology, Kepler learned of the Copernican system and became an advocate for it. As Tycho's assistant, Kepler was assigned the job of working on models of planetary motion. After Tycho's death, Kepler took over most of Tycho's records, and 4 years later, after trying 70 different combinations of circles and epicycles, he finally devised a combination for Mars that would predict its position—when compared to Tycho's observations—to within 0.13° (about one fourth the diameter of the Moon). Recall that before this, a typical error between predicted positions and observed positions was about 2°; however, the error of 0.13° still exceeded the likely error in Tycho's measurements. Kepler knew enough about Tycho's methods to know that an error of 0.13° in the data was too much, and he sought a model that would fit the data to the limits of its precision. This is a tribute both to Kepler's persistence and to his belief in the accuracy of Tycho's careful measurements.

Finally, Kepler decided to abandon the idea of circular orbits for the planets and to try other shapes. He tried various ovals, with the speed of the planet changing in different ways as it went around the oval. After 9 years of work, he found a shape that fit satisfactorily with the observed path of Mars. What's more, he found that the same basic shape worked not just for Mars, but for every planet for which he had data. The shape? An ***ellipse***.

> **ellipse** A geometrical shape of which every point (*P*) is the same total distance from two fixed points (the foci). (That is, in Figure 2-20, $PF_1 + PF_2 =$ constant = major axis.)

The Ellipse

At first glance, an ellipse is nothing more than an oval, but it is more specific than that; not every oval is an ellipse. **FIGURE 2-20** shows how to draw an ellipse. Each point where we placed a tack is called a focus of the ellipse. The plural of this word is "foci," so an ellipse has two foci.

Ellipses can be of various ***eccentricities (e)***, from $e = 0$ to $e = 1$. The eccentricity of an ellipse is defined as the ratio of the distance between the foci to the longest distance across the ellipse (the major axis). Essentially, the eccentricity tells us how "flat" the ellipse is. For example, if the tacks coincide, the ellipse would be a circle and its eccentricity would be zero (**FIGURE 2-21a**). If the tacks are a small distance apart, the ellipse would have a small eccentricity and might look like Figure 2-21b. If the

> **eccentricity of an ellipse** The result obtained by dividing the distance between the foci by the longest distance across an ellipse (the *major axis*).

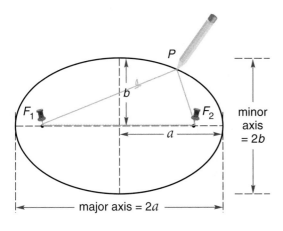

FIGURE 2-20 To draw an ellipse, start with two tacks, a pencil, and a string. Stick the two tacks into a board some distance apart and put a loop of string around them. Then use a pencil to stretch the string as shown. Finally, keeping the string taut, move the pencil around until you have completed the ellipse. Every point (*P*) on the ellipse is at the same total distance from the two fixed points F_1, F_2 (the foci); that is, $PF_1 + PF_2 =$ constant. Half of the major axis is called the semimajor axis (*a*) and is what we have previously referred to as the average distance from a planet to the Sun. The semiminor axis is usually denoted by *b*. The eccentricity can be calculated by $e = \sqrt{1 - (b^2/a^2)}$ or by $e = F_1F_2/2a$.

FIGURE 2-21 (a) An ellipse with zero eccentricity ($e = 0$), that is a circle. (b) An ellipse with a small eccentricity ($e = 0.4$). (c) A more eccentric ellipse ($e = 0.9$).

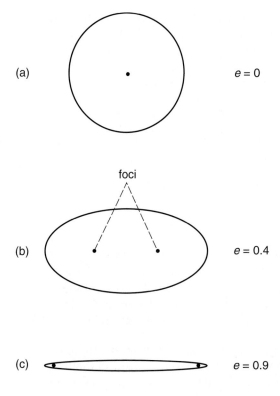

(a) $e = 0$

foci

(b) $e = 0.4$

(c) $e = 0.9$

tacks are farther apart or the string is shorter, the ellipse would have a greater eccentricity and might look like Figure 2-21c. Although different ellipses have different eccentricities, it is important to see that a definite rule governs the shape of an ellipse; that is, every point (P) on the ellipse is at the same total distance from the foci: $PF_1 + PF_2 =$ constant = major axis. Compare this with the definition of a circle, where every point on the circle is at the same distance from a fixed point, the circle's center. In this case, F_1 and F_2 coincide and thus $e = 0$.

An example of an elliptical shape you see every day is that of a circle seen at an angle. **FIGURE 2-22** shows a round object seen in perspective. The shape the artist drew to represent the rim of the basket correctly is an ellipse. The eccentricity of such an apparent ellipse is changed by changing your angle of viewing the rim. Viewing it from below, you see a circle, but by viewing it at an angle, you see its apparent shape become more eccentric.

Kepler's First Two Laws of Planetary Motion

Kepler published his first model of planetary motion in 1609 in his book, *The New Astronomy*. Today we summarize his findings for planetary motion into three laws, the first of which is as follows:

> **KEPLER'S FIRST LAW: Each planet's path around the Sun is an ellipse, with the Sun at one focus of the ellipse.**

FIGURE 2-23a shows an elliptical path of a planet around the Sun. We have exaggerated the eccentricity of the ellipse to make it more obvious. Most planets in our solar system move on nearly circular paths.

The second law tells us about the speed of a planet as it moves around its elliptical path. Kepler found that a planet moves faster when it is closer to the Sun and slower when it is farther away. He was able to make a more definite statement than this, however, one that would allow a calculation of the speed:

> **KEPLER'S SECOND LAW: A planet moves along its elliptical path with a speed that changes in such a way that a line from the planet to the Sun sweeps out equal areas in equal intervals of time.**

Suppose we consider the Earth moving in an elliptical path, such as that shown in Figure 2-23b. (The eccentricity of the Earth's orbit in this figure is greatly exaggerated.) Further suppose that point A in the figure represents the position of the Earth on December 19, and point B is its position on January 18. The yellow-shaded area (area X) represents the area "swept out" during this 30-day time period. Then suppose that on June 19 the planet is at point C. Kepler's second law tells us that in the next 30 days, the line from the Earth to the Sun must sweep out an area equal to X (area Y in the figure). Thus, the Earth can only get to point D by July 19. To understand this, recall that the area of a triangle is equal to half the product between its base and the height to that base. Even though we are not dealing with perfect triangles here, you see from the figure that the

To solve the problem of how to determine Mars' unknown orbit while riding on planet Earth, whose orbit was also unknown, Kepler selected observations of Mars on the same day in different years. This effectively froze the Earth in position. Then he did the same thing in different *Martian* years to map the orbit of Earth. Clever!

FIGURE 2-22 A circular shape, such as the rim of a basketball hoop, appears elliptical when viewed at an angle.

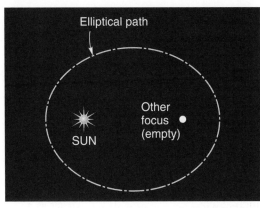

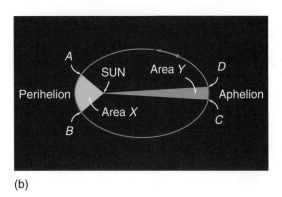

(b)

(a)

FIGURE 2-23 (a) Kepler's first law states that a planet moves in an elliptical orbit with the Sun at one focus of the ellipse. The orbits of actual planets are much more circular than the ellipse shown here. (b) Kepler's second law: A line from the Sun to a planet sweeps out equal areas in equal times. In the drawing, area X = area Y.

Earth is closer to the Sun at points *A* and *B* than at points *C* and *D*. The height of the "triangle" corresponding to area *Y* thus is greater than that for area *X* and, therefore, it must have a smaller "base" in order for both triangles to have the same area; that is, the path *CD* is shorter than *AB*. However, the Earth took the same amount of time to cover the paths *AB* and *CD*. We conclude then that the Earth's speed on its orbit is slower when it is farther from the Sun.

In this example, we considered the Earth and a time period of 30 days as it revolves around the Sun. The point of Kepler's second law, however, is that no matter what planet or time interval is selected, the area swept out during that time interval by the imaginary line connecting the planet and the Sun will be the same everywhere on the elliptical journey of the planet.

In essence, Kepler's second law describes how the orbital speed of a planet changes as the planet revolves around the Sun. A planet moves fastest when it is at the nearest point to the Sun (called the ***perihelion***), whereas it moves most slowly at its farthest point from the Sun (called the ***aphelion***).

perihelion/aphelion The closest/farthest point from the Sun on a planet's orbit.

Kepler's Third Law

Kepler went further than Ptolemy and Copernicus in his search for a working model for the heavenly motions. He also asked *why* the planets should have such motions. It was Kepler—not Isaac Newton, whom we discuss in the next chapter—who first hypothesized that there was a force that kept the planets near the Sun: a force we now identify as gravity. In addition to this force, however, Kepler felt that there must be a force sweeping the planets around the Sun. Just like the gravitational force, this force should become weaker with distance. That is why, he argued, the more distant planets move around the Sun more slowly. It was his consideration of this sweeping force that led him to his third law. Although no such sweeping force exists, it helped Kepler find a rule that later led Newton to discover the forces that *do* exist. Kepler's third law allows us to compare the speed of one planet along its path to that of another.

The distance in Kepler's third law—that is, the *semimajor axis*, or half the length of the major axis of the orbit—is very nearly the same as the planet's average distance from the Sun.

KEPLER'S THIRD LAW: The ratio of the cube of the semimajor axis of a planet's orbit to the square of its orbital period around the Sun is the same for each planet.

The third law is most easily understood when it is expressed in symbols. Let

a = semimajor axis of planet's elliptical orbit =
    (approximately) average distance of a planet from the Sun,

and

P = the planet's orbital period = the time it takes the planet to complete a
    full orbit around the Sun, relative to the stars (called the ***sidereal period***),

then

$$a^3/P^2 = C,$$

sidereal period The time it takes a planet to complete a full orbit around the Sun, relative to the stars.

HISTORICAL NOTE

Johannes Kepler

FIGURE B2-2 Johannes Kepler (1571–1630).

Johannes Kepler (**FIGURE B2-2**) was born in a small town in southern Germany in 1571. When he was 4, he contracted smallpox. The disease left him with poor eyesight, a condition that prevented him from becoming an observer and enjoying the astronomical sights reported to him by Galileo. Instead, his contributions to astronomy resulted from his great mathematical abilities and his insights to the observational data collected by others. Tycho hired Kepler in the hope that the mathematician would be able to verify Tycho's Earth-centered model, but Kepler apparently took the position with the idea of gaining access to the accurate data he needed to test his own ideas.

Kepler was a mathematician of some note; his work on determining volumes of objects by using approximations (inspired by trying to find the most convenient proportions for wine barrels) was an important early contribution to the development of calculus. Today we know Kepler for his three laws of planetary motion, but these laws seem to have been developed almost accidentally as he searched for other, deeper patterns in the heavens. For example, his first great endeavor was to try to find out why the planets are spaced apart as they are. When he was about 24 years old and working as a professor of mathematics, an idea occurred to him while he was lecturing. He was discussing the size of the largest circle that could be drawn inside a triangle and the smallest that could be drawn outside the triangle when he wondered if this could possibly be the basis for the spacings

of the planets. After working with various shapes, he decided that it was not two-dimensional figures that were significant, but three-dimensional solids. It was known that only five symmetric solids were possible (the 4-sided tetrahedron, 6-sided cube, 8-sided octahedron, 12-sided dodecahedron, and 20-sided icosahedron). That number is exactly what would be needed to fill the spaces between six planets. **FIGURE B2-3** shows a construction that illustrates the plan Kepler finally worked out. Kepler might have saved a lot of time if all of today's planets had been known then, for he would have had too many planets for the five solids.

Kepler's personal life was very difficult, and he was forced to cast horoscopes to support himself and his family. His consolation was that he believed his mathematical and scientific findings were important and would be recognized after he died.

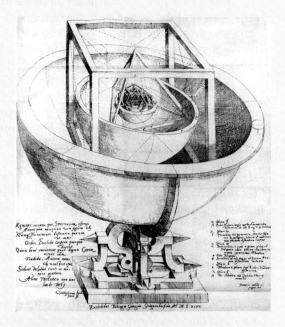

FIGURE B2-3 Kepler worked out a plan using the five regular three-dimensional figures to fix the spheres of the six known planets at their correct distances.

where C is a constant whose value depends on the physical units used for a and P. In using Kepler's third law, any units of measure can be used for the period and the distance. We choose units that make the calculation simple, expressing time in years and average distance in astronomical units (AU). Thus, for the Earth, $a = 1$ AU and $P = 1$ year. Applying Kepler's third law to the Earth we get

$$a^3/P^2 = (1 \text{ AU})^3/(1 \text{ year})^2 = 1 \text{ (AU)}^3/(\text{year})^2.$$

That is, the numerical value of the constant C in Kepler's third law is 1 for all planets in our solar system, as long as we measure the planet's orbital period in years and the semimajor axis of its orbit in astronomical units.

Kepler was very careful to use the sidereal period of a planet in his third law. This is the true orbital period of a planet as it revolves around the Sun, as seen by someone above the solar system, and it differs from its *synodic period*, which is the time it takes for the planet and Sun to be in the same configuration when seen from Earth.

synodic period The time between two successive identical configurations between a planet and the Sun, as seen from Earth.

EXAMPLE

Use the fact that Mars' orbital period is 1.88 years to calculate the average distance of Mars from the Sun.

SOLUTION The third law says that a^3/P^2 is also equal to 1 for Mars when the above units are used. After rewriting the equation, we substitute the value of Mars' period and solve for its distance from the Sun.

$$a^3/P^2 = 1 \ (AU)^3/ \ (year)^2,$$
$$a^3/(1.88 \ year)^2 = 1 \ (AU)^3/ \ (year)^2,$$
$$a^3 = 3.53 \ (AU)^3,$$
$$a = 1.52 \ AU.$$

This agrees with the measured value of the average distance of Mars from the Sun.

TRY ONE YOURSELF

Calculate the radius of Venus' orbit using the fact that its period of revolution around the Sun is 0.615 years. (Venus' orbit is very close to circular and, thus, its average distance from the Sun is essentially the radius of its orbit.)

TABLE 2-2 shows modern data for each of the planets known to Kepler. For completeness, it also includes data for Uranus, Neptune, and Pluto. In each case, Kepler's third law tells us that *to the limits of the accuracy of the data*, the distance cubed is equal to the sidereal period squared. (It is not completely accurate because planets are influenced not only by the Sun but also by the gravitational tug of the other planets.) But it is the synodic period, not the sidereal, that is easily measured from observations. How did Kepler find the sidereal period of a planet? We show how he did it in two calculations at the end of the chapter.

Kepler had fairly accurate data for the orbital periods of the planets. The actual distances to the planets from the Sun were unknown to him, but remember that the

TABLE 2-2			
Testing Kepler's Third Law			
Object	Distance from Sun (AU)	Sidereal period (years)	Distance cubed (AU³)	Sidereal period squared (yr²)
Mercury	0.387	0.241	0.0580	0.0581
Venus	0.723	0.615	0.378	0.378
Earth	1.000	1.000	1.000	1.000
Mars	1.524	1.881	3.540	3.538
Jupiter	5.204	11.862	140.9	140.7
Saturn	9.582	29.457	879.8	867.7
Uranus	19.201	84.011	7079.0	7057.8
Neptune	30.047	164.790	27,127	27,156
Pluto*	39.236	247.68	60,402	61,345
* Dwarf planet.			

In the first place, lest per-chance a blind man might deny it to you, of all the bod-ies in the universe the most excellent is the sun, whose essence is nothing else than the purest light, than which there is no greater star; it is...called king of the planets for his motion, heart of the world for his power, its eye for beauty, and which alone we should judge worthy of the Most High God.
Kepler

I measured the skies, now the shadows I measure. Skybound was the mind, earthbound the body rests.
Kepler's epitaph, as he composed it.

Copernican system allows us to calculate the *relative* distances to the planets. For ex-ample, it was known that Mars is 1.52 times farther from the Sun than the Earth is. This, however, is all that is needed to do calculations with Kepler's third law because 1.52 is the radius of Mars' orbit in AU. So Kepler didn't need to know the actual dis-tances. To the limits of the accuracy of Brahe's data, Kepler's third law fit.

2-10 Kepler's Contribution

Kepler's modification to the Copernican model brought it into conformity with the data; finally, the heliocentric theory worked better than the old geocentric theory. To achieve this fit to the data, however, it had been necessary to abandon the long-held idea of perfect circles in the heavens. Kepler's only reason for proposing ellipses for planetary motion was that they worked.

Our understanding of the solar system would be very unsatisfactory if it had re-mained where Kepler left it. At the time of Kepler, our scientific understanding of the world was just beginning to take major steps forward, and a man who was a contem-porary of Kepler—Galileo Galilei—provided what was needed for the next leap in understanding. The discussion of Galileo's contributions will be left to the next chap-ter, in which we show how they led, in turn, to the crowning achievements of Isaac Newton.

Conclusion

In this chapter we presented models of the heavens from antiquity to about 1600 AD (FIGURE 2-24), including the two major models of Ptolemy (geocentric) and Copernicus (heliocentric). Both models had simple explanations for the daily motion of the stars and for the annual motion of the Sun, but their explanations of planetary mo-tions were more complicated. Copernicus' model had an aesthetic advantage in that it explained all planetary motions in the same way, whereas Ptolemy's model had special rules for Mercury and Venus; however, when the models were compared in the most important of scientific criteria—fitting the data—the heliocentric system of Copernicus was not significantly better than the old geocentric system. To accept Copernicus' model would mean taking an entirely different view of the world, for his model moved the Earth from its central position and reduced it to being just one of the planets.

Kepler's revision of Copernicus' model solved its data-fitting problem by using ellipses rather than circles and by presenting simple rules for the speeds of planets. Using the criteria for scientific models, we judge Kepler's model a better one than

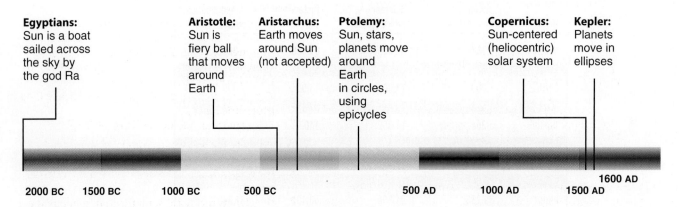

Egyptians: Sun is a boat sailed across the sky by the god Ra

Aristotle: Sun is fiery ball that moves around Earth

Aristarchus: Earth moves around Sun (not accepted)

Ptolemy: Sun, stars, planets move around Earth in circles, using epicycles

Copernicus: Sun-centered (heliocentric) solar system

Kepler: Planets move in ellipses

1600 AD

2000 BC   1500 BC   1000 BC   500 BC   500 AD   1000 AD   1500 AD

FIGURE 2-24 Models of the universe have changed over the past 4000 years with most changes coming in the last 500 years. Chapter 2 takes us only to about 1600 AD.

Ptolemy's. Kepler's revision suffered one flaw, however: the laws he proposed did not "make sense" in that they did not correspond to anything else in nature. They made the model fit the data, and they were reasonably "neat and clean," but they seemed to have been pulled out of thin air. Before the heliocentric system could be fully accepted, it would have to be tied successfully to other phenomena in our experience.

STUDY GUIDE

1. It is _____ to observe Venus in the sky throughout the entire night from central United States.
   A. sometimes possible
   B. always possible
   C. never possible

2. It is _____ to observe Jupiter in the sky throughout the entire night from a point on the Earth's surface.
   A. sometimes possible
   B. always possible
   C. never possible

3. If the plane of the Earth's equator were not tilted with respect to the ecliptic plane,
   A. the daylight period on Earth would be the same year-round.
   B. there would be no seasonal variations on Earth.
   C. Earth's poles would not experience 6-month-long nights.
   D. [All of the above.]

4. Which of the following choices is *not* a criterion for a good scientific theory?
   A. A theory should be aesthetically pleasing.
   B. A theory should be agreed on by all knowledgeable scientists.
   C. A theory should fit present data.
   D. [None of the above; all are criteria for a good theory.]

5. The Copernican model explained retrograde motion as due to
   A. planets moving along epicycles.
   B. planets stopping their eastward motion, moving westward a while, and then resuming their eastward motion.
   C. different speeds of the Earth and another planet in their orbits.
   D. [Both A and B above.]
   E. [None of the above; the Copernican model was unable to explain retrograde motion.]

6. Why was Copernicus forced to use epicycles in his model?
   A. To account for retrograde motion.
   B. To account for phases of the Moon.
   C. To predict accurately the position of a planet.
   D. [Both A and B above.]
   E. [All of the above.]

7. The observation that Mercury and Venus never get high in the night sky was explained by
   A. the geocentric model only.
   B. the heliocentric model only.
   C. both the geocentric and heliocentric models.

8. Which of the following choices could be calculated from the Copernican model but not from the Ptolemaic model?
   A. The length of the day on each planet.
   B. The actual distance from the Sun to each planet.
   C. The relative distance from the Sun to each planet.
   D. The approximate position of a given planet in the sky a few years in the future.

9. If you lived on Jupiter, which of the following planets might you see high overhead during the night (assuming that your sky was clear enough that you could see the stars)?
   A. Mercury and Venus only.
   B. Mercury, Venus, Earth, and Mars only.
   C. All of the planets except Mercury and Venus.
   D. All of the planets except Mercury, Venus, Earth, and Mars.

10. The Sun appears to move among the stars. The Copernican model accounts for this as being due to
    A. the Earth's rotation on its axis.
    B. the Earth's revolution around the Sun.
    C. the actual motion of the Sun against distant stars.
    D. the Earth changing speed in its orbit.
    E. different planets moving at different speeds.

11. If the Earth's diameter were double what it is, stellar parallax would be
    A. easier to observe.
    B. more difficult to observe.
    C. the same, for stellar parallax does not depend on the Earth's diameter.

12. Why did the existence of stellar parallax not convince people of Copernicus' time that his model was better than the geocentric model?
    A. Few people understood stellar parallax.
    B. Stellar parallax was understood, but ignored.
    C. Although most people understood the phenomenon, it was not clear that it provided evidence of the heliocentric theory.
    D. The phenomenon actually provides evidence for a geocentric theory.
    E. Stellar parallax had not yet been observed.

13. According to the heliocentric model, the reason the planets always appear to be near the ecliptic is that
    A. the ecliptic is only 23.5 degrees from the celestial equator.
    B. the planets revolve around the Sun in nearly the same plane.

C. compared with the stars, the planets are nearer the Sun.

D. the planets come much nearer to us than does the Sun.

14. The apparent motion of the planet Jupiter in the sky during a single night as seen by an Earthbound observer is
   A. from west to east, then backward a while, and then wet to east again.
   B. a west-to-east motion throughout the night.
   C. an east-to-west motion throughout the night.
   D. [None of the above, for no motion is observed.]

15. According to Kepler's laws, a planet moves
   A. at a constant speed through its orbit.
   B. fastest when nearest the Sun.
   C. fastest when farthest from the Sun.
   D. [None of the above, for there is no simple rule.]

16. The primary reason that it is hotter in Australia in December than in July is that
   A. the Southern Hemisphere is tilted toward the Sun in December.
   B. the Earth is closer to the Sun in December than in July.
   C. the Earth is moving faster in its orbit in December than in July.
   D. the Earth is on the hotter side of the Sun in December.
   E. [Both B and C above.]

17. If the Earth were in an orbit closer to the Sun, the
   A. day would be longer.
   B. day would be shorter.
   C. year would be longer.
   D. year would be shorter.
   E. [Two of the above are correct.]

18. Kepler's law of equal areas predicts that a planet moves fastest in its orbit when
   A. it is closest to the Sun.
   B. it is closest to the Earth.
   C. the Earth, Moon, and Sun are in a line.
   D. it is farthest from the Sun.
   E. [None of the above; the law makes no predictions as to speed.]

19. If there were another planet between Mercury and Venus,
   A. its "day" would be shorter than Earth's.
   B. its "year" would be shorter than Earth's.
   C. its speed in orbit would be less than Earth's.
   D. [Both A and C above.]
   E. [Both B and C above.]

20. From the law of equal areas, one can predict that the Earth
   A. moves the same distance in January as in July.
   B. moves faster when it is closer to the Sun.
   C. spins faster when it is closer to the Sun.
   D. [Both A and C above.]
   E. [Both B and C above.]

21. If a planet were found at a distance of 3 AU from the Sun, its sidereal period would be
   A. about 2.1 years.
   B. 3 years.

C. about 5.2 years.
D. 9 years.
E. about 0.3 years.

22. If a new planet were found with a period of revolution of 6 years, what would be its average distance from the Sun?
   A. About 2 AU.
   B. About 3.3 AU.
   C. 6 AU.
   D. About 9 AU.
   E. 36 AU.

23. Ptolemy's system of epicycles was used to explain the apparent
   A. daily motion of the stars.
   B. annual motion of the Sun around the sky.
   C. backward motion of the Moon through the sky.
   D. changing speeds and directions of the planets among the stars.
   E. motion of the Sun north and south of the celestial equator.

24. In science, what is the difference between a *theory* and a *hypothesis*?
   A. A hypothesis is more fully developed than a theory.
   B. A theory is more fully developed than a hypothesis.
   C. A hypothesis is based on a model, while a theory isn't.
   D. There is essentially no difference. The words are interchangeable.

25. Eratosthenes calculated the size of the Earth from
   A. angles to the Sun from locations a measured distance apart.
   B. angles to the Moon from locations a measured distance apart.
   C. its angular size and distance from the Sun.
   D. its orbital speed and distance from the Sun.
   E. its calculated rotational speed.

26. In what century did Ptolemy live?

27. What observation convinced Pythagoras that the Earth was spherical?

28. What is an epicycle and what did the use of epicycles enable us to explain?

29. Define *geocentric*.

30. List three criteria for a good scientific model.

31. Which of the criteria for a scientific model is the most important?

32. Explain how Eratosthenes measured the diameter of the Earth.

33. Identify or define: heliocentric, geocentric, astronomical unit, *De Revolutionibus*.

34. How did the Copernican model explain the daily motion of the heavens?

35. How did the Copernican model explain retrograde motion?

36. Why was Copernicus forced to use epicycles in his model?

37. Name a discovery made in the 19th century that the Copernican model fits (or can be made to fit), but the Ptolemaic model does not.

38. What rule did the Copernican model have concerning the speed of one planet compared with another?

39. How did each model explain why Mercury and Venus are never seen in the night sky?

40. Explain how to draw an ellipse, including how to draw one with more or less eccentricity.

41. State and explain Kepler's second law, the law of equal areas.

42. List the planets known in Kepler's time in order of their speeds in orbit. List them in order of their distances from the Sun.

43. Copernicus had stated that planets farther from the Sun move slower than nearer ones. What did Kepler's third law add to this statement?

---

1. Science never arrives at a final answer. As it answers one question, it just finds more. Why then should we continue doing science?

2. The ancient Greeks believed that "the universe is comprehensible" and that it can be described using mathematics. We consider this to be one of their most important contributions. What do you think we mean by saying that "nature is rational"?

3. What is meant by saying that one criterion for a good scientific model is its aesthetic quality?

4. Ptolemy's model placed the Sun on a sphere whose axis was not the same as the axis of the celestial sphere. Why was it necessary for the Sun's sphere to have a different axis?

5. Why did Ptolemy include epicycles in his model?

6. Explain why the Greeks used circles and spheres to account for heavenly motions.

7. What observation forced the Ptolemaic model to locate the center of Venus' and Mercury's epicycles on a line between the Earth and the Sun?

8. Name the three criteria for a good scientific model and discuss their relative importance.

9. Describe one observation that would disprove the Ptolemaic model and explain why it would conflict with the model.

10. We sometimes speak of the sky as being "the Heavens," and it is traditional in Christian circles to speak of a spiritual heaven as being up above us. Cite a pre-Christian historical reason for this link between the sky and heaven.

11. Explain the concept of Occam's razor by giving an example of its use in distinguishing which is the better of two hypotheses. (You may make up two hypotheses about some everyday event if you wish.)

12. Ptolemy observed parallax of the Moon against the distant stars. Explain how this can be observed. (Hint: It is not due to the motion of the Moon or of the Earth around the Sun.)

13. Was Ptolemy's model based on his own observations? Discuss the origin of the ideas in his model.

14. According to the Copernican model, what causes the Sun to change its position periodically from south of the equator to north of the equator and back again?

15. Compare the Copernican and Ptolemaic models as to their accuracy in fitting the data available in the 16th century.

16. What is Occam's razor and why is it important in considering the value of a scientific theory?

17. Explain why Copernicus was dissatisfied with Ptolemy's model.

18. Compare the Ptolemaic and Copernican models as to how each explains the Sun's apparent motion among the stars.

19. What observation led Tycho Brahe to develop his own model of the solar system? Describe his model.

20. It might be possible to prove a theory false, but it is never possible to prove it true. Explain.

21. When you read how each of the models explained retrograde motion, you may have found it easier to understand the explanation of the Ptolemaic model than the explanation of the Copernican model. If so, how can Copernicus' explanation for retrograde motion be called the simpler one?

22. Suppose that you have a friend across the Atlantic with whom you converse on the telephone. Describe a measurement that the two of you might make that would allow you to calculate the distance of the Moon. (Assume that you know how far apart the two of you are.)

23. When observed from Earth, Mercury and Venus have different motions than other planets. Describe this difference. (Hint: We can NOT observe directly that these planets are closer to the Sun than other planets.) Compare Copernicus' explanation for this difference with Ptolemy's explanation.

24. What is meant by the "eccentricity" of an ellipse? What is the eccentricity of a circle?

25. The Earth is closest to the Sun in January. Then why do we not experience our hottest weather in January?

---

1. Calculate the radius of Jupiter's orbit (in astronomical units) from data available to Copernicus. Data: Jupiter's sidereal period is 11.86 years, and quadrature occurs 87.5 days after opposition. (Hint: see the activity "The Radius of Mars' Orbit.")

2. Jupiter's semimajor axis is 5.2 AU. What are the smallest and largest distances possible between Earth and Jupiter?

3. A comet moves around the Sun on a highly eccentric orbit. Assume that its distance from the Sun at perihelion is 0.1 AU. The comet's period is 729 years. Use Kepler's third law to find the average distance of the comet from the Sun and the largest distance between the comet and the Sun.

4. Mercury's period of revolution is 88 days, or 0.24 years. Use Kepler's third law to calculate its average distance from the Sun in astronomical units.

5. The planet Uranus was discovered in 1781. The semi-major axis of its orbit is 19.2 AU. Calculate its period of revolution around the Sun.

6. A quarter at 50 feet spans an angle of about 0.1 degrees. About how many quarters could be placed side by side around a circle with a radius of 50 feet?

7. The eccentricity of Pluto's orbit is 0.24. Use a drawing of an ellipse to explain what this means quantitatively.

8. Suppose you live on planet X. One day the Sun is directly over a place 500 km to your south, while at your position the Sun's direction is 10 degrees off vertical. What is the radius of planet X?

9. The time between successive similar phases for the Moon is on the average about 29.53 days. (In Chapter 1, we defined this time period as a *lunar month* or the *synodic period* of the Moon.) If it takes the Moon an average of 1.2 hours more to go from first to third quarter than to go from third to first quarter, show that the angle Moon-Earth-Sun in Figure 2.7 is about 89.85 degrees.

10. **Synodic versus sidereal period.** FIGURE 2.25a shows the circular (as assumed by Copernicus) orbits of Jupiter and Earth. The time the Earth takes for a complete orbit with respect to the distant stars is its sidereal period (P_E), which is 1 year. The time Jupiter takes for a complete orbit is its sidereal year (P_{sid}). The time it takes Jupiter between successive similar alignments (for example, between **oppositions**, when the Sun and

Jupiter are on opposite sides of the Earth) is its synodic period (P_{syn}). Between two successive oppositions, the Earth covers an angle of $(360°/P_E) \cdot P_{syn}$, while Jupiter covers $(360°/P_{sid}) \cdot P_{syn}$. However, the Earth has covered a full 360 degrees more than Jupiter has. Therefore, $(360°/P_E) \cdot P_{syn} = (360°/P_{sid}) \cdot P_{syn} + 360°$. Show that $P_{sid} = P_{syn}/(P_{syn} - 1)$. Using the observed synodic period of Jupiter (398.9 days), show that its sidereal period is 11.86 years. Also show that the angle θ in the figure is about 33 degrees. (Watch for the conversion between days and years.) Copernicus used the same logic to find the sidereal periods of Mars and Saturn.

11. **Synodic versus sidereal period.** Figure 2-25b shows the circular (as assumed by Copernicus) orbits of Earth and Venus. The time the Earth takes for a complete orbit with respect to the distant stars is its sidereal period (P_E), which is 1 year. The time Venus takes for a complete orbit with respect to the distant stars is its sidereal period (P_{sid}). The time it takes Venus between successive similar alignments (for example, between **inferior conjunctions**, when the Sun and the Earth are on opposite sides of Venus) is its synodic period (P_{syn}). Between two successive inferior conjunctions, the Earth covers an angle of $(360°/P_E) \cdot P_{syn}$, while Venus covers $(360°/P_{sid}) \cdot P_{syn}$. However, Venus has covered a full 360 degrees more than Earth has. Therefore, $(360°/P_{sid}) \cdot P_{syn} = (360°/P_E) \cdot P_{syn} + 360°$. Show that $P_{sid} = P_{syn}/(P_{syn} + 1)$. Using the observed synodic period of Venus (583.9 days), show that its sidereal period is 224.7 days. Also, show that the angle θ in the figure is about 215 degrees. (Watch for the conversion between days and years.) Copernicus used the same logic to find the sidereal periods of Mercury.

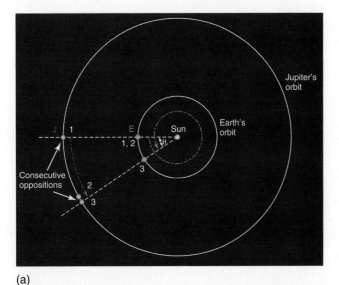

(a)

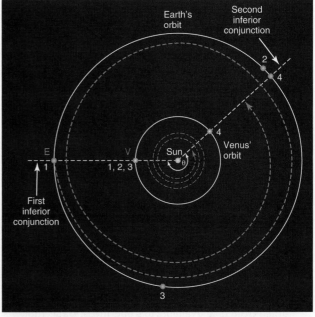

(b)

FIGURE 2-25 These diagrams show the relationship between the synodic and the sidereal period of the planet. The orbits are not in scale; however, for the cases shown, the corresponding positions of each planet (shown by numbers) are accurate.

1. Calculating the Earth's Radius

Let us assume that you would like to repeat Eratosthenes' calculation of the radius of the Earth. However, neither you nor your friend who volunteered to help you live at a place where the Sun is directly overhead at any day of the year. You decide to give it a try anyway. Both of you set a stick vertically in the ground and around noontime on the same day you measure the smallest shadow that the stick makes on the ground. You know the size of your stick and therefore, by drawing a scale of a right triangle (vertical stick and horizontal ground) you are able to determine the angle that the Sun's direction makes with the vertical. You also know the direct, north-south distance between your locations. Do you have enough information to find the size of the Earth? If yes, how would you calculate it?

2. The Radius of Venus' Orbit

In this activity, we measure the orbital radius of Venus. This is the method used by Copernicus to measure the orbital radius of the two **inferior planets**, Venus and Mercury (called this because their orbits are smaller than Earth's).

Mark a location for the Sun on the right side of a piece of paper, and draw an arc of a circle about halfway across the paper, representing the orbit of the Earth. The radius of this circle represents 1 AU. Make a scale drawing and assume that the Earth and Venus move in circular orbits. Then draw a line from the Sun horizontally across the paper and mark its intersection point with the Earth's orbit as the position of the Earth. Copernicus measured the angle between the Sun and Venus, when Venus was at its **greatest elongation**. This is the farthest from the Sun that Venus is seen to be in the sky from Earth. Therefore, when an inferior planet is at its greatest elongation, it means that the line connecting it with the Earth is tangent to the planet's orbit. The angle measured by Copernicus for Venus was 46 degrees.

Center a protractor on the Earth so that the Sun is at 0 degrees. Mark a point at 46 degrees and draw a line from the Earth to that point, extending it a bit further. Draw the perpendicular to this line from the Sun. This perpendicular distance is the radius of Venus' orbit. Using a compass you can now draw the orbit of Venus. Measure the radius of Venus' orbit; divide this by the Earth's distance to the Sun and compare your result with that in Table 2-1. Are you close?

3. The Radius of Mars' Orbit

Let us determine the radius of Mars' orbit from data available to Copernicus. To do so, we start by making a scale drawing, assuming that the Earth and Mars move in circular orbits. Mark a location for the Sun on one side of a piece of paper. As in **FIGURE 2-26a**, draw an arc of a circle about halfway across the paper to represent the orbit of the Earth. Then draw a line from the Sun horizontally across the paper. Suppose that when the Earth crosses your horizontal line, Mars is also crossing it, so the Sun and Mars are on opposite sides of the Earth. (Mars is then said to be in **opposition**.) Mars and the Earth are now moving in the direction shown in Figure 2-26a. Now suppose that Mars is observed night after night until it is at **quadrature**, which is defined as the position when Mars is 90 degrees from the Sun in the sky (seen from Earth), as shown in Figure 2-26c. This is observed to occur 106 days after opposition. This observation, along with the sidereal period of Mars (1.88 years), is all that we need to determine the radius of its orbit.

To draw your figure to scale, you must first calculate the angle through which the Earth moves in 106 days. This is obtained by dividing 106 days by 365 days and multiplying by 360 degrees. Draw a line from the Sun at the angle you have calculated (angle X in Figure 2-26b), and determine the position of Earth (point E) when Mars is at quadrature.

Because Mars is at quadrature when the Earth reaches point E, you should start at E and draw a line toward the right 90 degrees from the line to the Sun. This line points to the planet Mars. You don't know yet where Mars is located on this line. You can determine this by calculating the angle Mars moves through in 106 days. Use its period of 1.88 years to calculate this angle. (Because 1.88 years is 686 days, Mars moves 106/686 of a complete circle in 106 days.) Starting from the Sun, draw a line at the angle you have calculated

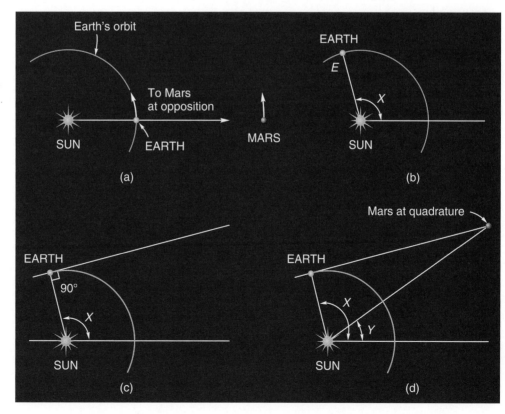

FIGURE 2-26 Steps involved in determining the radius of an outer planet's orbit. This figure is not drawn to scale, as yours must be.

(angle *Y* in Figure 2-26d). Because Mars lies along this line, it must be located where the two lines cross.

Measure the distance from the Sun to Mars, divide this by the Earth's distance to the Sun, and compare your result with that in Table 2-1. Are you close? This was the method used by Copernicus to measure the radius of not only Mars' orbit but also the orbital radius of every **superior planet**, that is, of every planet with an orbit larger than Earth's.

1. M. Caspar: *Kepler* (Dover, 1993).

2. O. Gingerich: *The Great Copernicus Chase, and Other Adventures in Astronomical History,* a collection of essays including the period of Copernicus and Kepler (Sky Publishing/Cambridge University Press, 1992).

3. "Observing the Occasion" by E. C. Krupp, a short biography of Tycho, in *Sky & Telescope* (December 1996).

4. "Copernicus and Tycho" by O. Gingerich, in *Scientific American* (December 1973).

5. A simple description of archaeoastronomy is given in "Archaeoastronomy: Past, Present, and Future" by A. Aveni (*Sky & Telescope*, November 1986, p. 456).

6. A recounting of Eratosthenes' experiment is given by Carl Sagan in *Cosmos* (Random House, 1980), in the chapter entitled, "The Shores of the Cosmic Ocean."

7. "From Aristarchus to Copernicus" by O. Gingerich, in *Sky & Telescope* (November 1983, p. 410).

8. An award-winning history of the development of astronomical thought is given by T. Ferris in *Coming of Age in the Milky Way* (Morrow, 1988).

9. The book *Introductory Readings in the Philosophy of Science, 3rd ed.* (edited by E. D. Klemke, R. Hollinger, D. Wyss Rudge, and A. D. Kline; Prometheus Books, 1998) has a number of interesting articles on science, including P. Thagard's "Why Astrology is a Pseudoscience."

EXPANDING THE QUEST

STARLINKS

Quest Ahead to Starlinks
http://physicalscience.jbpub.com/starlinks

Starlinks is this book's online learning center. It features **eLearning**, which contains chapter quizzes and other tools designed to help you study for your class. You can also find **online exercises**, view numerous relevant **animations**, follow a guide to **useful astronomy sites** on the Internet, or even check the latest **astronomy news** updates.

Gravity and the Rise of Modern Astronomy

3

ON JULY 20, 1969, HUMANS LANDED FOR THE FIRST TIME on another astronomical object. From the Moon's Sea of Tranquility they reported to mission headquarters in Texas, "Houston, Tranquility Base here. The Eagle has landed." Shortly thereafter, Neil A. Armstrong stepped onto the surface of the Moon saying, "That's one small step for a man, one giant leap for mankind." (Historians might note that although the words quoted here are what Armstrong meant to say, he actually left out the word "a.")

Armstrong and Edwin (Buzz) Aldrin, Jr., stayed on the Moon for 21.6 hours. When they left the lunar module, each wore a 90-kilogram spacesuit. On Earth this weighed about 200 pounds, but on the Moon it weighed a mere 33 pounds. During their 2.5 hours of walking on the Moon outside the lunar module, they deployed several instruments to measure various features of the Moon and unveiled a plaque that reads, "Here Men From Planet Earth First Set Foot Upon the Moon. July 1969 AD. We Came in Peace For All Mankind." As they left the Moon to rejoin Michael Collins in the command module and return to Earth, the plaque remained behind to express for ages to come a lofty ambition of the human race.

Apollo 11 astronaut Buzz Aldrin stands next to a lunar seismometer, looking back toward the lunar landing module.

All cross references to chapters, sections, figures, and tables pertain to the main text, *In Quest of the Universe, Sixth Edition. In Quest of the Solar System* contains Chapters 1–11 and 19 of the main text. *In Quest of the Stars and Galaxies* contains Chapters 1–5 and 11–19 of the main text.

The flight to the Moon and back was made possible by many people and many discoveries, some of which we discuss in this chapter. Perhaps the most basic discovery of all was that the laws of physics that describe motion on the Earth are the same laws that describe motion on the Moon and, indeed, everywhere else in the universe. Neil Armstrong's step onto the lunar surface was the last step in a sequence of events dating back over 300 years, with contributions from some of the greatest scientific minds of all time. Their work led to the exploration of space as well as to the rise of modern astronomy.

Although Kepler's laws succeeded in adding accuracy to the heliocentric system, it was the work of Galileo Galilei and Isaac Newton that led to the final triumph of that system. We usually credit the beginning of modern science to Galileo, while the work of Newton helped describe gravity and the causes of planetary motions. Newton formulated some of the fundamental laws of physics, and his work not only helped us understand the force of gravity, but also sparked the technological developments of the Industrial Revolution. In the 20th century, Albert Einstein changed our understanding of such basic concepts as space, time, and mass, refining Newton's laws to make them applicable to the vast regions and exotic objects that we find in a universe much larger and more wondrous than Newton envisioned.

In this chapter we look first at how Galileo's use of the telescope established the validity of the heliocentric system. Then we examine how Newton's new description of mechanical motion and his law of gravitation successfully explained the motions of celestial objects so that only a few unsolved problems remained. Those remaining problems were not solved until Einstein proposed an entirely new theory, the general theory of relativity.

3-1   Galileo Galilei and the Telescope

The contributions of Galileo Galilei to our understanding of the solar system consisted both of observations and theory. His observations were of particular value because he was the first person to use a telescope to study the sky. Galileo built his first telescope in 1609, shortly after hearing about telescopes being constructed in the Netherlands.

Several of Galileo's observations with his telescope affected the comparison between the geocentric and heliocentric theories, as we discuss later:

- Mountains and valleys on the Moon
- Sunspots—dark areas on the Sun that move across its surface
- More stars than can be observed with the naked eye
- Four moons of Jupiter
- The complete cycle of phases of the planet Venus (similar to the Moon's phases)

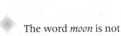 The word *moon* is not capitalized here. In this book, when talking in general about satellites of planets, we will use lowercase letters. Earth's satellite is named "Moon"; thus, we capitalize its name.

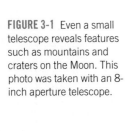

FIGURE 3-1 Even a small telescope reveals features such as mountains and craters on the Moon. This photo was taken with an 8-inch aperture telescope.

Observing the Moon, the Sun, and the Stars

The first three observations in the list provide no data that completely rule out either theory. However, they cast doubt on basic assumptions of the geocentric theory. The Ptolemaic idea of the perfection of the heavens was at the heart of the geocentric system, yet Galileo's telescope (like the small telescope used for FIGURE 3-1) revealed Earth-like features on the Moon, such as mountains and craters. Also, in the case of the Sun, the existence of dark spots did not fit at all with the idea of perfection in the celestial realm (**FIGURE 3-2**).

Galileo's telescope also revealed that the sky contains many more stars than had previously been imagined. Thomas Aquinas had incorporated the Ptolemaic model into Christian theology. Part of this idea was the centrality of humans, not only in position, but in importance. Passages in the first book of the Bible can be interpreted to indicate that the stars were put in the sky for the exclusive purpose "to shed light on the Earth" (Gen. 1:17). Why, then, do stars exist that are so dim that they cannot even be seen by the unaided eye? What is their purpose? The existence of these stars seemed to undermine the Ptolemaic model, and with it a literal interpretation of the Bible. For this reason, many people in Galileo's time simply refused to look through a telescope at the stars.

Jupiter's Moons

In January 1610, Galileo turned his attention to Jupiter. Near the planet, he saw three very faint "stars." They were all in line, two east of the planet and one west. As he continued to observe them night after night, he noted that there were four "stars," rather than three, and it became clear to him that they were not stars at all, but natural satellites—moons—of Jupiter. He concluded that they were objects that revolve around the planet, just as the planets themselves revolve around the Sun in the heliocentric model. (Today we call these four satellites the *Galilean moons* of Jupiter.)

The fact that they never appear north or south of the planet indicated to Galileo that their orbital plane is aligned with the Earth. **FIGURE 3-3** shows that someone viewing Jupiter's satellites from Earth would see them moving back and forth from one side of the planet to the other.

The Ptolemaic model held that everything revolves around the Earth, a special object in the model. Galileo's observation of Jupiter's satellites indicated otherwise.

FIGURE 3-2 Sunspots appear as dark spots on the Sun's surface. See Chapter 11 for more details on sunspots.

Galilean moons The four natural satellites of Jupiter that were discovered by Galileo. Their names, starting closest to Jupiter, are Io, Europa, Ganymede, and Callisto.

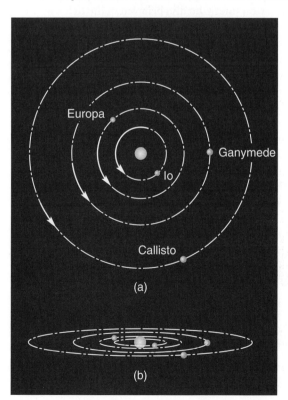

(a)

(b)

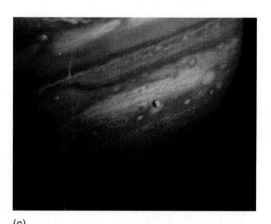

(c)

FIGURE 3-3 (a) The orbits of Jupiter's Galilean satellites as seen face on. (b) The same satellites and orbits seen almost edge on. From Earth, this is how this system appears. (c) This 1979 photo from the *Voyager 2* spacecraft shows the moons Io (left) and Europa (right) in front of Jupiter.

HISTORICAL NOTE

Galileo Galilei

Galileo Galilei (**FIGURE B3-1**) was born in Pisa, Italy on February 15, 1564, the same year that Shakespeare was born (and Michelangelo died). At age 12 Galileo went away to school, studying the usual course of Greek, Latin, and logic. He entered medical school when he was 17, but lost interest in it and took up mathematics. However, he was forced to leave school after 4 years, before graduating, because of a lack of money.

While in medical school, he invented a pendulum device for measuring pulse rates. After leaving school, he continued to study mathematics, applying it to the physics of motion and of liquids. At the age of 25, he was appointed professor of mathematics at Pisa and spent the next 2 decades as a college professor.

In his early years, Galileo had little interest in astronomy, but medical students whom he taught were required to learn some astronomy for use in medical astrology. He therefore became quite familiar with the Ptolemaic model. In 1597, however, he obtained a book published by Kepler concerning the Copernican theory. Galileo apparently preferred the Copernican theory from the time he learned of it, but kept these ideas secret (except from Kepler and a few others) to avoid controversy. Not until his publication of *Letters on Sunspots* in 1613, well after many of his telescopic discoveries had been made, did he openly espouse Copernicus' ideas.

Galileo's announcement of support for the heliocentric system started an uproar that would cause him—and society in general—much grief. One factor contributing to the opposition engendered by the *Letters* was that Galileo did not write them in Latin, the "international" language at the time in which most scholarly writings were written and which was understood by well-educated people in all countries. Instead, Galileo chose Italian, his native language and that of the ordinary literate person, because he was convinced that people other than scholars could understand his arguments and his evidence that the Copernican system was correct.

Primarily because of this controversy, in 1616 the Roman Catholic Church declared that the Copernican doctrine was "false and absurd" and issued a proclamation prohibiting Galileo from holding or defending it. Why did the church make such a strong statement? We must recall the times and the fact that the Ptolemaic theory and Aristotelian physics were a definite part of both religious doctrine and the general culture.

The Catholic Church was not alone in voicing opposition to the Copernican theory. Martin Luther, who was a contemporary of Copernicus, had called Copernicus "the fool who would overturn the whole science of astronomy." John Calvin asked, "Who will venture to place the authority of Copernicus above that of the Holy Spirit?" To appreciate why religious leaders were so concerned about this issue, we must realize that they considered the spiritual salvation of the individual of paramount importance—more important than answering the question of what is the best theory of the universe. They feared that the idea of a nongeocentric model might seem to undermine the supremacy of humans in God's plan, thereby confusing people and threatening their chance of salvation.

FIGURE B3-1 Galileo Galilei (1564–1642).

An Inquisition court sentenced Galileo, at the age of 70, to house arrest for life; it also banned his book *A Dialogue on the Great World Systems*, published in 1632, in which two fictional characters debate whether or not the Earth is at the center of the universe. Galileo was forced to abandon publicly "the false opinion" that the Sun is at the center of the universe, thus avoiding the fate of Giordano Bruno (a mystic and astronomer who also believed in the heliocentric model), who was burned at the stake 33 years earlier. In the end, the church was not able to stop the spreading of the new ideas, and in 1992, Pope John Paul II formally accepted that it was a mistake to condemn Galileo. He also proclaimed that knowledge obtained by science is soundly based on evidence and that this knowledge "reason can discover by its own power."

The important point in Galileo's story is not what the church did, but that similar things happen in cultures and societies around the world. (Actually, the Catholic Church supports the work of many professional astronomers and the Vatican Observatory, one of the oldest astronomical institutes in the world). There are many examples in our history, in many fields, where the status quo defends a specific ideology against new ideas that turn out to be true. It takes courage to defend one's ideas and to be open to the possibility of being proved wrong.

(a)

(b)

(c)

(d)

FIGURE 3-4 When viewed through a telescope, Venus exhibits a full set of phases, including (d) full, (c) gibbous, and (a and b) crescent. The planet's apparent size changes as it goes through its phases.

In the heliocentric system, everything revolves around the Sun. The Earth is just one of the planets, and Galileo's observation of a "solar system" in miniature does not contradict the model. In addition, the fact that Jupiter is able to move through space without leaving its satellites behind conflicts with the Aristotelian/Ptolemaic view, which held that if the Earth moved through space it would leave the Moon behind.

The Phases of Venus

The observation that Galileo found most convincing in supporting the heliocentric theory was that of a complete set of phases for the planet Venus. **FIGURE 3-4** shows four photographs of Venus at different times. It exhibits phases, similar to the Moon. To the naked eye, Venus appears as a dot of light, but Galileo's telescope revealed its phases. When the entire disk of Venus is seen lit, Venus is said to be in **full** phase (Figure 3-4d). (A full Moon is similar; we see the entire Moon lit.) When more than half but less than the entire disk of Venus is seen lit, it is said to be in a **gibbous** phase (Figure 3-4c). Venus is said to be in a **crescent** phase when less than half of its lit disk is visible (Figures 3-4a and b).

Repeated observations of Venus lead us to conclude that Venus never gets far from the Sun in the sky. It always appears either in the east shortly before sunrise or in the west shortly after sunset, and it is never seen high overhead at night. Ptolemy explained this by saying that the center of Venus' epicycle always remains on a line that joins the Earth and the Sun (**FIGURE 3-5a**).

What phases of Venus would we see according to the Ptolemaic model? In Figure 3-5a, the sunlit side of Venus never faces the Earth. At times we should be able to see

full (phase). The phase of a celestial object when the entire sunlit hemisphere is visible.

gibbous (phase). The phase of a celestial object when between half and all of its sunlit hemisphere is visible.

crescent (phase). The phase of a celestial object when less than half of its sunlit hemisphere is visible.

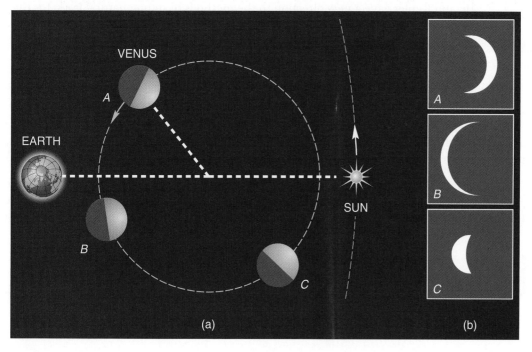
(a)   (b)

FIGURE 3-5 (a) Venus' motion according to Ptolemy. (b) This is how Venus would appear from Earth when it is at each of the three points shown in part (a).

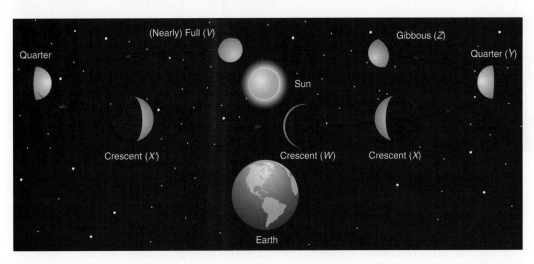

FIGURE 3-6 Venus at various places in its orbit, according to the Copernican, heliocentric model. This drawing shows how the model predicts that all phases are possible and correspond to what is actually observed. Compare the phases of Venus marked by *V*, *Z*, *X*, and *W* with those in Figure 3-4. Venus is brightest at positions *X* and *X'* as we explain in the Advancing the Model box on "Our Changing View of Venus" in Chapter 8.

Additional concepts such as **interior** and **superior planets**, **quadrature**, **greatest elongation**, and **opposition** are covered in *Activities 2 and 3* at the end of Chapter 2.

The Holy Spirit intended to teach us in the Bible how to go to Heaven, not how the heavens go.
Galileo

I do not feel obliged to believe that the same God who has endowed us with sense, reason, and intellect has intended us to forgo their use.
Galileo

part of its illuminated surface, but never much of it. Venus should never get beyond a crescent phase. Galileo, however, saw Venus in a gibbous phase, an observation that is unexplainable by the Ptolemaic model, but can be accounted for by the heliocentric model as shown in **FIGURE 3-6**. At point *X*, Venus would be seen in a crescent phase by an Earth-bound viewer. As it moves farther around and reaches point *Y*, it should appear in quarter phase. Then when it reaches point *Z*, it should appear in a gibbous shape, similar to the gibbous Moon.

We thus finally have an observation that gives us data for choosing the heliocentric system over the geocentric one. The heliocentric model explains all of the phases and therefore becomes the model of choice. In addition, it explains the correlation between the phases and the corresponding observed sizes of Venus. When Venus is in gibbous phase, it looks smaller than when it is in crescent phase. According to the heliocentric model, this happens because at gibbous phase Venus is farther from the Earth than when it is in crescent phase.

Galileo was a leader in breaking the bonds of Aristotelian and Ptolemaic thought. He and Kepler gave a new direction to the methods of stating natural laws, for their laws were stated in a manner that allowed measurement and testing by the methods of mathematics. In addition, Galileo referred constantly to experiments that would test his hypotheses. This was a new procedure in the study of nature. Even though experimentation was not a new idea (experiments were conducted by some ancient Greek scholars), the synthesis of Aristotle's ideas into religious belief by Aquinas resulted in an almost complete dependence on "authoritative" opinions. Today, a reliance on observation and experimentation, rather than on authority, is a cornerstone of science. Its beginning is usually credited to Galileo.

One problem remained after Galileo's observations: There seemed to be no logic supporting the laws of Kepler, except that they work. Galileo wrote that he had "opened up to this vast and most excellent science, of which my work is merely the beginning, ways and means by which other minds more acute than mine will explore its remote corners." Indeed, Isaac Newton, born the year Galileo died, was the person who synthesized Galileo's findings into one expansive theory of motion.

3-2   Isaac Newton's Grand Synthesis

Can a moving object slow down and stop by itself?

Today we summarize Newton's conclusions concerning force and motion in three laws, known as "Newton's laws of motion."

Newton's First Two Laws of Motion

The first of Newton's laws was built directly on conclusions (though incomplete) that Galileo had reached. Newton stated that an object at rest tends to stay at rest, whereas a moving object has a tendency to continue moving in a straight line at the same speed. A stationary object starts to move only when something causes it to move. A moving object changes direction or stops only because something causes it to change direction or stop. The tendency of an object to maintain whatever speed and direction of motion it has is called **inertia**, and Newton's first law is often called the *law of inertia*.

> **NEWTON'S FIRST LAW (THE LAW OF INERTIA): Unless an object is acted upon by a net, outside force, the object will maintain a constant speed in a straight line (if it were initially moving), or remain at rest (if initially at rest).**

Newton's first law is a basic observation about motion and cannot be directly proven or derived. Still, the law of inertia is a powerful tool that emphasizes cause and effect. It indicates that a force is needed to change the speed and/or the direction of an object's motion—that is, to **accelerate** it. Newton's second law quantifies and extends the first law. It tells us how much force is necessary to produce a certain *acceleration* of an object.

Consider the brick shown in FIGURE 3-7 and imagine that the wheels allow the brick to move without friction. The brick is at rest in part (a), and because it has inertia, it will stay at rest. In (b), you give the brick a push. While you are pushing, the brick accelerates. The amount of force you exert determines how great the acceleration is. As you increase the force, so does the brick's acceleration (c). This idea gives us the first part of Newton's second law: The acceleration of an object is proportional to the force exerted on it. "Proportional to" actually goes beyond what we deduced from our thought experiment. It tells us that twice the force will cause twice the acceleration.

Refer to FIGURE 3-8. Here we show one hand pushing on a frictionless brick as before and another hand pushing on a stack of two identical bricks. If the hands push with equal force, which hand will cause the most acceleration? It should be intuitive that the greater acceleration will be produced by the hand pushing on the single brick. It is important to see that friction has nothing to do with this. It is tougher to accelerate two bricks than one simply because the two have more inertia. In fact, two identical bricks have twice as much inertia as one brick. This is an idea we didn't discuss previously here: Some objects have more inertia than others. So before continuing with the second law, we must introduce the property of an object that determines its inertia—its *mass*.

An Important Digression—Mass and Weight

Mass, like inertia, is one of those terms that is used in everyday language but has a very specific definition in science. An object's **mass** is the measure of its inertia. Instead of saying that one object has twice the inertia of another, we normally say

inertia The tendency of an object to resist a change in its motion.

The "net" force acting on an object is the sum of all forces, taking into account both their magnitudes and their directions. Equal forces in opposite directions, acting on the same object, cancel one another.

accelerate To change the speed and/or direction of motion of an object.

acceleration A measure of how rapidly the speed and/or direction of motion of an object is changing.

The question of the origin of inertia has not yet been resolved. Newton believed that inertia is an intrinsic property of matter. Some today believe that it arises from the interaction of all matter in the universe, whereas others are exploring alternative ideas.

mass The quantifiable property of an object that is a measure of its inertia.

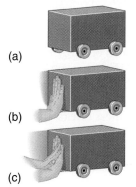

(a)

(b)

(c)

FIGURE 3-7 (a) The wheeled brick will accelerate (b) if a force is exerted on it. (c) If twice as much force is exerted on it, it will accelerate at twice the rate.

FIGURE 3-8 The same amount of force will give twice as much mass only half the acceleration.

HISTORICAL NOTE

Isaac Newton

To give an idea of the importance of Isaac Newton (FIGURE B3-2) to the history of thought, we quote the English poet Alexander Pope, who lived at the same time as Newton:

> Nature and nature's laws lay hid in night.
> God said, Let Newton be! and all was light.

Newton was born on Christmas Day, 1642, in a small village in England. His father, a farmer, had died before Isaac was born, and his mother decided that he was too frail to become a farmer. He was not a particularly good student, however, "wasting" much of his time tinkering with mechanical things, including model windmills, sundials, and kites that carried lanterns and scared the local people at night. When her second husband died, Isaac's mother called him home from school to help on the farm, where he spent most of his teenage years.

At the age of 19, Newton was admitted to Cambridge University. There he studied mathematics and natural philosophy (known today as science). In 1665, after Newton had received his degree and was serving as a junior faculty member, England was swept by the bubonic plague. This incurable disease killed more than 10% of the population of London in only 3 months, and those who could afford to do so escaped it by moving away from population centers. Newton returned to his mother's home at Woolsthorpe. There he worked feverishly on science for the next 2 years. These must have been two of the most productive years in the history of science, for during this time Newton made discoveries in light and optics, in force and motion, and in gravitation and planetary motion. He also devised a theory of color. To solve a problem in gravitation, he invented calculus. During this time he outlined what would become his major book, *Philosophiae Naturalis Principia Mathematica*, usually called *The Principia*. In later years, Newton wrote, "All this was in the two plague years of 1665 and 1666, for in those days I was in the prime of my age for invention, and

minded Mathematics and Philosophy more than at any time since."

Newton's manner of attacking a problem was simply to concentrate his mind on it with such intensity that he finally solved it. As a young man, Newton was not interested in publishing his work. He seemed to want to discover the mysteries of nature simply for his own curiosity, and his friends often had to persuade him to publish his findings. Besides, he did not like having to defend his views against criticism, and he wanted to avoid getting involved in arguments over who was the first to make certain discoveries. Nevertheless he *did* get involved quite fiercely in such disputes, including one with Gottfried Leibniz over which of them first developed calculus. This dispute lasted long after both had died. (Today they are both credited for discovering calculus independently.)

Newton had a very practical side as well as being a theoretician. He served as Warden of the Mint, and while in this office, he began the practice of making coins with small notches around their edges. (Check the quarters in your pocket.) This was done to discourage people from illegally shaving off and retaining the valuable metal of the coins.

Newton received many honors for his work and was given positions of authority. From 1703 until his death in 1727, he was president of the Royal Society, an organization of scientists. In 1705 he was knighted, the first scientist to receive this honor. When he died, Sir Isaac Newton was buried in Westminster Abbey after a state funeral.

FIGURE B3-2 Sir Isaac Newton (1642–1727) worked in many fields. In this painting, he is shown experimenting with a prism to investigate the nature of light.

---

that it has twice the mass. Mass is an intrinsic property of an object and remains the same independent of where the object is located in the universe.

Mass is *not* volume. Suppose that the brick in Figure 3-7 is a piece of styrofoam made to look like a brick. If you do not know this, however, you will be quite surprised when you push on it because it will accelerate much more than you expected. Why? Because a styrofoam brick has much less mass than a real brick, even though they both have the same volume.

Mass also is *not* weight. In our examples of pushing the frictionless bricks, the weight of the brick was not a factor. The brick's weight is the downward force experienced by the brick as a result of its gravitational interaction with the Earth. (See Section 3-4.) Weight did not oppose the pushing hand. This is a subtle distinction, but a very important one.

◆ Rotational inertia—the inertia of a spinning object—involves more than just mass and is discussed in Chapter 7.

Is mass the same thing as weight?

The worldwide unit of mass measurement is the kilogram. At the International Bureau of Weights and Measures near Paris is a platinum cylinder that has, by definition, a mass of *exactly* 1 kilogram. Even though a kilogram weighs about 2.2 pounds at the surface of the Earth, it is not correct to say that a kilogram is about equal to 2.2 pounds because a pound is a unit that expresses weight and a kilogram is a unit that expresses mass. The weight of an object depends on where the object is located in the universe.

Back to Newton's Second Law

Figure 3-8 shows one hand pushing on one brick and another hand pushing on two bricks. We stated that if the two applied forces were equal, the acceleration of the two bricks would be smaller. In fact, measuring the accelerations would show that if the forces are equal, the acceleration of the two bricks is exactly half that of the single brick, and the same force applied to three bricks results in one third the acceleration. Thus, for a given force, acceleration is inversely proportional to the mass being accelerated.

We have thus far discussed the relationships between force and acceleration and between mass and acceleration. We can sum them up in a single statement.

> NEWTON'S SECOND LAW: **A net external force applied to an object causes it to accelerate at a rate that is proportional to the force and inversely proportional to its mass.**

The second law is usually written as

$$\text{Force} = \text{mass} \times \text{acceleration or } F = ma.$$

The second law also makes it apparent that if the net external force is zero, there is no acceleration, which agrees exactly with Newton's first law.

Newton's Third Law

Newton's third law is simple to state:

> NEWTON'S THIRD LAW: **When object X exerts a force on object Y, object Y exerts an equal and opposite force back on X.**

This law seems very simple, but its implications are great. It is sometimes stated as this: "For every action there is an equal and opposite reaction." The implication is that it is impossible to have a single, isolated force. Forces always occur in pairs.

When you sit on a chair, you exert a force downward on the chair (**FIGURE 3-9**). The third law states that the chair exerts an equal force upward on you. We may call the force you exert on the chair the *action* force and the force of the chair on you the *reaction* force, but these labels can be assigned in an arbitrary way. The point is that one of these forces cannot exist without the other and neither "comes first."

The statement of Newton's third law indicates that two objects (called *X* and *Y* here) are *always* involved in the application of forces. A force cannot be exerted without an object to exert it and an object on which it is exerted. The word "object" here might refer to an individual atom, or it might refer to a collection of atoms; the collection might be a gas or a liquid as well as a solid object. For example, if you hold your hand out of the window of a moving car to feel the force of air resistance, it is the air that exerts a force on your hand. The third law tells us that your hand exerts a force on the air. This is not easy to see, but we know that your hand must deflect the air as it comes by and a force is necessary to do so. Calling the air an object may seem odd, but air is simply a collection of objects—atoms.

Is a kilogram a unit of weight?

A kilogram on the Moon weighs only about one third of a pound.

The concept of force is a fundamental one. It is reasonable to think of force as pushes and pulls, but this is subjective and not always accurate. We see the changes produced by a force and maybe the best way of describing force is as *the agent of change*.

In Newton's second law, force and acceleration are always in the same direction.

In using Newton's second law, we assume that the speeds involved are much less than the speed of light; otherwise, Einstein's theory of relativity comes into play.

Force of chair on body

Force of body on chair

FIGURE 3-9 When you sit, you exert a downward force on the chair, and the chair exerts an upward force on you. By Newton's third law, these forces are equal in magnitude and opposite in direction, but they do not cancel each other because they are exerted on different objects.

3-3   Motion in a Circle

Does accelerating an object always cause it to speed up or slow down?

You might try the Activity "Circular Motion" (at the end of the chapter) now.

centripetal force The force directed toward the center of the curve along which the object is moving.

Newton's second law says that a net external force acting on an object always produces an acceleration.

Consider what happens if you whirl a rock on a string. You are exerting a force on the rock toward the center of its circular motion as you pull inward on the string (**FIGURE 3-10a**). If you are careful, you might be able to make the rock go around you with a constant speed. Where is the acceleration that, according to Newton's law, must be produced by the force you are exerting? If the rock moves at a constant speed, the force is not causing a change in speed. Instead, it changes the *direction* of the rock's motion. Motion of an object in a circle at constant speed is an example of acceleration causing a change in direction; the acceleration is in the direction of the force—toward the center of the circle.

Parts (b) through (e) of Figure 3-10 show several potential paths for the rock when the string breaks. Based on Newton's laws, which of these paths is the way you expect the object to travel? Once the string breaks, there is no horizontal force (that is, parallel to the ground) on the rock. If this is the case, there can be no horizontal acceleration of the rock. According to Newton's first law, it thus will continue in a straight line (Figure 3-10d). When the rock was moving in the circle, the only horizontal force on it was exerted by the string. (The rock's weight pulls down, but we are considering only horizontal forces and horizontal motions.) The horizontal acceleration produced was therefore due only to the string.

A force is necessary to make something move in a curve. We give this force a name, calling it the **centripetal** (meaning "center-seeking") **force**. This is *not* a new kind of force, but rather the name we apply to a net force if that net force is causing something to move in a curve. For example, the centripetal force involved when you were whirling the rock was the force exerted by the string on the rock. The centripetal force on a car rounding a curve on a level roadway is the frictional force of the road acting on the tires in the direction toward the center of curvature. The roadway exerts a force on the car, keeping it in the curve. Do not think of centripetal force as another type of force, similar to a push, a pull by a string, or a friction force. It might be any of those three or other single forces or combinations of them; it is what we call the net force responsible for an object's motion along a curved path.

Now let us consider the force that causes the Moon to move around the Earth.

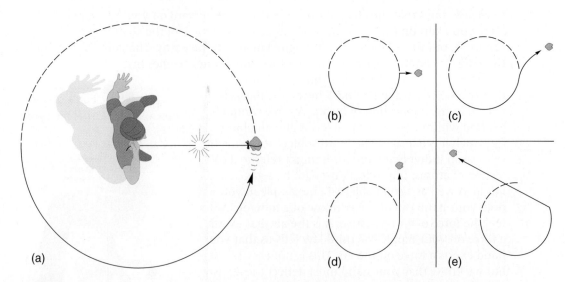

(a)   (b)   (c)   (d)   (e)

◯ **FIGURE 3-10**   (a) The string breaks as the rock is whirled in a circle. (b–e) Which way does the rock go after the string breaks?

3-4 The Law of Universal Gravitation

The importance to astronomy of Newton's three laws of motion becomes evident when they are combined with another of his accomplishments: the law of universal gravitation, or "the law of gravity."

Newton's first law states that an object continues at the same speed in the same direction unless some unbalanced external force acts on it. This law is applicable to objects on Earth, but what about objects in the sky? Aristotle believed that earthly laws do not hold in the heavens, but Newton sought to apply his laws of motion there, too. It was during his great, productive years of 1665 and 1666 that, in Newton's words, he "began to think of gravity extending to the orb of the Moon." The Moon follows a nearly circular path around the Earth, so if the laws of motion apply to the Moon, a force must be exerted on it toward the center of its circle—a centripetal force.

Newton hypothesized that there is a force of attraction between the Earth and the Moon and that this serves as the centripetal force. What causes this force? Newton didn't know, but he hypothesized that it is the same force that causes an object here on Earth to fall to the ground. The insight that all matter (hence the term "universal") interacts gravitationally developed gradually. Even Copernicus had made a similar, but incomplete, proposal. Newton's formulation of this idea is as follows.

THE LAW OF UNIVERSAL GRAVITATION: **Between every two objects there is an attractive force, the magnitude of which is directly proportional to the mass of each object and inversely proportional to the square of the distance between the centers of the objects.**

In equation form, this is

$$F = G\frac{m_1 m_2}{d^2},$$

where m_1 and m_2 are the masses of the two objects, d is the distance between their centers, and G is a constant number that depends on the units used.

According to Newton, *every* object in the universe attracts every other one. Because Newton's law of gravitation applies to every two objects, you may wonder why you are not gravitationally attracted to the person sitting next to you in the classroom. If we use Newton's law and substitute appropriate numbers for your mass, the mass of your classmate, and the approximate distance between you, we get a force that is about equal to the weight of a small flea. This tiny force, combined with the presence of frictional forces, explains why people are not attracted to each other (gravitationally, anyway). When one of the attracting objects is the Earth, however, one of the masses is enormous, and the force we experience in this case is what we call *weight* (FIGURE 3-11).

More than 100 years after Newton published his law, Henry Cavendish was able to measure the gravitational force between two ordinary objects in a laboratory. This experiment enabled him to determine the numerical value of $G = 6.67 \times 10^{-11}$ N · m²/kg². (1 N, one newton, is a unit of measurement for force. The weight of an average apple on the surface of the Earth is about 1 N.)

weight The gravitational force between an object and the planetary/stellar body where the object is located.

(a)   (b)

FIGURE 3-11 An object's mass doesn't change from one location to another, but its weight can change. An object that weighs 12 pounds on Earth (a) would weigh 2 pounds on the Moon (b), but its mass is the same.

EXAMPLE

Let us now apply Newton's law of gravity to compare the force required to keep the Earth orbiting the Sun, F_{ES}, to the one required to keep the Moon orbiting the Earth, F_{ME}. The masses of the three objects and their relative distances can be found in the appendices of this book.

SOLUTION

You might think that the force from the Sun on the Earth will be the greater one because the Sun has much more mass than the Moon; however, keep in mind that the gravitational force decreases as the *square* of the distance between two objects, and the Sun is much farther from the Earth than the Moon is. Applying Newton's law of gravity for both pairs we have

$$F_{ES} = G\frac{m_E m_S}{(d_{ES})^2} = 6.67 \cdot 10^{-11}\frac{\text{N} \cdot \text{m}^2}{\text{kg}^2} \cdot \frac{5.97 \cdot 10^{24}\,\text{kg} \times 1.99 \cdot 10^{30}\,\text{kg}}{(1.50 \cdot 10^{11}\,\text{m})^2} = 3.52 \cdot 10^{22}\,\text{N},$$

and

$$F_{ME} = G\frac{m_M m_E}{(d_{ME})^2} = 6.67 \cdot 10^{-11}\frac{\text{N} \cdot \text{m}^2}{\text{kg}^2} \cdot \frac{7.35 \cdot 10^{22}\,\text{kg} \times 5.97 \cdot 10^{24}\,\text{kg}}{(3.84 \cdot 10^{8}\,\text{m})^2} = 1.98 \cdot 10^{20}\,\text{N}.$$

Taking a ratio of the two expressions, we get

$$\frac{F_{ES}}{F_{ME}} = \left(\frac{m_S}{m_M}\right)\left(\frac{d_{ME}}{d_{ES}}\right)^2 = \left(\frac{1.99 \cdot 10^{30}\,\text{kg}}{7.35 \cdot 10^{22}\,\text{kg}}\right)\left(\frac{3.84 \cdot 10^{8}\,\text{m}}{1.50 \cdot 10^{11}\,\text{m}}\right)^2 \approx 180.$$

Indeed, the force from the Sun on the Earth is greater by about a factor of 180 than the force from the Earth on the Moon.

TRY ONE YOURSELF

At the surface of the Earth, a distance of about 6400 km from the Earth's center, an astronaut's weight is about 160 pounds. This is the attractive force on the astronaut from Earth. What is the value of this attractive force on the astronaut if he were at a height of 300 kilometers above the surface of the Earth? Taking account of your answer, how do you explain the fact that astronauts in the Space Shuttle, orbiting at 300 km above the Earth's surface, feel no weight?

According to Newton, the attractive force of gravity not only makes objects fall to Earth, but keeps the Moon in orbit around the Earth and keeps the planets in orbit around the Sun. Newton proposed that this law, along with his laws of motion, could explain the motions of the planets as well as the falling of objects on Earth. If he could explain the planets' motions, this would clear up the mystery of why Kepler's laws worked and bring the motions of the heavenly planets within the realm of our scientific understanding.

Arriving at the Law of Universal Gravitation

Kepler believed that the planets were pushed along their orbits by a force that spread out from the Sun and, therefore, decreased with distance.

How did Newton arrive at his formulation of the law of gravity? Kepler (in 1609) was the first to suggest that two objects placed in space would attract gravitationally and "come together . . . each approaching the other in proportion to the other's mass. . . ."

This is easy to test—in one case, anyway. A 10-kilogram object has twice the *mass* of a 5-kilogram object. The law of gravity predicts that a 10-kilogram object has twice the *weight* of a 5-kilogram object, and indeed it does. Weight is proportional to mass, so the law seems to work when one of the objects is the Earth. That is, the gravitational attraction between the Earth and an object is proportional to the object's mass (say m_1). According to Newton's third law, however, the object is also attracting the Earth with an equal and opposite force. Therefore, the gravitational force between the Earth and an object is also proportional to the Earth's mass (say m_2). Thus, the force is proportional to the product of the masses of the two objects, that is, $F \propto m_1 m_2$.

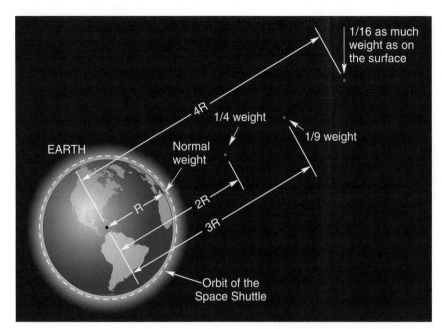

FIGURE 3-12 When an object gets farther from Earth, its gravitational force toward Earth decreases as the square of the distance to Earth's center. The weight of astronauts in the orbiting Space Shuttle is only about 5% less than their weight on Earth's surface. How do you explain the fact that they feel no weight?

Because of today's sensitive instruments, we can now measure differences in weight at different locations on Earth. As expected, the measurements confirm Newton's law of gravitation.

To test gravity's dependence on distance, we must be able to compare forces on objects at different distances from each other. The law states that the force is inversely proportional to the square of the distance between the objects' centers. For example, if the two objects are the Earth and your book and if you take your book to a location twice as far from the center of the Earth as it is now, the law says that the book will weigh only one fourth (which is $(1/2)^2$) as much. At three times as far from the Earth, it will weigh one ninth as much (**FIGURE 3-12**).

As we explain in Chapter 6, Ptolemy used parallax to determine that the distance from the Earth to the Moon is 27.3 Earth diameters, very close to today's value of 30.17 for the average distance to the Moon.

Newton, of course, was unable to change an object's distance from Earth's center significantly. He could have taken an object up on a mountain and measured its weight, but his theory predicted that the weight change in this case would be so small that it would not be measurable with the methods he had available. Instead, he used an object already in the sky: the Moon. And rather than comparing forces directly, he compared accelerations produced by the forces, as they should change in the same way that forces do.

Even though the Moon's orbit around the Earth is slightly elliptical, Newton assumed that the Moon circles the Earth at a distance of about 60 Earth radii under the influence of the centripetal force provided by Earth's gravity. He also knew that the Moon's sidereal period is about 27.3 days. As we show in one of the calculations at the end of the chapter, this information was enough for Newton to show that the Moon's centripetal acceleration is 0.0027 m/s^2, about $(1/60^2)$, or 1/3600 of the acceleration of gravity on Earth's surface, which is about 9.8 m/s^2.

The surface area of a sphere of radius R is $4\pi R^2$. Anything that diffuses outward from a central point, filling all space uniformly in all directions, therefore, becomes less concentrated by a factor of $1/R^2$.

We can reach the conclusion that gravity decreases inversely with the distance squared ($1/d^2$) if we assume that gravity "spreads out" from an object uniformly in all directions. Indeed, this outward diffusion has to pass through successive imaginary spheres centered on the object. Because the surface area of a sphere of radius d is $4\pi d^2$, as gravity spreads out from the object filling all space uniformly in all directions, it must become less "concentrated" by a factor of $1/d^2$.

3-5   Newton's Laws and Kepler's Laws

One had to be a Newton to see that the Moon is falling, when everyone sees that it doesn't.
Paul Valery (French poet and philosopher, 1871– 1945)

Kepler's laws were not derived from fundamental principles of nature; they were accepted because they seemed to work. Based on his laws of motion and his law of gravity, Newton was able not only to prove Kepler's laws but also expand beyond them. **TABLE 3-1** summarizes the types of motion as they were analyzed by Newton.

ADVANCING THE MODEL

Travel to the Moon

The technology to launch satellites was not developed until the second half of the 20th century. But 300 years earlier, Newton developed the required gravitational theory and predicted the possibility of artificial satellites. He presented the following argument for placing an object in orbit around the Earth, based on his laws of motion and his law of universal gravitation:

If a mountain could be found that extends above the Earth's atmosphere, and if a cannonball were fired from the summit, the ball's path would be somewhat as shown in **FIGURE B3-3**, line *A*. If a greater charge were used to shoot the cannonball, the ball might follow path *B*. Shoot harder yet and get path *C*. With careful adjustment (and a very powerful cannon!) the ball can be made to go around the Earth and return to where it started (line *D*). Then it will continue around again in the same path; the cannonball will be in orbit.

The gravitational aspects of a flight to the Moon are not vastly more complicated than those of Earth orbit. The mission begins with the spacecraft orbiting the Earth. As indicated in **FIGURE B3-4**, a rocket then launches the spacecraft out of Earth's orbit to begin its trip to the Moon. As the spacecraft gets farther from the Earth and nearer the Moon, the gravitational pull of the Moon becomes significant. The law of gravity tells us that every object in the universe attracts every other. At one point, the gravitational pulls toward Earth and Moon exactly cancel one another. Up until this point, the force pulling the craft toward the Earth has been greater than that pulling it toward the Moon, and the spacecraft has been slowing down. After passing that point, the pull toward the Moon is greater, and the spacecraft speeds up. Finally, when the spacecraft has reached the appropriate point in its journey (at about *X* in Figure B3-4), a rocket is fired to slow it down so that it remains in orbit around the Moon. Without this firing, the craft would have so much speed that it would swing right past the Moon. For most of the trip the astronauts feel weightless, as they are in free fall, coasting along with the craft.

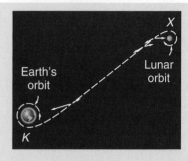

FIGURE B3-4 A craft sent to orbit the Moon starts out in Earth orbit. A rocket is fired at point *K* to send it toward the Moon. At point *X*, a rocket is fired to put it in Moon orbit. The arrows along the path illustrate the changing gravitational pull on the craft from the Earth and Moon. Distances are vastly out of scale here.

Once in orbit around the Moon, the lunar module disconnects from the command/service module that remains in orbit, fires its rockets to slow down, and descends to the Moon's surface. The lunar module does not have to be very large and does not need particularly powerful rockets. This is because the force of gravity on the surface of the Moon is only one sixth of that on the Earth; thus, the fall toward the Moon is not as fast, and the lift-off requires much less energy. In addition, the Moon does not have an atmosphere, so the frictional effects of an atmosphere need not be considered, as they do for takeoff and landing on Earth.

To leave the Moon, part of the lunar module is left behind when a rocket launches the small remaining craft into lunar orbit to reconnect ("dock") with the command module. Then, at the appropriate point, rockets fire again to send the craft out of lunar orbit and back toward the Earth.

Very little of the trip to and from the Moon is spent with the rockets firing. They are needed only to begin and end each portion of the trip and to make minor midcourse corrections. On one of the missions to the Moon, an astronaut spoke with his son by radio. His son asked who was driving. The reply was that Isaac Newton was doing most of the driving at the time.

The first manned *Apollo* flight took place in October 1968. In December of that year, three astronauts orbited the Moon, but did not land. Their mission provided the first whole-Earth photos ever made. In July 1969, astronauts Neil A. Armstrong and Edwin (Buzz) Aldrin, Jr., guided their lunar module "Eagle" onto the Moon at the Sea of Tranquility and uttered the famous words, "Tranquility Base here. The Eagle has landed." In all, 12 men visited the Moon between July 1969 and December 1972, when *Apollo* 17 blasted off from the Moon.

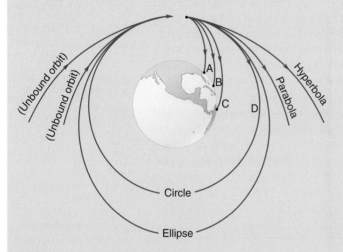

FIGURE B3-3 Newton drew a figure similar to this to illustrate how an artificial satellite would circle the Earth. Under the influence of gravity, the orbit of an object moving around a stationary body can be a circle, an ellipse, a parabola, or a hyperbola.

TABLE 3-1

Summary of Types of Motion

Motion	Type of Acceleration	Direction of Force and Acceleration
Linear	Change in speed	Forward (to speed up) Backward (to slow down)
Circular	Change in direction	Toward center of circle
Orbital	Change in speed and direction	Toward the object being orbited

As the poet Lord Byron (1788–1824) said of Newton, "When Newton saw an apple fall, he found in that slight startle from his contemplation . . . a mode of proving that the earth turn'd round in a most natural whirl, called 'gravitation.'" Indeed, Newton's major insight that gravity operated in the same way in the heavens as it did on Earth was the final blow to the Aristotelian ideas; the idea that the laws of nature as we uncover them on Earth apply in a similar fashion to the rest of the universe is now a basic assumption in astrophysics.

Newton took the Sun as the source of the force responsible for the motion of the planets. Using calculus, he showed that a planetary orbit will be elliptical if, and only if, the centripetal force varies as $1/d^2$ from the Sun, which is located at one of the foci. In other words, he proved Kepler's first law. He also showed that under the influence of the gravitational force between an object (such as a comet) and a stationary body (such as a star), the object's orbit could be elliptical (which includes circular), parabolic, or hyperbolic (Figure B3-3).

Newton showed that Kepler's second law results from the fact that the gravitational force acting on an object orbiting the Sun (or any other stationary body) always points toward the Sun. Without using math, here is how we can show that Newton's laws make Kepler's second law seem reasonable. Consider a planet moving on an (exaggerated) elliptical path around the Sun (**FIGURE 3-13**). When the planet is at position A, moving closer to the Sun, the gravitational force exerted on the planet by the Sun (shown by the arrow pointing toward the Sun) is not perpendicular to the motion of the planet; it is not simply a centripetal force. If it were a centripetal force, it would be exerted perpendicular to the motion of the planet, and it would have only one effect: to cause the planet to curve in its path. Instead, it has two effects: It causes the planet to curve, and it also causes the planet to speed up. This is because it pulls *forward* as well as sideways on the planet.

At location B in the figure, the planet is moving away from the Sun. Here the gravitational force pulls both backward and sideways on the planet, thus slowing it down as well as curving its path. We thus see that as the planet moves toward the Sun, it speeds up, and as it moves away, it slows down. This motion is incorporated in Kepler's second law.

Finally, using his laws of motion and his law of gravity, Newton modified Kepler's third law, which now can be used for any two objects orbiting each other as a result of their mutual gravitational attraction. Using standard units (meters for distance, seconds for period, and kilograms for mass), Newton's formulation of Kepler's third law for a **binary system** is

$$\frac{a^3}{P^2} = \frac{G}{4\pi^2} \cdot (m_1 + m_2),$$

where a = semimajor axis of the orbit, P = period of the orbit, m_1, m_2 = the masses of the two objects, and G = gravitational constant.

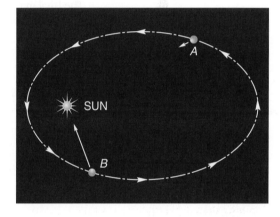

FIGURE 3-13 At point A, the planet is moving closer to the Sun, and the gravitational force on it causes it to speed up as well as curve from a straight line. At point B, the force of gravity slows the planet as well as curves its path.

binary system A system of two objects orbiting each other due to their mutual gravitational attraction.

In some cases, however, it is easier to measure the semimajor axis a in AU and the period P in years; then

$$\frac{a^3_{(AU)}}{P^2_{(yrs)}} = \frac{m_1 + m_2}{m_{Sun}}.$$

For the case of a planet (of mass m_1) orbiting the Sun (of mass $m_2 = m_{Sun}$), Newton's version of Kepler's third law actually simplifies to the original version by Kepler. This is the case because the mass of any planet is very small compared with the mass of the Sun; that is, $m_1 \ll m_{Sun}$, so the expression on the right side of the equation is essentially equal to one. For all objects orbiting the Sun, we thus get Kepler's original version: $a^3_{(AU)}/P^2_{(yrs)} = 1$.

EXAMPLE

Let us now examine the case of two stars orbiting each other. Suppose observations suggest that the orbital period of this system is 4 years and that the average distance between the stars is 8 AU. Substituting these numbers into Kepler's third law, we find

$$\frac{8^3}{4^2} = 32 = \frac{m_1 + m_2}{m_{Sun}}.$$

The total mass of both stars thus is 32 times the mass of the Sun. Without any additional information, we cannot find the mass of each star. We come back to this problem at the end of the next section.

TRY ONE YOURSELF
An artificial satellite is orbiting the Moon with period $P = 11.5$ hours (or 1.3×10^{-3} years) and at an average distance from the Moon's center of 5960 km (or 4×10^{-5} AU). What is the mass of the Moon? (The Sun's mass is about 2×10^{30} kilograms. Make sure you use the appropriate units for the period and average distance.)

3-6   Orbits and the Center of Mass

center of mass The average location of the various masses in a system, weighted according to how far each is from that point.

Newton's third law tells us that when one object exerts a gravitational force on another, the second object exerts an equal force back on the first. We thus should not expect that one of the objects just remains still whereas the other orbits around it. Instead, the two objects orbit about a point between them, called the ***center of mass*** of the objects. (Even though an object's center of mass is not always the same as its center of gravity, for almost all the cases we discuss the two points will be considered to be identical.) As a child, you probably played on a seesaw (or teeter-totter) with someone whose weight was greater than yours. This person had to sit closer to the pivot point to balance. In fact, both persons on a seesaw must position themselves in such a way that their center of mass is at the pivot point of the seesaw if they are to balance on it. Suppose person A weighs 50 pounds and person B weighs 150 pounds. Because person A's weight is three times smaller than B's, person A must sit three times farther from the pivot of the seesaw.

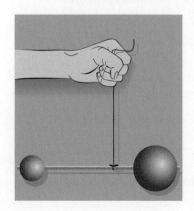

FIGURE 3-14 The two balls at the ends of the rod balance at the center of mass of the device, where the string is connected.

FIGURE 3-14 shows a large ball and a small ball connected by a rod and held by a string tied to the rod. The two balls balance because the string is connected at the center of mass

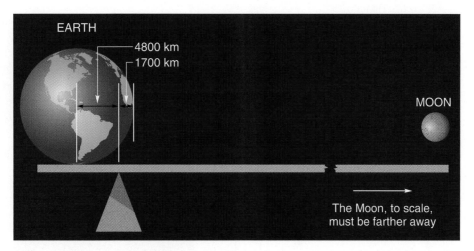

FIGURE 3-15 The center of mass of the Earth–Moon system is 4800 kilometers from the center of the Earth. If a model of the system were constructed to scale to sit on a weightless board, the construction would balance as shown.

of the objects. If this contraption were thrown into the air with a spinning motion—like a baton—it would spin around the center of mass. A planet and the Sun are somewhat like the two balls on the rod, but instead of being held together by a rod, they are held by the force of gravity. For the case of the Sun–Earth system, the Sun is 330,000 times more massive than the Earth and, thus, the center of mass of this system is essentially at the center of the Sun. However, for the case of the Earth–Moon system, the Earth's mass is only 81 times that of the Moon. Therefore, the Moon is 81 times farther from the center of mass of this system than the Earth. This means that the center of mass is about 4800 kilometers from Earth's center (380,000 kilometers from the Moon) or about 1700 kilometers below the Earth's surface (**FIGURE 3-15**).

It was the relationship between the distances of the Earth and Moon from the system's center of mass that allowed us to calculate the mass of the Moon. Today we know the Moon's mass more accurately by observing its gravitational effect on space probes that have flown by it, but until the space age, the method described above was the most accurate. In Chapter 12 (and the example later), we show that the center-of-mass method is what allows us to calculate the masses of stars.

Historically, the center of mass of the Earth-Moon system was determined by observing parallax of nearby planets due to Earth's motion around the center of mass.

EXAMPLE

In the previous example we found that the total mass of the binary system is 32 solar masses. Suppose observations suggest that one star (say star X) is three times farther from the center of mass than the other star (star Y). Star Y, therefore, has three times the mass of star X. As a result, we find that the mass of star X is 8 solar masses and the mass of star Y is 24 solar masses.

When an entire system moves under the influence of a net external force, it moves according to Newton's second law, behaving as if all of its mass were concentrated at the center of mass (**FIGURE 3-16a**). It thus is the center of mass of the Earth–Moon system that follows an elliptical path around the Sun (Figure 3-16b).

3-7 Beyond Newton

Science seeks to show that the various phenomena we observe are not independent of one another, but are instead manifestations of a relatively few basic principles. In the previous sections, we saw how Newton's work succeeded in describing planetary motions based on a few basic ideas on motion and gravity. In addition, Newtonian ideas explained a wide range of observations: In Chapter 6 we describe how they explain the tides and the Earth's precession; in Chapters 9 and 10 we mention that they

I do not know what I may appear to the world; but to myself I seem to have been only like a boy playing on the seashore, and diverting myself in now and then finding a smoother pebble or a prettier shell than ordinary, whilst the great ocean of truth lay all undiscovered before me.
Newton

FIGURE 3-16 (a) The center of mass of a thrown hammer follows a smooth path as the hammer rotates around that point. (b) Likewise, it is the center of mass of the Earth–Moon system that orbits the Sun in an elliptical path. (By holding a straight edge up to the paths drawn, you can see the Earth and Moon are always curving toward the Sun during the orbit. This is because the Sun's gravitational force on either object is stronger than the gravitational force from the other object.)

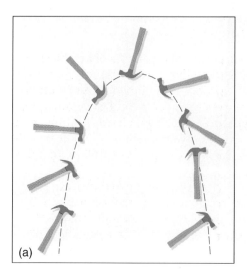

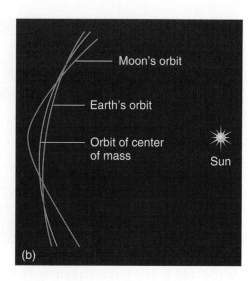

All things by immortal power,
Near or far,
Hiddenly
To each other linked are,
That thou canst not stir a
*   flower*
Without troubling a star.
Francis Thompson
(1859–1907)

were used to predict the existence of the planet Neptune, and the return of comet Halley, respectively.

From these examples, you might conclude (correctly) that Newton's laws are more fundamental than Kepler's. They explain motions in a measurable, mathematical manner; they fit the data and make predictions that can be checked, and they fit with other laws to make an overall theory that is a simple, unified package. (This package is called "Newtonian mechanics" or "classical mechanics," to distinguish it from the mechanics of Einsteinian relativity. **FIGURE 3-17** shows a timeline of how our ideas on gravity have changed through the years.)

Newton was not the first to consider laws that hold both for Earth and the heavens; the idea of the unity of nature began to take shape from the time of Galileo; however, it was Newton who first created a unified model. The word universe begins with the prefix "uni," meaning "one, single" (*unit, unify, unique,* and so on). Our modern science of astronomy could not exist without the basic understanding of the unity of nature. In addition, Newton's work confirmed the belief of the ancient Greeks that nature is explainable—that if we work hard at it, we have the ability to understand the many seemingly mysterious things that occur in nature.

As we found out in the 20th century, however, Newton's laws cannot be applied to all situations. In the next section we describe Einstein's special theory of relativity and his general theory of relativity, which go beyond Newtonian ideas to describe what happens at high speeds and in places where gravity is extremely strong. In Chapter 4, we discuss quantum mechanics, which better describes what happens inside an atom.

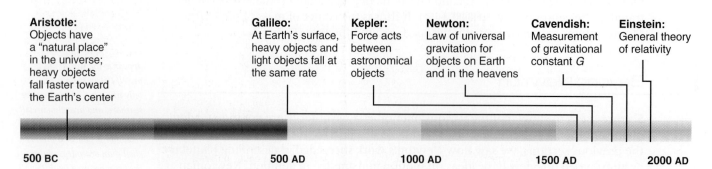

Aristotle:
Objects have a "natural place" in the universe; heavy objects fall faster toward the Earth's center

Galileo:
At Earth's surface, heavy objects and light objects fall at the same rate

Kepler:
Force acts between astronomical objects

Newton:
Law of universal gravitation for objects on Earth and in the heavens

Cavendish:
Measurement of gravitational constant *G*

Einstein:
General theory of relativity

500 BC   500 AD   1000 AD   1500 AD   2000 AD

FIGURE 3-17 Aristotle's explanation of gravity lasted for almost 2000 years, before it was replaced by scientific models based on observation and experiment.

3-8   General Theory of Relativity

Mass was defined earlier in this chapter as the measure of the inertia of an object. In stating the law of gravity, however, Newton proposed that mass is also the quantity that determines the strength of gravitational attraction. Why should the same quantity be the measure of two seemingly different physical properties? It is certainly not obvious that *inertia* (the resistance to a change in motion) should have anything to do with *gravitational attraction*. Yet the measures of these two properties are not just similar; experiments show that they are identical to at least one part in 1 trillion. Can this be simply a coincidence?

Scientists do not like such coincidences. They believe that if two things are so similar, there must be a reason. The attempt to explain this apparent coincidence is what led Albert Einstein to develop his general theory of relativity more than 200 years after Newton stated his law of gravity. The theory begins with a statement of the equivalence of gravity and acceleration, and fundamentally changes the way we look at the phenomenon of gravity.

The woman in FIGURE 3-18a has dropped a book on the surface of the Earth. In part (b), she drops the same book while in a spaceship that is far from Earth (and any other gravitational influences) and accelerating in the direction shown. If her spaceship is accelerating at the same rate that falling objects accelerate here on Earth, she will observe the book falling toward her feet in exactly the same way as it did on Earth's surface. She won't be able to tell the difference between "falling" caused by the ship's acceleration and falling caused by gravity. If she steps on a scale in such a spaceship, the scale will register the same weight as a scale on the Earth's surface. The ***principle of equivalence*** of Einstein's general theory of relativity tells us that there is no experiment whatsoever that she can do to distinguish between the two conditions. This principle forms the cornerstone of the theory in which the similarity of the effects of gravity and acceleration are explained by considering the curvature of space.

Space curvature is not easy to imagine because the world we experience is three dimensional, and we cannot imagine what dimension our world can curve into. We say that our world has three dimensions because we use three directions to specify the exact location of something. We can state these directions as north–south, east–west, and up–down. For example, you might state that to get from the library to your bed at home, you must go 4325 feet north, 5843 feet west, and 12 feet up to the second floor. Although other choices might be made for the three coordinates that specify the location of an object relative to your position, you must always give three pieces of information in our three-dimensional world.

Imagine some tiny two-dimensional creatures on a large, expanding balloon. The creatures have no height, and they only know of the two dimensions of their universe; they can perceive north–south and east–west, but have no conception of up–down. You might object that the surface of the balloon is not really two dimensional,

The spaceship is far enough from Earth (and other objects) that there is no observable gravitational force.

principle of equivalence
The statement that effects of acceleration are indistinguishable from gravitational effects.

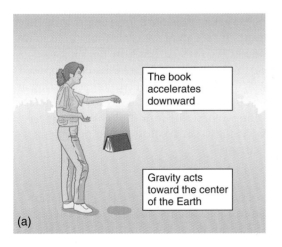

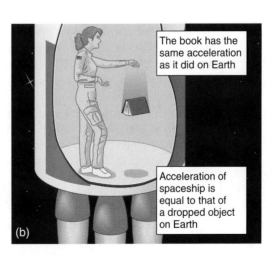

The book accelerates downward

Gravity acts toward the center of the Earth

The book has the same acceleration as it did on Earth

Acceleration of spaceship is equal to that of a dropped object on Earth

(a)   (b)

FIGURE 3-18 Whether on Earth or in a spaceship far from Earth and any other gravitational influences, accelerating at the acceleration due to gravity, the book will "fall" in the same way.

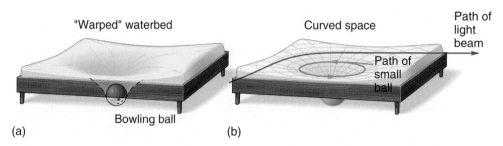

"Warped" waterbed

Curved space

Path of light beam

Path of small ball

Bowling ball

(a)   (b)

FIGURE 3-19 According to the general theory of relativity, matter curves the space around it. This curvature, in turn, influences the motion of objects in the vicinity. To visualize the curvature of three-dimensional space, in this figure we show how (a) a two-dimensional space (the surface of a waterbed) curves around a massive object (a heavy bowling ball), and how (b) the resulting distortion, in turn, affects the paths taken by light and other objects passing by. (Changes in the paths have been exaggerated for clarity.)

This idea is borrowed from *Flatland*, written by Edwin A. Abbott in 1884. This book is a classic illustration of multiple dimensions in space, but even more than that, it's great fun to read.

We will see later in the text that space curvature also explains black holes.

but if the balloon is large enough compared with the size of the creatures, they would not easily perceive the balloon's curvature. If one creature realizes that his universe is curved, how can he explain this idea to his contemporaries?

Saying that their universe is curved "downward" would have no meaning, because they have no conception of "up" and "down." We humans see that the balloon's surface is curved into the third dimension, but it would take a great stretch of the creatures' imagination to think of a third dimension, for it is not part of their everyday world.

By a similar analogy, we picture the curvature of *our* space. Suppose that the presence of a massive object causes space to be warped. We can picture space near the Sun as being warped analogous to the way the surface of a waterbed is warped by a bowling ball placed at its center (**FIGURE 3-19**). The waterbed's surface would be distorted so that a straight line following the surface would have to follow the distortion. Similarly, a beam of light passing near a massive object would follow the curvature of space caused by the massive object. Let us also imagine the motion of a small ball on this waterbed. Far from the distortion, the surface of the waterbed is fairly flat and thus the ball will move in a straight line; however, the closer the ball gets to the distortion, the more it will curve toward it. If the ball's speed is chosen appropriately with respect to the distortion, the ball may end up moving in an orbit around the sides of the distortion, in the same way a planet moves around the Sun. We can say that matter (in our case, the heavy bowling ball) tells space (the surface of the waterbed) how to curve and that the space curvature (or distortion) tells matter (the small ball) how to move.

3-9   Gravitation and Einstein

In Newton's view of gravity, interacting objects act at a distance; one object influences another across the empty space between them. If I let go of an apple that I hold in my hand, it *immediately* starts falling toward the Earth's surface, but how does it know when to start falling and how strong the Earth's influence is on it? Einstein proposed that instead of thinking of gravity as an attractive force between two objects acting at a distance, we think of space as being curved by the presence of mass so that objects move because of the curvature. In this section we describe four of the many experimental tests of the general theory of relativity; they clearly show that it provides a better description of gravity than Newton's theory.

Test 1: The Gravitational Bending of Light

As shown in Figure 3-19, according to the general theory of relativity, a light beam passing by a massive object will be deflected from its original straight path. This gravitational bending of light (which we discuss in more detail in Chapter 15) was predicted by Einstein in 1915, and observations made during a solar eclipse in 1919

HISTORICAL NOTE

Albert Einstein

The young Albert Einstein (FIGURE B3-5) found the formal, disciplinary schools of Germany at the end of the 19th century intimidating and boring, and he dropped out before completing high school. He studied at home, however, and learned to play the violin well enough that he became an accomplished violinist. His studies of geometry and science led him to conclude, at age 12, that the Bible is not literally true. That shock implanted in him a deep distrust of authority of any kind—a distrust that he carried with him throughout life.

On his second try, Einstein was granted admission to the Swiss Federal Institute of Technology in Zurich. While there, he often preferred to study on his own, reading the classical works of theoretical physics. He was granted a Ph.D. in 1900, but it was 2 years before he found regular work—as a patent examiner in the Swiss Patent Office.

Like Newton, Einstein's most revolutionary work was done during a short period in his early 20s. In 1905, Einstein published four important papers in the prestigious German physics journal *Annalen der Physik*. Although the paper for which he is best known is the one introducing special relativity, he was awarded the Nobel Prize for another, concerning the photoelectric effect of light. Many physicists quickly recognized the importance of his work, but it was not until 1909 that he was given a full-time academic position at the University of Zurich.

In 1903, Einstein married his college sweetheart, Mileva Maritsch, and they had two sons. Mileva's influence on Einstein's ideas is still debated. Their forced separation during World War I resulted in a divorce in 1919. Later that year, Einstein married his cousin Elsa.

In 1916, Einstein published his general theory of relativity, but his theories were slow to gain acceptance because of the lack of experimental verification. In 1919, however, the Royal Society of London announced that its scientific expedition to observe the solar eclipse of that year had verified Einstein's prediction of the gravitational deflection of starlight as it passes near the Sun. The international acclaim that followed changed Einstein's life, for he was suddenly considered a genius.

FIGURE B3-5 Albert Einstein (1879–1955) had several hobbies; one of his favorites was sailing.

Although Einstein continued his scientific work until he died in 1955, his fame also allowed him to exert influence in world affairs. Though a Jew and a critic of the political situation in Germany, he escaped the Nazis because he was visiting California when Hitler assumed power in 1933. Einstein renounced his German citizenship and never returned to his home country. The next year he became an American citizen.

It is ironic that Einstein's name is so closely linked to the atomic bomb. Although his theories predicted that mass could be converted to energy, which was the idea that led to the development of nuclear weapons, Einstein was an avowed pacifist who worked untiringly to prevent war, which he saw as the ultimate scourge of humanity. He argued that the establishment of a world government was the only permanent solution.

---

confirmed this prediction. Newton's theory predicts no gravitational effect on light, as light has no mass, and light is not observed to respond to gravity in our everyday world. The bending of light is observed only around very massive objects, and because Newton's theory had never been checked in such cases, no one realized that it makes incorrect predictions.

Suppose a total solar eclipse occurs while the Sun—as seen from Earth—is between two bright stars. (When the Sun is totally eclipsed by the Moon, stars can be seen in the sky.) FIGURE 3-20a shows the stars as they normally appear. During the eclipse, shown in part (b) of the figure, as light from the stars passes near the Sun before reaching Earth, it bends and, thus, it makes the stars appear slightly farther apart, as shown in part (c). The maximum predicted gravitational bending of light is very small, only 1.75 arcseconds, but the observations made during the 1919 expedition produced results in agreement with the theory, providing the first experimental support for the general theory of relativity.

Test 2: The Orbit of Mercury

In 1859, 14 years after predicting the existence of Neptune, Urbain Le Verrier reported that Mercury's elliptical orbit *precesses*; that is, it does not keep the same orientation in space. FIGURE 3-21 illustrates a greatly exaggerated precession, showing the

If we were to use Newtonian mechanics to describe the motion of a photon as a particle (of mass m and speed c) in a gravitational field, we would find that the photon's path is deflected by an angle that is half the size predicted by the general theory of relativity.

precession (of an elliptical orbit) The change in orientation of the major axis of the elliptical path of an object.

FIGURE 3-20 Light from the two stars is bent as it passes near the Sun, causing the stars to appear farther apart.

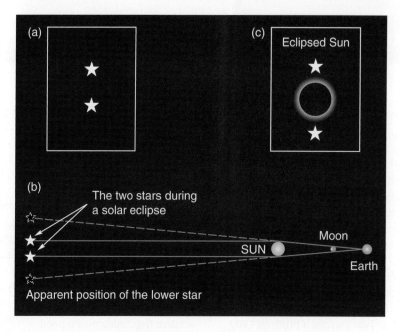

near point in the orbit (the perihelion) gradually sliding around the Sun. Mercury's orbit precesses at the rate of 574 arcseconds per century, which is less than a degree per century. Calculations show that Newton's theory of gravity can account for the basic effect because gravitational pulls by other planets, particularly Venus and Jupiter, would cause it. However, the total precession accounted for by these gravitational tugs amounts to only 531 arcseconds per century.

The unaccounted-for 43 arcseconds of precession per century presented a mystery for astronomers. One hypothesis held that there is another planet in the inner solar system that is responsible for the extra precession. This planet was named Vulcan. Extensive searches for it were carried out, but it was never found.

Einstein applied his general theory of relativity to the problem of the precession of Mercury's orbit and found that it accounted precisely for the 574 arcseconds of precession. He later wrote that, "for a few days, I was beside myself with joyous excitement."* Like Newton's theory, general relativity predicts the precession effect caused by the other planets, but unlike Newton's theory, it predicts additional precession due to properties of curved space. The Sun itself thus caused the extra 43 arcseconds of precession.

*The quotation is from Clifford M. Will, *Was Einstein Right?* (New York: Basic Books, 1993).

Additional Tests

gravitational wave
Ripples in the curvature of space produced by changes in the distribution of matter.

According to Einstein's theory of gravity, massive objects warp not only space around them but also time. Time slows down in the presence of gravity, and the greater the strength of gravity, the more time slows down. This effect (which we discuss in more detail in Chapter 15) was experimentally tested in 1960, and the results were in complete agreement with the general theory of relativity.

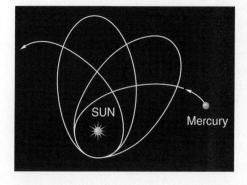

FIGURE 3-21 This illustration exaggerates the actual precession of the perihelion (the closest point to the Sun) of Mercury's orbit.

Finally, the general theory of relativity predicts the existence of *gravitational waves*. Consider Figure 3-19 and imagine that the massive object at the center of the curved space was made to move up and down. As a result, ripples will be produced in the curvature of space, and these waves will propagate outward at the speed of light. Even though we have not yet directly observed such waves, strong indirect evidence comes from observations of a binary system of two stars, as we discuss in Chapter 15. Newton's theory of gravity

ADVANCING THE MODEL

The Special Theory of Relativity

About 10 years before he developed the general theory of relativity, Einstein proposed the special theory of relativity, which is based on two postulates, the first of which states:

- All laws of physics are the same for all nonaccelerating observers, no matter what the speed of those observers.

People once thought that the laws of nature that govern objects here on Earth are different from those that rule the heavens. Then Newton showed that the same laws work for both—at least, the mechanical laws of force and motion. Einstein's first postulate completes the progression; he begins with the assumption that *all* laws, including those of electricity and magnetism, are the same everywhere. The first postulate abolishes the idea of absolute rest and therefore absolute motion. Motion is relative, as it is not possible to distinguish experimentally between two different uniformly moving observers.

Einstein's second postulate concerns the speed of light:

- The speed of light (*c*) is the same for all nonaccelerating observers, no matter what their motion relative to the source of the light.

Compare this behavior of light to that of ordinary objects in our experience. For example, when we catch a baseball, we see it coming at us faster if we are moving toward the thrower than if we are standing still. If the postulate is true, this doesn't happen for light. **FIGURE B3-6** shows an imaginative case in which people on fast-moving spaceships are measuring the speed of light that comes from an Earthling's flashlight. Technological advances since Einstein's time have allowed us to repeatedly confirm the second postulate; we now consider the constancy of the speed of light a law of nature.

On the basis of his two postulates, Einstein showed that Newton's laws of motion become increasingly inaccurate with increasing speed. The special theory of relativity predicts that (1) the observed length along the line of motion of a moving object becomes less than its length when measured at rest; (2) the observed passage of time becomes slower for the moving object; and (3) the observed inertia of an object becomes greater than its inertia when at rest. Special relativity shows that the three spatial dimensions and the time dimension are intimately bound, and thus it makes sense to describe them together as a four-dimensional continuum: *spacetime*. The framework of the special theory of relativity, however, is such that it does not include a consistent description of gravity; the theory's realm is flat spacetime. Gravitation, being a manifestation of the curvature of spacetime, is described by the general theory of relativity, which involves special relativity, the equivalence principle, and the local nature of physics.

One more conclusion based on the postulates of special relativity should be mentioned: mass (*m*) can be transformed to energy (*E*) and vice versa. The conversion between the two is governed by the equation $E = mc^2$. The equation was dramatically verified in 1945 by the first explosion of a nuclear bomb, for the energy of these bombs comes from the conversion of mass to energy. A more peaceful example of mass–energy conversion is provided by nuclear power plants, which produce electrical energy based on Einstein's theory. (We see in Chapters 11 and 12 that nuclear energy is the source of the energy of the stars.) The predictions of the special theory of relativity become significant only at speeds greater than those attained in everyday life. Every time a test has been conducted to check the theory, the theory has passed the test.

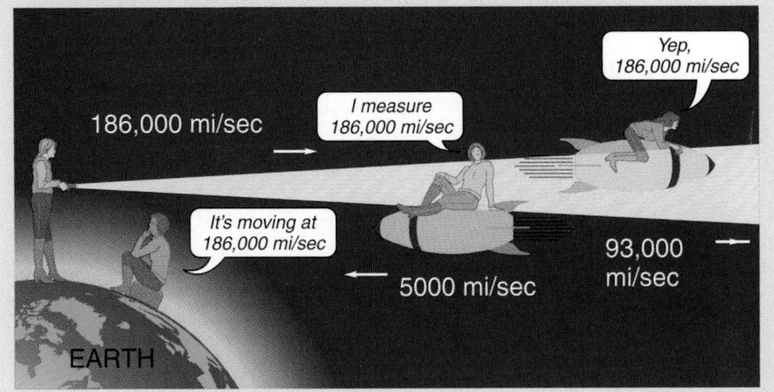

FIGURE B3-6 According to the special theory of relativity, the speed of light is the same regardless of the motion of the observer.

does not include any effect that gravity might have on time and does not predict the existence of gravitational waves.

All tests of Einstein's relativity theory have confirmed it, and the theory forms the basis of modern astronomy. We still talk about Newtonian ideas, however, for the same reason we still use a geocentric view of the world when we talk about "sunrises" and "sunsets." Newton's theory works very well in the realm of our everyday experiences and is more easily understood than Einstein's. You will see little mention of curved space in this text until the discussion of black holes, where Einstein's theory is essential.

Conclusion

correspondence principle
The idea that predictions of a new theory must agree with the theory it replaces in cases where the previous theory has been found to be correct.

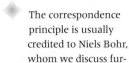 The correspondence principle is usually credited to Niels Bohr, whom we discuss further in Chapter 4.

There is a general principle in science concerning the replacement of old theories by new ones. The *correspondence principle* states that the predictions of the new theory must agree with those of the previous theory where the old one yields correct results. This is entirely reasonable. In the previous chapter, we discussed two primary models: Ptolemy's geocentric model and the heliocentric model, as presented by Copernicus and revised by Kepler. In this chapter, we discussed how Galileo's telescopic observations supported Kepler's results and how Newton expanded on the work of his predecessors by offering us a more unified view of the world around us.

We now find that Newton's ideas have been supplanted by Einstein's. Even though the theory of relativity does not seem to fit our everyday experiences, it is firmly established in science, and perhaps someday it will become part of our everyday thinking. It is also not only possible but likely that at some point in the future Einstein's theories will be supplanted by new theories that more accurately describe the universe.

STUDY GUIDE

1. Using his newly invented telescope, Galileo discovered all of the following except
   A. moons of Jupiter.
   B. phases of Venus.
   C. sunspots.
   D. stellar parallax.
   E. mountains on the Moon.

2. Which of the following planets can be seen (from Earth) in a crescent phase?
   A. Mercury.
   B. Venus.
   C. Mars.
   D. [Two of the above.]
   E. [All of the above.]

3. Which of the following observations by Galileo was most convincing in deciding between the two opposing theories of our planetary system?
   A. Moons of Jupiter.
   B. Mountains on the Moon.
   C. Phases of Venus.
   D. Sunspots.

4. Suppose you are riding as a passenger in a car when the car stops suddenly. What pushes you forward?
   A. Inertia.
   B. Your weight.
   C. The car's forward motion.
   D. Nothing pushes you forward.

5. If the same net force is applied to two different objects, one with a mass of 1 kilogram and the other with a mass of 2 kilograms,
   A. the 1-kilogram mass will have twice the acceleration of the other.
   B. the 1-kilogram mass will have more acceleration than the other, but not necessarily twice as much.
   C. both objects will have the same acceleration, for the force determines the acceleration.

6. If the mass of the Earth magically decreased with no change in its size, your weight (as you stand on the surface) would
   A. increase.
   B. not change.
   C. decrease.

7. According to Newton, the natural motion of an object is
   A. a circle.
   B. an ellipse.
   C. a straight line.
   D. retrograde motion.

8. If the distance between two objects is tripled, the gravitational force exerted by one on the other will be
   A. the same.
   B. one third as much.
   C. one ninth as much.
   D. three times as much.
   E. nine times as much.

9. The gravitational attraction between an object and the Earth
   A. stops just beyond the Earth's atmosphere.
   B. extends to about halfway to the Moon.
   C. extends about five sixths of the way to the Moon.
   D. extends to infinity.

10. The mass of the Moon is about 1/81 that of the Earth. If you were on the Moon, however, you would weigh 1/6 of your weight on Earth. Why do you weigh more than 1/81 of your Earth weight?
    A. The Moon is made of different materials than Earth.
    B. The Moon has a different density than Earth.
    C. The Moon has no atmosphere, whereas Earth does.
    D. The Moon is much smaller than Earth.
    E. [Both A and B above.]

11. Which statement best describes the relationship between Newton's laws and Kepler's laws?
    A. Newton proved that Kepler was wrong.
    B. Newton's laws and Kepler's laws are now considered equally valuable.
    C. Newton's laws are more fundamental and more powerful than Kepler's laws.
    D. Neither Newton's laws nor Kepler's laws have any modern applications.

12. When you whirl a rock around on the end of a string, centripetal force
    A. pulls outward on the rock.
    B. pulls inward on the rock.
    C. pulls outward on your hand.
    D. [Both A and B above, and the two forces balance.]
    E. [All of the above, for all of the forces are equal.]

13. Kepler's third law states that the ratio of the cube of a planet's semimajor axis to the square of the planet's period of revolution is equal to a constant. Newton found that the value of that constant depends on
    A. the sizes of the two objects.
    B. the masses of the two objects.
    C. the velocities of the two objects.

14. Newton checked his hypothesis concerning an inverse square law of gravitation by calculating
    A. the Moon's acceleration toward the Earth.
    B. the time required for the Moon to complete one orbit.
    C. the mass of the Earth.
    D. the mass of the Moon.
    E. [Both C and D above.]

15. The force of gravity is responsible for
    A. the weight of an object on Earth.
    B. the mass of an object on Earth.
    C. the tides.
    D. holding the Earth in its orbit.
    E. [All of the above.]

16. If the Earth's radius magically decreased with no change in its mass, your weight (as you stand on the surface) would
    A. increase.
    B. not change.
    C. decrease.

17. If the Earth were magically moved farther from the Sun but kept in orbit around it, the length of the year would
    A. increase.
    B. not change.
    C. decrease.

18. If the Sun's radius magically doubled but everything else remained the same, the Earth's orbital period would
    A. increase.
    B. not change.
    C. decrease.

19. If a small asteroid is found to orbit the Sun on the same orbit as Earth, the asteroid's orbital period will be
    A. about the same as Earth's.
    B. much greater than Earth's.
    C. much smaller than Earth's.

20. The principle of equivalence of the general theory of relativity tells us that
    A. effects of gravity are equivalent to effects of acceleration in the opposite direction.
    B. speeds measured in any location are equivalent.
    C. the speed of light is the same for all observers.
    D. Newton's laws are equivalent to Kepler's laws.
    E. Newton's laws are equivalent to Einstein's theories.

21. Which of the following choices confirmed a prediction made by the general theory of relativity?
    A. Observations of Jupiter's satellites.
    B. Observations of phases of Venus.
    C. Calculations of the orbit of Mercury.
    D. Calculations predicting the existence of Mars' moons.
    E. [None of the above; general relativity has not been successful in astronomical applications.]

22. The correspondence principle states that
    A. predictions made by a new theory must agree with those of the theory it replaces where the old theory fit the data.
    B. all predictions made by a new theory must agree with those of the theory it replaces.
    C. no predictions made by a new theory are expected to agree with those of the theory it replaces.
    D. effects of gravity are equivalent to effects of acceleration in the opposite direction.

23. List the following men in the order in which they lived: Copernicus, Aristotle, Ptolemy, Kepler, Galileo, Brahe. (Hint: You may have a "tie" between two of them.)

24. Most of Galileo's observations argued *against* the Ptolemaic system rather than *for* a heliocentric system. What was the exception?

25. Which of the phases of Venus could not be explained by the Ptolemaic model?

26. Define inertia and give an example of its action.

27. Which is the more fundamental quantity, mass or weight? Describe an observation that confirms your answer.

28. State Newton's three laws and cite an example of each.

29. How was the mass of the Moon (compared with Earth's mass) first determined?

30. What determines the magnitude (strength) of the gravitational force between two objects?

31. How did Newton use an astronomical object to check his hypothesized law of gravitation?

32. Explain how the law of gravitation accounts for the fact that planets move fastest when they are closest to the Sun.

33. Did Newton's laws conflict with Kepler's laws? Explain.

34. What provides the centripetal force to keep the Earth in orbit around the Sun?

1. What events took place during the century before Galileo that contributed to the revolutionary flavor of his times?

2. Neither the geocentric nor the heliocentric system made a direct prediction about whether Jupiter has moons. Why, then, did Galileo's discovery of moons have an impact on the choice of a model?

3. Figure 3-4 shows that Venus appears to be larger when it is in certain phases. Why does this occur?

4. Mars exhibits phases. What phase(s) would you expect to see in viewing Mars? In what phase(s) would Mars never appear when viewed from Earth?

5. Explain the distinction between mass and weight, showing that the difference is more than just a matter of what units are used.

6. Newton's laws tell us that no force is needed to keep something moving. Why, then, when we are driving on level ground, don't we turn off the engine of our car?

7. It seems presumptuous to call the law of gravitation "universal." What evidence did Newton have that the law applies beyond the Earth?

8. Explain why the work of Newton had implications beyond science.

9. Kepler held that the center of the Earth orbits the Sun in an elliptical path. Newton's laws tell us that this isn't exactly true. Explain.

10. Why do we teach Newton's law of gravitation even though general relativity is a more up-to-date explanation of the phenomena involved?

11. What causes weight?

12. Give an example of the correspondence principle in the case of the heliocentric theory replacing the geocentric theory.

13. Some people even today oppose the theory of evolution on religious grounds. Compare this position with the conflicts of Copernicus and Galileo with the church establishment.

1. Suppose that you move three times as far from the center of the Earth as you are now. By what factor will your weight change?

2. On Earth's surface, Big Al is about 6500 kilometers from its center. If the force of gravity on Big Al here is 250 pounds, how much will it be on him at a distance of 13,000 kilometers from the Earth's center?

3. Two planets (A and B) orbit a star S. Planet B is three times farther from the star than A is and has three times the mass of A. The force from the star on planet A is x. What is the force, in units of x, from the star on planet B?

4. In the preceding problem, suppose that the orbital period of planet A is two years and that it orbits the star at an average distance of 2 AU. What is the mass of the star compared to that of our Sun? (Assume the planets have small masses.) What is the orbital period of planet B?

5. Using data for Jupiter and its Galilean satellites (Io, Europa, Ganymede, and Callisto) given in the appendices of your book, show that these data agree with Newton's form of Kepler's third law.

6. The average center-to-center distance between the Earth and Moon is 3.84×10^8 m. The sidereal period of the Moon is 27.32 days. Assuming a circular orbit, show that the Moon moves on its orbit with a speed (v) of about 1020 m/s. (Hint: During one orbit, the Moon covers a distance equal to the orbit's circumference; also, speed is distance divided by time.) It can be shown that the centripetal acceleration of an object moving at constant speed on a circular orbit of radius r is given by v^2/r. Show that the Moon's centripetal acceleration is 0.0027 m/s^2 or $1/60^2$ of 9.8 m/s^2, which is the acceleration of gravity on Earth's surface.

1. Circular Motion

This activity should be done outside, away from anything breakable. Tie some object such as a shoe to the end of a fairly long (6 or 8 feet) string or rope. Now whirl the object in a horizontal circle around you. Feel the pull you must exert to keep the object in the circle. Whirl it faster. Do you have to increase the force you exert on the string?

Now let go of the string and carefully observe the path taken by the object. Forget the downward motion (caused by gravity) and concentrate on how the object travels horizontally. Figure 3-10 shows several potential paths for the object. Which of these did your object take?

2. Observing Venus

If you have a small telescope and Venus happens to be visible in the evening sky, observe the planet once a week for a month. In essence, you will be repeating Galileo's observations. Do your observations show that the observed shape of the planet is changing? If yes, can you tell if it is coming toward us or moving away from us?

The following books and articles will give you a sense of the impact that Galileo, Newton, and Einstein had on our understanding of the universe.

1. "Newton's Discovery of Gravity," by I. B. Cohen, in *Scientific American* (March, 1981).

2. "Newton's Principia: A Retrospective," by G. Christianson, in *Sky & Telescope* (July, 1987).

3. "How Galileo Changed the Rules of Science," by O. Gingerich, in *Sky & Telescope* (March, 1993).

4. J. Fauvel, et al., *Let Newton Be!* (Oxford University Press, 1988).

5. O. Gingerich, *The Great Copernicus Chase, and Other Adventures in Astronomical History*, a collection of essays including the period of Galileo and Newton (Sky Publishing/Cambridge University Press, 1992).

6. Clifford M. Will, *Was Einstein Right?* (Basic Books, 1986).

7. Lewis Carroll Epstein, *Relativity Visualized* (Insight Press, 1985).

8. George Gamow and Roger Penrose, *Mr Thompkins in Paperback* (Cambridge University Press, reissue edition, 1993).

9. George Gamow, *The Great Physicists from Galileo to Einstein* (Dover Publications, 1988).

10. Galileo Galilei, *Discoveries and Opinions of Galileo*, Stillman Drake, translator (Anchor Books/ Doubleday, 1957).

11. James Reston, *Galileo, audio cassette*, Jeff Riggenbach, narrator (Blackstone Audio Books, 1996).

12. Gale E. Christianson, *Isaac Newton and the Scientific Revolution. Oxford Portraits in Science* (Oxford University Press, 1998).

13. John L. Heilbron, *The Sun in the Church: Cathedrals as Solar Observatories* (Harvard University Press, 1999)

14. "Relativity Turns 100," by R. Panek, in *Astronomy* (February, 2005).

Quest Ahead to Starlinks
http://physicalscience.jbpub.com/starlinks

Starlinks is this book's online learning center. It features **eLearning**, which contains chapter quizzes and other tools designed to help you study for your class. You can also find **online exercises**, view numerous relevant **animations,** follow a guide to **useful astronomy sites** on the Web, or even check the latest **astronomy news** updates.

radio continuum (408 MHz)

atomic hydrogen

radio continuum (2.5 GHz)

molecular hydrogen

infrared

mid-infrared

near infrared

optical

x-ray

gamma ray

Light and the Electromagnetic Spectrum

4

The Milky Way seen at 10 wavelengths of the electromagnetic spectrum.

JAMES CLERK MAXWELL WAS BORN IN EDINBURGH, SCOTLAND IN 1831. His genius was apparent early in his life, for at the age of 14 years, he published a paper in the *Proceedings of the Royal Society of Edinburgh*. One of his first major achievements was the explanation for the rings of Saturn, in which he showed that they consist of small particles in orbit around the planet. In the 1860s, Maxwell began a study of electricity and magnetism and discovered that it should be possible to produce a wave that combines electrical and magnetic effects, a so-called electromagnetic wave. His analysis of this hypothetical wave showed that its speed would be 300,000 kilometers/second. Because this is the speed of light, Maxwell concluded that he had discovered the nature of light: Light is an electromagnetic wave.

In 1888, 9 years after Maxwell's death, radio waves were discovered by Heinrich Hertz and were shown to have properties similar to those of light. This verified Maxwell's prediction. The importance of Maxwell's work is indicated in the following quotation from the Nobel Prize winner Richard Feynman:

> *From a long view of human history—seen from, say ten thousand years from now—there can be little doubt that the most significant event of the 19th century will be judged as*

All cross references to chapters, sections, figures, and tables pertain to the main text, *In Quest of the Universe, Sixth Edition. In Quest of the Solar System* contains Chapters 1–11 and 19 of the main text. *In Quest of the Stars and Galaxies* contains Chapters 1–5 and 11–19 of the main text.

96

Maxwell's discovery of the laws of electrodynamics. The American Civil War will pale into provincial insignificance in comparison with this important scientific event of the same decade.

The Feynman Lectures on Physics. Vol. 2 (Reading, Mass.: Addison-Wesley Publishing Co., 1964).

An important part of your study of astronomy is to learn about the objects in our universe, but perhaps more important is to see how astronomy functions by learning *how* we know what we know about these objects. The only thing we obtain from them is the radiation they emit. This radiation, including not only visible light but many other types of radiation, carries to us a tremendous amount of information. To understand how astronomers analyze radiation to answer questions about celestial objects, it is necessary to learn something about radiation itself.

In this chapter, we examine the nature of light and show how we measure three major properties of stars: their temperatures, their compositions (that is, the chemical elements of which they are made), and their speeds relative to the Earth. Recall that we already have a tool, in the form of Kepler's third law (see Chapter 2), which allows us to find the total mass of a binary system. In addition, there are other tools, described in later chapters, which allow us to learn more about the stars (for example, how far away and how big they are).

4-1   The Kelvin Temperature Scale

Light transmits energy. We know this because we can clearly feel the warmth of sunshine. When we think of the Sun as an energy source (even though very different from other energy sources, such as a fireplace or a very hot iron bar), an obvious question to ask is how "hot" it is. What is its temperature? Even more important, what do we mean by the term "temperature," and how do we measure it? It should not be surprising that we started measuring temperatures long before we understood what temperature is. Check the accompanying Tools of Astronomy box where we describe the temperature scales currently in use.

The commonly used temperature scale in science is the **Kelvin scale**; its zero point corresponds to the lowest temperature possible (absolute zero), about $-273°C$. The intervals on the Kelvin scale are the same size as on the Celsius scale and, thus, in Kelvin temperature, the freezing point of water is 273 K and the boiling point is 373 K. (No degree symbol is included; the latter temperature is stated as "373 kelvin.") Figure B4-1 includes the Kelvin temperature scale on the right.

We now recognize that temperature is a fundamental quantity, as are mass and time. As such, it cannot be expressed in terms of other quantities; however, it is a good approximation for us to say that given the temperature of an object, such as a pot of water on a stove, we can calculate the average speed of each of its constituent particles; that is, temperature is a measure of their average **kinetic energy**. As the temperature of an object increases, each of its constituent particles moves faster, whereas as the temperature decreases, the particle speed also decreases (although not in a linear fashion). At absolute zero, we have a state of minimum atomic motion.

Kelvin temperature scale A temperature scale with its zero point at the lowest possible temperature ("absolute zero") and a degree that is the same size (same temperature difference) as the Celsius degree. $T_K = T_C + 273$.

kinetic energy An object's energy due to motion. For an object of mass m and speed v, its kinetic energy is equal to $(1/2)mv^2$.

EXAMPLE

A star's surface temperature is 6000 K. What is its temperature in degrees Celsius and Fahrenheit?

SOLUTION   In degrees Celsius, the star's temperature is $6000 - 273 = 5727°C$. In degrees Fahrenheit, it is

$$\frac{9}{5}(5727) + 32 = 10,341°F.$$

TOOLS OF ASTRONOMY

Temperature Scales

Around 1592, Galileo invented the first device indicating the "degree of hotness" of an object. The invention of the thermometer, in 1631, is credited to J. Rey, a French physician. The scale in common use in the United States is the Fahrenheit scale, but most of us are at least somewhat familiar with the Celsius temperature scale. The freezing point of water is defined as 0°C or 32°F, whereas the boiling point of water is defined as 100°C or 212°F. Because 180 divisions on the Fahrenheit scale correspond to 100 divisions on the Celsius scale, a Fahrenheit degree is smaller than a Celsius degree by a factor of 180/100 = 9/5. Thus, the two scales are related by

$$T_F = 32 + \frac{9}{5}T_C \quad \text{or} \quad T_C = \frac{5}{9}(T_F - 32).$$

The first two thermometers of Figure B4-1 compare the Fahrenheit and Celsius scales, from extremely low temperatures up to the boiling point of water, but let's think for a minute about what these temperature scales can and cannot tell us. What does it mean to say that object A is two (or a thousand) times hotter than object B, which is at 0°F? What is then the temperature of object A? The obvious answer is 0°F, but we know there is something wrong with this. (We could ask a similar question using 0°C.) Neither scale—nor, for that matter, any other scale that defines its zero mark arbitrarily—has anything to do with what temperature "is." Clearly, we would like to have an "absolute" temperature scale, one on which the zero mark is associated with the lowest temperature possible (absolute zero). Such a scale exists (**FIGURE B4-1**). It is called the **Kelvin** scale and is the one commonly used in science; it was first proposed by William Thompson (1824–1907), a British physicist and engineer who was made a baron in 1892, taking the title Lord Kelvin. He contributed important ideas across the entire range of physics.

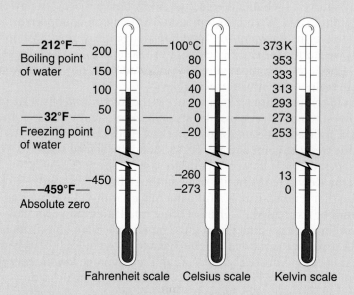

FIGURE B4-1 A comparison of temperature scales.

For a large enough Kelvin temperature, we can approximate its Celsius equivalent by the same number, while its Fahrenheit equivalent is approximately twice as large in value.

TRY ONE YOURSELF

Two objects A and B are identical except that B is at 0°C and A is twice "as hot" (meaning its temperature is twice as large). What is the temperature of object A? (Hint: You must work with the Kelvin scale for both objects.)

4-2   The Wave Nature of Light

Our understanding of the nature of light has changed several times over the years. Two lines of thought about light came to us from the ancient Greeks. The first suggested that light is a stream of extremely small, fast-moving particles and that our vision is the result of the interaction between our eyes and this stream. The second idea, suggested by Aristotle, pictured light as an "aethereal motion." Aristotle added *aether* as the fifth element to his four elements of nature (fire, air, earth, and water) and imagined that it fills all space. According to this idea, our vision is the result of movement of the aether produced by the object we perceive.

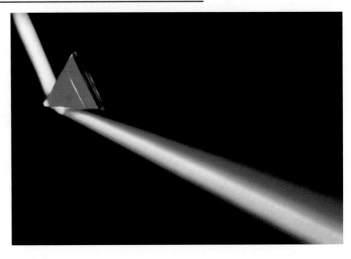

FIGURE 4-1 A prism separates white light into its component colors. This is shown by the beam emerging from the base of the prism and moving toward the bottom right in the image. In the case shown, however, there is an additional beam that emerges from the right of the prism and moves toward the upper right of the image. Why do we not see colors in this beam?

The first step away from Aristotle's idea was taken by Newton, who theorized that light consists of tiny, fast-moving particles. **FIGURE 4-1** shows a beam of white light passing through a glass prism. The emerging light is separated into colors—into a ***spectrum***. Newton showed that the prism does not add color to the light, as was previously thought, but rather that color is already contained in white light and that the prism merely separates the light into its colors. He showed this by using a second prism to recombine the colors produced by the first. In a separate experiment, Newton allowed one color of the spectrum created by a prism to pass through a second prism. Because each color of the spectrum remained unchanged by the second prism, he was able to show that color is a fundamental property of light. We see in later chapters that analysis of the spectrum of light from stars is extremely important in astronomy.

> **spectrum** The order of colors or wavelengths produced when light is dispersed.

Characteristics of Wave Motion

Light acts like a wave. **FIGURE 4-2** is a simplified drawing of a wave indicating that the distance between successive peaks (crests) of the wave is called the ***wavelength*** (λ). Waves you make by dipping your hand in a swimming pool might have a wavelength of about 2.25 inches.

A wave does not sit still, however. Imagine yourself fishing while sitting on a pier, watching waves pass underneath. As the waves move by, they cause the cork on the fishing line to move up and down. This indicates that the water itself moves up and down, rather than along the direction of the wave's motion. As the wave travels along the surface, the water's motion is *primarily* in the vertical direction and not along the direction of the wave. If you count the number of times the cork moves up and down, you might find that it moves through a complete cycle 30 times each minute. We say that the ***frequency*** of the cork's motion is 30 cycles per minute and therefore that the frequency of the wave is 30 cycles/minute. (Frequency is often reported in units of ***hertz***, abbreviated Hz, where 1 hertz = 1 cycle/second.)

Now suppose you measure the wavelength of the waves and find it is 20 feet. Because each wave, from crest to crest, is 20 feet long and 30 of these waves pass by you each minute, the waves must move at a speed of 600 feet/minute. We multiply wavelength by frequency to obtain the speed of the wave. In equation form,

$$\text{wave speed} = \text{wavelength} \times \text{frequency},$$

or, using symbols,

$$v = \lambda \times f,$$

where v = wave speed, λ = wavelength and f = frequency.

> **wavelength** The distance from a point on a wave, such as the crest, to the next corresponding point, such as the next crest.

> **frequency** The number of repetitions per unit time.

> **hertz** (abbreviated **Hz**) The unit of frequency equal to one cycle per second.

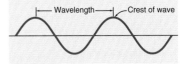

FIGURE 4-2 Wavelength is the distance between successive crests of a wave.

This equation applies to all types of waves, including light and sound waves. An application of its use for sound waves is shown in the next example. If you do the suggested exercise in this example, you will see that a sound of higher frequency has a shorter wavelength, as the speeds are the same. In the example, we use sound rather than light because sound waves have frequencies, velocities, and wavelengths within our everyday experience, whereas light waves do not. Now let's return to light and the spectrum produced when white light shines through a prism.

EXAMPLE

Sound travels at a speed of about 344 meters/second in air at room temperature. What is the wavelength of a sound that has a frequency of 262 Hz? (This is the frequency of the note C in music.)

SOLUTION   We start with the equation,

$$\text{wave speed} = \text{wavelength} \times \text{frequency, or}$$

$$344 \text{ m/s} = \lambda \times 262 \text{ cycles/s.}$$

We now solve the equation for the wavelength of the sound wave and do the calculation:

$$\lambda = \frac{344 \text{ m/s}}{262 \text{ cycles/s}} = 1.31 \text{ m.}$$

TRY ONE YOURSELF

What is the wavelength of a sound that has a frequency of 4000 Hz? (Hint: The speed of sound is the same for all frequencies, about 344 meters/second in air at room temperature.)

Light as a Wave

It only takes light 8.3 minutes to reach us from the Sun, but for faraway objects, we see them the way they were millions or even billions of years ago.

Can anything travel faster than the speed of light in a vacuum?

nanometer (abbreviated **nm**) A unit of length equal to 10^{-9} meter.

The color we see is not a property of the light itself but a manifestation of the system that senses it, that is our eyes, nerves, and brain.

White light is made up of light of many different wavelengths, all traveling at the same speed in a vacuum (and interstellar space is essentially a vacuum). The speed of light (c) is 300,000 kilometers/second (186,000 miles/second). If a light beam could be made to travel around the Earth's surface, it would circle the globe seven times in one second. According to Einstein's special theory of relativity (which we discussed in Chapter 3), the speed of light in vacuum is always measured as the same speed and is the fastest speed possible in the universe. Check the accompanying Advancing the Model box to find out how we measure this incredibly high speed.

We perceive light of different wavelengths as different colors. The wavelength of the reddest of red light is about 7×10^{-7} meters, or 0.0000007 meters. The wavelength decreases across the spectrum from red to violet, and the wavelength at the violet end of the spectrum is about 4×10^{-7} meters. In describing the wavelengths of visible light, a meter is much too long to be convenient, and thus, scientists use another unit—the **nanometer**. One nanometer (abbreviated nm) is 10^{-9} meters. So the shortest violet wavelength and the longest red wavelength are about 400 nm and 700 nm, respectively.

Using the equation $c = \lambda \times f$ to calculate frequencies of light waves, we obtain extremely high frequencies: 400 nm corresponds to 7.5×10^{14} Hz, and 700 nm corresponds to 4.3×10^{14} Hz.

The particular frequency or wavelength of light in vacuum determines its color; however, because color is so subjective and people are unable to distinguish between two very similar colors, scientists describe light by referring to wavelength in vacuum (or frequency, as the two quantities are related) rather than color (**TABLE 4-1**). They might mention the color in some cases, but this is to help us better picture the situation: Light can be described more accurately than by simply calling it "red" or "green."

4-3 The Electromagnetic Spectrum

The waves we see—visible light—are just a small part of a great range of waves that make up the **electromagnetic spectrum**. Waves somewhat longer than 700 nm (the approximate limit of red) are called *infrared* waves. FIGURE 4-3 shows the entire electromagnetic spectrum. The infrared region of the spectrum goes from 700 nm at the border of visible light up to about 10^{-4} meters, which is a tenth of a millimeter or 100,000 nm. Electromagnetic waves longer than that are called *radio* waves. Going the other way, from visible light toward shorter wavelengths, we first encounter *ultraviolet* waves and then *X-rays* and *gamma rays*.

It is important to emphasize that all of these types of waves (or rays, as certain portions of the spectrum are known) are essentially the same phenomenon. They differ in wavelength, and this causes some of their other properties to differ. For example, visible light is just that—visible. Ultraviolet is invisible to us, but it kills living cells and causes our skin to tan or burn. On the other hand, we perceive infrared as "heat" radiation, and yes, the radio waves in the spectrum are the same radio waves we use to transmit messages on Earth. They are handy for carrying messages containing sound and pictures (in the case of television) for several reasons, including the fact that they pass through clouds and bend around obstacles. All of these various waves are electromagnetic waves, just as visible light is. We give the various regions different names because of their properties and the uses we have for them.

The electromagnetic spectrum is important to astronomers because celestial objects emit waves in all the different regions of the spectrum. Visible light is a very small fraction of the entire spectrum. We humans tend to regard it as the important part, but this only reveals our limited outlook. Astronomers learn a great deal from the *invisible* radiation emitted by objects in the heavens.

TABLE 4-1

Approximate Vacuum Wavelength* Ranges for the Various Colors

Color	l (nm)
Violet	380-455
Blue	455-492
Green	492-577
Yellow	577-597
Orange	597-622
Red	622-720

* From now on, whenever we refer to the wavelength of light, we mean its wavelength in vacuum. As light travels from one medium to another, its speed and wavelength change but its frequency remains the same.

electromagnetic spectrum
The entire array of electromagnetic waves.

The waves are called "electromagnetic" because they consist of combined, perpendicular, oscillating electric and magnetic fields that result when a charged particle accelerates (see Figure B4-9b).

It is probably not worth memorizing these wavelengths of light waves, but it is handy to remember that the wavelength of visible light ranges from 400 (violet) to 700 (red) nanometers.

Is most of the electromagnetic spectrum made up of visible light?

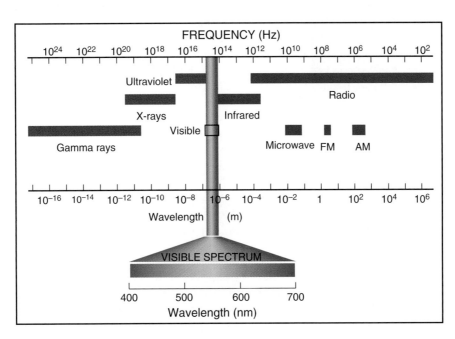

FIGURE 4-3 The electromagnetic spectrum is divided into several regions, depending upon the properties of the radiation. Region boundaries are not well defined. Notice the small portion of the spectrum occupied by visible light.

ADVANCING THE MODEL

Measuring the Speed of Light

Today we know that the speed of light c is 2.9979×10^8 m/s. How do we know that? You can't exactly time a light beam with a stop watch. Our understanding of light and its speed parallels the development of the technology that was used to measure this speed.

In the early 1600s, Galileo attempted to measure the speed of light by using his pulse-beat and comparing the time between opening his lantern while on a hilltop and seeing the light from his assistant's lantern from a distant hilltop. His attempts were not successful, but he correctly concluded that light is simply too fast to be measured by the slow human reaction.

Around 1675, the Danish astronomer Ole Roemer made the first accurate measurement of the value of c (FIGURE B4-2). Roemer had made many careful observations of Jupiter's moon Io and knew that Io's orbital period is about 1.76 days. As a result, he expected that he could predict Io's eclipses accurately. He was astonished to find that Io seemed to be behind its predicted position when the Earth was farther away from Jupiter (point A) and ahead when the Earth was closer (point B). Roemer correctly attributed this effect to the time required for light to travel from Jupiter to Earth. Light takes about 16.5 minutes to travel across the diameter of the Earth's orbit (2 AU); Roemer's actual measurement was 22 minutes. Using today's value for the AU (as it was not accurately known during Roemer's time) we find

$$c = \frac{\text{distance}}{\text{time}} = \frac{2 \times 1.5 \times 10^{11} \, \text{m}}{16.5 \times 60 \, \text{s}} = 3 \times 10^8 \, \text{m/s}.$$

(Roemer's actual calculation gave $c = 2.14 \times 10^8$ m/s.)

In 1849, French physicist Armand Fizeau measured the speed of light using the arrangement shown in FIGURE B4-3. His idea was to bounce a light beam between two mirrors, passing through a rotating toothed wheel in each direction. By choosing an appropriate rotation speed, the light beam can be made to pass through one gap on its way to the far mirror and the very next gap on its return to the observer. From the rotation rate, Fizeau calculated the time for the wheel to move from one gap between teeth to the next; this was the time during which light traveled from the wheel to the mirror and back. Dividing the distance by the time, Fizeau calculated $c = 3.15 \times 10^8$ m/s.

We can now routinely make accurate measurements of c in the laboratory. The speed of light in vacuum is now defined to be $c = 299,792.458$ km/s. Light has a smaller speed when going through other transparent media, but unless otherwise specified, we use $c = 300,000$ km/s.

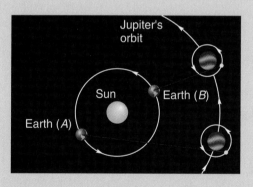

FIGURE B4-2 Roemer's method of measuring the speed of light. The scale of the orbits of Earth and Jupiter has been changed to exaggerate the effect.

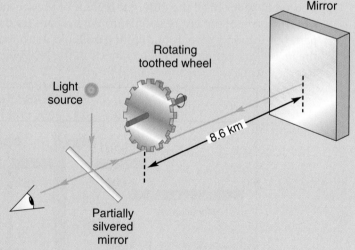

FIGURE B4-3 Fizeau's method of measuring the speed of light.

One of our problems with this invisible radiation is that most of it does not pass well through air and, thus, it does not reach the surface of the Earth. Our air is transparent to visible light and to part of the radio spectrum, but most of the rest of the electromagnetic spectrum is blocked to some degree. The chart in FIGURE 4-4 shows the relative absorbency of the atmosphere to various regions of the spectrum. Where the graph line is highest, the least amount of radiation gets through. Not much ultraviolet radiation, which damages living cells, penetrates to the surface.

Atmospheric absorption at various wavelengths

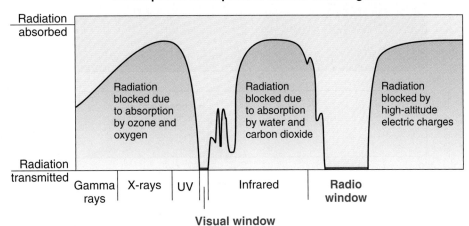

FIGURE 4-4 The height of the curve indicates the relative amount of radiation of a given wavelength blocked by the atmosphere from reaching Earth's surface. The atmosphere is transparent to two regions of the spectrum: visible light and part of the radio region.

Astronomers, however, wish to detect and examine these nonpenetrating radiations from space. They accomplish this by using balloons to carry detectors high into the atmosphere or by using artificial satellites to take detectors completely above the atmosphere. This will be covered in the next chapter.

Astronomers refer to *windows* in the atmosphere, saying that there is a visual window and a radio window. This means that our atmosphere allows radiation in these two regions of the spectrum to penetrate to the surface.

4-4 The Colors of Planets and Stars

How do we analyze the light spectrum from a celestial object to determine some of the object's properties? This analysis can be divided into two parts. First, we look at the overall spectrum, from which we typically determine the color of the object; however, when we look at the spectrum, we are really examining the actual wavelengths of light rather than just the color. The second analysis, discussed later in this chapter, involves examining individual regions of the spectrum.

Color from Reflection—The Colors of Planets

When you see a visible spectrum spread out on a screen, you see a particular color at a given location on the spectrum. This is because a wave whose wavelength is associated with that color is coming to your eye from that spot; however, the color we see in most objects does not correspond to a single wavelength. FIGURE 4-5a might be the spectrum of light from some lemons. It contains many different wavelengths of light; however, there is no violet or blue light, and the center of the spectrum is indeed in the yellow. Part (b) of the figure shows a graph that indicates the intensity of light of each wavelength. Where the graph is higher, the light of the corresponding color

Spectrum of light from lemons

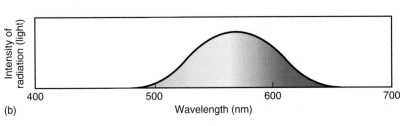

(a)

FIGURE 4-5 An example of reflected color. (a) If light reflected from the lemons were sent through a prism to reveal its spectrum, we would see that the lemons reflect mostly yellow light. (b) This graph indicates the relative intensity of the various wavelengths of light from the lemons.

(b)

FIGURE 4-6 An example of transmitted light. (a) The spectrum of light from the taillight of this particular car (right). The lightbulb emits white light, but the plastic cover over the bulb absorbs much of the light, letting pass some wavelengths in the red, orange, and yellow regions of the spectrum. (b) This graph indicates the relative intensity of the various wavelengths of light in this case.

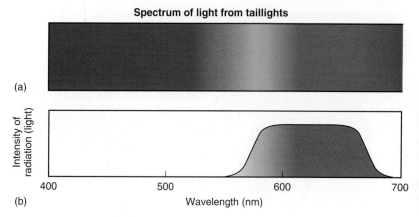

is brighter. Lemons have this spectrum because when white light strikes them, they absorb some of the wavelengths of the white light, especially blue and violet. They reflect only those we see in their spectrum (the wavelengths centered on the color yellow). This is what determines their color.

Planets have their colors because of a process like that described for the lemons. The rusty red color of Mars, for example, occurs because the material on its surface absorbs some of the wavelengths of sunlight and reflects a combination of wavelengths that looks rusty red to us.

FIGURE 4-6 shows the color spectrum and graph of intensity versus wavelength for the light from the red taillight of a car. The spectrum includes not only many wavelengths in the red part of the spectrum but also some orange. In this case, the bulb inside the taillight emits white light, but part of that light is absorbed by the plastic cover. The light that gets through the cover has the spectrum in the figure.

In the case of the Sun and other stars, light from the star is produced by emission within the star and some light is absorbed as it passes through the star's outer layer. The effect of this absorption on the color of the star is minimal, but nevertheless, the process is somewhat similar to that described for the taillight of the car. We will look at this in more detail later.

Color as a Measure of Temperature

The light emitted by the Sun and other stars can be compared with light coming from a lightbulb or from the element of an electric stove in an otherwise dark room. Consider what happens when you turn the burner of an electric stove to a low setting. It glows a dull red. The bottom curve in **FIGURE 4-7** is a graph of intensity versus wavelength for this case. This graph includes not only the visible portion of the spectrum, but quite a lot of the infrared. Recall that we experience infrared radiation as heat. In fact, the graph indicates that more infrared radiation is being emitted than visible radiation, and the graph reaches its peak in the infrared portion of the spectrum. Some red light is emitted, but very little light from the center and violet end of the visible spectrum.

Now turn up the heat on the stove. The burner begins to take on an orange glow. The second curve from the bottom in Figure 4-7 is a graph of intensity versus wavelength for this burner. Compare it with the bottom curve (the red-hot burner). First, the orange burner is emitting more radiation of all wavelengths. This should correspond to your experience, for you can feel that more infrared is being emitted and see that more light is coming from the burner. Second, the peak of the graph has moved over toward the visible portion of the spectrum, toward the shorter wavelengths.

This is about as far as you can go with an electric stove burner. If you have an object with a temperature you can control, you can increase its temperature so that the object emits most of its light in the yellow part of the spectrum. The third curve from the bottom is a graph of such an object. It actually corresponds to the Sun. The fourth

We want you to imagine a dark room because in a well-lit room, the lamp and stove element reflect light and you can see them even if they are turned off.

Astronomers usually call the graph of intensity versus wavelength for a star its *thermal spectrum*.

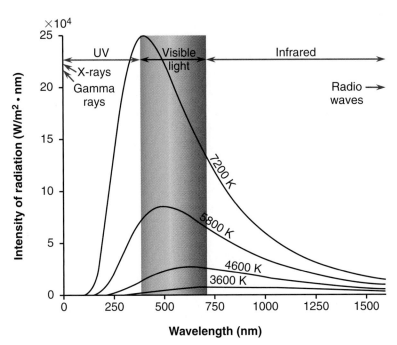

FIGURE 4-7 The bottom curve represents an object (at 3600 K with λ_{max} = 805 nm) that emits mostly infrared radiation. The next curve corresponds to an object (at 4600 K with λ_{max} = 630 nm) that emits most of its light in the orange part of the spectrum. The third curve from the bottom corresponds to the Sun (at 5800 K with λ_{max} = 500 nm). The 7200 K object would most likely seem "blue" because it emits most of its light in the violet part of the spectrum (λ_{max} = 403 nm). Recall that color is subjective.

blackbody A theoretical object that absorbs and emits all wavelengths of radiation, so that it is a perfect absorber and emitter of radiation. The radiation it emits is called **blackbody radiation**.

continuous spectrum A spectrum containing an entire range of wavelengths, rather than separate, discrete wavelengths.

Beta Cygni (named Alberio) is the second brightest star in the constellation Cygnus. It is the brighter of a closely spaced pair of stars with obvious color differences.

curve from the bottom of Figure 4-7 corresponds to an object of even higher temperature. You see that it emits most of its light in the violet part of the spectrum.

The temperatures indicated in Figure 4-7 are typical of surface temperatures of stars. As their sequence indicates, the higher the temperature of a star, the shorter the wavelength at which the star emits most of its energy. The relationship between an object's temperature (T) and the wavelength at which maximum emission occurs (λ_{max}) was discovered in 1893 by the German physicist Wilhelm Wien. **Wien's law** is expressed as

$$\lambda_{max(in\ nm)} = \frac{2.9 \times 10^6}{T_{(in\ K)}},$$

where λ_{max} is in nanometers and the temperature is in kelvin. You can check that this equation applies to the curves of Figure 4-7.

FIGURE 4-8 is an image of one of the coolest stars ever found. We often refer to stars by color as a quick way to indicate their temperature: A "white" star is hotter than a "red" star. In practice, of course, a "red" star does not appear red like a Christmas tree bulb, but it definitely has a red tint. To the unpracticed naked eye, color differences between stars are not at all obvious; however, if you have the opportunity to use a telescope to observe pairs of closely spaced stars, you can see this color difference easily.

The important point is that by examining the intensity versus wavelength curve for a star and using Wien's law, we can determine the star's surface temperature without ever visiting it!

Blackbody Radiation. Wien's law was derived from theoretical calculations about the radiation that would be emitted from an object that absorbs (or emits) all wavelengths completely. Such an ideal object is called a **blackbody**, and the radiation it emits is called **blackbody radiation**. A blackbody emits a **continuous spectrum**, which has a peak at a certain wavelength λ_{max}, but some energy is emitted at all wavelengths, as shown by the blackbody curves in Figure 4-7. It was found that although the stars are not perfect blackbodies, the theory applies very closely to them, as shown by **FIGURE 4-9**. This figure shows the intensity of the measured solar radiation outside the Earth's atmosphere from 300 to 830 nm; the Sun's

FIGURE 4-8 The *Hubble Space Telescope* imaged one of the lowest-temperature stars ever seen (upper right). It is a companion to the dwarf star at lower left. The surface temperature of the cool star may be as low as 2300 degrees Celsius. (The white bar is an effect produced by the camera.)

FIGURE 4-9 The Sun is almost an ideal blackbody at 5800 K. The measurements of the Sun's intensity were made above the Earth's atmosphere.

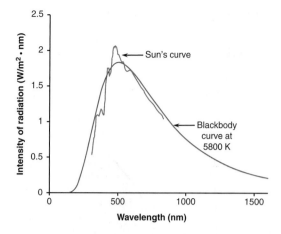

spectrum peaks in the visible part of the electromagnetic spectrum, where our eyes have become most sensitive. Superimposed to the data is a curve obtained by assuming that the Sun is a blackbody at 5800 K and that its intensity is measured outside Earth's atmosphere. Compare this curve with the 5800 K curve in Figure 4-7; as we discuss in Section 4-8, radiation from a source decreases in intensity as the inverse square of the distance from the source.

The Stefan-Boltzmann Law. Figure 4-7 indicates that the hotter an object is, the more radiation it emits. In 1879 an Austrian physicist, Josef Stefan, discovered the mathematical relationship between temperature and energy emitted, based on data published 14 years earlier. In 1884, another Austrian physicist, Ludwig Boltzmann, used a theoretical argument to show why the relationship occurs. This rule, now called the *Stefan-Boltzmann law*, is

$$F = \sigma T^4,$$

where T is the object's temperature on the Kelvin scale, F, called the energy flux, is the energy it emits per unit time per unit area, and σ (the Greek letter sigma) is a constant, called the *Stefan-Boltzmann constant*, that relates the two quantities. This law tells us that the energy flux of an object is directly proportional to its temperature to the fourth power. If the temperature of an object were to double, its energy flux thus would be 2^4, or 16, times greater than it was before.

As we see in Chapter 12, the Stefan-Boltzmann law allows us to find the radius of a star. (From observations, we can find a star's temperature and the total energy it emits each second. Recall the meaning of energy flux and the fact that the surface area of a sphere is proportional to the square of its radius.)

4-5   Types of Spectra

The spectrum of visible light shown in Figure 4-1 is a *continuous spectrum*. Such a spectrum is produced when a solid object (in this case the filament of a lamp) is heated to a temperature great enough that the object emits visible light. Not all spectra are of this type, as was discovered nearly 200 years ago.

Kirchhoff's Laws

In 1814, Joseph von Fraunhofer, a German optician, used a prism to produce a solar spectrum. He noticed that the spectrum was not continuous but had a number of dark lines across it (**FIGURE 4-10**). Fraunhofer had no explanation for these dark lines, but it was later discovered that they were the result of the sunlight passing through cooler gases (in the Sun's and the Earth's atmospheres). Then, in the mid-1800s, several

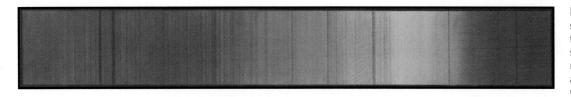

FIGURE 4-10 The Sun's spectrum may appear at first to be a continuous spectrum, but when it is magnified, we see dark lines across it where specific wavelengths do not reach us. Thus, the solar spectrum is an absorption spectrum.

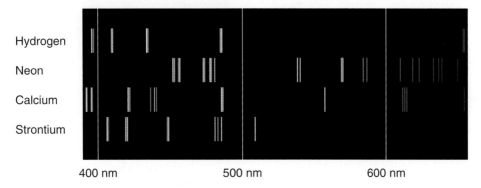

400 nm   500 nm   600 nm

FIGURE 4-11 The visible bright line spectrum of four different chemical elements. Each element emits specific wavelengths of light when it is heated. This applies not only to visible light (shown here) but also to infrared and ultraviolet. Such a spectrum is called an emission spectrum.

German chemists discovered that if gases are heated until they emit light, neither a continuous spectrum nor a spectrum with dark lines is produced; instead, a spectrum made up of *bright* lines appears. Further, they discovered that each chemical element has its own distinctive pattern of lines (**FIGURE 4-11**). This proved to be a very valuable way of identifying the makeup of an unknown substance and was soon developed into a standard technique that allows us to identify the chemical composition of matter.

In the 1860s, Gustav Kirchhoff formulated a set of rules, now called *Kirchhoff's laws*, which summarize how the three types of spectra are produced:

1. A hot, dense glowing object (a solid or a dense gas) emits a continuous spectrum (Figure 4-1).

2. A hot, low-density gas emits light of only certain wavelengths—a bright line spectrum (Figure 4-11).

3. When light having a continuous spectrum passes through a cool gas, dark lines appear in the continuous spectrum—a dark-line spectrum (Figure 4-10).

The dark lines that result when light passes through a cool gas (process 3) have the same wavelengths as the bright lines that are emitted if this same gas is heated (process 2).

Kirchhoff's laws tell us how to produce the various types of spectra, but the science behind the laws—the connection between the laws and the nature of matter—remained a mystery in the 19th century. The connection was finally made in 1913, when a young Danish physicist, Niels Bohr, proposed a new model of the atom.

Gustav Kirchhoff (1824–1887) was a German physicist and astronomer whose primary work was in the field of spectroscopy, the study of spectra.

4-6   The Bohr Model of the Atom

The atomic model accepted at that time was due primarily to the New Zealand physicist Ernest Rutherford. His model described the atom as having a **nucleus** with a positive electrical charge, circled by **electrons** with a negative electrical charge. Positive and negative electrical charges attract one another, and this electrical force holds the electrons in orbit around the nucleus. As an electron orbits the nucleus, it continuously changes direction and thus accelerates; however, according to classical theory,

nucleus (of atom) The central, massive part of an atom.

electron A negatively charged particle that orbits the nucleus of an atom.

HISTORICAL NOTE

Niels Bohr

Niels Bohr (1885–1962) was born into a very cultured Danish home. His father's interest in science led Niels to that subject, and he became known as a student who gave his utmost to every project—a reputation that continued throughout his life.

In 1922 Niels Bohr was awarded the Nobel Prize in physics "for his services in the investigation of the structure of atoms and of the radiation emanating from them." In his acceptance speech, he emphasized the limitations of his theory, and indeed, he seems to have been more aware of its limitations than other scientists who worked with the theory.

Niels' son Aage followed his father in the study of physics and won the Nobel Prize in 1975. One major project on which Niels and Aage Bohr worked together was the development of the atomic bomb. Niels' mother was Jewish, and after Hitler's army overran Denmark, Niels' family (**FIGURE B4-4**) fled their native land to avoid arrest. Niels and Aage came to the United States and helped with the Manhattan Project (the code name for the bomb development effort).

Bohr's contribution to science goes far deeper than the development of the Bohr model of the atom, as important as that is. His philosophical ideas on the nature of physical theory are perhaps his greatest contribution. Many important

FIGURE B4-4 Niels Bohr and his sons.

ideas of modern physics were clarified through the friendly arguments between Bohr and Einstein. Einstein would try to imagine situations in which the new ideas of quantum mechanics (the modern description of matter and energy) did not work, and Bohr would always figure out how quantum mechanics did explain these situations. It is a classic example of how science can progress through the informal discussions between scientists.

any charged particle moving in a curved path or accelerating in a straight-line path will emit electromagnetic radiation, thus losing energy. Therefore, what keeps the electrons from simply spiraling into the nucleus, leading to collapse of the atom? Bohr's model attempted to answer this question.

Before describing his model, we must emphasize that today we have a much more powerful model of matter and energy (called **Quantum Mechanics**); however, just like we still use Newtonian ideas to describe gravity even though we have the theory of relativity, we still use Bohr's model because many of its conclusions are essentially valid and it provides the basic conceptual tools in understanding atomic structure. Wherever appropriate, we mention important differences between the two models.

The ***Bohr atom***, as Niels Bohr's model is called, is based on three postulates:

1. Electrons in orbit around a nucleus can have only certain specific energies. To imagine different energies for the electrons, imagine electrons orbiting at different distances from the nucleus. The negatively charged electrons are being attracted to the positively charged nucleus; thus, to pull an electron farther away from the nucleus requires energy. Because only certain energies are possible, we speak of "allowed" orbits for the electrons. The element hydrogen has only one electron, but many possible energy levels and therefore many allowed orbits. The drawing in **FIGURE 4-12a** depicts the hydrogen atom with its electron in the lowest orbit (the ground state) and also shows, as an example, the next orbit this electron might have. The point of Bohr's first postulate is that the electron can have only specific energies and therefore specific orbits. This is far different from the solar system, where there are no limitations on possible positions of orbits.

Bohr atom The model of the atom proposed by Niels Bohr; it describes electrons in orbit around a central nucleus and explains the absorption and emission of light.

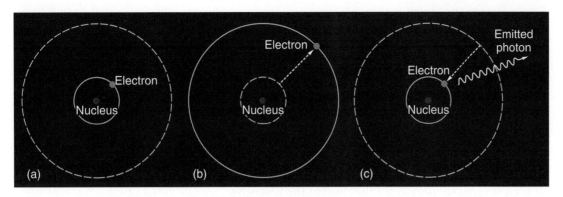

FIGURE 4-12 (a) The electron is in the lowest possible energy state—the ground state. (b) When the atom gains energy (perhaps by collision with another atom), the electron jumps up to a higher orbit. (c) The electron then quickly falls down to its original orbit and, as it does, emits energy in the form of a photon of electromagnetic radiation. The orbits are not drawn to scale.

Bohr simply assumed that an atomic electron in one of the allowed circular and stable orbits did not radiate any energy. (In the modern view of Quantum Mechanics, the wave-particle duality of light is extended to all entities, including matter. That is, there is a wave aspect associated with an electron, and therefore, there is a wavelength associated to a moving electron. This provides an interesting interpretation of Bohr's model of an atom; an allowed circular orbit has a circumference that is an integer multiple of the electron's wavelength. Also, according to Quantum Mechanics, we can calculate the probability that an electron of a given energy will be at some distance from the nucleus; Bohr's orbit for the same electron simply corresponds to the most probable distance.)

2. An electron can transit from one energy level to another, changing the energy of the atom. In terms of electron orbits, when energy is added to an atom, the electron moves farther from the nucleus. On the other hand, the atom loses energy when an electron moves from an outer orbit to an inner orbit. This lost energy leaves the atom in the form of electromagnetic radiation.

Our previous discussion of waves and light assumed that light acts as a simple long wave, similar to the waves you can make in a swimming pool. However, as we discuss in the Advancing the Model box on pages 112 to 113, light has a wave-particle duality. In the Bohr model, light is emitted not as continuous waves, but in tiny bursts of energy; each burst is emitted when an electron moves to an orbit closer to the nucleus. These tiny bursts of electromagnetic energy are called **photons**. The energy of the photon depends on the spacing between electron orbits.

3. The energy of a photon determines the frequency of light that is associated with the photon. The greater the energy of the photon, the greater the frequency of light, and vice versa. A photon of violet light thus has more energy than a photon of red light. The relevant equation is

$$E = hf,$$

where E = the energy of the photon, h = a constant (called *Planck's constant*), and f = the frequency of the light.

Emission Spectra

The Bohr model of the atom can be used to explain why only certain wavelengths are seen in the spectrum of light emitted by a hot gas. In its normal, lowest energy state, the electron of a hydrogen atom is in its lowest possible orbit, as indicated in Figure 4-12a. If this atom is given enough energy (perhaps by collisions with other atoms), the electron will jump to the next allowed orbit. Figure 4-12b illustrates this jump. An atom will not stay in its energized state long. Quickly, the electron falls down to a lower orbit, emitting a photon as it does, as shown in Figure 4-12c.

The wave particle duality of light is discussed in the corresponding Advancing the Model box on pages 112–113.

photon The smallest possible amount of electromagnetic energy of a particular wavelength.

Every physicist thinks he knows what a photon is. I spent my life to find out what a photon is and I still don't know it.
Albert Einstein

◆ The lowest energy state of an atom is usually called the *ground state*, and the energized states are called *excited states*.

The energy of this photon is exactly equal to the energy difference between the two orbits. Finally, because the energy of the photon determines the frequency of the radiation, the radiation coming from this atom must be of the corresponding frequency (and color).

We have described one atom emitting one photon. In an actual lamp that contains hot hydrogen gas, there are countless atoms gaining energy and countless atoms emitting photons as they lose energy. Different atoms will have different amounts of energy, depending on the energy they have absorbed (from a collision with another atom, for example). If a particular atom's energy corresponds to the electron being in the third orbit, the atom might release its energy in a single step, as shown in **FIGURE 4-13a**, or in two steps as shown in Figure 4-13b. That is, there are two different ways for an electron to move from the third to the first orbit.

Let us now assume that enough energy is available to cause electrons to move to even higher orbits. Using similar drawings, you should be able to show that there are four different ways for an electron to move from the fourth to the first orbit, eight different ways to move from the fifth to the first orbit, and so forth. Each jump of an electron for each of the steps involved in each different path corresponds to a certain specific energy and therefore to a certain specific frequency of emitted radiation. The electrons of some atoms will fall by some paths, and the electrons of other atoms will fall by other paths. As a result, radiation of several different frequencies will be emitted from the entire group of atoms. Not all frequencies will be emitted, however—just those that correspond to the electron jumps.

Hence, the spectrum from a heated, low-density gas is not a continuous spectrum. It contains only certain definite frequencies. We call such a spectrum an ***emission spectrum***—the bright line spectrum mentioned earlier.

Refer back to Figure 4-11, which shows the emission spectra of four elements. Each spectrum is different because the allowed energy levels of the atoms are different for each chemical element. No two chemical elements have the same set of energy levels, and thus, no two chemical elements have the same emission spectrum. This provides us with a valuable method of identifying elements, as each has a unique spectral "fingerprint." This process has some important applications here on Earth, but because in this book we are more interested in the stars, we look at a stellar application.

emission spectrum A spectrum made up of discrete frequencies (or wavelengths) rather than a continuous band.

Continuous and Absorption Spectra of the Stars

Kirchhoff's laws tell us that dark line spectra result from light with a continuous spectrum passing through a cool gas. Let us examine this in the case of the Sun.

The visible surface (the ***photosphere***) of the Sun emits a continuous spectrum. Even though the Sun, in most ways, is more like a gas than a solid, it produces a continuous spectrum rather than an emission spectrum. This is because as atoms are

photosphere The region of the Sun from which mostly visible radiation is emitted.

FIGURE 4-13 (a) The electron may fall directly from orbit 3 to orbit 1, emitting a single photon. (b) The electron may fall from orbit 3 to orbit 1 in two steps, emitting two photons whose energies sum to that of the single photon emitted in part (a). The orbits are not drawn to scale.

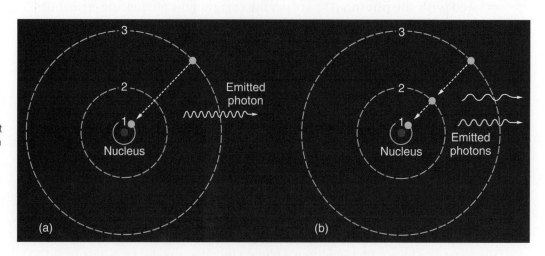

TOOLS OF ASTRONOMY

The Balmer Series

In Figure 4-11, notice the pattern to the spacing of bright lines in the hydrogen spectrum—they become progressively closer as they approach the blue end of the spectrum. The mathematical relationship that expresses this pattern was found in 1885 by Johann Jacob Balmer, a Swiss teacher. The lines visible in the figure are therefore called the **Balmer series** of spectral lines. This series served as the foundation for Bohr's work, and it is easily explained by Bohr's model of the atom. In FIGURE B4-5, the spacing of the lines depicting the orbits of the hydrogen atom represents the relative energy of each of the levels. According to Bohr's model, there is less difference between the energy levels as orbits get farther from the nucleus. The levels thus are spaced closer together toward the top of the figure.

Suppose that an electron is in the lowest energy level, the ground state. This electron will jump to a higher state when a photon of the appropriate energy strikes it. Look at the left side of FIGURE B4-6. Five arrows point upward from the ground state, representing electron jumps to higher orbits. On each arrow is printed the wavelength of the photon corresponding to the jump indicated. Each of these wavelengths is less than 400 nm, the shortest wavelength of visible light. They are in the ultraviolet region of the spectrum.

As a gas becomes hotter, more of its atoms have electrons in energy levels above the ground state. Suppose that hydrogen is at a temperature at which a significant number of its electrons are in energy level 2. The middle of Figure B4-6 shows the wavelengths of photons that would cause electrons at this level to jump to a higher level. The wavelengths of the photons that cause jumps from the second level are within the visible range—from 400 to 700 nm—and because energy levels are more closely spaced toward the top of the figure, the wavelengths toward the blue end of the spectrum are closer together. These wavelengths correspond to the wavelengths of the Balmer series.

The first set of wavelengths described previously, those in the ultraviolet region of the spectrum, form the *Lyman series*. As Figure B4-6 indicates, there is another series that falls in the infrared portion of the spectrum, called the *Paschen series*.

Thus far, we have discussed the absorption of photons to produce an absorption spectrum. As we noted in the text, the emission spectrum of hydrogen is produced when electrons fall from higher energies to lower. The spectrum shown in Figure 4-11 thus is an emission spectrum that resulted from electrons falling from higher energies to lower levels of the atom.

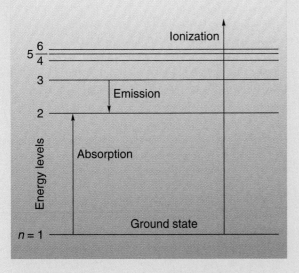

FIGURE B4-5 The energy levels of the hydrogen atom (not drawn to scale). The levels are progressively closer in energy as they are farther from the nucleus. The length of each arrow corresponds to the energy involved in the process. An atom is *ionized* when an electron absorbs enough energy and escapes.

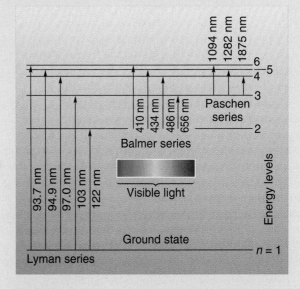

FIGURE B4-6 Electrons that transit from the ground state to higher energy levels do so by absorbing photons of the wavelengths shown. Electron jumps for the first three series of the hydrogen atom's spectrum are represented here.

ADVANCING THE MODEL

Evidence for the Wave-Particle Duality of Light

In the 1660s, English scientist Robert Hooke challenged Aristotle's particle model of light for the first time. He proposed a simple wave theory in which light was the manifestation of fast oscillations in the aether. He reasoned that this model better explained the color patterns observed in thin transparent films, such as soap bubbles. On the other hand, Newton favored the particle model for light and suggested that light has a dual nature and that it is a stream of particles that can induce oscillations in the aether. Newton, however, misunderstood how waves behave. He expected that if light were a wave, it should clearly spread out after passing through an opening, instead of producing the observed narrow beams. In his opinion, the particle model for light better explained how light propagates along straight lines and why shadows are sharp. The problem is that light *does* spread out as Newton expected, but this can be seen only if the opening is of the same approximate size as the wavelength of the light, which is extremely small (at least for visible light, which was what he used at the time).

Both the particle and wave models for light explained two of the most common phenomena we observe. Light reflects—it bounces like a tennis ball off a flat surface. Also, light refracts—it changes direction when going from one medium (such as air) to another (such as water). The particle model of light, however, gained acceptance for more than a century, mainly because of Newton's reputation and authority.

Around 1801, Thomas Young first proposed a simple but convincing experimental test to distinguish between the two competing models. Young's experiment was to shine light of a specific wavelength (from a single source) through two narrow parallel slits, close to each other. The light then fell on a screen a certain distance away (**FIGURE B4-7**). Young observed a pattern of light and dark bands on the screen, called interference fringes, which he explained by assuming that light is a wave. As light waves go through the slits, they *diffract* (they spread out) in a series of crests and troughs. When the waves meet, they add up algebraically as shown in **FIGURE B4-8**. Bright (dark) fringes appear on the screen in places where the waves interfere constructively (destructively). Young used this *double-slit experiment* (actually he used pinholes instead of slits) to measure the wavelength for violet (400 nm) and red light (700 nm).

The experiments of the French physicist Augustin Fresnel (1788–1827) helped put the wave model of light on a firm mathematical basis. The interference of light clearly

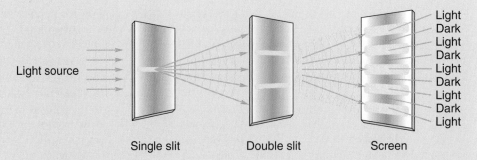

FIGURE B4-7 Young's double-slit experiment. The colors used for the light areas on the screen are for illustration purposes only and correspond to decreasing intensity of the light.

showed its wave nature, but what was the nature of the waves? In the early 1860s, the Scottish mathematical physicist James Clerk Maxwell (1831–1879) succeeded in unifying what was then known about electricity and magnetism into four equations, which today are named after him. He found that his equations described the existence of *transverse* waves (such as the up-and-down waves traveling on a stretched rope) that combine electromagnetic effects and move at a speed that is almost the same as the then known speed of light. To understand this, consider a small sphere carrying a uniform charge Q. If we place a tiny positive test-charge q_0 anywhere in the space around Q, it will experience a force of a specific direction and strength. We say that there is a *field* of force around charge Q. To represent visually this electric field, we draw continuous lines that are tangent to all of the "specific directions of force," as shown in **FIGURE B4-9a**. Now consider what will happen if we make charge Q oscillate; if you visualize the field lines as made of rubber bands, you can see that the oscillation will set up waves propagating outward. Indeed, the changes in the strength of the field that occur due to the motion of charge Q propagate outward like

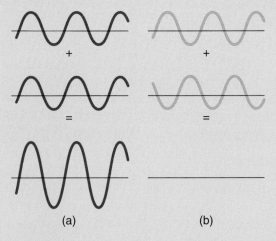

FIGURE B4-8 (a) Waves interfering constructively. (b) Waves interfering destructively.

ADVANCING THE MODEL

Evidence for the Wave-Particle Duality of Light (Cont'd)

waves with a speed equal to c. Maxwell found that magnetic fields can be formed not only by moving charges but also by changing electric fields, and that electric fields form not only by charges but also by changing magnetic fields (Figure B4-9b). Maxwell wrote that "we can scarcely avoid the inference that light consists in the transverse modulations of the same medium which is the cause of electric and magnetic phenomena." Nine years after Maxwell died, the German physicist Heinrich Hertz (1857–1894) produced radio waves in his lab, which confirmed all the properties of light known at the time. At the close of the 19th century, the wave model for light seemed to be very successful; however, some unanswered questions still remained. The most important ones dealt with the continuous spectrum of blackbody radiation and with the emission and absorption spectra.

In 1900, while trying to explain the continuous spectrum of blackbody radiation, the German physicist Max Planck (1858–1947) discovered an equation that fit the blackbody curve. In an effort to understand its meaning, Planck was forced to accept that electromagnetic waves can only have discrete (*quantized*) energy values that are integer

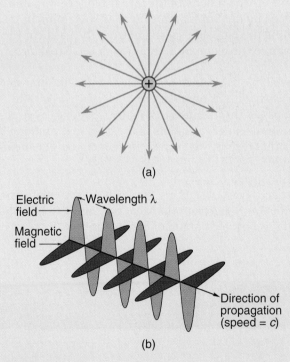

(a)

Electric field — Wavelength λ
Magnetic field

Direction of propagation (speed = c)

(b)

FIGURE B4-9 (a) Electric field lines around a positive charge. If the charge were negative, the lines would be pointing toward the charge. Keep in mind that this is a three-dimensional field. The field is stronger where the concentration of lines is denser. (b) Electromagnetic radiation (from radio waves to X-rays) consists of oscillating electric and magnetic fields, perpendicular to each other, moving through space at the speed of light.

multiples of a minimum energy value, called a *quantum* of energy. This quantum of energy is given by hf, where f is the frequency of the wave and h is Planck's constant. Today we consider h to be a fundamental constant of nature, like the gravitational constant G and the speed of light c. It is also the basis of *Quantum Mechanics*, which is the modern description of matter and energy; however, the quantum idea did not fit at all with the prevailing wave model of light.

The quantum nature of light was taken seriously by Einstein, who in 1905 used it to explain a very puzzling phenomenon, known as the *photoelectric effect*. When light of a certain frequency shines on a metal surface, electrons are ejected from it with a range of energies. (This is the basis of today's photocells in door openers, bar code scanners, and hundreds of other applications.) However, the maximum kinetic energy of these electrons does not depend on how bright the source of light is! According to the wave model of light, if we increase the rate of light energy falling on the surface, individual electrons will absorb more energy and will be emitted with larger kinetic energies. This is not what is observed. Increasing the brightness of the light source results in more electrons being emitted, but it doesn't affect their maximum kinetic energy. In addition, there is a minimum frequency that the light must have in order for any electrons to be emitted. This is also in contrast to the wave model for light, according to which if we wait long enough, the electrons will be able to absorb enough energy to be emitted, independent of the frequency of the light source.

Einstein's explanation was based on the assumption that light energy striking the metal surface is quantized in small bundles. These bundles behave as if they are a stream of massless particles that today we call *photons*. The energy of each photon is equal to hf. Convincing evidence that light indeed shows its particle nature when it interacts with matter was provided by the American physicist Arthur Compton in 1922. He measured the change in the wavelength of X-ray photons as they were scattered by free electrons (an observation now called the *Compton effect*) and found it to be exactly as predicted by the theory of a collision between two particles.

What, then, is the nature of light? It seems that light has a *wave-particle duality*. When it propagates through space, light can be described by a wave model (as shown by the double-slit experiment). When it interacts with matter, light can be described by a particle model (as shown by the photoelectric effect and the Compton effect). We cannot say whether light *is* particle or wave. This is not an either/or situation; light seems to be both particles and waves and thus is probably neither.

pushed together, their energy levels are broadened (as if slightly different orbits are allowed). As atoms become more and more tightly pressed together, their energy levels begin to overlap so that a full range of orbital energies is possible. An entire range of photon energies thus is emitted by the atoms, and instead of separate, distinct spectral lines appearing in the spectrum, an entire range of frequencies appears.

Before the light from the Sun gets to us on Earth, it must pass through the relatively cooler atmosphere of the Sun as well as through the atmosphere of the Earth. The Sun does indeed have an atmosphere and, as we will see in a later chapter, its atmosphere is much deeper than Earth's. As the light passes through these gases, atoms of the gases absorb some of it. This absorption of energy raises an atom's energy level, but because only certain specific energy levels are possible, only certain amounts of energy can be absorbed by the atom. This results in the reverse of what we had before: instead of an atom emitting a photon as it releases energy, it absorbs a photon as it absorbs energy. Just as a hot, low-density gas emits photons of certain energies, the same gas when in a cooler atmosphere absorbs photons of the same energies.

FIGURE 4-14a represents the emission spectrum of some element. Figure 4-14b shows the ***absorption spectrum*** that results when white light is passed through the cool gas of this same element. The dark lines of the absorption spectrum correspond exactly to the bright lines of the emission spectrum.

You might point out that after the cool gas has absorbed radiation, it must re-emit it. Shouldn't this then cancel out the absorption? No. As **FIGURE 4-15** shows, the re-emitted light is sent out in all directions. Certain frequencies of the light that was originally coming toward the Earth thus are scattered by the atmosphere of the Sun. This results in less light of those frequencies reaching us, and we observe an absorption spectrum.

The spectrum of the light that is re-emitted is an emission spectrum. During a total eclipse of the Sun, light from the main body of the Sun is blocked out, and astronomers can see the light emitted by the Sun's atmosphere and examine its emission spectrum. It was by examining the emission spectrum from the gas near the Sun that astronomers first discovered the element helium (from the Greek *helios*, the Sun).

The Sun, and other stars as well, has various chemical elements in its atmosphere (that is, different kinds of atoms, each with its own pattern of electron orbits and spectral lines). As the white light passes through this gas, many frequencies are absorbed, corresponding to the various chemical elements of the gas. By examining the complicated absorption spectrum that results, we are able to deduce what elements are present in the star's atmosphere.

We thus answer a question that, just a century ago, was thought to be unanswerable. We now know what the surface layers of stars are made of! As you might appreciate from the complexity of the Sun's spectrum (look back at Figure 4-10), the analysis is fairly complicated, but it is now a common one in astronomy.

We discuss the Sun in Chapter 11.

absorption spectrum A spectrum that is continuous except for certain discrete frequencies (or wavelengths).

Absorption spectra are the dark line spectra of Kirchhoff's laws.

The elements in the Earth's atmosphere also absorb radiation, the frequencies of which depend on what elements are in our atmosphere, but we know what those elements are and can take them into account.

FIGURE 4-14 (a) The emission spectrum of an element; (b) the absorption spectrum of the same element. The absorbed wavelengths in (b) are the same as the emission lines in (a).

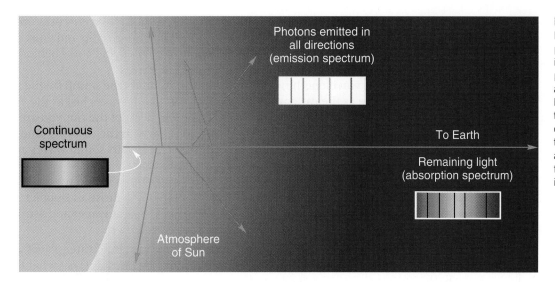

FIGURE 4-15 Consider a beam of light leaving a point on the Sun and moving toward the Earth. As it passes through the Sun's atmosphere, some wavelengths are absorbed and then re-emitted in random directions. This results in the solar spectrum being an absorption spectrum. Light from the Sun's atmosphere is an emission spectrum.

4-7   The Doppler Effect

Have you ever stood near a road and listened to the sound of a siren on a car as it sped by you? Recall how the sound changed when the siren passed, going from a higher pitch down to a lower pitch. This phenomenon has a very important parallel in astronomy. To understand it, we first consider water waves.

FIGURE 4-16a is a photo of waves spreading from a disturbance on the surface of water. The waves move away from the source in a regular way and appear the same in all directions from the source. Figure 4-16b was made by moving a vibrating object toward the right as it makes waves on water. Look at the difference between the waves in front of and behind the moving source. Four important points can be made about this case, although only one of them is apparent in the photo.

1. Even though the source of the waves is moving, the waves still travel at the same speed in all directions. The source's motion does not push or pull the waves. The source just disturbs the liquid, and the disturbance moves away at a speed that depends only on the liquid's characteristics—water, in this case.

2. The wavelengths of the waves in front of the moving source are shorter than they would be if the source was stationary, and the wavelengths behind the moving source are longer. For waves traveling at the same speed, the wavelength is

(a)

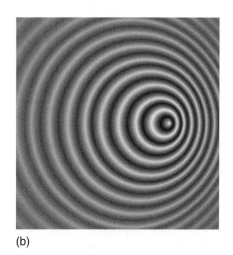

(b)

FIGURE 4-16 (a) Waves spread evenly from their source. (b) If the source is moving to the right, the waves are compressed in front of the source and stretched behind it.

inversely proportional to the frequency. This means that if there were corks on the water, the corks in front of the moving source would bounce up and down with a greater frequency than if the source were not moving, and corks behind the moving source would bounce with a lower frequency.

This change in wavelength is the same effect that causes the sound frequency of a siren to change as the car passes. The siren emits sound waves just as the vibrating object produces water waves. When you are in front of the car, you are in the region of shorter wavelength and higher frequency. In the case of sound, high frequency means high pitch. After the siren has passed, its sound waves are stretched in wavelength, causing you to hear a lower pitch.

This effect, which we see here both in water waves and in sound waves, is called the ***Doppler effect***, named after the Austrian physicist Christian Doppler (1803–1853), who first explained it. Before we apply it to light waves and astronomy, let's continue the list of important things to know about it.

3. The sound does indeed get louder as a siren approaches (and the water waves get higher as the vibrating object approaches a bobbing cork), but this is *not* what the Doppler effect is about. The Doppler effect refers only to the *change in wavelength* (and therefore frequency) of the wave.

4. The frequency does *not* get higher and higher as the source approaches at a uniform speed. Rather, the frequency is observed to be higher than the source's (but constant in value) as the source approaches and lower than the source's (but constant in value) as it recedes.

The Doppler Effect in Astronomy

The Doppler effect also occurs for electromagnetic waves, which is what we receive from the stars. This means that if an object is coming toward us, the light we receive from it will have a shorter than normal wavelength, and from an object moving away we will receive a longer than normal wavelength. The reason you can't observe this for moving objects here on Earth is that the amount of the shortening and lengthening of the wave depends on the speed of the object *compared with the speed of the wave*.

For water waves, which move fairly slowly, you observe the Doppler effect even for a slow-moving wave source. In the case of sound, which has a speed much greater than water waves, you notice the Doppler effect only for fairly fast objects. A car going 70 miles/hour is moving at about 10% of the speed of sound.

In the case of light, you don't perceive the Doppler effect for a car traveling at 70 miles/hour because it is moving at only one ten-millionth the speed of light. To describe the Doppler effect for light, consider a spaceship with a lamp emitting green light. If the spaceship is moving away from us at a great enough speed, the wavelength of the light we see from the lamp is stretched so that the light appears red. The light is ***redshifted***. If the spaceship is approaching, the light is ***blueshifted***.

The spaceship example does illustrate the Doppler effect, but it is very misleading in an important respect: Except for very distant galaxies, objects in the heavens do not move with speeds great enough to actually change their colors appreciably. The redshift or blueshift caused by the Doppler effect is very small in most cases. If the spectra of stars were continuous spectra, there would be few cases in which the Doppler effect could be used to detect motion. It is the spectral lines (usually the absorption lines) that make the Doppler effect such a powerful tool.

As an example, imagine that we record the spectrum of hydrogen gas in the laboratory (**FIGURE 4-17a**). Figure 4-17b represents the spectrum of a star having only hydrogen in its atmosphere (which is unrealistic, but this is a simplified example). The absorption lines in the star's spectrum do not align exactly with the emission lines of the laboratory spectrum. Instead, the absorption lines are shifted slightly toward the red. This indicates that the star is moving away from us.

There are three major differences between our example and a measurement of a real star: First, a real star's spectrum has many more spectral lines. Second, the Doppler shift is almost always much smaller than that indicated in the example. When we later show photos of actual spectra, they will usually be a magnified por-

Doppler effect The observed change in wavelength of waves from a source moving toward or away from an observer.

By *normal* we mean the wavelength of the emitted light as measured by someone traveling with the source.

redshift A change in wavelength toward longer wavelengths.

blueshift A change in wavelength toward shorter wavelengths.

(a)

(b)

FIGURE 4-17 (a) Emission spectrum of hydrogen when the source is stationary with respect to the observer. (b) Absorption spectrum of hydrogen for a receding source (at 0.7% of the speed of light). Note the shift of all lines toward redder colors.

tion of a small part of the visible spectrum. Third, astronomers do not normally use color film in recording the spectrum. This may seem odd, but color is not easy to describe accurately, whereas wavelength is. The wavelengths of absorption lines can be measured very accurately and compared with the lines from a laboratory spectrum to determine the existence and the precise amount of the Doppler shift.

The Doppler Effect as a Measurement Technique

Thus far, we have described the Doppler effect as a method for detecting whether a star is moving toward or away from us, but it is more powerful than this. From measurements of the *amount* of the shifting of the spectral lines, we can determine the **radial velocity** of the star relative to Earth. The radial velocity is the star's velocity toward or away from us and must be distinguished from its **tangential velocity**, which is its velocity across our line of sight.

FIGURE 4-18 distinguishes the two velocities. To measure tangential velocity, we must look for motion of the star across our line of sight (like the yellow car in Figure 4-18), and this motion can be detected only for relatively nearby stars. To measure radial velocity, we use Doppler shift data and the following equation:

$$\frac{\Delta\lambda}{\lambda_0} = \frac{\upsilon}{c},$$

where $\Delta\lambda = \lambda - \lambda_0$ = wavelength difference, λ = observed wavelength, λ_0 = wavelength of spectral line from stationary source, υ = radial velocity of object, and c = velocity of light.

If we solve this equation for what we are usually calculating—that is, the velocity of the object—we obtain

$$\upsilon = c\left(\frac{\Delta\lambda}{\lambda_0}\right).$$

When an object is approaching the observer, the wavelength difference is negative and so is the radial velocity. When an object is moving away from the observer, the wavelength difference is positive and so is the radial velocity.

An object's velocity is a quantity that gives us the object's speed and its direction of motion.

radial velocity Velocity along the line of sight, toward or away from the observer.

tangential velocity Velocity perpendicular to the line of sight.

This equation applies if the object is moving much slower than the speed of light, as is the case for most galaxies other than the very distant ones. For distant galaxies, a slightly more complex equation is needed, as relativistic effects become important.

EXAMPLE

The wavelength of one of the most prominent spectral lines of hydrogen is 656.285 nanometers (this is in the red portion of the spectrum). In the spectrum of Regulus (the brightest star in the constellation Leo), the wavelength of this line is observed to appear greater by 0.0077 nanometers. Calculate the speed of Regulus relative to Earth, and determine whether it is moving toward or away from us.

FIGURE 4-18 The red car has a radial (to-or-fro) velocity with respect to the observer, while the yellow car has a tangential (side-to-side) velocity. The Doppler effect cannot be used to detect tangential velocity.

SOLUTION   First, the data indicate that the wavelength of the line in the spectrum of

Regulus is *longer* by 0.0077 nanometers. This means that the star is moving *away* from us. To determine its speed, we substitute the given values in the equation that relates Doppler shift and speed:

$$\upsilon = c\left(\frac{\Delta\lambda}{\lambda_0}\right) = (3.0\times10^8 \text{ m/s})\times\left(\frac{0.0077 \text{ nm}}{656.285 \text{ nm}}\right)$$

$$= 3500 \text{ m/s} = 3.5 \text{ km/s}.$$

Regulus thus is moving away from the Earth at a speed of about 3.5 kilometers/second. This is about 2 miles/second, or 8000 miles/hour, a typical speed for nearby stars. Because the Earth's speed in its orbit around the Sun is about 30 kilometers/second, the Earth's motion would have to be taken into account in making the measurement. Our calculation assumed that the Earth was between the Sun and Regulus so that the Earth had no motion toward or away from the star.

TRY ONE YOURSELF
The nearest star to the Sun that is visible to the naked eye is Alpha Centauri. With Earth's motion removed, the 656.285 nanometers line of hydrogen has a wavelength of 656.237 nanometers in Alpha Centauri's spectrum. Calculate the radial velocity of this star relative to the Sun, and tell whether it is moving toward or away from the Sun.

The annual variation in the Doppler shift observed in stellar spectra provides evidence for the Earth's motion around the Sun. The evidence was not available, of course, when the question was controversial.

Police radar measures motorists' speeds by bouncing waves from their cars and detecting the Doppler shift in the reflected waves.

binary star (system) A pair of stars gravitationally bound so that they orbit one another.

Other Doppler Effect Measurements

The use of the Doppler effect is not limited to measuring the speeds of stars. Other applications include the following:

1. Measuring the rotation rate of the Sun. Galileo was the first to observe that sunspots move across the Sun, thus providing a method to measure the Sun's rotation rate. The Doppler effect gives a second method. The measurement is done by examining the light from opposite sides of the Sun. Light from the side moving toward us is blueshifted and light from the other side is redshifted.

2. In a similar manner, the rotation rates of distant stars, other galaxies, planets, and the rings of Saturn can be measured. The fact that the light in the case of planets has been reflected by the planet (or rings) rather than emitted by it does not matter; the light is still shifted by the Doppler effect. In fact, the rotations of Mercury and Venus were first revealed by reflected radar waves.

3. Many stars are part of a two (or more) star system in which the stars orbit one another. If their orbits happen to be aligned so that each star moves alternately toward and away from us, we can detect this motion by the blueshift and redshift of each star's spectrum. When we discuss such **binary stars** in Chapter 12, we will see that they are very important in our quest to learn more about stars.

Relative or Real Speed?

The speed measured by the Doppler effect is the object's speed *relative to the speed of the Earth*. We, therefore, must take the Earth's speed into account in any calculation made with the Doppler effect, but even then we are only measuring the speed of the object with respect to the Sun. What about the object's *real* speed? There is no such thing because all speeds are relative to something. When you say that a car is moving at 50 miles/hour, you mean "relative to the surface of the Earth." When you walk up the aisle of a moving plane, you have one speed relative to the seated passengers, another relative to the Earth, another relative to the Sun, and so on.

This understanding of the relativity of motion is called *Galilean relativity* (or *Newtonian relativity*). Einstein's theory of relativity goes much further than this.

Just as all motion is relative, all nonmotion is relative, too. When we say that something is at rest, we *usually* mean that it is at rest relative to the Earth. It is meaningless to say that something is absolutely at rest. Thus, there is no loss of meaning to our finding the velocity of an object relative to the Earth and then correcting for the Earth's motion around the Sun. All motion is relative.

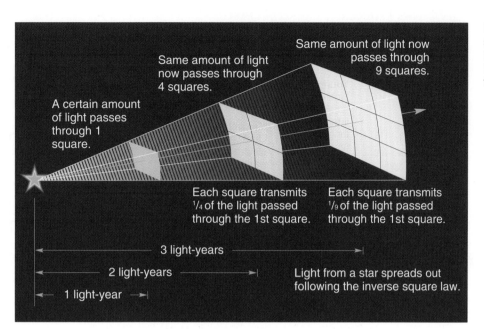

FIGURE 4-19 As light spreads from a star, its intensity decreases as the inverse square of the distance.

4-8 The Inverse Square Law of Radiation

Everyone has experienced the fact that the intensity—that is, the brightness—of light decreases when we move farther from its source. The decrease in intensity follows an ***inverse square law***, a relationship that states that radiation spreading from a small source decreases in intensity as the inverse square of the distance from the source.

We can reach the conclusion that light intensity (just like gravity) decreases inversely with the distance squared ($1/d^2$) if we assume that light "spreads out" from its source uniformly in all directions. Indeed, this outward diffusion has to pass through successive imaginary spheres centered on the source. Because the surface area of a sphere of radius d is $4\pi d^2$, as light spreads out from the source filling all space uniformly in all directions, it must become less "concentrated" by a factor of $1/d^2$ (FIGURE 4-19).

In Chapter 12, we discuss how we use this inverse square law to find how much energy a star really emits if we know its distance from us and its apparent brightness.

inverse square law Any relationship in which some factor decreases as the square of the distance from its source.

The inverse square law applies only to small sources of light, but this requirement is easily met in most astronomical cases.

Conclusion

One of the major differences between Astronomy and other disciplines is that the subject of our studies, the universe, is beyond our control. Stars are born and die continually, galaxies collide and, most important of all, when we look at an object we see it as it was when it emitted the light we now see. One of the means to learn about the cosmos is by unlocking the information in the radiation we collect. In this chapter, we examined the evolution of our ideas on light since antiquity (FIGURE 4-20). We also discussed how the spectra of stars are used in two very different ways. First, the spectra are treated as continuous spectra, and we examine their graphs of intensity versus wavelength. By measuring the wavelength of the maximum intensity of radiation, we can determine the temperature of a star's surface. In doing this analysis, we ignore the absorption lines. Recall that these lines are very narrow, and their presence does not change the overall pattern of the blackbody curve.

Second, we examine the absorption lines within the spectrum. These lines allow us to determine the chemical composition of stars. In addition, when we measure the Doppler effect, the shift in the lines allows us to calculate the radial speeds of stars relative to Earth, as well as the speeds of rotation and revolution of celestial objects.

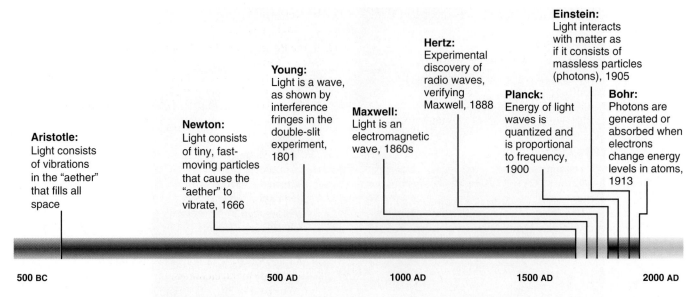

FIGURE 4-20 Over the years, explanations for the nature of light went back and forth between wave models and particle models. As we saw in the Advancing the Model box on pages 112–113, today we talk about the wave-particle duality of light; it propagates through space as a wave but interacts with matter as a stream of particles.

STUDY GUIDE

RECALL QUESTIONS

1. The frequency of visible light falls between that of
   A. infrared waves and radio waves.
   B. X-rays and cosmic rays.
   C. ultraviolet waves and X-rays.
   D. short radio waves and long radio waves.
   E. ultraviolet waves and radio waves.

2. Infrared radiation differs from red light in
   A. intensity.
   B. wavelength.
   C. its speed in a vacuum.
   D. [All of the above.]
   E. [None of the above.]

3. The frequency at which a star emits the most light depends on the star's
   A. distance from us.
   B. brightness.
   C. temperature.
   D. eccentricity.
   E. velocity toward or away from us.

4. In an infrared photo taken on a cool night, your skin will appear brighter than your clothes.
   A. Correct.
   B. Wrong, it depends upon the color of your clothes.
   C. Wrong, your clothes will appear brighter.
   D. Wrong, they will be equally bright.

5. Light waves of greater frequency have
   A. shorter wavelength.
   B. longer wavelength.
   C. [Either of the above; there is no direct connection between frequency and wavelength.]

6. Which of the following choices does not have the same fundamental nature as visible light?
   A. X-rays.
   B. Sound waves.
   C. Ultraviolet radiation.
   D. Infrared waves.
   E. Radio waves.

7. As an electron of an atom changes from one energy level to a higher energy level by absorbing a photon, the total energy of the atom
   A. increases.
   B. decreases.
   C. remains the same.

8. Which of the following choices is produced when white light is shined through a cool gas?
   A. An absorption spectrum.
   B. A continuous spectrum.
   C. An emission spectrum.
   D. [All of the above.]
   E. [None of the above.]

9. The solar spectrum is which of the following?
   A. An absorption spectrum.
   B. A continuous spectrum.
   C. An emission spectrum.
   D. [All of the above.]
   E. [None of the above.]

10. The spectrum of light from the Sun's atmosphere *seen during a total solar eclipse* is
    A. an emission spectrum.
    B. an absorption spectrum.

   C. a combination of emission spectrum and absorption spectrum.
   D. a continuous spectrum.
   E. a combination of all three types of spectra.

11. Analysis of a star's spectrum *cannot* determine
   A. the star's radial velocity.
   B. the star's tangential velocity.
   C. the chemical elements present in the star's atmosphere.
   D. [More than one of the above.]

12. According to the Doppler effect,
   A. sound gets louder as its source approaches and softer as it recedes.
   B. sound gets higher and higher in pitch as its source approaches and lower and lower as its source recedes.
   C. sound is of constant higher pitch as its source approaches and of constant lower pitch as its source recedes.
   D. [Both A and B above.]
   E. [Both A and C above.]

13. The Doppler effect causes light from a source moving away to be
   A. shifted to shorter wavelengths.
   B. shifted to longer wavelengths.
   C. changed in velocity.
   D. [Both A and C above.]
   E. [Both B and C above.]

14. We can determine the elements in the atmosphere of a star by examining
   A. its color.
   B. its absorption spectrum.
   C. the frequency at which it emits the most energy.
   D. its temperature.
   E. its motion relative to us.

15. Which list shows the colors of stars from coolest to hottest?
   A. Red, white, blue.
   B. White, blue, red.
   C. Blue, white, red.
   D. Red, blue, white.
   E. Blue, red, white.

16. The Doppler effect is used to
   A. measure the radial velocity of a star.
   B. detect and study binary stars.
   C. measure the rotation of the Sun.
   D. [Two of the above.]
   E. [All of the above.]

17. Sound waves cannot travel in a vacuum. How, then, do radio waves travel through interstellar space?
   A. They are extra-powerful sound waves.
   B. They are very high frequency sound waves.
   C. Radio waves are not sound waves at all.
   D. The question is a trick, for radio waves do *not* travel through interstellar space.
   E. Interstellar space is not a vacuum.

18. The energy of a photon is directly proportional to the light's
   A. wavelength.
   B. frequency.
   C. velocity.
   D. brightness.

19. In the Bohr model of the atom, light is emitted from an atom when
   A. an electron moves from an inner to an outer orbit.
   B. an atom gains energy.
   C. an electron moves from an outer to an inner orbit.
   D. one element reacts with another.
   E. [Both A and B above.]

20. The intensity/wavelength graph of a "blue-hot" object peaks in the
   A. infrared region.
   B. red region.
   C. yellow region.
   D. ultraviolet region.

21. The emission spectrum produced by the excited atoms of an element contains wavelengths that are
   A. the same for all elements.
   B. characteristic of the particular element.
   C. evenly distributed throughout the entire visible spectrum.
   D. different from the wavelengths in its absorption spectrum.
   E. [Both A and D above.]

22. Each element has its own characteristic spectrum because
   A. the speed of light differs for each element.
   B. some elements are at a higher temperature than others.
   C. atoms combine to form molecules, releasing different wavelengths depending on the elements involved.
   D. electron energy levels are different for different elements.
   E. hot solids, such as tungsten, emit a continuous spectrum.

23. The speed of sound is 335 meters/second. What is the wavelength of a sound that has a frequency of 500 cycles/second?
   A. 0.67 meters.
   B. 1.49 meters.
   C. 165 meters.
   D. 835 meters.

24. Suppose the speed of a water wave is 12 inches/second and the wavelength is 4 inches. What is the frequency of the wave?
   A. 48 cycles/second.
   B. 18 cycles/second.
   C. 8 cycles/second.
   D. 3 cycles/second.
   E. 1/3 cycles/second.

25. Define *wavelength* and *frequency* in the case of a wave.

26. Which has the higher frequency, light of 400 nanometers or light of 450 nanometers?

27. Approximately what are the least and greatest wavelengths of visible light?

28. Name six regions of the electromagnetic spectrum in order from longest to shortest wavelength. (Do not list the colors of visible light as separate regions of the spectrum.)

29. Which parts of the electromagnetic spectrum penetrate the atmosphere?

30. Define the following terms: electron, nucleus, photon.

31. State the three postulates on which the Bohr model of the atom is based.

32. How is the energy of the photon related to the frequency of the light?

33. Why do the various elements each have different emission spectra?

34. Name the three different types of spectra, and explain how each is produced.

35. Explain how we know what elements are in the atmospheres of the stars.

36. Carefully explain what the Doppler effect is. For light waves, does the Doppler effect tell us anything about the intensity of the light in front of and behind a moving light source?

37. Distinguish between the way a spectrum is used to determine the temperature of a star and the way it is used to determine the star's composition and/or motion toward or away from us.

1. Why do we have three different temperature scales instead of one? What are the advantages (if any) of each?

2. Two people are having fun in the ocean. One is surfing, and the other is floating up and down. Their different motions could be used to illustrate two quantities in the equation $v = \lambda \times f$. Which two?

3. The period of a wave is defined as the amount of time required for the wave to complete one cycle. What, then, is the relationship between the period and the frequency of a wave?

4. It is especially easy to get a sunburn when skiing high in the mountains. How does the high altitude contribute to the danger of sunburn?

5. Why can't we hear radio waves without using the device that we call a "radio"?

6. Nowadays, radio waves (both for radio and TV) are striking your body all the time, but you are not allowed to get too many X-rays in one year. Why is this so? After all, both radio waves and X-rays are a part of the electromagnetic spectrum.

7. What does an object do to light to give the object its color? How does this differ from the color of an object that emits its own light?

8. Draw a graph of intensity versus wavelength for a red star and another for a white star. Describe two ways in which the graphs differ.

9. The text explains that cool stars are redder than hot ones. However, couldn't the red color result instead from the Doppler effect if the star is moving away from us? Explain.

10. If absorption lines were to be included in the graph of intensity versus wavelength for a star, how would this change the graph?

11. How would Figure 4-10 be different if it were the spectrum of a red star?

12. How do we know what chemical elements are in the atmosphere of the Sun? Explain.

13. What is the Doppler effect and why is it important to astronomers?

14. Figure 4-18 greatly exaggerates color differences for the two cars. If it showed a car coming toward you, what color would the car be, using the same exaggeration?

1. If the speed of a particular water wave is 12 meters/second and its wavelength is 3 meters, what is its frequency?

2. The speed of sound in air at room temperature is about 344 meters/second. What is the wavelength of a sound wave with a frequency of 256 Hz?

3. Express 500 nanometers in meters.

4. The 656.285 nanometers line of hydrogen is measured to be 656.305 nanometers in the spectrum of a certain star. Is this star approaching or receding from the Earth? Calculate its radial speed relative to the Earth.

5. If a certain star is moving away from Earth at 25 km/s, what will be the measured wavelength of a spectral line that has a wavelength of 500 nm for a stationary source?

6. Suppose the Kelvin temperature of blackbody X is three times as great as that of blackbody Y. Compare the energy released by equal areas of the two objects.

7. Use Wien's law to determine the wavelength at which an object at body temperature (98.6°F) emits most of its energy. (Hint: Make sure you find the corresponding temperature in kelvin.) What kind of radiation is this? Can you see it?

8. Two thermometers, one marked in °F and the other in °C, are placed in the same container. At what temperature will both thermometers read the same?

9. Matter falling toward a black hole gets compressed and heated to about 10^6 K. In what part of the electromagnetic spectrum would you search for black holes?

10. A star has five times the surface area of the Sun, but its surface temperature is smaller by a factor of three. Does this star emit more energy than the Sun or less? Explain.

11. (a) We said that light takes 8.3 minutes to reach us from the Sun. Show that this is the case. (b) When we are observing an object that is 5000 light years away, do we see it as it is now? Explain.

12. Compare the light we receive from a 100-watt lightbulb that is 30 feet away with the light from a similar bulb that is 10 feet away.

Doppler Effect Measurement

Figure 4-17a shows a hydrogen emission spectrum (in the visible part of the spectrum) when the source is stationary with respect to the observer, whereas Figure 4-17b shows a corresponding hydrogen absorption spectrum for a receding source. In general, spectra are much more complicated, including many lines from different elements; however, let us imagine that we have isolated the hydrogen lines in this spectrum. In addition, consider the spectrum as a "map," exactly as is. We would like to calculate the speed of the source, using the Doppler effect.

For a map to be useful, it must include the correct scale. The lines in Figure 4-17a correspond to the following wavelengths from right to left: 656.3 nm (red), 486.1 nm (blue-green), 434.0 nm (violet), and 410.1 nm (violet). Measure the distances between the lines in millimeters (mm), and calculate the scale with units nm/mm.

Now compare the position of the lines between the two spectra. For example, by how many millimeters has the 656.3-nanometer line shifted to the right (toward longer wavelengths)? Using the scale you found in the previous step, translate this shift (redshift) into nanometers. This is the change between the observed and laboratory-based wavelength for this hydrogen line ($\Delta\lambda$). Using $\lambda_0 = 656.3$ nanometers and the Doppler equation, show that the speed of the receding source is 0.7% of the speed of light.

1. "The Duality in Matter and Light," by B.G. Englert, M. O. Scully, and H. Walther, in *Scientific American* (December, 1994).

2. "The Truth About Star Colors," by P. C. Steffy, in *Sky & Telescope* (September 1992, p. 266).

3. "Unlocking the Chemical Secrets of the Cosmos," by O. Gingerich, in *Sky & Telescope* (July, 1981).

4. "The Electromagnetic Spectrum," by H. Augensen and J. Woodbury, in *Astronomy* (June, 1982).

5. "Beyond the Rainbow," by J. B. Kaler, in *Astronomy* (September, 2000).

6. "Starlight Detectives," by A. W. Hirshfeld, in *Sky & Telescope* (August, 2004).

Quest Ahead to Starlinks
http://physicalscience.jbpub.com/starlinks

Starlinks is this book's online learning center. It features **eLearning**, which contains chapter quizzes and other tools designed to help you study for your class. You can also find **online exercises**, view numerous relevant **animations**, follow a guide to **useful astronomy sites** on the Web, or even check the latest **astronomy news** updates.

ACTIVITIES

EXPANDING THE QUEST

STARLINKS

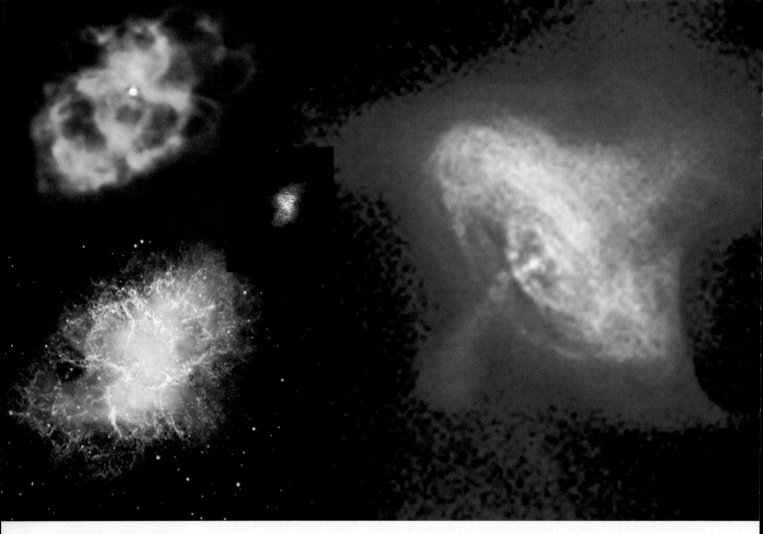

Telescopes: Windows to the Universe

5

The Crab nebula as seen in ultraviolet (inset) and (clockwise from the right) in X-ray, visible, and radio radiation.

LOOKS CAN BE DECEIVING. DURING THE PAST FEW DECADES, astronomers have come face to face with this old saying. Up until the middle of the 20th century, almost all of our knowledge of the sky came from visible light waves. Therefore, photographs of the sky showed what our eyes see, just adding detail because telescopes can detect very dim light and magnify the image. Still, it seemed that "what we see is what we get."

Then along came detectors of radiation in other parts of the spectrum. The chapter-opening photo shows four views of the Crab nebula, which is the leftover debris of a star that exploded almost a thousand years ago (1054 AD). The bottom left image was taken with an optical (visible-light) telescope. The larger image was taken by the *Chandra X-Ray Observatory*, which is in orbit around the Earth. It shows the spinning, top-like structure that may provide the energy for the glowing gases seen in visible light. Which image shows reality? The answer is "both, and more!" for the same picture taken in radio or ultraviolet radiation looks different still. We must conclude that our eyes don't tell us the whole story. In astronomy, what we see is just part of what we get.

All cross references to chapters, sections, figures, and tables pertain to the main text, *In Quest of the Universe, Sixth Edition. In Quest of the Solar System* contains Chapters 1–11 and 19 of the main text. *In Quest of the Stars and Galaxies* contains Chapters 1–5 and 11–19 of the main text.

To understand the universe, we rely heavily on observations. As with any branch of science, we need observational data to test existing theories or develop new ones. The distances between us—the observers—and the objects of our studies are vast, and the cosmos is very dynamic in nature. All we can do is passively collect the radiation sent from these objects and try to decipher the information it carries. The most useful tools for collecting radiation are telescopes.

Galileo Galilei was the first to use a telescope to study the heavens systematically, as we discussed in Chapter 3. Much was learned before Galileo's time by naked-eye observations, but Galileo's telescope changed astronomy—and our outlook on the universe—tremendously. As we make telescopes larger and take them into space, above the distortions of the Earth's atmosphere, we realize that we are just beginning to learn about the mysteries of our universe.

The past century has brought a multitude of telescopic tools, and our view of the skies has expanded to all parts of the spectrum well beyond visible light. We begin this chapter by describing some properties of light that are important to the understanding of visible-light telescopes. Then we discuss the use of such telescopes. Finally, the chapter concludes with a look at some of the nonvisible-light telescopes that are so indispensable to modern astronomy.

O telescope, instrument of much knowledge, more precious than any scepter!
—Kepler, in a letter to Galileo

5-1 Refraction and Image Formation

The discussion of light thus far has concentrated on its wave properties. The effect of these properties is examined later in this chapter, but for the moment, we concentrate not on the nature of light, but on the path that light travels. That path is usually very simple, for light travels in a straight line as long as it remains in the same (uniform) *medium*. It may, however, change direction upon entering a second medium. FIGURE 5-1 shows a ray of light passing through a wedge of glass. We see that the light travels in a straight line before and after passing through the glass and that it travels in a straight line inside the glass; however, the ray bends when it passes through each surface of the glass. The phenomenon of the bending of a wave as it passes from one medium into another is called **refraction**.

Two factors determine the amount of refraction that occurs when light crosses from one material into another. The first factor is the relative speeds of light in the two materials. In Chapter 4, we stated that the speed of light in vacuum is about 3×10^8 m/s. In air, light's speed is just slightly slower. In glass, however, light travels at about 2×10^8 m/s, depending on the type of glass. Because of the change in speed, a ray of light may bend significantly when going from air into glass.

For a simple analogy, consider what happens when you roll a pair of toy wheels connected by an axle from a smooth wooden floor onto a carpet. If the wheels cross the interface between the two surfaces perpendicularly (face-on), then they will slow down but will continue moving in the same direction. If, on the other hand, they cross the interface between the two surfaces nonperpendicularly, the wheel that first hits the carpet will slow down while the other wheel, still on the smooth floor, will continue moving at its original speed. As a result, the axle will turn; it will change direction. In a similar way, light changes direction when it crosses the interface between two different media nonperpendicularly (FIGURE 5-2a). Figure 5-2b shows an everyday example of refraction.

To understand the second factor in refraction, consider FIGURE 5-3a, which shows several rays of light (that came from beyond the left side of the page) passing through a lens-shaped piece of glass. Notice that the rays that strike the surface of the glass (the interface between the two media) at a glancing angle (farther from the perpendicular) bend more than those rays that hit the glass more "head-on." The central ray strikes the glass perpendicularly and does not bend at all. This is a general rule: the larger the angle between the ray of light and the perpendicular line to the interface, the more the light bends on passing through the surface.

The word *medium* refers, in a generic way, to the "material" though which light travels, such as air, glass, water, or the almost perfect vacuum in the far reaches of the universe.

refraction The bending of light as it crosses the boundary between two materials in which it travels at different speeds.

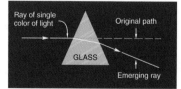

FIGURE 5-1 When light passes through a wedge of glass, it is bent from its straight-line path. The smaller the angle of the wedge, the less the bending. (The "wedge" is really a prism, but we are concentrating here on the bending of the light rather than its separation into colors.) The bending occurs both when the light enters the glass and when it emerges.

FIGURE 5-2 (a) As a light beam crosses the interface between two different media (such as from air into water, which is optically denser), it bends because of the change in the speed of propagation (as shown in Figure 5-1). This is similar to the change in direction when a pair of toy wheels travels from a smooth floor onto a rough carpet; however, if we consider two light rays instead of two wheels, both rays change direction at the interface and the distance between them increases by an amount that depends on the type of media. (b) The pencil in a cup of water appears bent because of the refraction of light. As the light rays from the submerged part of the pencil rise toward the observer, they bend at the interface.

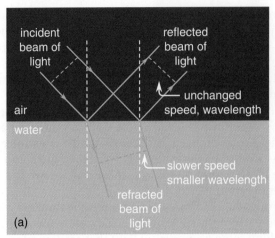

image The visual counterpart of an object, formed by refraction or reflection of light from the object.

focal point (of a converging lens or mirror) The point at which light from a very distant object converges after being refracted or reflected.

focal length The distance from the center of a lens or a mirror to its focal point.

Refraction of light in raindrops is a part of the process that gives us the rainbow.

Refer again to Figure 5-3a. The surfaces of the lens have just the right curvature to cause all of the rays of light shown in the figure to pass through the same spot. We could even block some of the incoming light rays (for example, the one labeled *X*) and still get an **image** of the object, although a bit dimmer. This is important later in this chapter in our discussion of a class of telescopes.

To see how a lens forms an image, imagine that each ray in the figure came from a distant star, which we treat as a point. A point light source emits light rays in all directions; if this source is far from our telescope, only a few of these rays will enter the telescope, and they will be essentially parallel to each other. If we put a piece of paper at the point where the light rays come together, all the light from that star that passes through our lens will come to a single point on the piece of paper. In fact, the rays of light coming from other stars will likewise come to a focus on the paper, forming an *image* of that area of the sky (Figure 5-3b).

The **focal point** of a lens is that point where light from a very distant object comes to a focus. This is the point where the rays converge in Figure 5-3.

The **focal length** of the lens is the distance from the center of the lens to the focal point. Depending on the curvature of their surfaces, different lenses have different focal lengths.

Refraction has important consequences for observations involving the Earth's atmosphere. When light passes from the (almost perfect) vacuum of space into Earth's atmosphere, it continuously refracts as it moves through the air. Each thin layer in the atmosphere is under different conditions and therefore behaves as a different medium. As a result, objects appear higher in the sky than they really are (**FIGURE 5-4a**), and the amount of refraction depends on the object's altitude above our horizon. We must take this into account when measuring positions of celestial objects. This phenomenon also explains why the Sun looks "flattened" when it is closer to the horizon. Its lower edge, being closer to our horizon, is shifted upward more than its upper edge (Figure 5-4b).

FIGURE 5-3 (a) A lens bends incoming rays of light toward a single point. When the incoming rays are parallel to the axis (as shown here), they cross at the lens' focal point. (b) Light rays from two different stars are focused into two separate images.

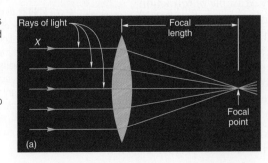

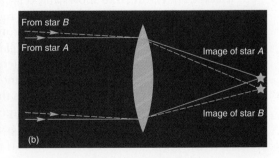

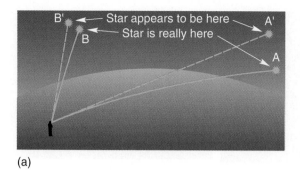

(a)

(b)

FIGURE 5-4 (a) A simplified picture of atmospheric refraction. The position of the observed object seems to be higher in the sky than it really is. (b) The Sun appears flattened at sunset because the lower edge is refracted higher than the upper edge. As shown in part (a), the amount of refraction depends on the altitude above the horizon. As a result, the Sun appears more flattened the closer it is to the horizon.

5-2 The Refracting Telescope

Lenses are at the heart of the refracting telescope. (The reflecting telescope will be discussed later in this chapter.) The simplest use of a telescope (in principle, anyway) is as a lens for a camera (FIGURE 5-5a). The camera is using the long-focal length lens of the telescope in place of its regular lens (Figure 5-5b). The telescope lens simply replaces the regular camera lens and brings the image to a focus on the film. As we see later, telescopes can be mounted so that they can track stars across the sky and allow astronomers to take long time exposure photographs of the heavens. Long time exposures allow us to photograph much fainter objects than can be seen by simply looking through a telescope. The use of a telescope with a camera thus is much more important to a professional astronomer than its use for direct viewing.

Figure 5-5c shows how a small telescope is used for direct observation. The primary lens—the lens through which the light passes first—is called the *objective lens*, or simply the *objective*. This lens brings the light to a focus at the focal point, which is where the film is placed when the telescope is used with a camera. This is where the image is located. For direct viewing, a second lens, the *eyepiece*, is added just beyond the focal point. This lens simply acts as a magnifier to enlarge the image.

objective lens (or objective) The main light-gathering element—lens or mirror—of a telescope. It is also called the primary lens.

eyepiece The magnifying lens (or combination of lenses) used to view the image formed by the objective of a telescope.

Chromatic Aberration

A prism separates white light into its colors because different wavelengths of light are refracted different amounts. Except for the ray of light that goes straight through the center of a lens, rays go through a lens in much the same way as they go through a prism. Light passing through a lens separates into colors, and this causes the lens to

(a)

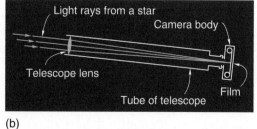

(b)

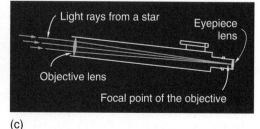

(c)

FIGURE 5-5 (a) A simple way to use a telescope for photography is to let the telescope serve as the camera's lens. (b) The image is focused directly on the film by the telescope's main lens. (c) When a telescope is used for direct viewing, the eyepiece magnifies the image formed by the objective lens. This image is formed at the objective's focal point.

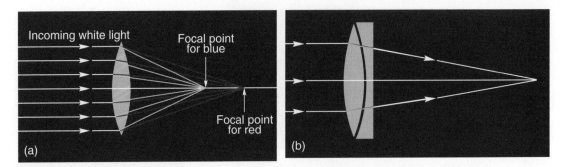

FIGURE 5-6 (a) Chromatic aberration. A lens exhibits a prism effect, separating white light into its colors. The lens therefore focuses each color at a different place. Only red and blue are shown here. (b) A lens can be corrected for chromatic aberration by the proper combination of lenses of different types of glass. Here the second lens brings the separated colors back together at the image.

have a slightly different focal length for each wavelength of light (**FIGURE 5-6a**). As a result, there is no single place where an image is exactly in focus. If the film of a camera is placed at the point where the red light focuses, the other colors will be out of focus, and the result will be an image with a fuzzy, bluish edge. This phenomenon, called ***chromatic aberration***, occurs not just in telescopes, but in regular cameras as well. Fortunately, it can be corrected, at least in part.

The amount of color separation that occurs when light passes through a lens depends not only on the curvature of the glass but also on the type of glass. Some kinds of glass separate the colors more than other kinds. Telescope and camera manufacturers use this fact to correct for chromatic aberration. In all but the cheapest toy-store telescopes, the objectives of refracting telescopes are made of two lenses instead of one. As Figure 5-6b shows, the second lens has reverse curvature from the first. This curvature is not enough to undo all of the converging effect of the first lens, however, and the light is still brought to a focus. The second lens is made of a different type of glass than the first, and although it does not cancel out the bending of the light, it does cancel out most of the color separation. Such a combination of lenses is called an ***achromatic lens***.

chromatic aberration The defect of optical systems that results in light of different colors being focused at different places.

achromatic lens (or achromat) An optical element that has been corrected so that it is free of chromatic aberration.

5-3   The Powers of a Telescope

When most people think of a telescope's power, they think of magnification. Magnification, however, is the least important of the three major powers of a telescope (the other two being the light-gathering power and the resolving power). After all, magnifying a blurry image will not show any more details. Again, we describe small telescopes, but the same ideas apply to large telescopes used by professional astronomers. In fact, the reasons for using such large telescopes are related to the powers of telescopes.

Do telescopes just magnify things?

Angular Size and Magnifying Power

The ***angular size*** of an object is the angle between two lines that start at the observer and go to opposite sides of the object. **FIGURE 5-7a** shows this angle for the case of someone looking at the Moon. (Angular size is defined very similarly to angular separation in Chapter 1. Angular size measures the apparent size of one object, whereas angular separation measures the apparent angular distance between two objects.) The angular size determines, for example, how big the image of the Moon is on the retina of your eye.

For a telescope (as well as binoculars and several other optical instruments), ***magnifying power*** or ***magnification*** is defined as the ratio of the angular size of an object when it is seen through the instrument to the object's angular size when seen with the naked eye. Figure 5-7b shows the angular size of the Moon as seen through

angular size (of an object) The angle between two lines drawn from the viewer to opposite sides of the object.

magnifying power, or **magnification** (of an instrument) The ratio of the angular size of an object when it is seen through the instrument to its angular size when seen with the naked eye.

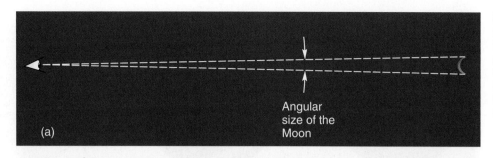

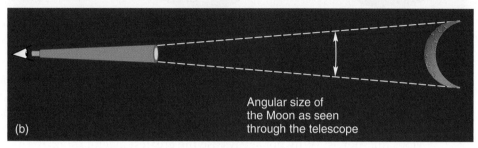

FIGURE 5-7 (a) The Moon seen by the naked eye. (b) In a telescope, the angular size of the Moon is apparently larger. The magnification of the telescope is the ratio of the two angular sizes.

a particular telescope. You might estimate from the angles in the figure that this telescope has a magnification of about four; that is, the telescope magnifies the object four times.

The magnification of a particular telescope depends on the focal lengths of the objective and the eyepiece. It can be calculated using the following formula:

$$\text{magnifying power} = \frac{\text{focal length of objective}}{\text{focal length of eyepiece}} \quad \text{or} \quad M = \frac{f_{\text{obj}}}{f_{\text{eye}}}.$$

This means that the greatest magnification can be achieved by having a long-focal length objective and a short-focal length eyepiece.

EXAMPLE

The telescopes used in an introductory astronomy laboratory have objectives with focal lengths of 1250 mm. One eyepiece has a focal length of 25 mm. What is the magnification produced by this telescope? What is the angular size of the Moon as seen through this telescope, using the 25-mm eyepiece? (The naked-eye angular size of the Moon from Earth is about 0.5°.)

SOLUTION   To calculate the magnification, we use the equation that relates it to focal lengths:

$$M = \frac{f_{\text{obj}}}{f_{\text{eye}}} = \frac{1250 \text{ mm}}{25 \text{ mm}} = 50.$$

The magnifying power of the telescope thus is 50. We say that the magnification is 50 times, or 50×.

Now, because the Moon's angular size as seen by the naked eye is 0.5° and the telescope magnifies the Moon 50 times, the Moon's angular size in the telescope will be 50 × 0.5° = 25°.

TRY ONE YOURSELF
What is the magnification produced by a telescope having an objective with a focal length of 1.5 meters when it is being used with an eyepiece that has a focal length of 12 millimeters. (Hint: In doing the calculation, you must express the two lengths in the same units: either meters or millimeters.)

FIGURE 5-8 Increasing the magnification decreases the field of view and makes the image darker. The entire telescopic view is shown in each case.

(a) 50×   (b) 100×   (c) 250×

As we hinted in both the example and the suggested problem, you might use an eyepiece of a different focal length. Indeed, it is a minor matter to change the eyepiece of a telescope. An eyepiece costs relatively little so it is common to have a few different eyepieces for a telescope to have different magnifications available. The obvious question is, "Why not always use the greatest magnification?" There are several answers to this question.

The first answer is that as the magnification increases, the ***field of view*** of the telescope decreases (**FIGURE 5-8**). Field of view refers to how much of the object is seen at one time. This is entirely reasonable, for if the object appears larger, not as much of it will be contained within the view of the telescope. When viewing the Moon, for example, you may wish to see all of it rather than just a portion. If so, you need an eyepiece with a long focal length, thus producing less magnification.

The other reasons why we do not always use the greatest magnification relate to the other powers of a telescope.

field of view The actual angular width of the scene viewed by an optical instrument.

Light-Gathering Power

Notice in the three views in Figure 5-8 that the more magnified the image, the darker it is. To see why this occurs, consider the region of the Moon we see in view (c). The light from this region was concentrated in a small segment of the image shown in view (a), but it is now more spread out in the more magnified view. That is, the same light that covers only part of the image in view (a) has to illuminate the entire image in view (c). The image is therefore darker. There are two ways to make the image brighter and still retain this magnification. If we are taking a photograph, a longer time exposure can be used. When we do this, we allow the light's effect to accumulate on the film over a longer time, and the film becomes more exposed. The other way is to capture more light from the Moon in the first place. This can only be done by using a larger objective, which brings us to the second power of a telescope: the ***light-gathering power***, which is often the most important power.

light-gathering power A measure of the amount of light collected by an optical instrument.

The light-gathering power of a telescope refers to the amount of light it collects from an object. Most objects that are observed with a telescope are very faint, and to obtain an image, we need to capture as much light from them as possible. This is true whether we are using the telescope for photography or for direct viewing.

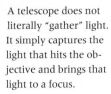 A telescope does not literally "gather" light. It simply captures the light that hits the objective and brings that light to a focus.

The major way to gather more light is to use a telescope with a larger objective. The amount of light that strikes the objective simply depends on its area. This is the primary reason that it is desirable to have a telescope with a large-diameter objective and why the size of the objective is one of the features specified when discussing a telescope. For example, we might describe a particular telescope as a refractor with a 12-centimeter objective of focal length 140 centimeters; however, although we state the *diameter* of the objective, it is the *area* that is important. Because the area of a circle is proportional to the square of the diameter, we must be careful in making comparisons of light-gathering power.

━━━━━ EXAMPLE ━━━━━

How does the light-gathering power of a telescope with a 6-inch objective lens compare with that of a telescope with a 9-inch objective?

SOLUTION   The area of a circle depends on the square of its diameter and, thus, the squares of the diameters of the telescopes must be compared to compare their light-gathering power. We set up a ratio of the squares of the two diameters:

$$\frac{9^2}{6^2} = \frac{81}{36} = 2.25.$$

The light-gathering power of the larger telescope thus is more than twice that of the smaller.

TRY ONE YOURSELF
Compare the light-gathering power of a small telescope, with an objective lens of diameter 10 centimeters, with that of a fully dark-adapted human eye with a pupil of diameter 5 millimeters.

The desire to be able to see fainter and fainter objects in space has led us to make larger and larger telescopes. Before discussing large telescopes, however, another advantage of size, the third and last power, must be considered.

Resolving Power

One property of light that we assume in our everyday life is that it travels in straight lines (unless it reflects from a mirror or is refracted at an interface). If we could not assume this, we would be unsure whether an object we see is in front of us or behind us. Yet there are exceptions to this rule: Light does not always travel in straight lines. The exceptions are usually unimportant, but look at FIGURE 5-9. This is a magnified photograph of the shadow of a regular household screw. The shadow was made by holding the screw a few meters from a screen and illuminating it with a small, bright light source. The edges of the shadow are not distinct; there are light and dark fringes near the edges.

In this case, the light passing near the edge of the object (the side of the screw) has "spread out" slightly. The effect is small and is difficult to see in everyday life. We do not discuss the reason that light acts this way except to point out that water waves behave similarly; they bend around corners. This phenomenon is called *diffraction*.

The amount of diffraction that occurs when light passes through an opening depends on the wavelength of the light and the size of the opening. The longer the wavelength, the more diffraction, and the larger the opening, the less diffraction.

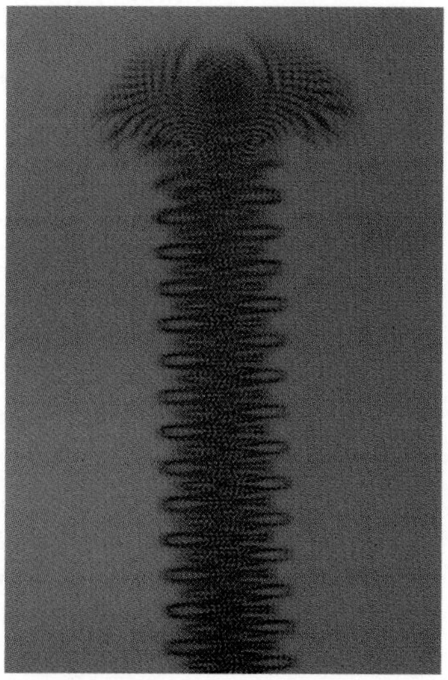

In a telescope, the objective itself forms the opening through which the light passes. The effect is small, but it is there: Light that should bend regularly and accurately according to the laws of refraction fans out a slight amount. This results in the image not being exactly clear. The image of a star that should appear as a single point is blurred into a small spot. If we increase the magnification, we simply make the blurred spot bigger and fainter.

FIGURE 5-10a is a photograph of the Big Dipper and indicates what seems to be a single star in the Dipper's handle. If you have fairly good eyes and look at the Dipper in a clear dark sky, you can see that this star is not one, but two stars (named Mizar and Alcor). We say that your eyes and the clarity of

diffraction The spreading of light upon passing the edge of an object.

FIGURE 5-9 If an extremely small light source is used, the shadow of an object will have light and dark fringes at its edge. This is due to diffraction of light. (A laser was used to make this photo, but laser light is not necessary to produce the effect.)

FIGURE 5-10 (a) Mizar and Alcor, stars in the handle of the Big Dipper, are seen as a single star by many people, but can be resolved into two stars by those with better eyesight. (b) Even a small telescope resolves Mizar into two stars (lower right). Alcor is at the upper left.

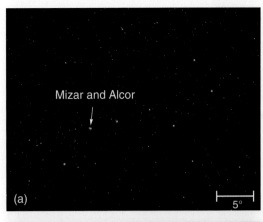

resolving power (or resolution) The smallest angular separation detectable with an instrument. It thus is a measure of the instrument's ability to see detail.

the sky allow you to *resolve* the pair of stars. Someone with poorer eyesight may be unable to resolve the pair. Figure 15-10b shows what you would see if you looked at this pair of stars with a small telescope. The photograph shows three stars, two of them very close together. In fact, the two stars that are close together were seen as one star (Mizar) when viewed with the naked eye; it was the brighter of the naked-eye pair. The fainter of the naked-eye pair, Alcor, is the third star in the photograph, at the opposite side of the field of view. The telescope is able to resolve the group of stars into three.

The ***resolving power*** of an instrument is the smallest angular separation between two stars that can still be resolved as two by the instrument. Resolving power thus is described in terms of an angle.

What is it about a telescope that determines its resolving power? Naturally, the quality of the lenses is a major factor, but even with perfect optical components, the resolving power of a telescope is limited by diffraction. Because less diffraction occurs with a large objective, the maximum resolving power can be achieved by a telescope with a large-diameter objective.

Assuming that diffraction is the only limitation, a telescope's resolving power can be calculated by using the expression

$$\theta \approx 2.5 \times 10^5 \times \frac{\lambda}{D},$$

where θ = angular resolution in arcseconds, λ = wavelength of the light used, in meters, and D = diameter of the telescope objective, in meters.

EXAMPLE

What is the resolving power of a human eye?

SOLUTION   Assuming that a fully dark-adapted human eye has a diameter of 5 millimeters = 5×10^{-3} meters, for visible light of wavelength λ = 600 nanometers = 6×10^{-7} meters, we get

$$\theta \approx 2.5 \times 10^5 \times \frac{\lambda}{D} = 2.5 \times 10^5 \times \frac{6 \times 10^{-7}}{5 \times 10^{-3}} = 30'' = 0.5' = 1/120 \text{ of a degree}.$$

TRY ONE YOURSELF
The *Hubble Space Telescope* has a 2.4-m primary mirror. (The expression for the resolving power of a telescope works for both refracting and reflecting telescopes.) Show that its resolving power for visible light of 600 nanometers is about 0.06" or 1/60,000 of a degree. This is about the equivalent of the angle subtended by a quarter from 80 kilometers (50 miles) away.

Based on size alone, the largest telescopes should have a resolving power far greater than the best backyard telescope, but in fact, the lack of clarity of the Earth's atmosphere becomes a major factor in limiting the resolution of large telescopes. This lack of clarity is caused by turbulence of the air and air pollu-

FIGURE 5-11 M-13 is a cluster of more than 100,000 stars about 23,000 light-years away orbiting the center of our Galaxy. The image at top right was obtained using adaptive optics (resolution of about 0.06 arcseconds), whereas the image of the same field at bottom right was obtained without adaptive optics under exceptional conditions (seeing of about 0.26 arcseconds). The field of view is about 20 arcseconds.

tion—the latter caused by modern civilization or simply by dust. Recall that light is refracted as it passes from one material into another if there is a difference in its speed in the two materials. In fact, light travels at slightly different speeds in air at different temperatures and densities. Our atmosphere always contains some amount of turbulence, and this causes air at various temperatures to move across the line of sight of a telescope. This results in the image moving slightly in nearly random directions, causing the image of a point source to become blurred. This blurring and twinkling of the image caused by the Earth's atmosphere is called **astronomical seeing**, and it is commonly measured by the diameter of the seeing disk (**"seeing"**). This is the diameter of the fuzzy image obtained while observing a point-like source (such as a star) through the Earth's atmosphere and corresponds to the best angular resolution that can be achieved. Atmospheric seeing places a limit on the resolution of even the largest telescope. The largest telescopes on Earth have a practical resolving power of about 0.5 arcseconds, improving to about 0.25 arcseconds on the best nights. To improve on the limit the Earth's atmosphere places on resolving power, astronomers place telescopes in space or use new observing techniques (such as adaptive and active optics and interferometry, which we discuss later in this chapter). **FIGURE 5-11** shows how such techniques improve the quality of an image.

This atmospheric turbulence is what causes the *twinkling* of the stars when they are viewed with the naked eye. Most planets do not seem to twinkle because their angular size is larger than the scale of the atmospheric turbulence, and the distortions are averaged out over the size of the image.

astronomical seeing The blurring and twinkling of the image of an astronomical light source caused by Earth's atmosphere.

5-4 The Reflecting Telescope

An inwardly curved mirror will bring rays of light to a focus, just like a lens (**FIGURE 5-12a**), allowing us to use such a mirror as the objective of a telescope. Figure 5-12b shows the arrangement devised by Isaac Newton. A small flat mirror is arranged in front of the objective mirror to deflect the light rays out to the eyepiece or camera body.

The largest refractor in existence is the 40-inch diameter telescope at Yerkes Observatory at Williams Bay, Wisconsin (**FIGURE 5-13**). Reflecting telescopes are made much larger than this, however, and they are less expensive. Astronomers prefer to build and use reflecting telescopes instead of refracting ones for the following reasons.

seeing The best possible angular resolution that can be achieved.

An inwardly curved mirror is said to be *concave* and, thus, the objective mirror of a telescope is concave. A mirror with an outward curvature—such as the passenger-side mirror on many cars—is said to be *convex*.

1. To be achromatic, a refractor requires two lenses. This means that four surfaces of glass have to be shaped correctly. A front-surface mirror, on the other hand, has

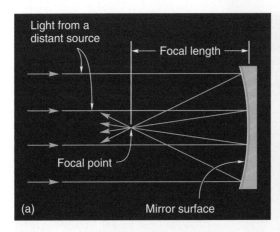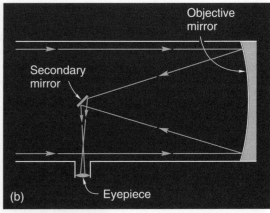

FIGURE 5-12 (a) A curved mirror can bring incoming light rays to a focus. Again, the focal point is defined as the point where incoming rays that are parallel to the axis of the mirror converge. (b) The Newtonian focal arrangement places a small flat mirror in the path of the reflected rays so that they are bounced off to the side and into the eyepiece (or camera or other instrument).

Newtonian focus The optical arrangement of a reflecting telescope in which a plane secondary mirror is mounted along the axis of the telescope to intercept the light reflected from the objective mirror and reflect it to the side.

FIGURE 5-13 The 40-inch diameter Yerkes Observatory refractor. The telescope tube is nearly 20 meters (60 feet) long, indicating that the focal length of the objective is about that long. The telescope was completed in 1897. The floor of the observatory can be raised so that the eyepiece is within reach.

only one critical surface, thus simplifying the construction of the objective. How "perfect" must the surface be? We typically want to keep any deviations from the desired shape of the surface to be less than about $\lambda/20$. For example, if we are observing in the visible part of the spectrum and set $\lambda = 500$ nanometers, any deviations must be kept less than 25 nanometers. Such a small deviation for, say, a 10-meter diameter surface, corresponds to requiring that the largest mountain—deviation from a flat surface—on Earth be smaller than about 5 centimeters (2 inches) in height.

2. It is impossible to correct lenses completely for chromatic aberration. When light reflects from a mirror, however, all wavelengths reflect in exactly the same way; thus, reflectors automatically eliminate chromatic aberration problems.

3. Because a reflector's mirror is front surfaced, the light does not pass through the glass of the mirror. The glass thus does not need to be as perfect as that used for a refractor. It is difficult (and therefore expensive) to make large pieces of glass without tiny air bubbles or other imperfections. In addition, a fraction of the light going through the lens of a refracting telescope gets absorbed, and this is especially true for short-wavelength light. Reflecting telescopes do not have this problem, as light simply reflects off the mirror.

4. For a refracting telescope, light must pass through the objective lens; thus, the lens must be supported from its edges. However, as the size of the lens is increased in order to collect more light, its weight increases, and gravity will deform its shape. In addition, this deformation will change with the orientation of the telescope. For a reflecting telescope, we can support the mirror's weight by a support structure, in a honeycomb form, behind the reflecting surface. This also allows the use of a computer-controlled active system of pressure pads that can slightly change the mirror's shape to accommodate for distortions caused by gravity or the atmosphere.

For all of these reasons and others, a large reflector is much less expensive and more practical than a large refractor. As a result, all really large telescopes are reflectors; however, even reflectors have some image distortions.

Large Optical Telescopes

A reflecting telescope with an eyepiece arrangement like that in Figure 5-12b is called a *Newtonian telescope*, or is said to have a **Newtonian focus**. The Newtonian focus is a common one for small telescopes. **FIGURE 5-14** shows an arrangement common in

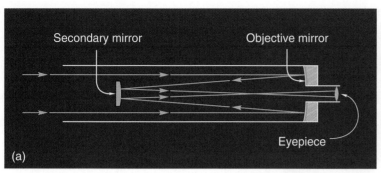

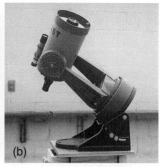

FIGURE 5-14 (a) In the Cassegrain focal arrangement, the secondary mirror is curved outward (convex), and the light is reflected back through a hole in the objective or primary mirror. (b) This is a 5-inch diameter Cassegrain telescope.

large telescopes, the **Cassegrain focus** (invented by G. Cassegrain, a French optician who lived at the time of Newton). The eyepiece or camera body is at the back of the telescope. In this arrangement, the secondary mirror is not a flat mirror but has an outward curvature. The effect of the curved secondary mirror is that an objective that actually has a short focal length can be given a longer effective focal length and thus can be contained in a short telescope.

In very large telescopes, observing is often done at the **prime focus**. This is the point where the light from the objective mirror comes to a focus, the focal point. FIGURE 5-15 shows the 200-inch (5-meter) Hale telescope on Mount Palomar. The observer's "cage" is at the prime focus of the instrument, located at the top end of the telescope. The astronomer can sit in the cage and be carried around with the telescope as it moves. The observer's cage does block some light to the objective mirror, but only a small fraction of it.

Losing some light because of reflection from the objective mirror back along the direction of incoming light is a drawback of reflecting telescopes; however, if the secondary mirror or the observer's cage is small enough, only a fraction of the incoming light is lost. Consider the case of a 5-meter objective mirror and a 1.5-meter cage. The amount of light received is proportional to the area of the objective mirror, while we lose an amount of light proportional to the area of the cage. Both areas are proportional to the squares of their diameters and, thus, we lose about $1.5^2/5^2$ or about 10% of the incoming light.

FIGURE 5-16 shows another design, the **Coudé telescope**. This is a useful design for the case when large and heavy equipment, such as a spectroscope, is used for observing through the telescope. The equipment is set at the focal point, which is outside the main telescope tube and thus does not exert a weight on the telescope that must be compensated for. This design also increases the focal length substantially, which can be very useful for high-resolution observations.

New telescope technology has produced telescopes larger than the Hale telescope. The Tools of Astronomy box on p. 139 describes how a laboratory at the University of Arizona is making mirrors in a radically different way than ever before. The world's most powerful single telescope is currently the Large Binocular Telescope on Mount Graham, Arizona. The telescope (the result of a collaboration between institutions in the United States, Italy, and Germany) consists of two 8.4-meter mirrors on a common mount, making it equivalent in light-gathering power to a single 11.8-meter mirror (because $2 \times 8.4^2 = 11.8^2$).

Cassegrain focus The optical arrangement of a reflecting telescope in which a convex secondary mirror is mounted so as to intercept the light reflected from the objective mirror and reflect the light back through a hole in the center of the primary.

prime focus The point in a telescope where the light from the objective is focused. This is the focal point of the objective.

Coudé focus The optical arrangement of a reflecting telescope in which two mirrors are used to reflect the light coming from the objective to a remote focal point.

FIGURE 5-15 The Hale telescope on Mount Palomar. The objective mirror is right above the two people pointing at it. The observer's cage is at the top left of the image.

FIGURE 5-16 In the Coudé design, light reflects off three mirrors before it exits the telescope.

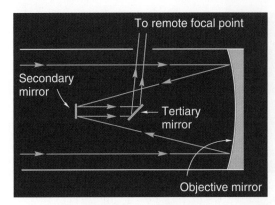

*And now our hundred inch...
I hardly dare to think what
this new muzzle of ours may
find. Come up, and spend
that night among the stars.*
—British poet Alfred Noyes,
in *Watchers of the Skies*, written during the first observations with the 100-inch
Mt. Wilson telescope, Nov.
1, 1917.

active optics A tech-
nology that works by
"actively" keeping a
telescope's mirror at its
optimal shape against
environmental factors
such as gravity and wind.
It works on timescales of
a second or more.

A new generation of extremely large ground-based telescopes is under study. The first such telescope to start construction is the Giant Magellan Telescope, to be completed by 2016 at a site in northern Chile. It will have seven 8.4-meter mirrors and thus the light-gathering power of a single 22-meter telescope.

To achieve the best viewing conditions, it is advantageous to locate telescopes high in the mountains in dry, clear climates. This greatly reduces but does not eliminate the blurring effects of the atmosphere. These effects are reduced even further by using new techniques, as described in the following section.

Active and Adaptive Optics

To collect more light, the objective mirror must be as large as possible, but the problem with making a large mirror is that it tends to bend and sag when the telescope moves. As a result, the mirror's changing surface destroys image quality. To avoid this problem, new large reflecting telescopes use very thin mirrors together with an array of actuators (little motor-driven pistons) behind the mirror that keep it in an optimal shape. This new technology is called ***active optics*** and works as follows: Sensors on the mirror send information about the mirror's shape to a computer, which then drives the actuators to keep the mirror in perfect shape as it moves or is affected by the wind.

The first fully active telescope was the 3.6-meter New Technology Telescope of the European Southern Observatory (ESO); it entered into operation in 1989 at La Silla, Chile, and served as a test case for ESO's Very Large Telescope (VLT) (**FIGURE 5-17a**). The mirrors of VLT's four 8.2-meter telescopes are flexible, and the image they produce is constantly monitored by a computer. To keep the image sharp under different conditions, the computer controls about 200 actuators that push and pull on the back of each mirror to adjust its shape. When the four telescopes work in combined mode, their total light-gathering power is that of a 16.4-meter single telescope.

Active optics is also used for the Keck telescope (operating since 1993) on the 4200-meter summit of Mauna Kea in Hawaii. The Keck I telescope (Figure 5-17b)

FIGURE 5-17 (a) The four telescopes of the VLT in Paranal, Chile. VLT's useful wavelength range extends from the near ultraviolet up to 25 μm in the infrared. (b) The 36 mirrors of the Keck I telescope are reflecting the underside of the prime focus cage, where the 1.4-meter secondary mirror is mounted. The hole at the center of the objective is for the Cassegrain focus. (c) A yellow laser beam is coming out of one of the four telescopes of the VLT. The Milky Way is shining on the left. One of the Magellanic Clouds, a satellite galaxy to our own, is visible on the right.

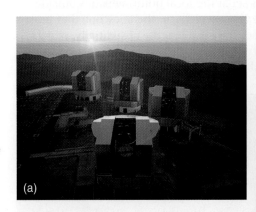

has a 10-meter aperture consisting of 36 hexagonal mirrors; each mirror is 1.8-meters across, weighs 880 pounds, is mounted separately, and is controlled by a computer.

Active optics works on scales of a second or more and differs from *adaptive optics*, which works on much shorter scales to compensate for the effects of atmospheric distortion. An adaptive optics system includes a sensor that measures atmospheric distortion on a timescale of a few milliseconds, a deformable mirror that is located in the path of the light from the source, and a computer that controls and reshapes the mirror so that the distortion is eliminated. But how does the computer know what the appropriate shape of the mirror should be? The computer must know both the distorted and undistorted image of the observed object.

One method to accomplish this is to use as a reference a bright star that is near the position of the observed object. Light from both the reference star and the observed object is affected similarly by astronomical seeing; however, we know that the undistorted image of a star should typically be a point that shows no details. From the distorted and undistorted images of the reference star, the computer calculates the effects of atmospheric distortion and uses that information to correct the image of the observed object. The problem with this method is that it is limited to only the study of objects that are near bright reference stars. An alternative method involves the use of a laser beam to generate an artificial star; for example, the naturally occurring layer of sodium atoms about 90 km (56 miles) above the Earth's surface can be excited by laser light at 589 nm and thus serves as a reference artificial star almost anywhere in the sky (Figure 5-17c).

Active and adaptive optics are now used in many of the world's large telescopes.

Telescope Accessories

Telescopes have many other astronomical uses besides obtaining images of celestial objects. Three of these applications and their corresponding accessories are described here.

- Even though a camera can be attached to a telescope to take photos, photographic film is not a very efficient detector of light, for only about 5% of the light hitting the film causes the chemical reaction that results in an image. For this reason, astronomers often use various electronic light detectors, particularly the *charge-coupled device (CCD)*.

    This device, about the size of a postage stamp (FIGURE 5-18a), is divided into small squares, each capable of detecting the intensity of the light that hits it. Just as a black-and-white newspaper photo (or the screen of an electronic game) is made up of many individual pixels, a CCD may have more than 4 million pixels. Figure 5-18b is a highly magnified CCD image in which individual pixels are visible.

    The intensity of light collected in each pixel of a CCD is stored as a number in a computer. The data can then be used to show what the object would "look like" in a regular photograph, or they can be used to produce a false-color image that reveals some other aspect of the object (Figure 5-18c). For example, a false-color image may illustrate the intensity of radiation from the object by showing each brightness level as a different color. Most modern astronomical "photos" are actually CCD images, not film-based photographs. CCDs have revolutionized the way astronomers count photons. These devices not only can count almost all the incident photons but also can detect a wide range of wavelengths. When viewing objects of very different intensities simultaneously, CCDs can differentiate between them. Finally, CCDs have a linear response; that is, the signal increases linearly with the number of incident photons.

- One important measurement made of celestial objects is the intensity of light that is received at various wavelengths, a procedure called *photometry*.

    In the past, this was done with a device similar to the light meter of a camera, but today, a CCD is normally used. The measurement can be made by placing filters in front of the light detector that allow only the wavelength of interest to pass through.

adaptive optics A technique that improves image quality by reducing the effects of astronomical seeing. It relies on an active optics system and works on timescales of less than 0.01 seconds.

charge-coupled device (CCD) A small semiconductor chip that serves as a light detector by emitting electrons when it is struck by light. A computer uses the pattern of electron emission to form images.

The devices used in digital cameras are similar to CCDs.

photometry The measurement of light intensity from a source, either the total intensity or the intensity at each of various wavelengths.

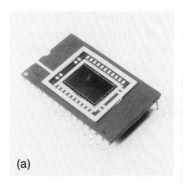

(a)

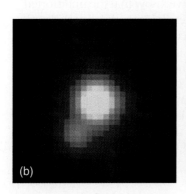

(b)

(c)

FIGURE 5-18 (a) A CCD (charge-coupled device) is a rectangle of the semi-conductor silicon. This one contains nearly 164,000 electric circuits that detect the intensity of light striking them. A computer analyzes the resulting data and produces an image. (b) This CCD image has been magnified so much that the individual pixels are obvious. (c) This false color image of the Moon not only emphasizes slight natural color differences, but also compresses the spectrum from the ultraviolet to the near infrared so that it all shows as visible. The image was made from images from the spacecraft *Galileo*.

spectrometer An instrument that measures the wavelengths present in electromagnetic radiation. (A **spectrograph** is a spectrometer that produces a photograph of the spectrum.)

diffraction grating A device that uses the wave properties of electromagnetic radiation to separate the radiation into its various wavelengths.

- In Chapter 4, we pointed out that a tremendous amount of information about celestial objects is obtained by spectral analysis—the examination of light that has been separated into its various wavelengths. To obtain data for such an analysis, a **spectrometer** is connected to a telescope. This instrument uses a prism—or, more commonly, a **diffraction grating**—to separate light into its colors. A spectrometer produces either a photograph of the spectrum or numerical data about the intensity of light at various wavelengths. The data can then be converted into a graph of intensity versus wavelength, as shown in **FIGURE 5-19**.

5-5   Radio Telescopes

Thus far the discussion has concentrated on optical telescopes—telescopes that gather visible light. Besides visible light, the type of radiation from space that best penetrates the atmosphere is radio waves. In 1932, Karl Jansky, working on radio transmission for Bell Laboratories, noticed that static received by his antenna originated in the Milky Way. (See the Historical Note box on page 143.) When better radio receivers were designed (largely during World War II), astronomers were able to pinpoint the sources of celestial radio waves, and the field of radio astronomy was born.

Two problems arise in examining radio waves from space. First, the intensity of radio waves from a star is much less than the intensity of visible light. Second, because the wavelengths of radio waves are a million times longer than the wavelengths of visible light, there is a corresponding de-

FIGURE 5-19 This drawing shows two different ways of representing a spectrum. The dark absorption lines in the photograph correspond to the dips in the intensity versus wavelength graph.

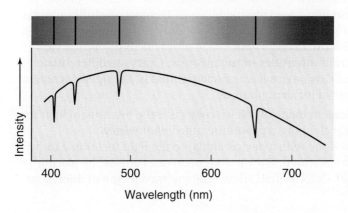

Intensity →

Wavelength (nm)

400   500   600   700

TOOLS OF ASTRONOMY

Spinning a Giant Mirror

The 5-meter (200-inch) mirror for the Hale Telescope was cast at the Corning Glass Works in 1934. It was nearly 60 years before a mirror this large was made again in the United States. Then, in 1992, the Steward Observatory Mirror Laboratory at the University of Arizona in Tucson transformed 10 tons of glass into a 6.5-meter mirror blank.

The Steward Laboratory is making mirrors in a radically different way than ever before. In the past, mirrors were made by melting glass to form large pieces with flat surfaces. Then, after the glass hardened, the working surface of the glass was carefully ground into a curve to form a concave mirror surface. For the 6.5-meter mirror, that would have meant starting with more than twice the weight of the finished mirror, with 12 tons of glass then to be ground away, taking a year of extra work. Under the stands of the University of Arizona football stadium, technicians of the Steward Laboratory form the mirror surface by spinning the mirror as the glass melts (**FIGURE B5-1**).

Rotating the furnace—a technique called spin-casting— saves time and money in the production of large telescope mirrors. It shapes a natural curve in the surface of the molten glass in the same way that swirling a liquid in a glass causes a curved surface. The curvature created this way is close to the mirror's final parabolic shape, the rotation speed determining the amount of curvature. To create the 6.5-meter mirror, the furnace was spun at 7.4 revolutions per minute. This may seem slow, like a merry-go-round or carousel, but remember that the mirror was 6.5 meters—21 feet—across.

The new mirrors use much thinner glass than in the Hale telescope mirror, making them much lighter. The 6.5-meter mirror has 64% more light-gathering area than the Hale telescope mirror, but weighs half as much. The glass facesheet of the mirror surface averages only a little more than an inch thick; strength is provided by a honey-comb structure cast all in one piece with the facesheet and extending back nearly 30 inches at the edge of the mirror and half that at the center.

FIGURE B5-1 In the spin-casting technique, molten glass is placed in a rotating furnace to form a mirror with a curved upper surface. Increasing the rotation rate increases the amount of curvature.

Another advantage to spin-casting is a deeply curved, parabolic surface with a focal length much shorter than conventional mirrors. The shorter focal length means that the mirrors require a much shorter, less expensive enclosure.

The first 6.5-meter Steward Observatory mirrors have been installed at the Multiple Mirror telescope on Mount Hopkins in southern Arizona and at the Magellan Project telescope on Las Campanas, Chile. Two 8.4-meter mirrors were also installed on the Large Binocular Telescope on Mt. Graham, Arizona.

crease in the resolution of images made with radio waves. (Diffraction is greater with longer wavelengths.)

Both of these problems are solved in the same way—by making radio telescopes extremely large. The spherical mirror of the world's largest radio telescope, located in Arecibo, Puerto Rico, is fixed on the ground; however, its various antennas, suspended on a track above the mirror, enable it to receive radio emissions from different parts of the sky. Shown in **FIGURE 5-20**, it is a telescope constructed by perforated aluminum panels supported by steel cables stretched across a natural bowl between hills. This telescope is 300 meters (1000 feet) in diameter and scans the sky as the telescope moves along with the Earth. Changes in the direction from which it detects radio signals can be achieved by moving its antenna, which hangs from the cables suspended above the bowl.

FIGURE 5-20 The radio telescope near Arecibo, Puerto Rico, is the world's largest. Its radio detectors are part of a 900-ton platform suspended above the reflecting surface.

Radio telescopes are similar in principle to optical reflecting telescopes and to the satellite dishes we use to receive television signals from Earth satellites. In each case the reflector simply directs waves to a small detector located at the reflector's focal point. You can see the supports for the detector in Figure 5-20.

At Green Bank, West Virginia, the first of a modern generation of radio telescopes was completed in 2001. In a regular radio telescope, the detector and its supports not only block out a small portion of the waves, but they also cause diffraction effects. The detector of the new Green Bank Telescope (GBT) is located off to the side (**FIGURE 5-21**). This is done by constructing the reflecting surface in the shape of part of a much larger reflector. The surface of the GBT is not symmetrical. It is a 100- by 110-meter section of an imaginary giant 208-meter symmetric radio telescope and consists of 2004 separate panels, each of which is computer controlled. This adaptive capacity, the first use of adaptive optics for radio telescopes, permits the surface to be adjusted as the metal supports flex because of the telescope's motion. The resulting surface accuracy allows the GBT to be useful at shorter wavelengths than would otherwise be possible.

The image formed by a radio telescope is not a normal photograph. The radio telescope simply detects the intensity of radio signals from the area of the sky toward which it is pointed. One way to display and examine the data received is to plot a graph of the intensity of the radiation as the telescope moves across a small portion of the sky. **FIGURE 5-22a** shows what such a plot might look like. A more complete image of the radio-emitting object can be obtained by scanning the radio telescope back and forth across the celestial object and feeding the data into a computer that is programmed to represent the various intensities of the radio waves as different colors. Parts (b) and (c) of Figure 5-22 were produced in this manner and indicate the intensity of radio waves from a small portion of the sky. **FIGURE 5-23** illustrates the different appearance of the same object (a galaxy) when seen in visible light, infrared, and radio waves.

5-6 Interferometry

The resolution formula is

$$\theta = 2.5 \times 10^5 \times \frac{\lambda}{D}.$$

FIGURE 5-21 The GBT is the first radio telescope to take advantage of adaptive optics. A computer controls each of the panels making up the surface of the reflecting dish. The GBT is the world's largest fully steerable radio telescope. The weight of the moving part of the telescope is 16 million pounds (7300 metric tons). The GBT is 485 feet (148 meters) tall—taller than the Statue of Liberty.

The resolution of a telescope depends on both the diameter of the telescope and the wavelength of the radiation. The greater the diameter of the telescope, the better the resolution (assuming "seeing" is not a factor), but the greater the wavelength, the poorer the resolution. Even though radio telescopes are very large, radio waves are so long that the best resolution from a single radio telescope is of the order of a number of arcminutes. A radio source the size of a star thus would still appear in a radio telescope to be a blur as large as half the diameter of the Moon.

The solution to this problem lies in the fact that a giant radio telescope would retain the same resolution if only two portions of its outer surface were being used, as shown in **FIGURE 5-24a**. Astronomers take advantage of this idea by combining two radio telescopes so that they act as one: In a sense, the two telescopes substitute for two portions of the outer part of a giant telescope, as in Figure 5-24b. In this way, they are able to obtain resolutions equal to that of a single large telescope.

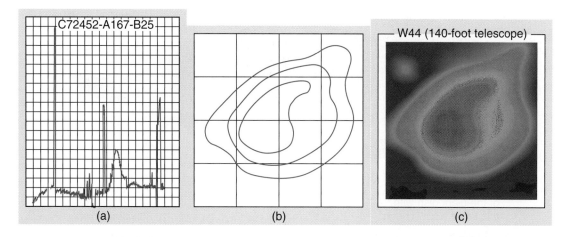

FIGURE 5-22 (a) A typical graph made by one scan of a radio telescope across a source. (b) A contour map can be made from a number of such scans. The strength of the radio signals is greatest in the center. (c) Color has been added to the contour to produce a false color image.

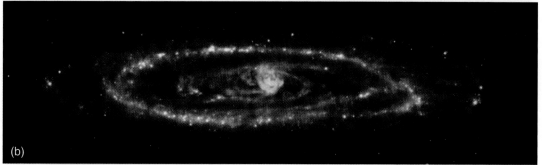

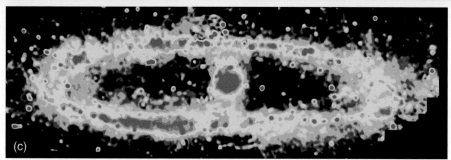

FIGURE 5-23 (a) The Andromeda galaxy, shown here in visible, is a close neighbor to our Galaxy. (b) This view of Andromeda from NASA's *Spitzer* infrared telescope combines images taken at 24 microns (blue, showing the warmest dust), 70 microns (green), and 160 microns (red, showing the coolest dust). (c) A radio image of Andromeda shows that most radio waves are emitted from its spiral arms and its center.

Using two telescopes to act as one is not a simple matter. To understand the problem, refer to **FIGURE 5-25**. Radio waves from a distant source are shown striking the dish of a radio telescope. They are reflected so that a single wave gets to the detector (located at the focal point of the dish) at the same time from all areas of the dish. This feature must be retained when two (or more) radio telescopes are used as one. Waves from each single dish must be combined in the correct relationship. We say that the waves from the telescopes "interfere" with one another when they combine; therefore, the technique of linking two (or more) telescopes so that they act as one is called ***interferometry***. If the telescopes are close enough that they can be directly con-

interferometry A procedure that allows several telescopes to be used as one by taking into account the time at which individual waves from an object strike each telescope.

FIGURE 5-24 Two small radio dishes (b) can be made to have the same *resolution* as a large radio telescope (a) that has a diameter equal to the distance between the two small ones. The *strength* of the signals detected will be much greater in the case of the single large telescope.

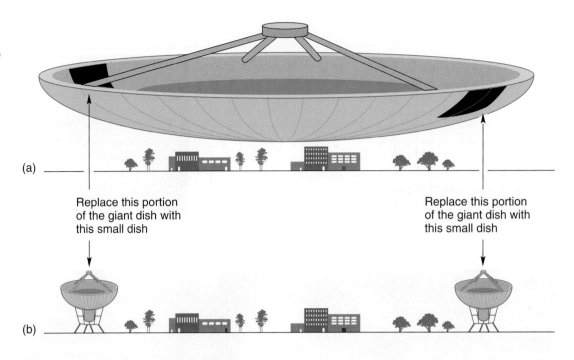

(a)

Replace this portion of the giant dish with this small dish

Replace this portion of the giant dish with this small dish

(b)

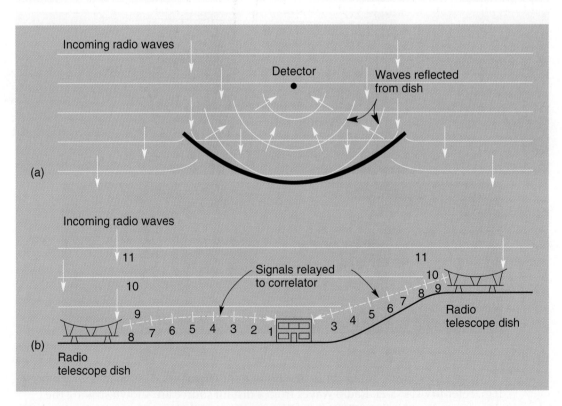

FIGURE 5-25 (a) All portions of an incoming wave reach the detector of a radio telescope at the same time. (b) If two radio telescope dishes are to function as a single telescope, the waves must likewise reach the detector at the same time, or at least the time difference must be corrected for. (This drawing shows an obvious and oversimplified case of waves whose difference in reception times must be accounted for by the correlator.) As the Earth rotates, it causes a change in the difference in path from the source to the telescopes; as a result, the signals from a point radio source alternately arrive in phase and out of phase producing interference fringes. Also, if the source has a finite angular size, then different parts of the source are at different distances from the telescopes. High-speed computers and appropriate software are needed to obtain images from interferometric data.

HISTORICAL NOTE

Radio Waves from Space

In 1927, Karl Jansky received his B.A. in physics from the University of Wisconsin. After 1 year of graduate study, he was hired by Bell Laboratories to do research on radio communications. By this time, the first telephone service across the Atlantic had opened, and the telephone signal was sent by radio. Static was a common problem, however, and Jansky was assigned the task of designing and building an antenna to find the source(s) of the static (FIGURE B5-2). He had no specific experience in radio or electrical engineering, but after much study and some dead ends, he built a 100-foot-long antenna with which he was able to detect weak radio signals and to determine the direction from which they came.

In January 1932, Jansky wrote in his monthly report that he detected "...a very steady continuous interference— the term 'static' doesn't quite fit it. It goes around the compass in 24 hours. During December this varying direction followed the sun." As the early months of 1932 passed, he found that the direction from which the signals came seemed to move around, getting farther from the Sun. He anticipated that after the summer solstice, the apparent source would move closer to the Sun again, but instead, it continued to move around the sky during the year. By August 1932, Jansky decided that the static was coming from a fixed place among the stars; it was some type of "star static."

Jansky's star static was the first detection of radio waves from space. The source of the waves was the center of our Galaxy. Jansky had been slow to recognize the celestial nature of the source in part because he did not know astronomy well. Astronomers, in turn, were slow to recognize the significance of his work because they were unfamiliar with electronics and radio, and they did not imagine that celestial objects emit radio waves. Radio astronomy did not grow quickly after Jansky went on to other things, but today it is a major branch of astronomy and provides us with information about the heavens that could not be learned in other ways.

FIGURE B5-2 Jansky's "merry-go-round" antenna could rotate to pinpoint the source of the radio signal.

nected, then the signals from the telescopes can be directly sent to the correlator. If the telescopes are widely separated, then extremely accurate atomic clocks are used for synchronization, and the signals from the telescopes are recorded on magnetic tape and then sent to the correlator.

In the New Mexico desert is an array of radio telescopes used for observations by interferometry. FIGURE 5-26 is a photograph of part of this Very Large Array. The telescopes ride on a double pair of railroad tracks arranged in a Y shape so that they can be moved and the arrangements changed.

The farther apart the telescopes, the better the resolution that can be obtained by interferometry. To achieve a longer baseline—distance between telescopes—some telescopes across the world are being used as part of a single array. For example, the Very Long Baseline Array consists of 10 radio telescopes across the United States (from Hawaii to the Virgin Islands), whereas the European Very Long Baseline Interferometry (VLBI) program includes an array of 18 radio telescopes that span the globe from Arecibo to China and from Finland to South Africa. Radio telescopes in space can be used as part of the VLBI program. Such long baseline programs achieve resolutions of fractions of milliarcseconds. An angle of one milliarcsecond is less than the angular diameter of a dime at 1000 miles!

Interferometry is also being employed in the newest optical telescopes. ESO's VLT is actually four separate telescopes, rather than mirrors, that focus light at the same point.

CHARA, at Georgia State University, has built an optical stellar interferometer, an array of six telescopes of only 1-m aperture each. They are placed in a Y-shaped array contained within a 400-m diameter circle on Mount Wilson, CA. Because each telescope is relatively small, this system costs much less than most new telescopes while producing a limiting resolution of 0.2 milliarcseconds in the visible range.

FIGURE 5-26 The Very Large Array is spread over 15 miles of the New Mexico desert.

To use the four telescopes as one, the signals from each must be combined using techniques similar to those originally developed for radio telescopes. Because the wavelength of visible light is much shorter than radio waves, matching waves from different sources is more critical for visible light waves. Therefore, optical interferometry is more difficult than radio interferometry. Using the four 8.2-meter telescopes together with four 1.8-meter auxiliary telescopes, the VLT can reach an angular resolution of less than a milliarcsecond in the visible, 50 times better than the *Hubble Space Telescope*, and provide astrometry at 10-microarcseconds precision, equivalent to that of a 130-meter telescope!

5-7 Detecting Other Electromagnetic Radiation

Visible light and radio waves pass through our atmosphere, and these two portions of the electromagnetic spectrum were the first that astronomers used in their quest to understand the heavens. (Examine again Figure 4-4, which shows atmospheric absorption at different wavelengths.) But celestial objects emit radiation over the entire range from radio waves to gamma rays, and modern astronomy has tools that study each region of the spectrum. (See the following Tools of Astronomy box where we describe a number of space telescopes.)

In the range of wavelengths from about 1200 to 40,000 nm (40 μm) (called the *near infrared*) are several narrow wavelength regions whose radiation penetrates to the surface of the Earth. Because water vapor is the chief absorber of radiation in the infrared, infrared observatories are located on mountains where the air is dry. The extinct volcano Mauna Kea is an ideal location for infrared telescopes, and two major observatories are located there.

> These wavelengths are called "near infrared" because they are the part of the infrared region that is near the visible portion of the spectrum.

Radiation in the far infrared region—wavelengths greater than about 40 μm—is emitted by cooler celestial objects, such as planets and newly forming stars, and is very valuable in the study of these objects. Far infrared does not penetrate our atmosphere as deeply as shorter wavelengths, and we must therefore locate our instruments higher to detect it.

As we move from infrared to wavelengths shorter than visible—shorter than about 400 nanometers—we come upon ultraviolet, X-rays, and gamma rays. Ozone is the chief absorber of most of this radiation, and our atmosphere has a layer of ozone at altitudes between about 20 and 40 kilometers. Therefore, telescopes designed to detect this range of radiation must be located in space.

In building X-ray telescopes we cannot use mirrors because X-ray light is very energetic, and unless it hits the surface of the mirror at grazing angles it gets absorbed by the mirror (**FIGURE 5-27**).

FIGURE 5-27 This drawing shows how the mirrors in X-ray telescopes collect and focus light.

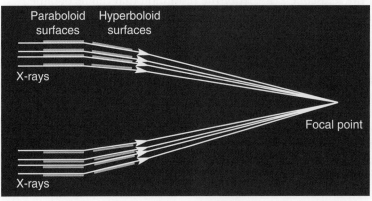

Just imagine what happens when you throw stones toward the surface of a pond; only the stones thrown at grazing angles will skip across the surface. That is why telescopes such as the *HST* cannot be used to observe objects in the X-ray part of the spectrum.

TOOLS OF ASTRONOMY

Space Telescopes

Infrared Telescopes

The SOFIA (Stratospheric Observatory For Infrared Astronomy) project involves a modified Boeing 747-SP aircraft carrying a 2.7-meter reflecting telescope at altitudes as high as 12,000 meters (45,000 feet), above 99% of the atmosphere's water vapor. The telescope's wavelength range is between 0.3 μm and 1600 μm. SOFIA, a joint NASA/German Space Agency effort, began operations in early 2009–2010 winter and is expected to be fully operational in 2014 and have a 20-year lifetime.

The *Spitzer Space Telescope* was launched in August 2003. It is a 0.85-m infrared telescope that observes the universe in the range of 3 to 180 μm. The telescope has a shield to protect it from the Sun and the Earth's infrared radiation, and it is cooled to a temperature only 5.2°C above absolute zero, by a tank of 360 liters of liquid helium, so that it can observe infrared signals from space without interference from its own heat. Infrared light can penetrate the regions in space that are filled with dense gas and dust clouds, giving us a unique view of star-forming regions, planetary systems, galactic centers, and cool objects too dim to be detected at visible wavelengths. **FIGURE B5-3** shows a *Spitzer* image of the "mountains of creation" in the W5 star-forming region.

The GALEX Ultraviolet Telescope

NASA's *Galaxy Evolution Explorer (GALEX)*, launched in 2003, is an orbiting 0.5-m space telescope that observes galaxies in

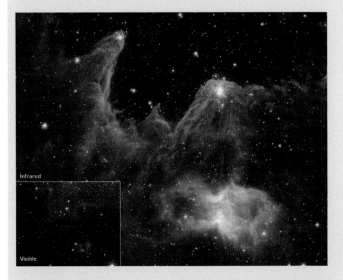

FIGURE B5-3 An image composite comparing *Spitzer's* infrared image of the W5 star-forming region to a visible-light picture of the same region taken by the Digitized Sky Survey (inset). The "mountains of creation" reveal dust glowing in the infrared because they have been warmed up by the stars inside them.

FIGURE B5-4 A composite image of the nearby spiral galaxy M101. The *GALEX* image (in blue) shows the presence of young stars (only 10 million years old) in the galaxy's spiral arm. The two Digital Sky Survey images in visible light trace stars that have been living for more than 100 million years (in green) or more than a billion years (in orange).

UV light (130–300 nm). *GALEX* is looking for clues on how the earliest galaxies evolved and probes the causes of star formation during the early period of our universe. **FIGURE B5-4** is an example of *GALEX*'s impact on the study of galaxy evolution by showing how we can trace the evolution of star formation in a galaxy such as M101.

X-Ray Telescopes

NASA's *Chandra X-ray Telescope* started operating in 1999 and has four nested mirror shells as shown in Figure 5-27. (The name "Chandra" was the nickname of the Indian physicist S. Chandrasekhar, who shared a Nobel Prize for his theoretical work on the structure and evolution of stars.) It is one of the most sophisticated observatories ever built and has an unusual elliptical orbit that takes it more than a third of the way to the Moon; its closest approach to Earth is 10,000 km. *Chandra* is designed to observe X-rays from high-energy regions of the universe, such as hot gas in the remnants of exploded stars. *Chandra*'s chapter opening image of the Crab Nebula is an example of how X-ray images provide different information than visible-light images of the same object.

ESA's *X-ray Multi-Mirror Satellite (XMM-Newton)*, launched in 1999, does not have the resolution of *Chandra* but can "see" further back in the universe's history because it can collect more light. It has 58 nested mirror shells, and its 0.3-m optical monitor with UV capabilities gives *XMM-Newton* a multi-wavelength capacity (**FIGURE B5-5**). Together with the joint US/Japan *Suzaku X-ray Telescope* (launched in 2005), these telescopes are already revolutionizing our un-

TOOLS OF ASTRONOMY

Space Telescopes *(Cont'd)*

FIGURE B5-5 A UV image of the M81 galaxy obtained by *XMM-Newton's* Optical Monitor. It covers a region of one quarter of a degree square; M81 is at least 22,000 light-years across.

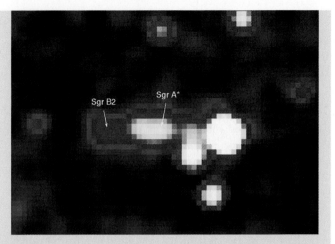

FIGURE B5-6 This is a false color image of the central region of the Milky Way. Sgr A* indicates the position of the supermassive black hole and Sgr B2 that of a molecular cloud at a projected distance of 350 light-years away. The cloud is currently being exposed to gamma rays emitted by the black hole 350 years ago during one of its active stages. Other gamma-ray emitters are also shown in the image.

derstanding of supernova remnants, active galactic centers, galaxy clusters, and black holes.

Gamma-Ray Telescopes

Since its 1991 launch and up to its deorbiting in 2000, the *Compton Gamma Ray Telescope* has provided high-quality data on subjects ranging from solar flares to supernova explosions to accreting black holes of stellar mass. The *Compton Observatory* was designed to map the gamma ray sky to solve some of the outstanding questions that earlier missions had posed and—perhaps most importantly—to watch for the unexpected. One of its most important findings was that gamma-ray burst sources are distributed in a manner consistent with a cosmological population.

The *International Gamma-ray Astrophysics Laboratory (INTEGRAL)*, launched in October 2002, is a collaboration involving ESA, NASA, and Russia. It carries four scientific instruments: an imager (giving sharp images of gamma-ray sources), a spectrometer (measuring gamma-ray energies), an X-ray monitor, and an optical camera (for identifying the sources). All four instruments observe the same region of the sky at the same time, allowing for a clear identification of the gamma-ray sources. *INTEGRAL* has already made enormous contributions to our understanding of extremely energetic phenomena such as supernova explosions and com-

pact objects such as black holes and neutron stars. **FIGURE B5-6** is an *INTEGRAL* observation of the central region of our Galaxy and reveals that the supermassive black hole (Sgr A*) at the center was much more active in the past.

The *Fermi Gamma-ray Telescope*, which was launched in 2008, is the result of an international collaboration between NASA, the U.S. Department of Energy, and institutions in Japan and Europe. It will investigate gamma-ray bursts, the composition of dark matter, and the acceleration mechanism responsible for the high speeds of outflows (jets) from black holes.

The Hubble Space Telescope

The *Hubble Space Telescope (HST)* is mainly an optical telescope but can observe across the spectrum from the near infrared to the near ultraviolet regions (115–2500 nm). The *HST* (**FIGURE B5-7**) has a modular design so that on subsequent shuttle missions astronauts can replace faulty or obsolete parts with new and/or improved instruments. This was fortunate because soon after its 1990 launch astronomers discovered that the objective mirror of the *HST* was slightly misshaped. The telescope has performed flawlessly after a repair mission in late 1993, during which corrective optics packages were installed. During the fourth repair mission in 2009, astronauts installed new gyros and batteries (which extend *HST*'s life through 2014), a new wide-field camera (which improves *HST*'s sensitivity by 10–30 times), a new spectrograph, and made necessary repairs to existing equipment.

TOOLS OF ASTRONOMY

Space Telescopes *(Cont'd)*

FIGURE B5-7 The crew of the Shuttle mission STS-82 took this photo of the *Hubble Space Telescope* after they had released it.

The *HST* has a 2.4-meter primary mirror and is roughly cylindrical in shape, 13.1 meters end to end and 4.3 meters in diameter at its widest point. Radiation enters the telescope through an opening below the open door at upper left in Figure B5-7.

It takes about 95 minutes for the *HST* to complete one orbit around the Earth. Only part of this time is spent observing, with the remainder spent on "housekeeping" functions, such as receiving command loads and sending data to Earth, turning the telescope to find a new target or avoid the Sun or Moon, and similar activities. To keep the telescope operating efficiently, commands are sent to the *HST* several times a day.

The *HST* is a unique instrument, a cooperative program of NASA and ESA, and an invaluable resource for all astronomers. The following are but a few of its contributions: It has uncovered evidence for the existence of supermassive black holes and evidence in support of the Big Bang theory, uncovered details of the processes involved in the formation of planets, stars, and galaxies, and is helping astronomers calculate the age of the universe.

The successor to the *HST* is the *James Webb Space Telescope (JWST)*. It is scheduled to launch in 2013 for a 5- to 10-year mission. Its range is from visible green light through the invisible midinfrared (0.6–28 μm) and has a 6.5-m primary mirror. The *JWST* will see objects 400 times fainter than those currently observed by large infrared telescopes. Its goal is to observe the first stars and galaxies in the universe and to give us a better understanding of the dark matter problem.

The shortest wavelength in the electromagnetic spectrum corresponds to gamma radiation. As a result, gamma rays are too energetic to be reflected by mirrors, making gamma-ray detection an integral part of astronomy only since the 1960s. New detectors, using advanced solid-state technology, are slowly catching up in resolution with detectors at other wavelengths, giving us an increasingly sharper view of the universe at high energies.

Conclusion

In this chapter, we showed how refraction and reflection of light allow us to gather radiation from dim stellar objects and focus it to form an image. We saw that the powers of a telescope include not only magnification, but also light-gathering power and resolving power. This analysis showed the importance of large telescopes and led to a discussion of reflecting telescopes, which can be made much larger than refractors.

The importance of nonoptical telescopes in our quest to understand the universe cannot be underestimated. They permit us to observe objects that are invisible to the eye yet emit vast quantities of electromagnetic energy. New and different telescopes—including space telescopes—have provided us with information that was impossible to obtain by other means. The Chapter 4 opening images clearly show the different appearances an object—in this case our entire Galaxy—can have when observed at different wavelengths. These differences provide us with an enormous amount of information that was not available from ground-based observations centered on visible light. As later chapters show, entirely new celestial objects have been discovered in recent years by the new generation of telescopes. Undoubtedly, telescopes of the future will continue to bring us new and unexpected results and open whole new areas of exploration in astronomy.

Galileo's telescope began a revolution in astronomy nearly 400 years ago. Today the *HST* and other new telescopes are producing a comparable revolution. We do indeed live in exciting times.

STUDY GUIDE

1. The best site for an optical telescope is a place where the air is
   A. thin and dry.
   B. thick and dry.
   C. thin and moist.
   D. thick and moist.

2. Which of the following features determines the light-gathering power of a telescope?
   A. The diameter of the objective.
   B. The focal length of the objective.
   C. The focal length of the eyepiece.
   D. [Two of the above.]

3. Which of the following features determines the resolving power of a telescope?
   A. The diameter of the objective.
   B. The focal length of the objective.
   C. The focal length of the eyepiece.
   D. [Two of the above.]

4. Which of the following features determines the magnifying power of a telescope?
   A. The diameter of the objective.
   B. The focal length of the objective.
   C. The focal length of the eyepiece.
   D. [Two of the above.]

5. When the magnification of a telescope is increased by changing eyepieces,
   A. the apparent angular size of the object is increased.
   B. the field of view is decreased.
   C. the brightness of the object is decreased.
   D. [All of the above.]
   E. [None of the above.]

6. Which of the following telescopes has the greatest light-gathering power?

Telescope	Focal Length of Eyepiece	Focal Length of Objective	Diameter of Objective
A	24 mm	150 cm	12 cm
B	6 mm	100 cm	8 cm
C	18 mm	125 cm	20 cm
D	12 mm	90 cm	6 cm
E	12 mm	100 cm	10 cm

7. Which of the telescopes in question 6 has the greatest magnification?

8. The objective of most radio telescopes is similar to the objective mirror of a reflecting optical telescope
   A. in being concave in shape.
   B. in its approximate diameter.
   C. in being made of Pyrex glass.

9. The field of view of a telescope is
   A. the range of distance from the telescope over which it is in focus.
   B. the particular object being viewed by the telescope.
   C. the range of practical magnifying powers for the telescope.
   D. the range of wavelengths that can be detected by a particular telescope.
   E. the actual angular width of the scene viewed by the telescope.

10. Why are achromatic lenses used in optical telescopes?
    A. They reduce diffraction.
    B. They reduce color fringing.
    C. They produce greater magnification.
    D. They allow more light-gathering power.

11. Which of the following choices puts a limit to the useful magnification of a given telescope?
    A. Diffraction of light.
    B. Redshift of distant objects.
    C. The limit to how well lenses can be made.
    D. Reflection of light from parts of the telescope.

12. The resolving power of a telescope is a measure of its
    A. magnification under good conditions.
    B. overall quality.
    C. ability to distinguish details in an object.
    D. [All of the above.]

13. The light-gathering power of a telescope is determined by
    A. the telescope's magnification.
    B. the diameter of the objective of the telescope.
    C. the clarity of the sky.
    D. [All of the above.]
    E. [None of the above.]

14. Radio telescopes need not have finely polished surfaces because
    A. we are not interested in detail in the radio image.
    B. the speed of radio waves is less than that of light.
    C. the speed of radio waves is greater than that of light.
    D. radio telescopes can be used during the day.
    E. radio waves have longer wavelengths than light waves.

15. The primary purpose of a typical radio astronomer's work is to
    A. look for signals from other beings.
    B. send out radio waves to other beings.
    C. send out radio waves to be reflected back from stars and galaxies.
    D. receive radio waves sent out by radio sources.
    E. [All of the above.]

16. The eyepiece of a telescope is primarily used
    A. to collect as much light as possible.
    B. as a magnifier.
    C. as a prism to break light into its component colors.

17. A 40-inch telescope has _____ times the light-gathering power of a 10-inch telescope.
    A. 4
    B. 8
    C. 16
    D. 40
    E. [Either A, B, or C, above, depending on the eyepiece used.]

18. The Keck telescope is
    A. a large ground-based optical telescope.
    B. an orbiting telescope.
    C. a single radio telescope.
    D. an array of radio telescopes.

19. The Hubble telescope is
    A. a large ground-based optical telescope.
    B. an orbiting optical telescope.
    C. a single radio telescope.
    D. an array of radio telescopes.
    E. an orbiting infrared telescope.

20. The Arecibo telescope is
    A. a large ground-based optical telescope.
    B. an orbiting optical telescope.
    C. a single radio telescope.
    D. an array of radio telescopes.
    E. an orbiting infrared telescope.

21. Interferometry is used to increase
    A. magnifying power.
    B. resolving power.
    C. light-gathering power.
    D. [All of the above.]
    E. [None of the above.]

22. Which of the following effects is reduced by using a larger telescope?
    A. Diffraction.
    B. Refraction.
    C. Reflection.
    D. Chromatic aberration.
    E. [All of the above.]

23. Define refraction, chromatic aberration, and diffraction.

24. How is an achromatic lens made? Describe the defect it corrects.

25. Define magnifying power. Why do we not always use the highest magnification available?

26. What is meant by light-gathering power, and what is it about a telescope that determines its light-gathering power?

27. Explain what is meant when we say that a certain telescope can "resolve" a particular pair of stars.

28. Sketch a Newtonian reflecting telescope, showing the relative positions of the primary mirror, the secondary mirror, the eyepiece, and the image produced by the objective.

29. About how large is the largest refracting (visible light) telescope? The largest reflecting (visible light) telescope? Explain why one type can be made larger than the other.

30. Why do we not have to correct the objectives of reflectors for chromatic aberration?

31. Why must radio telescopes be made so large? About how large is the largest?

32. Why are most research telescopes located on mountains?

33. What is interferometry, and what is its advantage?

34. What is the primary advantage of locating a telescope in space?

35. What is spectroscopy?

1. Define and distinguish between reflection and refraction, and explain how each phenomenon is used in telescopes.

2. Consider the following telescopes:

Telescope	Focal Length of Eyepiece	Focal Length of Objective	Diameter of Objective
A	24 mm	150 cm	12 cm
B	6 mm	100 cm	8 cm
C	18 mm	125 cm	20 cm
D	12 mm	100 cm	10 cm

   Which telescope has the greatest light-gathering power? Which one has the greatest magnification? What is the magnification of that telescope?

3. What determines the resolving power of a telescope? The magnification? The light-gathering power?

4. In the drawings showing light rays that come from very distant objects, the rays are represented as being parallel. If two rays come from a single point, how can they ever be parallel? A drawing may be helpful.

5. If the magnification of a telescope can be changed by changing eyepieces, why don't we use an eyepiece that will magnify objects thousands of times?

6. If radio telescopes use the principle of reflection, why do they not require a shiny reflecting surface?

7. Describe some special features of two recently constructed telescopes.

8. Why are the telescopes of the Very Large Array arranged so far apart? It would seem more convenient to have them in a tight cluster.

9. Write a report on the future plans that NASA has for space observatories. (You might begin with articles in *Sky & Telescope* or *Astronomy*. Also, visit NASA's Web page.)

10. Why are all large telescopes reflectors?

11. Some of the largest telescopes in the world are located on Mauna Kea in Hawaii, at an altitude of 4.2 km. What are some of the factors that astronomers consider when deciding the location for a telescope?

12. Suppose you and your neighbor each have a satellite dish to receive television signals from a satellite, but the signals from each dish are weak. Suppose that you decide to solve your problem by combining the signals from the two dishes to produce a signal twice as strong. Why won't this work well?

1. Suppose you have lenses of the following focal lengths: 30, 10, and 3 centimeters. If you wish to construct a telescope of maximum magnification, which two lenses would you use? Which would be the objective and which the eyepiece? What magnification would this telescope produce?

2. The pupil of your eye is the opening through which light enters. The maximum diameter of the pupil of a human eye is about 0.5 centimeters. How does the light-gathering power of two eyes compare with that of a telescope with a 10-centimeter objective?

3. The telescope at Mount Pastukhov in Russia has a 6-meter objective. Compare its light gathering power with that of the 5-meter Hale telescope on Mount Palomar.

4. To have four similar mirrors with the same light-gathering power as one 16-meter circular mirror, what must be the diameter of each of the four?

5. The *HST* has a 2.4-meter objective mirror. If we would like to use it to observe Pluto in the visible, could we distinguish any of Pluto's features? (Diameter of Pluto = 2390 kilometers. Closest distance of Pluto from Earth ≈ 38 AU, at which distance Pluto's angular size is about 0.08 arcseconds.)

6. In the text, we discussed the concept of using a space radio telescope in conjunction with ground-based radio dishes in order to increase the baseline. An 8-m space radio telescope is at a distance (from Earth's surface) of about 21,000 kilometers at apogee and 560 km at perigee. Observations are made at 18, 6, and 1.3 centimeters. What is the best possible angular resolution that we can achieve?

1. Making a Telescope

To build your own simple telescope, you will need two cardboard tubes that fit tightly together. You may even use kitchen paper tubes, as long as one tube fits tightly inside the other and you can roll the small tube in and out of the larger one. You will also need two converging lenses: the first, the objective, should be the size of the opening of the large tube, and the second, the eyepiece, should be about half the size of the opening of the smaller tube. Both lenses should have their flat sides facing out of the tubes and their curved sides facing in. The objective can be held in place with a piece of tape wrapped around the tube. To keep the eyepiece in place, you may use a foam ring (a piece of foam that fits inside the smaller tube and has its central region cut so that it plays the role of the eyepiece holder). By moving the smaller tube, you can focus on a specific object. Find the magnification of the refracting telescope you just built and use it to observe objects near and far.

2. Making a Spectroscope

To make a simple spectroscope you will need a small piece of diffraction grating, some black construction paper, tape, scissors, and a box. (Plastic diffraction grating is available for a few dollars from Edmund Scientific Co., http://www.scientificsonline.com.) The simplest box that works for this purpose is a toothpaste box, which has at each end a "tongue" that folds in and a pair of "ears" that cover the opening. Cut off the ends of the ears so that there is an opening about half an inch wide between them. Cut a piece of the black construction paper so that after rolling it and inserting it in the box it fits loosely inside. Also, cover the inside face of each tongue of the box with this paper by using some tape. The purpose of the paper is to create a dark interior, reducing extra reflected light. Punch a hole at the center of one tongue, and cover this hole with the small piece of diffraction grating by taping the edges of the grating on the inside face of the tongue. On the other tongue you will create a thin slit by making two parallel and closely spaced cuts along the tongue (from the tip of the tongue to the body of the box). Close the box. Tape around the edges of the tongues, and your spectroscope is ready. Use it to look at any light source by putting the end with the grating close to your eye. Compare the spectra you see when looking at an ordinary incandescent light bulb, a fluorescent light source, a mercury vapor streetlight and the sky. (**Warning: Never look directly at the Sun.**)

1. The book *Observing the Universe: A Guide to Observational Astronomy and Planetary Science* (2004, Cambridge Univ. Press) introduces a range of useful techniques and skills for those wanting to do observational work.

2. The article "Untwinkling the Stars" by R. Fugate and W. Wild in *Sky & Telescope* (May, 1994, p.24 and June, 1994, p.20) discusses adaptive optics.

3. The book *Great Observatories of the World* by S. Brunier and A-M. Lagrange (2005, Firefly Books Ltd.) includes a history of and the technology used in 56 professional astronomical observatories.

4. If you are interested in buying a telescope, you may want to read the article "Buying the Best Telescope" by A. Dyer in *Sky & Telescope* (December, 1997).

5. "2MASS: Unveiling the Infrared Universe," by M. Skrutskie, in *Sky & Telescope* (July 2001).

6. "Seeing in Infrared," by A. Swafford, in *Astronomy* (August 2002).

7. "Telescopes 101," by M. E. Bakich, in *Astronomy* (June 2005).

8. "Giant Telescopes of the Future," by R. Gilmozzi, in *Scientific American* (May 2006).

9. "Hubble's Top 10," by M. Livio, in *Scientific American* (July 2006).

10. "Window on the Extreme Universe," by W. B. Atwood, P. F. Michelson, and S. Ritz, in *Scientific American* (December 2007).

11. "Next Light: Tomorrow's Monster Telescopes," by J. Lowe, in *Sky & Telescope* (April 2008).

Quest Ahead to Starlinks
http://physicalscience.jbpub.com/starlinks

Starlinks is this book's online learning center. It features **eLearning**, which contains chapter quizzes and other tools designed to help you study for your class. You can also find **online exercises**, view numerous relevant **animations**, follow a guide to **useful astronomy sites** on the Web, or even check the latest **astronomy news** updates.

EXPANDING THE QUEST

STARLINKS

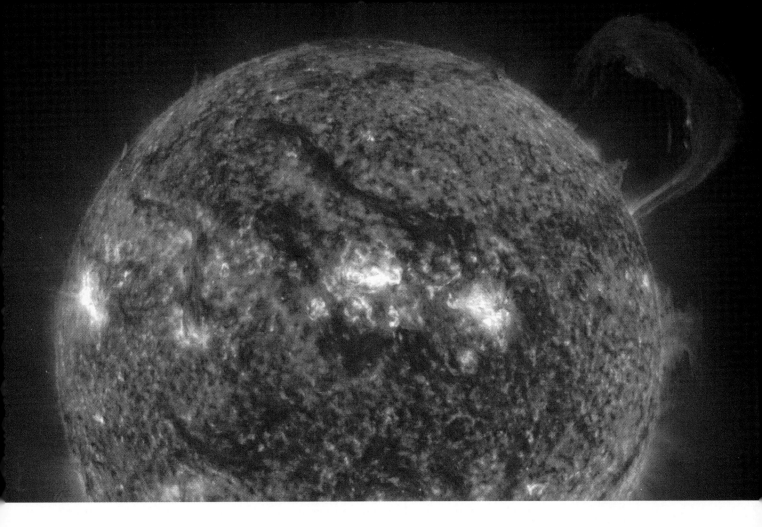

The Sun

THE CHAPTER-OPENING PHOTOGRAPH, WHICH WAS TAKEN FROM SPACE, shows prominences on the Sun's surface, giving us a hint of the tremendous energy within. Over the years, many well-known physicists tried to explain where all of this energy comes from. In 1854, Hermann von Helmholtz hit on the idea of gravitational contraction as the source of the Sun's heat. He calculated that the Sun could have existed for 25 million years this way, which was not long enough to match the findings of geologists of the time concerning the age of the Earth. In 1862, Lord Kelvin, one of the most imposing figures in science, refined the calculations to say that the Sun could have been shining for 500 million years, but no more. By this time, Darwin's theory of evolution was being debated all over Europe, and for evolution to be correct, the Earth (and Sun) had to be much older than this. Kelvin, however, obtained the same result in several versions of the same calculation, always basing his work on gravitational contraction. His only concession to the possibility of error was to say, "I do not say there may not be laws which we have not discovered."

There is such a law: radioactivity. Henri Becquerel announced his discovery of radioactivity in 1896, touching off a flood of research activity. One result was the work of young Ernest Rutherford, who showed that radioactive materials could produce large amounts of energy. This energy could provide additional heat in the Earth and

Prominences on the Sun are huge eruptions of gas that often form arches along magnetic field lines. The entire Earth could easily fit beneath one of these arches.

All cross references to chapters, sections, figures, and tables pertain to the main text, *In Quest of the Universe, Sixth Edition. In Quest of the Solar System* contains Chapters 1–11 and 19 of the main text. *In Quest of the Stars and Galaxies* contains Chapters 1–5 and 11–19 of the main text.

FIGURE 11-1 The Sun is the ultimate source of oil (including that pumped by this rig) and of all energy on Earth except nuclear and geothermal energy.

the Sun that would mean these objects are far older than Kelvin's results—billions of years older, in fact. Here is Rutherford's description of the address he gave on this topic at a meeting of England's Royal Institution:

I came into the room, which was half dark, and presently spotted Lord Kelvin in the audience and realized that I was in for trouble at the last part of my speech dealing with the age of the earth, where my views conflicted with his. To my relief, Kelvin fell fast asleep, but as I came to the important point, I saw the old bird sit up, open an eye and cock a baleful glance at me! Then a sudden inspiration came, and I said Lord Kelvin had limited the age of the earth, provided no new source [of energy] was discovered. That prophetic utterance refers to what we are now considering tonight, radium! Behold! the old boy beamed upon me. (From A.S. Eve, *Rutherford*, Cambridge, UK: Cambridge University Press, 1939, p. 107.)

In fact, we now believe that the source of energy in the Sun is not due to the energy of radioactive decay, but comes from reactions of nuclei in the hot, dense core. We discuss these reactions in this chapter.

In our quest to understand our universe, we started by first examining the basic tools we have for deciphering the information we get from celestial objects. Then we used these tools to understand the formation and evolution of our own planet and that of the remaining objects in our neighborhood—the solar system. We now study the Sun, the most important member of our solar system. Without it, life on Earth would not be possible. It is also the closest laboratory for studying how other stars evolve and, thus, it is a logical next step in our quest to understand the physical universe.

Although the Sun is the celestial object of most importance to life on Earth (**FIGURE 11-1**), it is just an ordinary star. The cosmic importance of the Sun is limited to the fact that it is the central object of the planetary system in which we live. Many other stars are bigger and brighter. Many are more interesting and unusual, and most will far outlive the Sun.

We begin this chapter with a brief overview of the major properties of the Sun. Then we examine the source of the Sun's energy, along with various theories for the production of that tremendous energy. Finally, we describe the Sun in more detail, starting at its center, where energy production takes place, and proceeding outward.

11-1 Solar Properties

FIGURE 11-2 If the Earth were at the center of the Sun, the Moon would orbit about half-way to the Sun's surface.

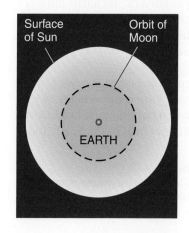

Surface of Sun

Orbit of Moon

EARTH

As viewed from Earth, the Sun has an average angular diameter of 31'59", just barely less than 32 minutes of arc. By taking 1.50×10^8 kilometers as the average distance from Earth to Sun and by using the small-angle formula (the relationship between angular size, distance, and actual size discussed in Section 6-1), we can calculate that the Sun's diameter is 1.39×10^6 kilometers, about 110 times Earth's diameter and about 10 times Jupiter's. **FIGURE 11-2** illustrates the great size of the Sun. (Activity 1 at the end of the chapter shows how you can measure the diameter of the Sun.)

From Kepler's third law (as revised by Newton), we calculate that the mass of the Sun is 1.99×10^{30} kilograms. This is about 333,000 times the mass of the Earth! The Sun's average

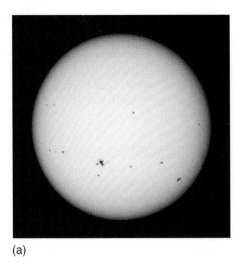

(a)

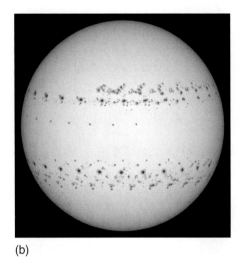

(b)

FIGURE 11-3 (a) Sunspots appear as dark spots on the Sun's surface. (b) This composite image shows the Sun's visible surface for most days of August 1999; the same sunspots appear many times as the Sun's rotation carries them across its surface.

density, then, is 1.41 g/cm³, about the same as the density of Jupiter. This is just the first of several similarities we will see between the Sun and some of the planets in our solar system.

Nearly 400 years ago, when Galileo first used a telescope to view the Sun, he observed dark spots moving across its face (**FIGURE 11-3a**). Figure 11-3b is a series of photographs of the motion of such *sunspots*. Galileo concluded from sunspot motions that the Sun rotates with a period of more than a month. Today we know that the Sun exhibits differential rotation; it rotates with a period of 25 days at its equator and nearly 35 days near its poles. Recall from Chapter 9 that the equatorial regions of the Jovian planets also rotate faster than their polar regions.

> **sunspot** A region of the Sun's surface that is temporarily cool and dark compared with surrounding regions.

11-2 Solar Energy

The Sun emits energy in all portions of the electromagnetic spectrum. A valuable piece of information about the Sun is the rate at which it emits its energy, or total power output. Fortunately, this is not too difficult to determine.

We start by measuring the rate at which solar energy strikes the Earth's atmosphere. This determination was made long ago by measuring the amount of energy from the Sun that strikes an area on the Earth's surface and then correcting for the energy absorbed by the atmosphere. Today it is done most accurately from satellites above the atmosphere. Measurements show that solar energy strikes the upper atmosphere of the Earth at the rate of about 1370 watts per square meter. This value is used in the following example to show that the Sun's *luminosity*—the energy the Sun radiates into space per second—is 3.85×10^{26} *watts*. (If the Sun were a lightbulb, this would be its "wattage.") This is an awesome amount of *power*. The energy that the Sun releases *in 1 second* is about the same as the energy released by the simultaneous explosion of 4 trillion atomic bombs! The solar energy received by the surface of the Earth, assuming we could collect and harness it efficiently, is enough to cover the energy needs of the entire world population 10,000 times over.

The example also illustrates nicely why the inverse square law applies to radiation from the Sun (or any distant object). The surface area of a sphere that is centered on the Sun depends on the square of the sphere's radius. A sphere twice the distance from the Sun thus has a surface area four times as great, and only one fourth as much energy strikes a square meter on the surface of this more distant sphere.

> **luminosity** The rate at which electromagnetic energy is emitted.

> **watt** A unit of power that corresponds to a specific amount of energy each second. (Imagine how dim a 1-watt light bulb would be.)

> **power** The amount of energy exchanged per unit time.

The Source of the Sun's Energy

It has been estimated that a 1% change in solar luminosity would result in a temperature change on Earth of 1°C or 2°C (about 2°F to 4°F). When we consider that the last major ice age on Earth resulted from a temperature decrease that averaged only

5°C across the planet, we see how critical it is that the Sun maintain a uniform rate of energy production. In reality, this rate is not uniform; the amount of solar energy that strikes the Earth varies with time, the variations over the centuries being of the order of fractions of a percent. The more we study the interaction between the Sun's energy output and our planet, however, the more it seems that even these small variations have consequences for life on Earth. What is the source of this energy, which must have remained approximately constant far into the past?

EXAMPLE

Calculate the energy output of the Sun in watts, given that solar energy strikes the Earth at the rate of 1370 watts/meter² and that the Sun is 1.496×10^8 kilometers from Earth.

SOLUTION   First, because we have expressed the area on Earth in square meters, we should also express the Earth-Sun distance in meters:

$$1.496 \times 10^8 \text{ km} \times \frac{1000 \text{ m}}{1 \text{ km}} = 1.496 \times 10^{11} \text{ m.}$$

Imagine a sphere around the Sun at the Earth's distance. We must calculate how many square meters are on that surface (**FIGURE 11-4**). To do this, use the equation for the area of a sphere, with 1.496×10^{11} meters being the radius of the sphere in the calculation:

$$\text{area of sphere} = 4\pi r^2 \quad (r = \text{radius of the sphere})$$

$$= 4 \,(3.14)\,(1.496 \times 10^{11} \text{ m})^2 = 2.81 \times 10^{23} \text{ m}^2.$$

Each of these square meters receives 1370 watts of solar power. Therefore,

$$\text{total solar power} = (1370 \text{ watts/m}^2) \times (2.81 \times 10^{23} \text{ m}^2)$$

$$= 3.85 \times 10^{26} \text{ watts.}$$

TRY ONE YOURSELF
How many watts of power strike one square meter of Mars' surface? (Mars is 2.3×10^{11} meters from the Sun.) Solve this by using the total solar power calculated in the example and the area of a sphere at Mars' distance.

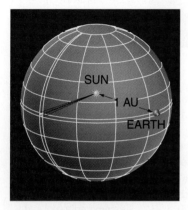

FIGURE 11-4 An imaginary sphere is shown drawn around the Sun at the distance of the Earth. It is a simple matter to calculate the area of such a sphere and to determine the solar energy that strikes each square meter of it.

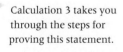 Given the power striking the Earth's surface, you could use the inverse square law of radiation, discussed in Section 4-8, to calculate this number.

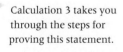 Calculation 3 takes you through the steps for proving this statement.

Before the 20th century, several hypotheses had been suggested to explain the source of the Sun's energy. All have now been rejected, based on additional data. For example, it was proposed that chemical reactions (such as the burning of a fuel) are the source of the Sun's energy. We now know that this cannot be the case, simply because if the Sun were made of a fuel such as coal or oil, it would burn out in a few centuries at the rate that it is releasing energy.

In the mid-19th century, Hermann von Helmholtz and Lord Kelvin proposed that the source of the Sun's energy is a very slow gravitational contraction. Such contraction compresses the gases inside the Sun, raising their temperature. This is similar to the air in a bicycle tire getting warmer when you compress it with a pump. When the Sun's gases got hot enough, they started radiating energy out into space. The calculations of Helmholtz and Kelvin showed that gravitational contraction could have produced the Sun's energy output with a reduction in the Sun's diameter of only a few tens of meters per year—so slight that it would not have been enough to notice in recorded history. Assuming that the Sun was formed from a large diffuse cloud, however, they calculated that gravitational contraction could not have started more than a few hundred million years ago. This time period seemed long enough in the 19th century, as the Earth was thought to be much younger. Their theory seemed to be a good one; it fit the available data.

Then, in the 20th century, geologists discovered that the Earth's age is not a few hundred million years, but rather a few *billion* years—10 times longer. The contraction theory had to be abandoned, and the source of the Sun's energy was again an open question.

In the first decade of the 20th century, as a result of Einstein's special theory of relativity, we started considering mass and energy as interconvertible. That is, one can be transformed into the other. In the late 1920s, it was hypothesized that this process could be the source of energy in the Sun. Then during the 1930s, Hans Bethe at Cornell worked out the theory that today explains how the Sun has produced its tremendous power for the past 4 to 5 billion years and how it will continue this production for another similar period of time.

Solar Nuclear Reactions

Recall from Chapter 4 that the Bohr model of the atom proposed that the atom consists of a nucleus surrounded by orbiting electrons (**FIGURE 11-5**). That nucleus is our focus now. An atom's nucleus makes up about 99.98% of the mass of the atom and consists of two kinds of particles: **protons** and **neutrons**.

Protons have a positive electrical charge and neutrons have no electrical charge. The number of protons in the nucleus determines what element the atom is. For example, if the nucleus of an atom contains 1 proton, that atom is necessarily an atom of hydrogen. If it contains 2 protons, it is helium; if 6, carbon; if 92, uranium. A nucleus of hydrogen, on the other hand, is not limited to a specific number of neutrons. Although most hydrogen nuclei contain no neutrons, some have one neutron, and a few have two.

It is important to distinguish nuclear reactions from chemical reactions. Chemical reactions, which we encounter in everyday life, involve atoms changing the ways in which they are combined with other atoms. When we burn paper, for example, the paper's carbon atoms combine with oxygen atoms from the air, producing carbon dioxide molecules. **FIGURE 11-6** illustrates this reaction and emphasizes that this is a *chemical bond*, formed by the sharing of electrons between the carbon atom and each of the oxygen atoms. The forces involved here are electromagnetic in nature. Forces responsible for the structure of the nuclei are not involved in chemical reactions.

Nuclear reactions, on the other hand, involve forces between nuclear particles; orbiting electrons are not part of these reactions. There are many types of nuclear

The relationship between the amount of energy (E) that can be created from a certain amount of mass (m) is $E = mc^2$. The conversion factor c is the speed of light in vacuum.

Bethe was awarded the Nobel Prize in 1967 for his theory that the source of the energy from stars is thermonuclear reactions from which hydrogen is converted into helium.

proton The positively charged particle in the nucleus of an atom.

neutron The nuclear particle with no electric charge.

Do not confuse the terms *proton* and *photon*; they are very different. Refer to Section 4-6 for a description of photons.

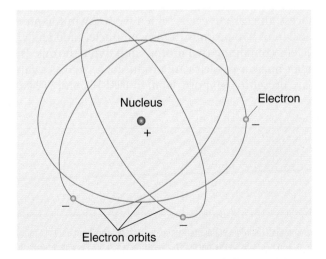

FIGURE 11-5 The Bohr atom, with electrons (which have a negative charge) circling the positive nucleus. If the atom were this size, the nucleus would still be too small to see. Drawings such as this one are used only as visualization tools. The orbits shown for the electrons correspond to the most probable distances from the nucleus at which electrons are observed.

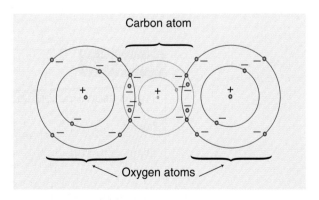

FIGURE 11-6 A carbon dioxide molecule (like all molecules) is held together by chemical bonds, in this case involving the sharing of electrons between two oxygen atoms and one carbon atom. Nuclear forces do not come into play in the bonding. Drawings such as this one are used only as visualization tools.

fusion (nuclear) The combining of two nuclei to form a different nucleus.

reactions, but only one will be of interest to us: the *fusion* reaction. In a nuclear fusion reaction, two nuclei combine to form a larger nucleus. They "fuse."

The core of the Sun is too hot to allow for complete atoms to exist. Instead, nuclei and electrons are separate from one another and bounce around at great speeds. The primary source of energy in the Sun (and in all stars during most of their lifetimes) is a series of nuclear fusion reactions in which four hydrogen nuclei are fused to form one helium nucleus. In the process, a small fraction of the mass of the nuclei is changed into energy. This is where Einstein's theory comes into play. Let's look at the process.

Most hydrogen nuclei consist simply of one proton. Before the fusion reaction, we have four hydrogen nuclei, and after the reaction, there is one helium nucleus. Let us subtract the mass of one helium nucleus (6.6447×10^{-27} kg) from the mass of four hydrogen nuclei (each having a mass of 1.6726×10^{-27} kg):

$$\begin{aligned} \text{mass of 4 hydrogen nuclei} &= 6.6905 \times 10^{-27} \text{ kg} \\ - \text{ mass of 1 helium nucleus} &= 6.6447 \times 10^{-27} \text{ kg} \\ \hline \text{difference} &= 0.0458 \times 10^{-27} \text{ kg.} \end{aligned}$$

The difference between the mass of the original matter and the resulting matter is very small, not only in terms of the actual amount (less than 10^{-28} kg per reaction), but also in that the "lost" mass (which has been completely converted to energy) is only seven tenths of 1% of the original mass of four hydrogen nuclei. Not even 1% of the mass is changed into energy. In fact, the energy produced by a trillion such fusion reactions is only enough to lift up this book by about a foot.

If the energy produced per fusion reaction is so small, how does the Sun produce an output of 3.85×10^{26} watts? The answer lies in the huge number of fusion reactions occurring in the Sun's core every second—about 10^{38} reactions per second. This implies that nearly 5 million metric tons of matter must be completely converted into energy each second. This involves the transformation of some 626 billion kilograms of hydrogen to about 622 billion kilograms of helium every second. Although this is a tremendous amount of matter by human standards, it is almost insignificant when compared with the Sun's total mass. If the Sun were originally pure hydrogen, it would take about 100 billion years for the Sun to convert its entire mass to helium at the present rate of consumption. (As we discuss later in the chapter, only the inner portion of the Sun—about 30% of its mass—is involved in the reaction and is converted to helium.)

In practice, the process by which hydrogen is converted into helium incorporates three steps, called the *proton–proton* chain. (This chain is the main fusion process in the Sun, responsible for 98.5% of the energy production. The remaining 1.5% is produced by a different process, the carbon cycle, which we study in Chapter 14.) The reactions start with a fusion of two protons (hydrogen nuclei) and end with the production of a helium nucleus containing two protons and two neutrons. During the process, two other smaller nuclear particles are produced, as well as a gamma ray. The solar fusion reactions are presented in more detail in **TABLE 11-1** and **FIGURE 11-7**.

Energy released as gamma rays (γ) interacts with electrons and hydrogen nuclei and heats the Sun's interior. This heating supports the Sun from collapsing under its own gravity.

proton-proton (p–p) chain The series of nuclear reactions that begins with four protons and ends with a helium nucleus.

TABLE 11-1

The Proton-Proton Chain

Reaction	Explanation
$^{1}_{1}\text{H} + ^{1}_{1}\text{H} \rightarrow ^{2}_{1}\text{H} + e^{+} + \nu_{e}$	Two protons combine to produce a deuterium nucleus ($^{2}_{1}\text{H}$; the 2 indicates the total number of protons and neutrons in the nucleus), a positron (e^{+}, which is a positively charged electron), and an electron neutrino (ν_{e}).
$^{2}_{1}\text{H} + ^{1}_{1}\text{H} \rightarrow ^{3}_{2}\text{He} + \gamma$	A deuterium nucleus joins with another proton to produce a helium nucleus ($^{3}_{2}\text{He}$, containing two protons and one neutron) and a gamma ray (γ).
$^{3}_{2}\text{He} + ^{3}_{2}\text{He} \rightarrow ^{4}_{2}\text{He} + 2 \cdot ^{1}_{1}\text{H}$	Two helium nuclei fuse to form the common type of helium ($^{4}_{2}\text{He}$) and two protons.

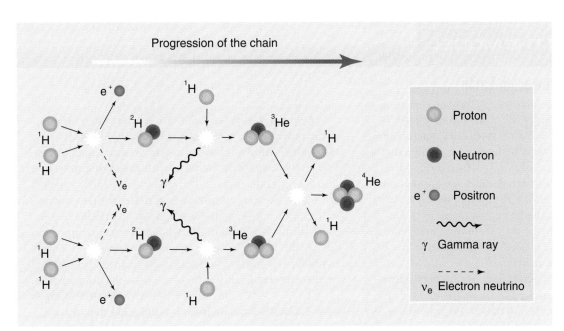

Progression of the chain

Proton

Neutron

e⁺ Positron

γ Gamma ray

νₑ Electron neutrino

FIGURE 11-7 The proton-proton chain begins with four protons and ends with a helium nucleus. The four protons are shown at left combining in separate reactions to produce two deuterium nuclei (each with a proton and a neutron).

Look at the first reaction in the table. Here, two hydrogen nuclei fuse to form the nucleus of another type of hydrogen, called *deuterium*, which has a neutron in its nucleus along with the proton. In addition, two other particles are formed, and these two particles fly away from the reaction at great speeds. One of them, the positive electron, or *positron*, combines with an electron and completely annihilates into gamma-ray photons. The other particle, the electron *neutrino*, escapes from the Sun and does not cause significant heating within the Sun. This particle is discussed later in this chapter.

In reality, only about 85% of the Sun's energy is produced by the chain described previously. The remaining 15% is produced by two other, similar chains, where the third step involves the temporary formation of the element beryllium before finally producing a helium nucleus. These additional two chains are important because they produce electron neutrinos with larger energies than the chain shown in Figure 11-7. In the next section, we discuss the importance of these neutrinos in our understanding of the interior of the Sun.

Hydrogen exists throughout the world. Yet it does not, on its own, fuse into helium. The reason for this is that all nuclei have a positive electric charge and, therefore, they repel one another. This electrical repulsion force acts over great distances, at least compared with the very short distances over which nuclear forces act. This means that the particles are unable to get close enough together for the attractive nuclear forces to take over unless they happen to be moving toward one another at a great speed (**FIGURE 11-8**). Because the temperature of a gas is determined by the speed of its particles, the particles of hot hydrogen are more likely to fuse than those of cool hydrogen. Significant fusion occurs only in high-temperature matter.

In addition, as we have seen, a great number of fusions of hydrogen nuclei must occur each second to produce the Sun's power. We thus know that the density of matter must be extremely high in the region of the Sun where fusion occurs. To see how and where this happens, let us investigate the Sun's internal structure.

deuterium A hydrogen nucleus that contains one neutron and one proton.

positron A positively charged electron emitted from the nucleus in some nuclear reactions.

neutrino An elementary particle that has little rest mass and no charge but carries energy from a nuclear reaction.

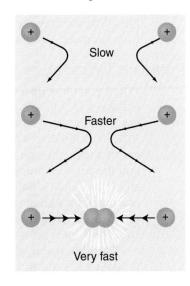

Slow

Faster

Very fast

FIGURE 11-8 Nuclei moving toward one another at too slow a speed will be repelled because of their positive charges; however, if they are moving fast enough, electrical repulsion will not be strong enough to prevent them from colliding and fusing.

ADVANCING THE MODEL

Fission and Fusion Power on Earth

The dream of unlimited energy has been with us at least since the beginning of the Industrial Age, and humans wondered at the tremendous power of the Sun long before that. We now know that nuclear fusion reactions are the source of the Sun's energy, and just a few decades ago humans harnessed this energy (if *harnessed* is an appropriate word here) in the hydrogen bomb. The bomb's name comes from the fact that it uses a form of hydrogen (deuterium) as its fuel. The nuclear reaction in the H-bomb is similar to that in the Sun; it produces helium from hydrogen.

Peaceful uses of fusion power have not yet been developed, although much research has taken place over the last 6 decades and is still in progress. A hint at the problems of controlling fusion can be seen by considering the tremendous temperatures and pressures that are necessary to produce fusion in the centers of stars. On the other hand, such reactions have an essentially unlimited supply of fuel—deuterium—and therefore, controlled energy production from fusion is a very attractive goal.

Although fusion is much more common than fission in the universe as a whole, humans developed fission first. Fission involves the release of energy when a large nucleus is broken into two medium nuclei, with mass being converted into energy in the process. The atomic bomb (poorly named, for it uses the energy of the nucleus rather than the energy of the outer atom) was the first application of fission power. Since the development of the A-bomb, we have learned to control this reaction, and today we use the energy of fission to produce electricity in nuclear power plants. In contrast to the ready availability of fuel for fusion, the uranium that must be used for fission power is definitely limited. Many people, scientists and nonscientists alike, question the wisdom of building and using fission power plants. Fission power must be viewed as only a temporary solution to the problem of finding a long-range source of energy on Earth. (See the references at the end of this chapter for more information on nuclear fission and fusion.)

11-3   The Sun's Interior

Obviously, we cannot examine the interior of the Sun directly; however, astronomers have learned much about its interior by computer modeling, based on observations of the Sun's surface and our knowledge about the behavior of matter at the temperatures and pressures necessary to sustain fusion reactions. We know that at temperatures as high as the Sun's, matter must exist as a gas rather than as a solid or a liquid, and this fact makes the analysis easier, for gases are much simpler than liquids or solids. As is discussed later, the temperature on the surface of the Sun is about 5800 K, and it increases greatly below the surface. At these temperatures, solids and liquids cannot exist, and most electrons are stripped away from their nuclei. As a result, most of the material of the Sun's interior consists of free nuclei and free electrons. The behavior of this material, however, is similar to that of a simple gas, and thus, we must first study properties of gases. The properties of importance to us are temperature, pressure, and ***particle density***.

particle density The number of separate atomic and/or nuclear particles per unit of volume.

Pressure, Temperature, and Density

When you blow up a balloon, the gas inside exerts a pressure on the rubber of the balloon and supports it against its tendency to contract. The pressure exerted by the gas is the result of collisions of the individual gas molecules with the rubber surface. FIGURE 11-9a illustrates this. Each molecule, as it strikes the rubber wall and rebounds, exerts a tiny force on the wall. Although we cannot detect each individual bounce and the corresponding force, the overall force exerted by the gas is simply the total of all of these individual tiny forces. To see this, imagine tiny grains of sand being fired at a board by a great number of sand throwers, as in Figure 11-9b. The force exerted on the board by a single grain of sand might seem negligible, but the overall result could be a force great enough to cause the board to move.

This discussion sometimes refers to force and other times to pressure. ***Pressure*** is defined as the amount of force exerted per unit area and might be expressed as

pressure The force per unit of area.

newtons per square meter or pounds per square inch. When we think of a single grain of sand or a single atom rebounding from something, it thus is more natural to speak of the force exerted by the particle. On the other hand, when we think of many sand grains or many atoms striking over a large area, we speak of the force exerted on each unit of area, or the pressure.

What then determines the pressure of a gas? There are two factors: the speed of the molecules of the gas and their particle density. To see that each of these factors is important, think again of the gas molecules in the balloon. If we heat this gas while keeping the balloon's volume constant, then the molecules will start moving faster. Each collision with the inside surface of the balloon will be more violent, exerting more force on the wall of the balloon. In addition, a greater speed results in more collisions. For these two reasons, the total force exerted on one square centimeter of the wall will be greater; that is, the pressure exerted by the gas will be greater. On the other hand, if the gas is cooled, it will exert less pressure. We thus can conclude that the pressure and temperature of a gas are related. More rigorous analysis shows that in fact a direct mathematical relationship exists between them, so that one changes proportionately to the other.

To appreciate the effect of particle density on pressure, imagine that twice as much gas at a given temperature is somehow put into the balloon without allowing it to expand. If this is done, twice as many molecules will be striking the inside walls of the balloon, exerting twice as much pressure. This is what causes the balloon to expand when more gas is added without restraining the volume. We thus can conclude that pressure and density of a gas are related. In fact, a direct mathematical relationship exists between them so that one changes proportionately to the other.

Pressure, density, and temperature thus are interrelated. If one changes, one or both of the others must change. Now let's turn back to the Sun and consider how these factors determine the character of the Sun's interior.

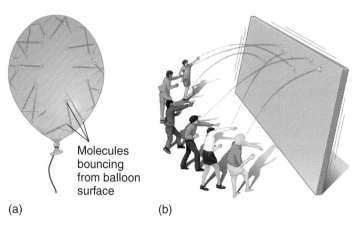

Molecules bouncing from balloon surface

(a)   (b)

FIGURE 11-9 The walls of a balloon (a) are held out by numerous collisions by molecules inside the balloon, in the same way that a great number of tiny sand throwers (b) could exert a force on a board and topple it backward.

One newton is a unit of force (about 0.225 lb, the weight of an average apple.)

Under ideal conditions, pressure is proportional to density times temperature.

Hydrostatic Equilibrium

The force of gravity holds the solar material to the Sun just as the force of gravity holds the atmosphere of the Earth near its surface. In this respect, the Sun is merely a big ball of gas. Earth's atmosphere is denser near the surface, not simply because the force of gravity is greater there than higher up, but also because the pressure exerted by the gas above compacts the lower layers. Gases lower in the atmosphere have to support the gases above. The same logic applies to the Sun. At any particular depth below the Sun's surface, the pressure of the gas at that point must be enough to support the gas above. It therefore is convenient to think of the Sun as having layers, like the various layers of an onion, as shown in **FIGURE 11-10a**. Keep in mind, however, that in the Sun there are no distinct boundaries between layers, but rather a continuous change as we move toward or away from the center.

Because the Sun is in a state of equilibrium (that is, neither noticeably contracting nor expanding), the pressure downward on any thin layer must be equal to the pressure exerted upward on that layer (Figure 11-10b). Knowing the total mass of the Sun, we can calculate the weight of gas above any particular layer and thus the pressure that is needed within the layer to support the gas above.

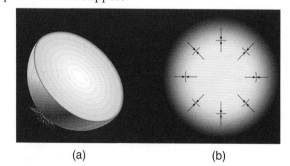

(a)   (b)

FIGURE 11-10 (a) You can think of the Sun as consisting of multiple layers, like those of an onion, except that there is no distinct boundary between the Sun's layers. (b) Within the Sun, the pressure upward on any layer must be the same as the pressure downward.

hydrostatic equilibrium
In a star or a planet, the balance between the downward pressure caused by the weight of material above a thin layer and the upward pressure exerted by material below.

◆ The density at the Sun's center is about 150 g/cm³, approximately 20 times the density of iron. The temperature there is about 15.6 million K.

The equilibrium conditions in the Sun are known as *hydrostatic equilibrium*. The name is almost self-explanatory: "Hydro" refers to the fluid state. This is basically just a more complex case of the situation we discussed with the inflated balloon. In that case, the stretched rubber holds the air inside in a compressed state. As long as the outward pressure of the compressed air inside is enough to support the inward pressure of the rubber, equilibrium is maintained.

Because the gas at the center of the Sun is supporting the weight of the gas all the way out to the surface, we should expect great pressures at the center. In fact, the pressure there is calculated to be about 2.5×10^{11} times that on the surface of the Earth. This tremendous pressure pushes protons close enough together that hydrogen fusion can take place. Only near the center of the Sun are the temperature and density of hydrogen great enough to support fusion. The solar core, where fusion is taking place, extends out to perhaps 25% of the radius of the Sun.

The fusion reactions in the core provide a heat source that must be taken into account when calculating conditions within the Sun. As we discussed earlier, when a gas is heated, it tends to expand. A balloon expands when its temperature rises, but then it stabilizes at a new equilibrium condition. Likewise, the Sun exists in a state of equilibrium, with the force of gravity balanced by forces tending to expand the gas.

To see how hydrostatic equilibrium works, imagine that the Sun could somehow be compressed artificially. Under compression, the pressure within the Sun would increase. The fusion rate would then increase, raising the temperature and pushing the Sun back out to another equilibrium position. As our discussion of the life cycle of stars in Chapter 14 explains, once the energy production of a star slows and the core cools, contraction of the core begins; however, as long as energy production is stable, the star remains in equilibrium.

To see what happens to the energy produced in the core of the Sun, we must look at energy transport within the Sun.

Energy Transport

conduction The transfer of energy in a solid by collisions between atoms and/or molecules.

convection The transfer of energy in a gas or liquid by means of the motion of the material.

We observe that energy is radiated from the Sun's surface; however, the fusion reactions in the Sun occur at its core, where the temperatures and pressures are greatest. The energy produced at the core must then be transported out to the surface in one (or more) of the three possible methods: conduction, convection, and radiation. These same three processes occur here on Earth and, indeed, everywhere in the universe.

If you put one end of a spoon on the burner (or in the flame) of a kitchen stove and hold the other end, you'll feel your end of the spoon gradually getting warmer. Energy is being transferred by vibration through the metallic crystal structure of the spoon. This method of transfer is called *conduction*. Imagine the atoms near the end of the spoon in the fire. As that end heats, the atoms vibrate at greater speeds (**FIGURE 11-11**); however, these atoms exert forces on adjacent atoms of the metal and cause those atoms to start vibrating faster. Gradually, the increased vibration spreads up the spoon until the atoms at the other end are also vibrating more rapidly than they were. In transferring energy by conduction, atoms do not move from one region to another, but vibrational energy—thermal energy—is transferred.

Conduction requires that the particles of the substance be in close contact, as are the atoms in a solid. In a star, this is not the case, except in some extremely dense stars, and thus, conduction is not a significant factor in transporting energy from within the Sun.

Convection occurs when the atoms of a warm fluid (liquid or gas) move from one place to another. Put your hand about a foot above a hot stove burner. You will feel hot air rising from the burner. This takes place because heated air is less dense than cooler air and, therefore, the hot, less dense air rises. The result is that energy is transferred upward from the stove by the motion of the hot gas. In forced-air central heating/cooling systems, the motion of the hot (or cool) air is caused by fans. On Earth, convection currents (thermals) enable hang gliders and gliding birds to travel long distances, carried along with the rising air.

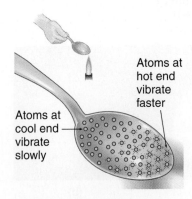

Atoms at hot end vibrate faster

Atoms at cool end vibrate slowly

FIGURE 11-11 The fast-vibrating atoms at the end of the spoon in the fire cause atoms next to them to vibrate faster. This continues until atoms at the far end are also vibrating fast, meaning that this end also becomes hot.

In a star, convection between adjacent layers is significant only when the temperature difference is great compared with the pressure difference. In the case of our Sun, this condition is met only in the region within about 200,000 kilometers of the surface (**FIGURE 11-12**). In this region, convection constantly mixes the solar material as hot gas rises and cooler gas descends. Deeper within the Sun, convection is almost inconsequential, and mixing does not occur to a great degree.

The final method of energy transfer is by *radiation*. If you hold the palm of your hand exposed to the stove burner, you can tell that your hand is being heated by another method besides rising hot air. To emphasize this, hold your hand off to the side, where it is not in the stream of hot air, and you will still feel your hand being heated. Radiation of energy occurs in all portions of the electromagnetic spectrum. Its effect on another object depends only on whether the receiving object absorbs the particular wavelengths of radiation emitted by the radiator. The air of your kitchen, for example, is transparent to most electromagnetic radiation produced by the stove burner, and thus, it does not absorb the radiation and is not heated by it directly. Your hand, however, being opaque to the radiation, absorbs it and is heated by it.

Inside the Sun and most stars, radiation is the principal means of energy transport. If the Sun were transparent, the electromagnetic radiation produced in the core would travel outward at the speed of light and reach the surface in about 2 seconds. In reality, the material in the Sun's radiative zone is too hot for atoms to exist and thus the radiation coming from the core scatters as it encounters free electrons and atomic nuclei. On the other hand, the material in the convective zone is cool enough for some atoms to exist (such as carbon, nitrogen, oxygen, calcium, and iron) and, thus, the radiation coming from the core gets absorbed and then reemitted, absorbed, reemitted, and so on, with a typical distance between successive absorptions of 1 centimeter (**FIGURE 11–13**). The reemissions occur in random directions, and as a result, energy that began perhaps as a gamma ray photon resulting from the proton–proton chain travels a very circuitous path and may take hundreds of thousands of years to reach the surface. This seems an impossibly long time for something traveling at the speed of light, but keep in mind both the multitude of scatterings, absorptions, and reemissions taking place and the extreme length of the roundabout path taken by the energy. As a result, any information about the Sun's core carried by the photons produced there is lost to us. This is similar to what happens when you forward a message by fax, using a fax you received from a friend, who did the same thing before you, and so on. The end result is that the final recipient will most likely not be able to read the message.

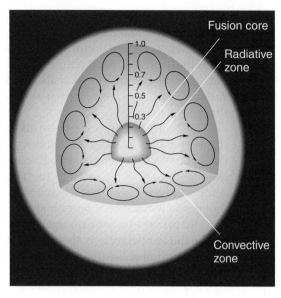

FIGURE 11-12 This figure illustrates the relative thicknesses of the three energy-transport zones of the Sun.

radiation The transfer of energy by electromagnetic waves.

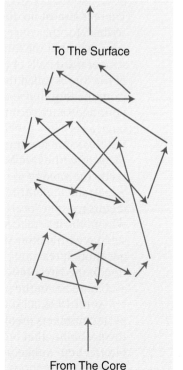

FIGURE 11-13 A photon is absorbed and reemitted numerous times as it travels from the Sun's core. Each reemission is in a random direction.

Can we tell what happens in the Sun's core by analyzing the light produced there?

Figure 11-12 shows the various portions of the interior of the Sun: the core, the radiative zone, and the convective zone. The temperature at the Sun's core, where the nuclear reactions occur, is about 15.6 million K, and the density is about 150 times that of water. Both temperature and density decrease as we move toward the radiative zone, which starts at about 25% of the distance from the center to the Sun's surface. The temperature here is about 7 million K, and the density is about 20 times that of water.

The radiative zone extends to about 70% of the distance to the surface, where the temperature is about 2 million K and the density is 0.2 times that of water. Across the thin layer between the radiative and convective zones (occupying the region from 69 to 71% of the distance to the surface), there is a change in the chemical composition and also the flow speed of the material. These changes in flow speeds are thought to generate the Sun's magnetic field.

The convection zone occupies the region within 200,000 km from the surface, where the temperature is about 5800 K and the density 2×10^{-7} times that of water. As we mentioned before, the presence of atoms in this layer makes it harder for the radiation from the core to escape; as a result, the trapped heat makes the material unstable and gives rise to convection currents, which carry energy quickly to the Sun's surface. After the energy of the Sun reaches the surface, it is again radiated outward. The energy from the Sun is released primarily as ultraviolet, visible, and infrared radiation, but it also comes in two less familiar forms: charged particles and neutrinos. As we explain in Section 11-5, the charged particles flowing outward from the Sun have observable effects on Earth. Neutrinos, however, are very difficult to detect because the probability of them interacting with matter they pass through is very low. This presents astronomers with a major problem.

Nevertheless, neutrinos can tell us about the *current* conditions in the Sun's core because they move very close to the speed of light in vacuum (c). (Until late in the 20th century, neutrinos were thought to be massless and thus moving at c. The first evidence for neutrino mass was seen in 1998. The latest experiments suggest that neutrinos have a very small mass, less than a billionth the mass of a proton; as such, their speed is very close to c.)

Solar Neutrinos and the Standard Solar Model

standard solar model
Today's generally accepted theory of solar energy production.

◆ Recall from Figure 11.7 that the proton–proton chain produces two electron neutrinos for each helium atom produced.

There is almost no doubt about the fundamental ideas of the solar model just described, for the concepts of pressure and density are well understood, and we are confident that the Sun's energy is produced by nuclear fusion; however, we are less sure of the details of the workings of the Sun's interior. The generally accepted theory of the Sun is called the ***standard solar model***; it predicts that so many neutrinos flow from the Sun that about 65 billion of them pass through every square centimeter of your body each second. These neutrinos do not affect your body because neutrinos have a very low probability of interaction with whatever matter they pass through, but this same low interaction rate makes them difficult for astronomers to detect. Measuring the number of solar neutrinos that reach the Earth allows us to check directly the validity of the standard solar model.

To shield neutrino detectors from cosmic rays and natural radioactivity, the detectors must be located far underground. Otherwise, the other radiation would overwhelm the few reactions caused by neutrinos. In addition, because neutrinos react so seldom with matter, a neutrino detector must contain a large amount of material to get enough reactions to detect. The world's first solar-neutrino detector began operation in the late 1960s; a number of other experiments have been conducted since, and more are in the planning process.

Until 1998, all solar neutrino experiments were finding 30 to 60% of the expected numbers predicted by the standard solar model. These results did not rule out the possibility that our theories concerning the detection of neutrinos were in error, but it is very unlikely that all experiments were wrong, as they all used very different detection techniques and were thoroughly tested. Also, the neutrinos produced in the Sun's core have a wide range of energies. The different detectors were sensitive to different energy ranges, and the measured neutrino deficit depends on the energy of the neutrino. The discrepancy between the observed and predicted numbers of neutrinos was commonly referred to as the *"solar neutrino problem."*

This discrepancy gave scientists two choices. The first choice was that the standard model was not correct. Many modifications of the model were proposed, but none gave a satisfactory solution to the "solar neutrino problem." Also, the model has been so successful in helping us understand the Sun, and therefore the stars in general, that in the absence of a better model there was no compelling reason to abandon it. When experimental data contradict the predictions of a theory, the theory could be wrong; however, it is also possible that there is some "new physics" that once understood could become an integral component of the existing theory. (Refer to our discussion on science, geocentrism, and heliocentrism in Chapter 2 and the evolution of our ideas on light, matter, and gravity covered in earlier chapters.)

The second choice was that during their 8-minute flight from the Sun's core to Earth, solar neutrinos oscillate between different types of neutrinos that the experiments at the time could not detect. (Indeed, according to the standard model of particle physics, neutrinos come in three types related to three different charged particles: the electron, and its lesser known relatives, the muon and the tau.) This is precisely what physicists Stanislaw Mikeyev, Alexei Smirnov, and Lincoln Wolfenstein proposed; however, for neutrinos to oscillate they must have mass. The so-called MSW theory was put to the test in experiments around the world.

As described in the Tools of Astronomy box on Solar Neutrino Experiments, the first evidence for neutrino oscillations was seen in 1998 by the Super-Kamiokande experiment. In early 2002, results from the Sudbury Neutrino Observatory (SNO) showed with great certainty that solar neutrinos change their type en route to Earth and that the total number of neutrinos observed by SNO is in excellent agreement with the number predicted by the standard solar model. Strong support for neutrino oscillations came in late 2002 from the KamLAND project and in 2006 by the MINOS experiment.

Neutrino oscillations are real, and they seem to explain the solar neutrino problem; however, so far we can only measure about 0.005% of the total number of neutrinos emitted by the Sun. The remaining neutrinos are difficult to detect because they are at lower energies. Additional experiments have been proposed in a concerted effort to solve this very serious problem.

There is at least one other possibility. The prediction of the total number of neutrinos produced by the Sun is based on the energy we receive from the Sun; however, as we have seen, the solar energy that comes to us as electromagnetic radiation requires millions of years to get from the core of the Sun to the photosphere where it is radiated away. The energy we receive had its beginning millions of years ago, but solar neutrinos from the core reach us in only about 8 minutes. Could it be that the core of the Sun has decreased its energy production since it released the electromagnetic energy we are now receiving? If so, then the change in solar activity will not have an effect on Earth until thousands or millions of years from now, when it will cause another ice age.

Understanding solar neutrinos is a very important problem in astronomy. At stake is not just the accuracy of a model for the Sun. If the standard solar model is accurate, which now seems to be the case, then we can be confident about our understanding of not only our Sun but of all stars. This obviously has ramifications for our understanding of galaxies and the overall evolution of the universe. The latest results suggest that neutrinos do have mass, and thus, being the most numerous entities in the universe other than photons, they contribute to a small degree to its overall mass and influence its evolution.

There is something wrong either with the Sun or with the neutrinos—or with what we think we know about them. John Bahcall, co-leader of the Homestake mine neutrino experiment

11-4 Helioseismology

In 1962, scientists discovered that the Sun is vibrating. Doppler shift measurements indicate that parts of the ***photosphere*** move up and down about 10 km with a period of about 5 minutes. Since then, many other vibration frequencies have been

photosphere The visible "surface" of the Sun. The part of the solar atmosphere from which mostly visible light is emitted into space.

TOOLS OF ASTRONOMY

Solar Neutrino Experiments

The world's first solar-neutrino detector (FIGURE B11-1a) began operation in the late 1960s in the Homestake Gold Mine in Lead, South Dakota. This detector, operated by a group from Brookhaven National Laboratory led by Raymond Davis, Jr., and John Bahcall, used 378,000 liters (100,000 gallons) of perchloroethylene (C_2Cl_4), a dry-cleaning fluid containing chlorine. When a solar neutrino strikes a chlorine atom with enough energy, a reaction occurs, transforming the chlorine into a radioactive isotope of argon that has a half-life of 35 days. The accumulated argon atoms were collected from the tank and counted, thereby revealing the number of neutrinos involved in the reactions. The results of experiments conducted from 1970 to 1995 suggest that, on average, one radioactive argon atom was created in the tank every 3 days. This corresponds to about 30% of the number of neutrinos predicted by the standard solar model.

In an effort to confirm or improve these results, other solar-neutrino experiments followed. The Kamiokande experiment operated from 1986 to 1995 in Kamioka, Japan. This detector had the ability to provide information on the direction of the neutrinos' travel. It confirmed that neutrinos do indeed come from the Sun, giving direct evidence that nuclear reactions occur in the Sun. Like the Homestake experiment, it found fewer neutrinos than theory predicts (about 50%).

The two detectors discussed thus far were sensitive only to high-energy neutrinos, which account for a very low percentage of the total neutrino production. Two other neutrino detectors, known as SAGE (Soviet American Gallium Experiment) and GALLEX, employ the element gallium, which is able to detect the low-energy neutrinos produced by the dominant reactions in the Sun's core. In these experiments, the interaction between solar neutrinos and gallium atoms produces radioactive germanium atoms, which can be collected and counted. SAGE (1990 to 2001) was housed in a tunnel under a mountain in the northern Caucasus, near a town called Neutrino City, and detected about 55% of the number of neutrinos predicted by the standard model. Statistically more substantial results came from the GALLEX experiment (1991 to 1997), which was located in a tunnel in Italy; it detected about 60% of the predicted neutrinos.

The Super-Kamiokande experiment, a joint Japan–U.S. collaboration and a continuation of the Kamiokande experiment, started operating in 1996 (Figure B11-1b). Super K is sensitive to electron neutrinos and muon neutrinos and can distinguish between them. In 1998 it found evidence for neutrino oscillations and thus for the possibility that neutrinos may have mass. It also found about 50% of the predicted neutrinos.

The Sudbury Neutrino Observatory (SNO), in Sudbury, Canada, is a Canada-Britain-U.S. collaboration that began operating in 1999 (Figure B11-1c). By early 2002, SNO scientists were able to use the unique properties of heavy water (where the hydrogen has an extra neutron in its nucleus) to measure the total number of neutrinos of all three types reaching their detector. They found that the number of electron neutrinos observed is only about one third of the total number reaching the Earth. This shows with great certainty

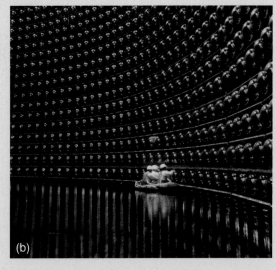
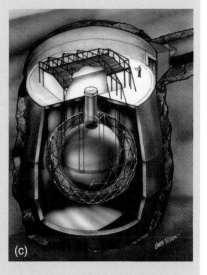

FIGURE B11-1 (a) The Homestake neutrino experiment located nearly a mile under the hills of South Dakota. (b) Scientists use a boat to check the detectors in the Super-Kamiokande neutrino experiment, which uses 50,000 tons of ultrapure water. (c) An artist's view of the SNO. The SNO detector consists of 1000 metric tons of ultrapure heavy water surrounded by 9600 light sensors, which detect tiny flashes of light emitted as neutrinos are stopped or scattered in the heavy water. The detection rate is about 1 neutrino/hour. It is the size of a 10-story building, 2 kilometers underground in Inco's Creighton Mine near Sudbury, Ontario.

TOOLS OF ASTRONOMY

Solar Neutrino Experiments *(Cont'd)*

that solar neutrinos change their type en route to Earth and arrive as a mixture of electron-, muon-, and tau-neutrinos. These results imply that neutrinos have mass and that the mass differences between the three types can be calculated. These masses are minute, and for all practical purposes, neutrinos can still be thought of as traveling at the speed of light in vacuum. The total number of neutrinos observed by SNO is also in excellent agreement with calculations of the nuclear reactions powering the Sun, suggesting that the standard solar model is quite accurate in describing the inner workings of the Sun. This agreement is very impressive if we consider the fact that the predicted number of neutrinos depends on the 25th power of the temperature at the center of the Sun. We can now calculate this temperature to better than 1%.

Strong support for neutrino oscillations came in late 2002 from the KamLAND project. This detector is located at the Kamioka mine in Japan and measures antineutrinos (the antimatter equivalent of neutrinos) from all 17 nuclear power plants in the country. Because matter and antimatter are mirror images of each other, studying antineutrinos produced by the fission reactions in the power plants should be the same as studying neutrinos produced by the fusion reactions in the Sun's core. In this case, the amount of nuclear material generating these neutrinos is well known, and thus, physicists can observe neutrino oscillations without making any assumptions about the properties of the source of the neutrinos. The results of this experiment clearly showed that neutrino oscillations are real.

In the MINOS experiment, operational since 2005, a beam of muon neutrinos travels underground from Fermilab (in Batavia, Illinois) to a particle detector in Soudan, Minnesota, 735 km away. In 2006, scientists announced that only 52% of the muons were detected, a clear indication of neutrino disappearance and therefore neutrino mass.

Additional experiments have been proposed in a concerted effort to detect the low-energy neutrinos generated in the Sun's core. For example, the main goal of the Borexino experiment, in Gran Sasso, Italy, is to observe neutrinos of specific energy created during the nuclear chain reactions involving beryllium. According to the standard solar model, this is the second most important neutrino production reaction after the basic proton–proton chain. The flux of these neutrinos is also predicted more accurately and is about a thousand times greater than the flux measured by Super-Kamiokande and SNO.

discovered, all taking place at the same time. **FIGURE 11-14a** is a computer simulation illustrating a high-frequency vibration. In this figure, the blue color represents regions where material is expanding at a given instant and orange represents areas where it is contracting. On the surface, the orange areas thus are moving inward and the blue areas outward. If this were a movie, in about 2 minutes the colors would reverse. The

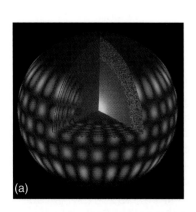

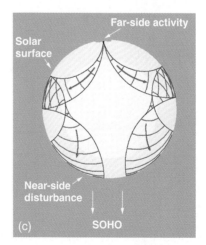

FIGURE 11-14 (a) A computer model of solar resonance that produces the observed vibrations of the photosphere. (b) An image of the actual solar surface showing oscillations; blue areas are moving toward us, and red areas are moving away from us. The motions are mainly radial (inward and outward), as shown by the fact that the signal is strongest near the center of the imaged disk of the Sun and weakest near the edge. (c) The *SOHO* satellite sees "through" the Sun by observing the disturbances at its near side as a result of the pressure waves generated by the activity (for example, a sunspot) at its far side.

pulsations are caused by waves, similar to sound waves, produced by pressure fluctuations in the turbulent convective motions of the Sun's interior. When the waves reach the photosphere, they reflect back toward the interior. These inward moving waves refract because of the changing physical conditions and eventually return to the surface. These trapped sound waves set the Sun vibrating in millions of different patterns. The combination of waves coming out and going back in produces a resonating effect, like a "gong."

Just as geologists use earthquakes to study the interior of the Earth, *helioseismology*—the study of vibrations of the Sun—is giving us insight into the Sun's interior. The Global Oscillation Network Group (GONG) has established a six-station network of telescopes around the world that obtain nearly continuous observations of the Sun's oscillations. The direct evidence we now get about the Sun's interior allows us to better test our theories of stellar structure and evolution and to find the role that magnetic fields play in the Sun's behavior. Also, a more accurate measurement of the helium abundance of the Sun put limits on cosmological models of the early universe and helped falsify a suggested explanation of the solar neutrino problem. Figure 11-14c shows how helioseismology has been used to see "through" the Sun in an effort to predict strong solar activity. In addition, it has provided us with information about flows from the Sun's equator toward the poles in the top 25,000 km of the convection zone, with the return flows being deeper in the zone. Finally, the rotation rate in the Sun was found to vary with depth and latitude; most of the variation occurs just below the base of the convection zone where we now think the Sun's magnetic field is generated.

helioseismology The study of the propagation of pressure waves (similar to sound) in the Sun.

11-5 The Solar Atmosphere

FIGURE 11-15 is a combination of six photos of the Sun, each taken by a different method and in a different portion of the spectrum. The yellow "pie wedge" at the top right position shows the Sun in visible light—the way it appears in a "regular" photograph. In this case we see what we call the Sun's "surface," the photosphere, although the Sun does not have a surface in the sense that the Earth does. The Sun's photosphere is the part of the Sun from which we receive visible light.

Figure 11-15 shows that we receive different information about an object depending upon the wavelength used in taking a photograph. The photo (particularly the "pie wedge" to the left) shows that there is material beyond the Sun's surface that is not visible to our eye. This is the solar atmosphere. It is convenient to divide the atmosphere into three regions: the photosphere, the chromosphere, and the corona. We now discuss each in turn.

Photosphere = "sphere of light."

limb (of the Sun or Moon) The apparent edge of the object as seen in the sky.

The Photosphere

The photosphere is a very thin layer (about 400 kilometers thick), meaning that we can see to that depth. This thickness implies that some of the light we receive from the Sun comes from one depth within the photosphere and other light comes from other depths. When we look at an edge (the *limb*) of the Sun, we see that it appears to be "darker" than the center of the solar disk (**FIGURE 11-16a**).

This limb darkening occurs because we see to a lesser depth as a result of observing the Sun at a grazing angle. Figure 11-16b shows this effect, which is important in that it allows us to analyze light from different depths within the photosphere and therefore to determine the temperature at different depths.

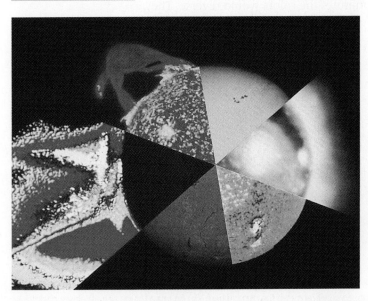

FIGURE 11-15 This figure was made by combining as separate wedges six different photos of the Sun. Each photo was made using a different portion of the spectrum.

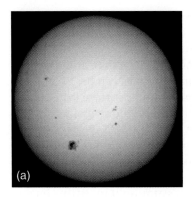

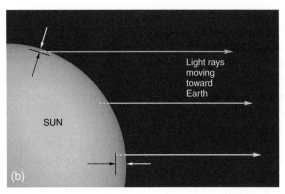

FIGURE 11-16 (a) The solar disk, the photosphere, is visible. Notice the limb darkening. (b) The light we receive from the center of the disk of the Sun originated at a greater depth than the rays we receive from near the edge. The three light rays travel an equal distance inside the photosphere; therefore, the one from the limb must originate in the upper photosphere, which is relatively cooler and thus glows less brightly. **Warning: Never look directly at the Sun**.

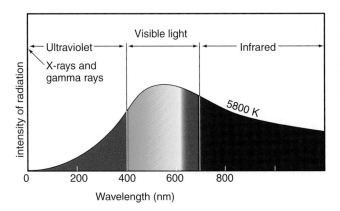

FIGURE 11-17 The intensity/wavelength graph of light from the Sun reaches a peak about the center of the visible portion of the spectrum.

We learn that the photosphere varies in temperature from about 6500 K at its deepest to about 4400 K near the outer edge. Overall, the light we receive from the photosphere is representative of an object whose temperature is about 5800 K. An intensity/wavelength graph of the radiation from the Sun (FIGURE 11-17) peaks near the center of the visible spectrum.

The pressure of the outer photosphere (calculated as for the inner layers, from knowing the gravitational force there and the amount of material above each layer) is only about 0.01 the pressure at the surface of the Earth. Knowing the temperatures and gas pressures of the photosphere, we calculate the density of particles there to be only about 0.0005 of the density of air at sea level on Earth (even though the gravitational field there is 28 times what it is at the Earth's surface).

When we observe the base of the photosphere (FIGURE 11-18a), we see irregularly shaped bright areas surrounded by darker areas, a constantly changing patchwork with individual regions appearing and disappearing with a period of a few minutes. Recall from the discussion of the intensity/wavelength diagram in Section 4-4 that a hotter object emits more radiation than a cooler object of the same size. The brighter areas of the Sun are brighter simply because they are hotter. This *granulation* of the

The solar spectrum peaks near the center of the electromagnetic region visible to us; as a result, it is reasonable to think that this is not a coincidence but that our eyes evolved so as to use the available radiation efficiently. However, see the reference at the end of this chapter on this issue.

granulation Division of the Sun's surface into small convection cells.

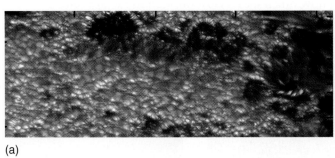

(a)

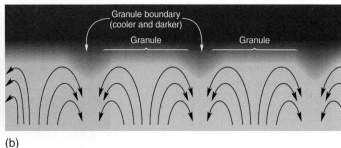

(b)

FIGURE 11-18 (a) This image (taken in 488 nm) of an active region near the solar eastern limb (top of the image) shows granules and sunspots (the dark areas). The tick marks are 1000 km apart. The three-dimensional nature of the photosphere is clearly seen in the elevated structures around the dark "floors" of the sunspots. Numerous bright "faculae" are also visible; these are granular structures slightly hotter than the surrounding photosphere and associated with strong magnetic fields. (b) Granules are seen where hot material from below the photosphere rises. Where it descends after cooling slightly, we see the darker edges of the granules. The flow can reach speeds of up to 15,000 miles per hour (6.7 km/s), producing sonic "booms" and other noise that generates waves on the Sun's surface.

Sun's surface is the result of convection. Granules are areas where hot material (the light areas) is rising from below and then descending (the dark surroundings). Figure 11-18b illustrates the effect and demonstrates that the photosphere is a boiling, churning region. A granule is about 1000 kilometers (600 miles) across, and thus, each granule covers an area about the size of Texas. Supergranules can be 35,000 kilometers across. Supergranules appear to move across the Sun's surface faster than the Sun rotates; however, this is an illusion similar to "the wave" done by fans at a sporting event.

Using methods described in Chapter 4, we find the chemical composition of the photosphere to be about 78% hydrogen and 20% helium (by mass). The remaining 2% consists of some 60 elements. All of these elements are known on Earth and occur in about the same proportions on Earth as in the Sun's atmosphere, with a few exceptions. The exceptions are of two types: (1) elements such as helium that have masses so low that they would have escaped Earth if they were once here in abundance and (2) elements found on Earth but whose characteristic spectra are such that they would not be detectable in the solar spectrum if the elements were as rare in the Sun as they are on Earth. As we pointed out in Chapter 7, this similarity of composition between objects as different as Earth and the Sun is not accidental, but results from the way the Sun and the solar system formed.

It is the composition of the Sun's atmosphere—not of the entire Sun—that we deduce from the solar spectrum. From our knowledge of nuclear fusion, we know that helium must be more abundant in the core of the Sun than in the atmosphere. Overall, the Sun is theorized to be about 73% hydrogen and 25% helium (by mass); this leaves only about 2% for the remainder of the elements.

The Chromosphere and Corona

The **chromosphere**, a region some 2000 kilometers thick lying beyond the photosphere, is not normally observable from Earth. It was first reported in the 17th century during a solar eclipse. It appears as a bright red flash, lasting only a few seconds, when the Moon has just covered the photosphere. During solar eclipses from 1842 to 1868, it was examined in more detail. Its spectrum was observed to be a bright line (or emission) spectrum because in viewing it we are seeing light from a hot gas with the dark sky behind it. Because the chromosphere is so much dimmer than the photosphere, it is only observed at the time of an eclipse, when the brighter portions of the Sun are blocked out (**FIGURE 11-19a**).

FIGURE 11-20 is a photograph of the chromosphere that was taken at a wavelength that allows us to see its structure. The **spicules** that can be seen shooting upward into the corona typically reach a height of 6000 to 10,000 kilometers and last from 5 to 20 minutes. Spicules occur periodically, every 5 minutes or so, at the same location. They are caused by sound waves of the same period that leak into the Sun's atmosphere and develop into shock waves that push these jets of plasma upward. Spicules play a very important role in how much mass is lost through the solar wind.

Today, the chromosphere and the region beyond it, the **corona**, can be observed by the use of a telescope that produces an artificial eclipse of sorts. With this instru-

The composition of the photosphere by volume is mainly hydrogen (91%) and helium (8.9%).

About 92% of the atoms of the Sun are hydrogen and 8% are helium. The Sun's composition changes slowly over time as hydrogen is converted to helium in the core.

chromosphere The region of the solar atmosphere between the photosphere and the corona.

Chromosphere = "sphere of color."

spicule A narrow jet of gas that is part of the chromosphere of the Sun and extends upward into the corona.

corona The outermost portion of the Sun's atmosphere.

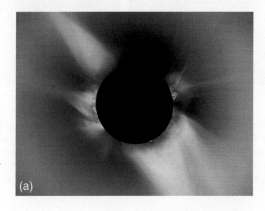 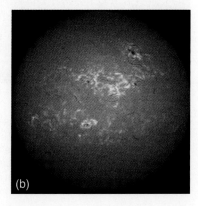

FIGURE 11-19 (a) At the higher temperatures in the chromosphere, hydrogen emits light that gives off a reddish color (H-alpha emission). This colorful emission can be seen in prominences that project above the limb of the Sun during total solar eclipses. (b) This is an image of the Sun taken with a filter that isolates H-alpha emission.

ment we can observe the Sun's atmosphere at various depths. We learn that as one moves outward from the photosphere, the temperature increases instead of diminishing, as we would expect. It is as high as 30,000 K in the outer portions of the chromosphere and continues to increase beyond the chromosphere into the corona, where it may reach 2 million K. This change in temperature occurs rapidly in a transition region of about 300 km between the chromosphere and the corona.

Most of the radiation emitted in these regions is in the X-ray portion of the spectrum rather than in the visible. (You can show this by using Wien's law.) As a result, you might think that these regions would be extremely bright because of their high temperature; however, the chromosphere and corona have a low density of matter and, thus, hardly any matter is available to glow. The corona's density is less than one trillionth that of the Earth's atmosphere. A simple example is to consider what would happen if you were to put your hand inside a hot oven. Even though the temperature is the same for the walls and air in the oven, you would get burned much easier by touching the walls—where the matter density is large and thus provides lots of thermal energy.

The reason for the high temperatures within the chromosphere and corona is only now becoming clear. After all, the corona cools rapidly, losing its energy as radiation into space and, thus, something must be pumping energy into it from below. The prevailing theory has been that the high temperature is the result of sound waves that are produced within the convective regions of the Sun and intensify as they pass outward until they are absorbed in the chromosphere and corona, heating these regions. Recent data, however, call this explanation into question, and it is now accepted that the heating is caused by an interaction between the Sun's magnetic field and its differential rotation. Continuous changes in the Sun's magnetic field result in pressure changes at the base of the field, which allow enormous jets of gas to escape from the Sun's interior into the solar atmosphere. As a result, some of the trapped, energetic sound waves (described in Section 11-4) escape the photosphere and transfer their energy into the chromosphere and corona. FIGURE 11-21, taken by NASA's *TRACE* spacecraft in 1999, shows "solar moss," a sponge-like feature that is associated with regions where there is strong magnetic activity. The moss consists of very hot gas (about 1 million K), occurs in patches as large as 20,000 kilometers (12,000 miles) in extent, and sometimes reaches 5000 kilometers (3000 miles) high above the Sun's visible surface. This observation gives us a glimpse of how the magnetic field of the Sun becomes increasingly organized as we move from the photosphere into the corona. The transition region is very dynamic.

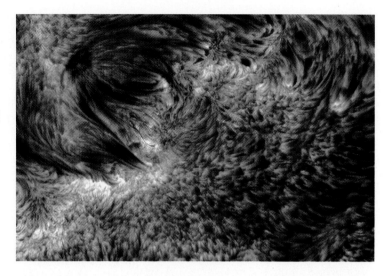

FIGURE 11-20 The short, dark features in the right half of this image are spicules, jets of gas shooting upward into the corona at speeds of 14 km/s (30,000 miles/hour) and then dying down within a few minutes. This 2003 image of an area toward the limb of the Sun was taken through a filter in H-alpha light (656 nm). Also visible in the upper left are some small sunspots connected by magnetic loops. The diameter of the spicules is about 500 km, and the entire area measures about 60,000 by 43,000 km on the Sun.

The word *corona* comes from the Latin word for crown.

A telescope designed to photograph the atmosphere of the Sun (when there is no eclipse) is called a *coronagraph*.

Launched in 2006, *Hinode* (Japanese for "sunrise") is studying the Sun's magnetic field and its role in powering the solar atmosphere. The mission is a collaboration between Japan, U.S. and U.K.

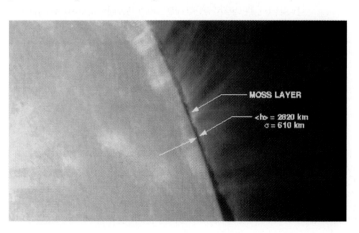

FIGURE 11-21 The "solar moss" (observed in the extreme UV) by the *Transition Region and Coronal Explorer* (*TRACE*) spacecraft.

MOSS LAYER

⟨h⟩ = 2820 km
σ = 610 km

FIGURE 11-22 This is a composite photo of the Sun and its corona taken in March 1988. The surface of the Sun is a combination of X-ray and visible-light images. The photo of the corona was taken during the March 1988 eclipse. Note the irregularity of the corona and how it streams outward to form the solar wind.

prominence The eruption of solar material beyond the disk of the Sun.

Yesterday I got a prominence photo good enough to prove the success of the method, and the result is that I am just now feeling pretty neat. George Ellery Hale, upon taking a photo of the Sun in the light from calcium vapor, in 1891.

solar wind The flow of charged particles from the Sun.

coronal hole A region in the Sun's corona that has very little luminous gas.

An additional process that contributes to the heating of the chromosphere and corona is *magnetic reconnection*; this occurs when oppositely directed magnetic field lines interact, releasing the energy stored in the magnetic field. Observations by *TRACE, SOHO*, and *Hinode* are providing new insights into how the Sun's magnetic field controls the release of mass and energy into the solar atmosphere.

The corona, a region extending for millions of kilometers from the Sun, has been observed during total solar eclipses for centuries, although many people used to claim that it was just an optical illusion caused by the sudden dimming of light as the Sun is eclipsed. **FIGURE 11-22** shows its extremely irregular appearance.

The photograph on the first page of this chapter and **FIGURE 11-23** show spectacular occurrences in the Sun's atmosphere. These are **prominences**, eruptions of solar material up into the chromosphere and corona. Some of these are relatively slow moving and remain fairly stable for as long as a few days. They may reach as high as thousands of kilometers above the photosphere. Some move much more quickly, ejecting material from the Sun at speeds up to 1500 km/s and reaching heights of nearly a million kilometers. Prominences are often associated with sunspots and the solar activity cycle to be discussed in the next section.

When we divide the Sun into regions, we must remember that the boundaries between them are artificial, for we have named them and distinguished between them on the basis of certain selected properties. For example, we consider that the process of energy transport is important, and we therefore talk of a radiative zone and a convective zone. If we emphasized another property, we might not make a division between these two parts of the Sun at all. In addition, although the boundaries between various regions appear sharp in our drawings, in reality, they are not as well defined. This is especially true of the outer limits of the corona, where the coronal material becomes the solar wind.

The Solar Wind

The **solar wind** is a continuous outflow of charged particles from the Sun, mostly in the form of protons and electrons. **FIGURE 11-24a** is an X-ray image of the Sun, in false color, showing the hottest, most dense regions as bright, and the cooler, less dense regions as dark. Such images indicate that X-ray emission is not uniform and that the active regions change appearances on a time scale of hours to days. The large dark area is called a **coronal hole** because it has very little luminous gas.

Coronal holes correspond to regions where the magnetic field lines are open, thus providing a corridor for charged particles to escape into space, generating the solar wind. The high coronal temperatures result in the wind particles escaping the Sun's strong gravity. These particles stream through space, taking mass away from the Sun, about 6×10^{16} kilograms every year. This corresponds to only a tiny fraction of the Sun's total mass.

Figure 11-24b shows that the solar wind is not uniform and that its speed is higher over coronal holes. Near the Earth, the solar wind normally travels at about 400 km/s and has a density of about 2 to 10 particles per cubic centimeter. Recall from Section 10-4 that one effect of the solar wind is that it causes comet tails to point away from the Sun as its particles sweep comet material along with them.

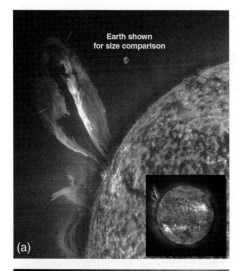

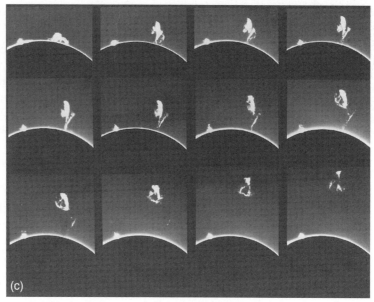

FIGURE 11-23 (a) A solar prominence. The Earth could easily fit under one of the loops shown in the picture. (b) Gas erupts in all directions from the Sun's surface in this photograph taken by the *TRACE* satellite. (c) This sequence of photos, taken from space, shows how this particular prominence progressed as charged particles were pushed from the Sun by its magnetic field.

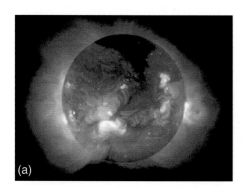

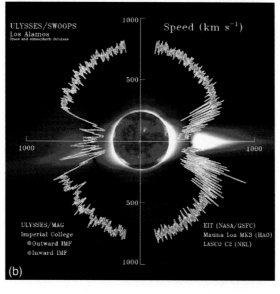

FIGURE 11-24 (a) An X-ray image of the Sun, in false color, taken by the *Yohkoh* satellite. The large, dark area (corresponding to a cooler, less dense region) is a coronal hole. (b) A composite image of the solar wind and corona, in false color, with additional information on the solar magnetic field and speed of the solar wind. Data and images were taken by the *Ulysses* and *SOHO* spacecraft (ESA/NASA missions) and correspond to the 1994 period of sunspot minimum.

Perhaps a more dramatic effect of the solar wind is the auroras seen near the poles of the Earth. Auroras (refer back to Figure 6-25) result when the solar wind creates tears in the Earth's magnetosphere, allowing energy and charged particles to enter. This is the driving force for space weather activity around Earth. Where the Earth's magnetic field lines converge toward the surface of the Earth, the electrically charged particles strike the molecules of the upper atmosphere and cause them to emit the beautiful, eerie glow we call an aurora.

Recall the photo of an aurora on Saturn (Figure 9-21) and on Jupiter (Figure 9-8).

11-6 Sunspots and the Solar Activity Cycle

Observations of dark spots on the Sun were reported by the Chinese as early as the 5th century BC. It is sometimes possible to see very large sunspots with the naked eye if the Sun is viewed when it is very near the horizon. Europeans did not report sunspots until Galileo saw them with his telescope, perhaps because the Europeans did not have observers as astute as the Chinese, or perhaps—after Aristotelian thought was adopted—because Aristotle had proclaimed that the Sun was flawless. (We tend not to see what we disbelieve.)

In the late 18th century, Alexander Wilson hypothesized that sunspots were places where we were seeing through the outer surface of the Sun and into a cooler interior. William Herschel, the discoverer of Uranus (see the Historical Note in Chapter 9), even thought that the interior of the Sun might be cool enough to support life. Today's spectroscopic measurements of solar temperatures reveal that sunspots are indeed about 1500 K cooler than the surrounding photosphere. They are still very hot, however. At the central part of a sunspot the temperature may be as low as 3900 K. Using the Stefan-Boltzmann law (see Section 4-4), we find that the radiation emitted by a sunspot is $(5800/4300)^4 \approx 3$ times less than that from the surrounding photosphere, and as a result, the sunspot appears dark. Sunspots are temporary phenomena, lasting anywhere from a few hours to a few months.

The explanation for sunspots involves the magnetic field of the Sun, which can be measured using a technique discovered late in the 19th century. For an object in a magnetic field, the field can cause each emission line of the object's spectrum to split into two or more lines, and the strength of the magnetic field can be determined from the extent of the splitting. The splitting can be measured in the spectrum of light from individual parts of the Sun and is an important tool in studying the Sun.

Sunspots often appear in pairs, aligned in an east–west direction. Early in the 20th century it was found that the magnetic field in a sunspot is about 1000 times as strong as the magnetic field of the surrounding photosphere. In addition, we find that sunspot pairs have opposite magnetic polarities, one being north and the other south.

Sometimes the Sun contains a great number of sunspots, and sometimes few or no sunspots are seen. In 1851, Heinrich Schwabe, a German chemist and amateur astronomer, discovered that there is a fairly regular cycle of change in the number of sunspots and that the cycle lasts about 11 years. He found that although individual spots do not last long, about the same number are found on the Sun at any one part of its cycle. The cycle varies somewhat in period, but averages about 11 years between repetitions. FIGURE 11-25a is a graph showing how the number of sunspots has changed since 1880. The extended sunspot record shown in Figure 11-25b suggests that the Sun went through a period of inactivity during 1645–1715, when very few sunspots were seen on its surface (the Maunder minimum). Solar observations during this period were extensive enough that the lack of observed sunspots was well documented. This period corresponds to the "Little Ice Age" climatic period on Earth, and there is evidence that similar periods of inactivity existed in the more distant past. There is clearly a connection between the changes in solar activity and our climate on Earth.

Modeling the Sunspot Cycle and the Sunspots

At a sunspot maximum, most spots occur about 35 degrees north or south of the equator. Then, as the cycle progresses, the spots are seen closer and closer to the equator. By the time they reach the equator, the cycle is at a minimum, and new spots are beginning to form again at greater latitudes. If we plot the location of the spots as time goes by, we get the pattern shown in FIGURE 11-26, called a *butterfly diagram* for obvious reasons. It is important to point out that a given sunspot does not move from higher to lower latitudes. (The lifetime of a sunspot can be up to a few months, much shorter than the 11-year solar cycle.) Instead, the diagram tells us that as time passes and old sunspots die out, new ones form closer to the Sun's equator.

◆ The darkest region of a sunspot may be as large as 30,000 km in diameter; this is about twice the Earth's diameter.

◆ The splitting of spectral lines by strong magnetic fields is called the *Zeeman effect* after the Dutch physicist who discovered it.

◆ When a large number of sunspots appear on the Sun's surface, its luminosity decreases by about 0.1%.

The reality of the [sunspot] minimum and its implication of basic solar change may be but one more defeat in our long and losing battle of wanting to keep the Sun perfect, and if not perfect, constant, and if not constant, regular. Why the Sun should be any of these when other stars are not is probably more a question for social than for physical science.
John Eddy, solar physicist, 1976.

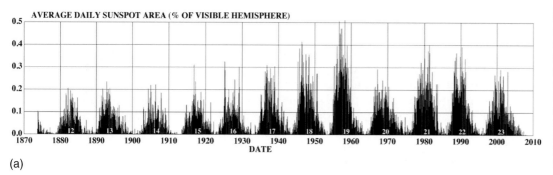

(a)

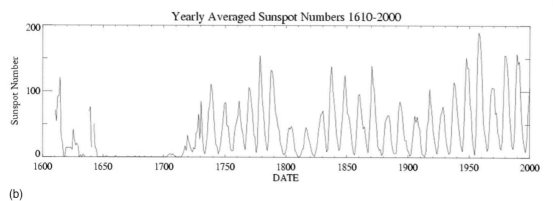

(b)

FIGURE 11-25 (a) The number of sunspots varies with a period of about 11 years, but there is a great difference in the maximum number during each cycle. It also appears that some cycles are double-peaked. (b) The Maunder minimum (1645–1715) corresponds to a period of inactivity for the Sun and the "Little Ice Age" for Earth.

DAILY SUNSPOT AREA AVERAGED OVER INDIVIDUAL SOLAR ROTATIONS

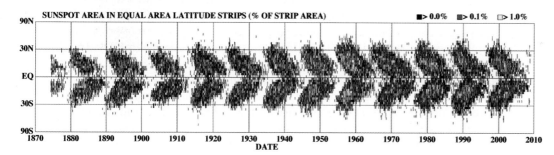

FIGURE 11-26 This plot, called a butterfly diagram, shows the location and relative number of sunspots as the years pass. When each set of butterfly wings forms, the sunspots are at higher latitudes. Then, as new sunspots are formed, they move closer to the equator. Also, about every 11 years there is a period of very few sunspots.

The leading modern hypothesis explains the existence of sunspots and their 11-year cycle as being due to patterns of magnetic field lines, which are generated by the flow of the hot ionized gases in the thin boundary layer between the radiative and convective zones. The Sun's rotational velocity suddenly changes at this boundary, and this velocity shear drives the formation of the solar magnetic field. It is thought that groups of these lines form "tubes" threading through the Sun. When the tubes first form, they are relatively straight and buried deep within the Sun as shown in FIGURE 11-27a. The differential rotation of the Sun, however, causes the lines to wrap around the Sun, as shown progressively in parts (b), (c), and (d) of Figure 11-27. As the tubes become more and more twisted around the Sun, they are forced to the surface by convection. When they break through, we see a pair of sunspots, one with a north magnetic pole and one with a south magnetic pole.

An interesting feature of the 11-year cycle is that during one cycle, the leading sunspot of each pair in a given hemisphere of the Sun is a north magnetic pole and the trailing sunspot is a south pole. (In the other hemisphere, the opposite is true.) Then, during the next cycle, the pattern reverses, with the leading sunspot in that hemisphere being a south pole. The Sun's overall magnetic field also reverses from one cycle to another. During each cycle, the polarity of a leading sunspot in a given hemisphere is the same as the polarity of the Sun's magnetic pole for that hemisphere

FIGURE 11-27 It has been proposed that tubes of magnetic field lines form below the Sun's surface. The Sun's faster rotation near its equator then twists the tubes around the Sun (a–d). The insets show the concentrated magnetic field lines breaking through the solar surface resulting in sunspots.

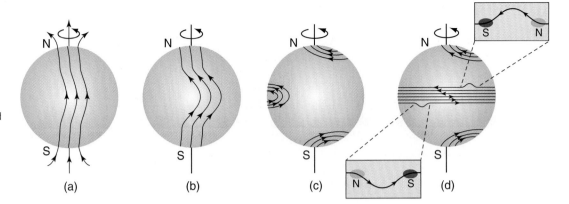

(a)   (b)   (c)   (d)

The Japan/US/UK solar X-ray observatory *Yohkoh* was launched in 1991. Its mission ended in 2001 and it burned during re-entry in 2005. The ESA/NASA *So*lar and *H*eliospheric *O*bservatory (*SOHO*) satellite was launched in 1995, and its mission was extended until December 2009.

(as shown in Figure 11-27d). The entire magnetic cycle of the Sun thus has a 22-year period. The reversal of the Sun's magnetic poles happens at the middle of every 11-year sunspot cycle, during the period of sunspot maximum. Currently, the Sun's north magnetic pole points through the Sun's southern hemisphere, and it will do so until the year 2012. FIGURE 11-28 illustrates the relationship between sunspots and the Sun's magnetic field.

What causes the Sun's field to flip every 11 years? Observations made with the *Yohkoh* and *SOHO* satellites show that giant loops of hot plasma, extending into the Sun's corona, link each of the Sun's magnetic poles to sunspots of opposite polarity (trailing sunspots) near the Sun's equator. As flows from the equatorial regions of the Sun to the magnetic poles transport opposite magnetic flux, the Sun's magnetic field steadily weakens. At the same time, the leading sunspots at both hemispheres migrate toward the equator, and when they meet, their opposite magnetic polarities cancel each other out. (Recall from Section 11-4 that helioseismology has allowed us to study the flows of differentially rotating magnetic bands inside the convection zone. Magnetic bands at high latitudes migrate toward the poles, and bands from low latitudes migrate toward the equator.) At the height of sunspot maximum, the magnetic poles change polarity and the Sun's magnetic field begins to grow in a new direction. During the flip, the field is very weak and uneven across the Sun's surface, which in turn becomes very active, shooting bubbles of hot gas and energy in every direction. At the time of sunspot minimum, the Sun's magnetic field resembles that of a bar magnet (just like Earth's) and is about 100 times stronger than Earth's (or about as strong as a refrigerator magnet).

Observations made by *SOHO* are helping us understand the stormy areas on the Sun's surface where sunspots appear (FIGURE 11-29). We mentioned before that the magnetic field is very concentrated in sunspot regions, but we know from experience, playing with magnets, that magnetic fields of similar polarity repel each other. The same should be happening in a sunspot region, and as a result, the sunspot should dissipate quickly. Instead, we observe sunspots that last for weeks. How is this

FIGURE 11-28 (a) A visible-light photo of the Sun, showing sunspots. (b) A magnetic map of the Sun on the same day shows where the magnetic field is strongest on the Sun.

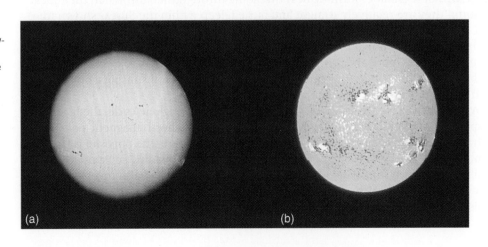

(a)   (b)

possible? We know that the strong magnetic field below a sunspot behaves like a "plug" that stops the normal upward convective flow from the hot solar interior. As a result, the material above the plug cools. This includes the observable sunspot region, and that is why it looks darker than its surroundings, but as this material cools, it becomes denser. According to the *SOHO* observations, this material then plunges downward at up to 5000 km/hour, drawing the surrounding plasma and magnetic field inward toward the center of the sunspot in the process. This increases the strength of the magnetic field, which in turn prevents more energy from reaching the solar surface. As the cooling material above the plug sinks, it draws more plasma and magnetic field inward, setting up a cycle that can last as long as the field is strong enough. As Figure 11-29 suggests, the region below the plug is hot, and this is just the opposite of the conditions at the surface. It also seems that the observed outflows at the surface are confined to a very narrow layer. The *SOHO* observations give us a better understanding of the overall structure of sunspots, but we still do not have a clear picture of the details, especially at the roots of the sunspots.

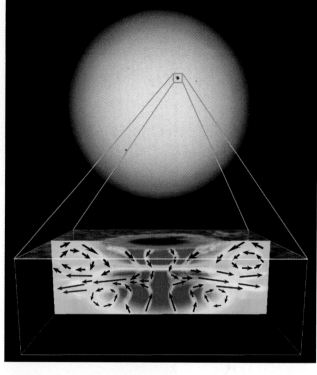

FIGURE 11-29 An artist's version of the region below a sunspot. Hot regions are shown in red, cool regions in dark blue.

Sunspots can occur in large groups that cover an area tens of times the diameter of the Earth. As shown in Figure 11-18, the observed structure of a sunspot includes a dark area where the magnetic fields are locally vertical surrounded by a lighter area where the fields are locally horizontal. The release of magnetic energy can cause violent solar eruptions such as solar flares and **coronal mass ejections**, which we study next. The main sources for such activity are large clusters of sunspots. With the help of helioseismology, astronomers were able to observe strong circulation patterns near the Sun's surface that play a role in holding the clusters together. Measuring such winds may prove to be a powerful tool in predicting solar flare activity.

Solar Flares and Coronal Mass Ejections

The turbulent magnetic field of the Sun is responsible for the prominences discussed earlier, giving prominences their unique shapes, such as those in the chapter-opening photograph and Figure 11-23. It also causes the colossal flareups called **solar flares** that normally occur during sunspot maxima. Lasting from a few minutes to a few hours and reaching up to 100,000 kilometers in length, a solar flare can release the equivalent energy of a few million of our largest nuclear weapons (**FIGURE 11-30**). According to the model presented in the last section, these flares occur when a great number of twisted tubes of magnetic field lines release their energy at once through

coronal mass ejection An event in which hot coronal gas is suddenly ejected into space at speeds of hundreds of kilometers per second.

solar flare An explosion near or at the Sun's surface, seen as an increase in activity such as prominences.

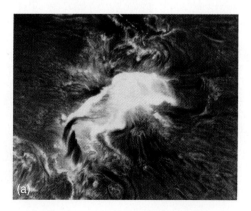

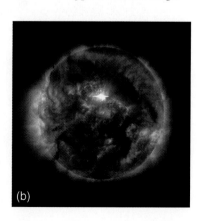

FIGURE 11-30 (a) An X-ray image of a solar flare. (b) A powerful flare observed by the *SOHO* spacecraft on July 2002. The flare was associated with an Earth-directed full-coronal mass ejection.

FIGURE 11-31 A coronal mass ejection. This is one frame of an ejection in progress, observed in 1997 by *SOHO* using a coronagraph. This produces an artificial eclipse of the Sun by placing an "occulting disk" over the image of the Sun, allowing us to observe the Sun for long periods of time.

the photosphere. They occur near sunspots, usually along the dividing line between areas of oppositely directed magnetic fields. In just a few seconds, flares can heat solar material to tens of millions of degrees and accelerate solar particles to very high speeds. In the case of the largest flares, these particles can reach the Earth in less than an hour. They are responsible not only for spectacular auroras but also for disruptions of Earthly radio transmissions.

Radio disruption occurs when particularly energetic particles of the solar wind strike a layer of the Earth's atmosphere called the ionosphere. The ionosphere plays a part in radio transmission because it reflects radio waves back down to the Earth's surface. Normally, the Earth's magnetic field prevents particles of the solar wind from reaching the ionosphere by deflecting and trapping them, but the high-energy particles emitted by a solar flare are able to penetrate to, and disrupt the ionosphere. As a result, communication technologies using radio waves, including the Global Positioning System, can be seriously impacted.

Coronal mass ejections (**FIGURE 11-31**) are often associated with solar flares and prominences but can also occur in their absence. Such ejections and flares might be just different aspects of the same phenomenon. We know they are related, with some

FIGURE 11-32 A coronal mass ejection showing twisted magnetic field lines. The field acquires this twist (or helicity) beneath the solar surface.

events showing more eruptive behavior, whereas others show more flare behavior. These mass ejections occur more frequently when the Sun is most active. They are huge bubbles of gas threaded with magnetic field lines that are ejected from the Sun over the course of several hours. The mass associated with these ejections can be in the billions of tons (more mass than Mt. Everest), and the speed of the charged particles can be up to 2000 km/s.

Beautiful twisted structures can be seen in coronal mass ejections (**FIGURE 11-32**). The twists in the magnetic field probably originate below the solar surface, and just like twisted coils of spring metal, these twisted structures contain energy that is used to blast the material into space.

We know that the active regions on the Sun's surface from which coronal mass ejections originate consist of a great number of loops filled with plasma. Such loops are shown in Figure 11-23b and in **FIGURE 11-33a**. The plasma that fills the coronal loops is not static. Instead, observations from the *TRACE* and *SOHO* satellites suggest that

FIGURE 11-33 (a) A *Yohkoh* image of a loop in soft X-rays. The width of the arrow's tip represents the size of the Earth. Such loops are about 50,000 kilometers in size and 450,000 kilometers long (equivalent to 40 Earths side by side). (b) A *SOHO* image of the loop in part (a) as it erupts into space.

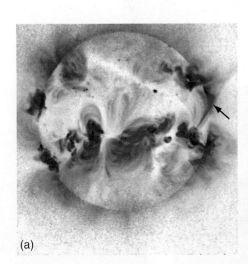

(a)

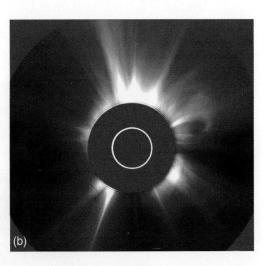

(b)

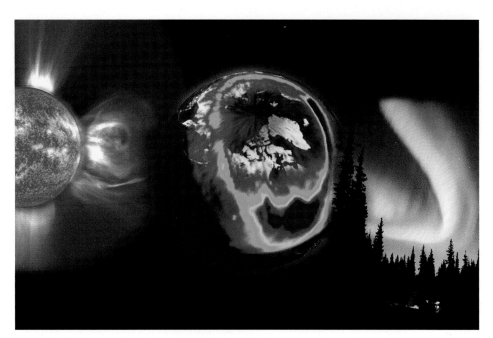

FIGURE 11-34 A coronal mass ejection, an aurora across the United States as seen from space on July 14, 2000, and an aurora display in Alaska are shown in this composite image representing the three most visible elements of space weather.

the plasma moves in the loops at great speeds, probably because of uneven heating at their bases.

Coronal mass ejections disrupt the flow of the solar wind and produce disturbances that strike the Earth with sometimes catastrophic results (**FIGURE 11-34**). When the disturbances reach Earth, they can create a major disruption in the Earth's magnetic field, which can knock out communication satellites and cause power surges. One such power surge overload resulted in a collapse of Quebec's hydroelectric system, cutting off six million people for 9 hours in March, 1989. In 1997, a TV communication satellite was lost because of a coronal mass ejection.

Do solar flares and coronal mass ejections have other effects on the Earth and on the lives of those of us who live on this planet? As we learn more about the Sun, we may find that what seem like small quirks in the Sun's behavior are actually of major importance to life on Earth.

We mention just two possible connections. First, when high-speed protons from the Sun bombard the Earth's upper atmosphere they break up molecules of nitrogen gas and water vapor. The broken molecules form nitrogen and hydrogen oxides, which then react with ozone and reduce its amounts. Even though the overall effect of such a bombardment is minimal (on the order of 1% of the total ozone), it is important to understand and separate the natural effects on ozone from the human factor. Second, cycles of drought at the Yucatan peninsula and cycles of cooling and warming at other parts of the Earth, which can be tracked through radioactive carbon-14 dating, seem to be correlated with a known 206-year cycle in solar intensity. Small variations in the Sun's energy output, if sustained over a long period of time, could have catastrophic climate effects. The observations seem to suggest that some mechanism in our climate is amplifying these natural variations.

During the solar maximum that lasted from 2000 to 2003, solar storms knocked out power in the northern parts of Canada and Sweden and destroyed a satellite used to verify credit card payments at gas stations in the United States.

As we explained in the discussion of the formation of the solar system, the solar wind had a major function in determining the nature of today's solar system.

Conclusion

Like our ancestors, we recognize the importance of the Sun to life on Earth. We also try to understand the changes we observe and how these changes influence our planet. The Sun's luminosity varies on time scales from milliseconds to billions of years. We think that some of these variations affect our climate, but we do not yet know how. Similarly, the ultraviolet light and X-rays emitted by the Sun heat up our atmosphere. The solar wind emanating from the Sun affects the Earth's magnetic field, pumps energy into the radiation belts, and can cause power surges. As we become more dependent on satellites, we will increasingly feel the effects of space weather and the need to predict it.

To astronomers, the Sun takes on even more importance because it is by far the closest star. Because astronomers must understand stars if they are to understand the workings of the universe, the Sun becomes critical in such a study. Many physical processes occurring elsewhere in the universe can be examined in detail on the Sun. Solar astronomy teaches us much about stars, planetary systems, galaxies, and the universe itself.

STUDY GUIDE

RECALL QUESTIONS

1. The Sun's energy is generated by
   A. gravitational contraction.
   B. nuclear fission.
   C. hydrogen fusion.
   D. helium fusion.
   E. chemical reactions.

2. The layer of the Sun that is normally visible to us is the
   A. corona.
   B. chromosphere.
   C. photosphere.
   D. core.
   E. solar wind.

3. Sunspots are areas on the Sun that are
   A. hotter than their surroundings.
   B. cooler than their surroundings.
   C. brighter than their surroundings.
   D. [Both A and B above.]
   E. [Both B and C above.]

4. The energy produced in nuclear reactions in the Sun results from
   A. friction as the nuclei crash together.
   B. heat produced from the electrical effects of the reactions.
   C. the increase in mass of the particles due to the reactions.
   D. the decrease in mass of the particles due to the reactions.

5. Why is a high temperature needed for energy production in the core of the Sun?
   A. Hydrogen will not combine with oxygen at a low temperature.
   B. Energy is needed to overcome electrical repulsion.
   C. Electrons will not recombine at low temperatures.
   D. The force of gravity is greater at high temperatures.
   E. Speeds are less at high temperature, and thus, there is more time for reactions between nuclei.

6. We know that the Sun's energy does not result from a chemical burning process because
   A. of the Doppler effect.
   B. of the redshift.
   C. the Sun would have burned up already.
   D. [Both B and C above.]
   E. [Both A and B above.]

7. The two forces producing hydrostatic equilibrium in the Sun to determine its size are
   A. electrical forces and gravity.
   B. nuclear forces and gravity.
   C. electrical forces and gas pressure.
   D. electrical forces and nuclear forces.
   E. gravity and gas pressure.

8. As the Sun "burns,"
   A. its total mass decreases very slightly.
   B. its total mass increases very slightly.
   C. its energy decreases, but the Sun's mass remains the same.
   D. energy is produced, but the Sun's mass remains the same.
   E. [None of the above.]

9. During a total solar eclipse, the Sun's atmosphere becomes visible. Why?
   A. It is brighter during an eclipse because of light reflected from the Moon.
   B. The light reemitted after absorption becomes visible because the brighter Sun is blocked out.
   C. The atmosphere becomes hotter during an eclipse.
   D. [The statement is not true.]

10. The total luminosity of the Sun can be calculated from its
    A. rotation period and temperature.
    B. rotation period and diameter.
    C. diameter and distance from the Earth.
    D. diameter and the solar energy at Earth's distance.
    E. distance from Earth and the solar energy detected at Earth's distance.

11. Two factors that determine the pressure of a gas are
    A. the speed of the molecules and the particle density.
    B. the speed of the molecules and the gas' temperature.
    C. nuclear reactions in the gas and its temperature.
    D. chemical reactions in the gas and its temperature.
    E. [Both C and D above.]

12. At any particular level within the Sun, the pressure outward is
    A. less than the pressure inward.
    B. equal to the pressure inward.
    C. greater than the pressure inward.
    D. [No general statement can be made.]

13. Granulation of the photosphere is a direct result of
    A. heat conduction.
    B. convection.
    C. heat radiation.

14. A prominence is
    A. a cool spot on the Sun.
    B. the ejection of material from the photosphere.
    C. a fairly permanent bulge on the photosphere's surface.
    D. a reaction within the Sun's core.

15. The solar wind extends
    A. about to Mercury's orbit.
    B. about to Venus' orbit.
    C. almost to Earth's orbit.
    D. far beyond the Earth's orbit.

16. The 11-year cycle of sunspots corresponds to
    A. the period of change in the magnetic field of the Sun.
    B. the rotation period of the Sun near the equator.
    C. the rotation period of the Sun near the poles.
    D. the revolution period of Jupiter.
    E. [None of the above.]

17. Which of the following is the thinnest layer of the Sun?
    A. corona.
    B. chromosphere.
    C. photosphere.
    D. radiative layer.
    E. convection layer.

18. Solar energy strikes the Earth at the rate of 1380 watts/m². It strikes a sphere that is 2 astronomical units from the Sun at a rate of
    A. 345 watts/meter².
    B. 690 watts/meter².
    C. 1380 watts/meter².
    D. 2760 watts/meter².
    E. 5520 watts/meter².

19. The primary source of energy for the Sun is a series of nuclear reactions in which
    A. four hydrogen nuclei fuse to form a helium nucleus.
    B. a helium nucleus fissions to form four hydrogen nuclei.
    C. uranium nuclei fission to form several other elements.
    D. two nuclei fuse to form uranium or plutonium.
    E. oxygen nuclei combine to form more massive nuclei.

20. To begin nuclear fusion in a star, high temperatures are required to overcome the
    A. nuclear force between the protons.
    B. nuclear force between the electrons.
    C. electrical force between the neutrons.
    D. electrical force between the protons.

21. The photosphere is
    A. the layer of the Sun where energy is created from mass.
    B. the outermost layer of the Sun.
    C. the layer of the Sun that we see when observing the Sun in visible light.
    D. the layer of the Sun in which we see granulation.
    E. [Both C and D above.]

22. When four hydrogen nuclei fuse to form a helium nucleus, the total mass at the end is _____ the total mass at the beginning.
    A. less than
    B. the same as
    C. more than

23. The sun emits its most intense radiation in which region of the electromagnetic spectrum?
    A. Radio.
    B. Infrared.
    C. Visible.
    D. Ultraviolet.
    E. X-ray.

24. Nuclear theory predicts that we should detect
    A. fewer neutrinos than we do.
    B. just the amount of neutrinos that we do, confirming the theory.
    C. more neutrinos than we do.

25. What is meant when it is said the Sun has "differential" rotation?

26. Describe some evidence that shows that the source of solar energy cannot be chemical reactions.

27. Distinguish between chemical and nuclear reactions, giving an example of each.

28. What is it about the nucleus of an atom that distinguishes one atom from another?

29. Name the chemical element that is consumed and the element that is produced in the Sun. What produces the energy when the change occurs?

30. In what physical state is most of the material in the interior of the Sun?

31. Define and explain *hydrostatic equilibrium*.

32. How do we know how great the pressures are at certain depths below the surface of the Sun?

33. List the three methods of heat transfer, giving an example of each. What method(s) is important in which region(s) of the Sun?

34. If all electromagnetic radiation travels at the speed of light, why does radiated energy take so long to get from the center of the Sun to the surface?

35. Describe the thickness and temperature of the photosphere.

36. If the chromosphere and corona are so hot, why are they not brighter than the photosphere?

37. According to present theory, what causes sunspots?

1. Why do nuclear reactions occur only at the center of the Sun?

2. Explain, without using any math, how we measure the total energy output of the Sun.

3. What evidence do we have that the Sun's energy does not result from the burning of fossil fuels?

4. If the mass of the Sun decreases because of nuclear fusion, why don't we see the Sun decreasing in size?

5. Distinguish between force and pressure as defined in science, and give an example of units in which each can be expressed.

6. Describe the relationship among pressure, density of particles, and temperature in a gas.

7. What would happen to the Sun's core if the rate of fusion reactions decreased suddenly?

8. How do we know the pressure deep within the Sun?

9. What method of energy transport causes the handle of a poker to get warm when the business end of the poker rests in a fire?

10. We see the Sun not as it is now, but as it was 8 minutes ago. The energy we detect, however, began millions of years ago. Discuss the implications this has on what we mean by the word "now." How does this provide a possible explanation for the neutrino problem?

11. What was the solar neutrino problem, and why are astronomers so interested about solar neutrinos?

12. List and describe the layers of the Sun's atmosphere.

13. Historically, what has been the value of an eclipse in studying the Sun?

14. Until recently, the solar neutrino problem was an example of an unsolved problem concerning the Sun. Another such problem exists in explaining temperatures within the solar atmosphere. Explain the problem.

15. If the magnetic field of the Sun reverses every 22 years, what is meant when we refer to the "northern" hemisphere of the Sun?

1. Calculate the angular size of a 15,000-km sunspot as seen from Earth.

2. Verify that the average density of the Sun is about 1.4 times greater than that of water.

3. If 1 kg of coal is burned in 1 second, it will produce 2 million watts of power. How many kilograms would have to be burned each second to produce the Sun's energy output? If the Sun were made of coal, how much time would pass before it would burn out at its present rate of energy production? (In fact, oxygen would be needed for the burning—nearly three times more mass of oxygen than of coal—so only one fourth of the Sun's mass would be coal.)

4. Use Wien's law (see Section 4-4) to calculate the wavelength that corresponds to the peak of the Sun's blackbody curve. Use 5800 K as the temperature of the Sun's photosphere. What color does this wavelength correspond to? Why does the Sun appear yellow to our eyes?

5. The average power requirement for the United States is about 6×10^{12} watts. With our current technology we can collect solar power at the surface of the Earth of about 200 watts/meter². If we were to build one huge circular solar collector, what would its radius be in miles? (Hint: The area of a circle of radius R is πR^2.) Do you think this idea could lead to a solution for our energy needs?

1. Measuring the Diameter of the Sun

With simple equipment, you can measure the size of the Sun with a fair degree of accuracy. All you need is a sunny day, a piece of cardboard, a ruler, and the knowledge that the Sun is 150,000,000 kilometers from Earth. An activity in Chapter 1 discussed observing an eclipse by pinhole projection; we use pinhole projection here, also.

Punch a small hole (perhaps one eighth inch) in a piece of cardboard and hold the cardboard so that the Sun shines through the hole onto a surface behind it. (Refer to Figure 1-46.) You may have to adjust the size of the hole to get an image bright enough to see clearly, and you might use additional cardboard to shield your screen from reflected sunlight.

Making sure that the screen is perpendicular to a line from the pinhole, measure the diameter of the image of the Sun and the distance from the pinhole to the screen. Now, as shown in **FIGURE 11-35**, the following ratio applies:

$$\frac{\text{diameter of Sun}}{\text{distance to Sun}} = \frac{\text{diameter of image}}{\text{distance from screen to image}}$$

Use this equation to calculate the diameter of the Sun.

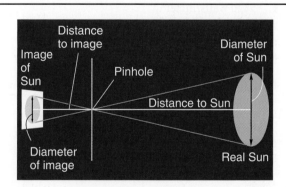

FIGURE 11-35 This drawing, obviously not to scale, illustrates the relationship between distances and sizes of the Sun and its image. Since the triangle at the left is similar to that at the right, the ratio of distance to size is the same for each.

To get a feel for the accuracy of your measurement, make several measurements with the screen at different distances. How closely do your various measurements agree? What is the largest source of error in this procedure? How does the value you obtained compare to that found in the Data Page?

2. Observing Sunspots

Sunspots were first observed by the naked eye, as described in the text, but such a method is not recommended. You would probably have to search the setting Sun for long periods of time over many years before you saw your first sunspot, and staring even at the setting Sun might damage your eyesight.

A more realistic way to view sunspots is with a telescope. **Do not, however, look directly at the Sun with a telescope**. You can obtain a solar filter to put over the front of the telescope but it is even better to use the telescope to project an image of the Sun on a screen, as described later here.

First, a caution: the intensity of sunlight is so great that you run a risk of damaging your telescope. One way to decrease this risk is to cover the objective lens with a piece of cardboard with a hole cut in it smaller than the lens. Tape it down so that it won't fall off and it will block out some of the light. If your telescope has a finderscope, you should cover it by taping a piece of cardboard (without a hole) over its objective lens. This will not only prevent eye damage to

someone who, out of habit, looks through it, but it will prevent the Sun from burning out the finderscope's crosshairs.

In using a telescope to view the Sun, set it up by first focusing on a distant object (**NOT THE SUN**). Then pull the eyepiece out just slightly. Now is the time to cover the finderscope and partially cover the objective. Point the telescope exactly at the Sun. This is not as easy as it sounds, and the best way is to move it until the shadow of the telescope tube is smallest. **DON'T LOOK THROUGH THE FINDERSCOPE**. When this is done, you should be able to see a spot on a screen held behind the eyepiece. Focus by moving the screen and eyepiece to various positions until you get the view you want. A cardboard shield around the telescope will shadow your image from direct rays and improve your view. If you have a star diagonal (which reflects the image off to the side), use it. Trace the image of the Sun and any spots you see. Then repeat your observation in a day or two and look for motion of the sunspots.

Figure 1-45 (in Activity 6 in Chapter 1) shows a small telescope being used to project the Sun's image during an eclipse.

Have fun, but be careful.

You can find information about current missions to study our Sun at NASA's homepage (http://www.nasa.gov). The "ultimate" page on neutrinos can be found at http://cupp .oulu.fi/neutrino//. An annotated bibliography for nuclear fission can be found at http://alsos.wlu.edu.

1. "How Stars Shine," by J. Trefil, in *Astronomy* (January, 1998).

2. "Where Are the Solar Neutrinos?" by J. Bahcall, in *Astronomy* (March, 1990).

3. "*SOHO* Reveals the Secrets of the Sun," by K. Lang, in *Scientific American* (March, 1997).

4. "The Sun-Climate Connection," by S. Baliunas and W. Soon, in *Sky & Telescope* (December, 1996).

5. "On the Solar Spectrum and the Color Sensitivity of the Eye," by D. K. Lynch and B. H. Soffer, in *American Journal of Physics* (vol. 67, No. 11, November 1999).

6. "Solving the Solar Neutrino Problem," by A. B. McDonald, J. R. Klein, and D. L. Wark, in *Scientific American* (April 2003).

7. "The Paradox of the Sun's Hot Corona," by B. N. Dwivedi and K. J. H. Phillips, in *Scientific American* Special Edition (September 2003).

8. "The Fury of Solar Storms," by J. L. Burch, in *Scientific American* Special Edition (September 2004).

9. "The Stellar Dynamo," by E. Nesme-Ribes, S. L. Baliunas, and D. Sokoloff, in *Scientific American* Special Edition (September 2004).

10. "The Mysterious Origins of Solar Storms," by G. D. Holman, in *Scientific American* (April 2006).

Quest Ahead to Starlinks
http://physicalscience.jbpub.com/starlinks

Starlinks is this book's online learning center. It features **eLearning**, which contains chapter quizzes and other tools designed to help you study for your class. You can also find **online exercises**, view numerous relevant **animations**, follow a guide to **useful astronomy sites** on the Internet, or even check the latest **astronomy news** updates.

12 Measuring the Properties of Stars

Mauna Kea Observatories, Mauna Kea, Hawaii. The summit (4200 meters high) houses the world's largest observatory for optical, infrared, and submillimeter astronomy.

ASTRONOMER SIDNEY WOLFF AND HER HUSBAND WERE INVOLVED in the building of the observatory on Mauna Kea, a 14,000-foot high mountain in the Hawaiian Islands, in the late 1960s. In an essay in *The Scientist*, she writes:

> We had a wonderful time in those early days, developing a site without even such basic amenities as a source of water; a site where all power had to be generated locally because there was no power line, a site where blizzards raged in winter and where even in summer temperatures dipped to freezing every night. But we learned. We learned about altitude sickness and the best strategies for forcing our bodies to acclimate. We learned first aid so we could cope with accidents, since professional help was hours away. We learned how to handle heavy machinery and how to maintain generators. We learned more about telescope gears and worm drives and how to repair scored gears than we ever wanted to know. Nearly every one of us who was involved in those early days can tell—loves to tell—stories of being nearly trapped on the mountain during a blizzard, of hiking to the summit because the road was blocked by snow, of climbing to the top of the dome to remove snow so that not a moment of observing was lost. It was a great adventure, an adventure that surely I had not envisioned when I planned a life of research alone in my office.*

*Sidney C. Wolff, National Optical Astronomy Observatories.

All cross references to chapters, sections, figures, and tables pertain to the main text, *In Quest of the Universe, Sixth Edition. In Quest of the Solar System* contains Chapters 1–11 and 19 of the main text. *In Quest of the Stars and Galaxies* contains Chapters 1–5 and 11–19 of the main text.

Only recently in human history have we even become aware of the astonishing fact that each of the thousands of stars we see is another sun similar to the one that rules our sky. This realization makes obvious the immense distances to those stars; just imagine how far away our Sun would have to be to appear as dim as one of these stars. Can we hope to learn much about such faraway objects? As we explained in Chapter 4, spectral analysis can be used to determine both the temperature of a star and the chemical composition of its atmosphere. Galileo and Newton would have been amazed that we can learn such things; however, temperatures and chemical compositions are only the beginning. In this chapter we discuss how parallax allows us to calculate the distances to many stars and how, after we know their distances, we can determine other properties, including their luminosities, motions, sizes, and masses. We describe relationships among the various properties that give us clues as to why one star differs from another, how stars are formed, how their lives progress, and how they die. We delay discussion of the life cycle of stars until the next chapter, turning our attention now to how we measure those properties of stars that divulge information about their life cycles.

The stars, while differing the one from the other in the kinds of matter of which they consists, are all constructed upon the same plan as our sun, and are composed of matter identical at least in part with the materials of our system.
William Huggins, 1863

Ye Stars! Which are the poetry of heaven.
Lord Byron (1788–1824)

12-1   Stellar Luminosity

When we speak of the brightness of a star, we must be careful to distinguish between its apparent brightness (**FIGURE 12-1**) and its **luminosity**. In Section 11-2 we explained how we can calculate the luminosity of the Sun if we know the Sun's distance and the amount of solar radiation striking a given area of Earth in a certain amount of time. Suppose two stars differ in apparent brightness so that different amounts of light reach the Earth from the two stars. The cause for this might be any combination of three things: (1) one star may be inherently brighter than the other (its luminosity may be greater); (2) one star may be closer, making it appear brighter; or (3) there may be more interstellar material absorbing light from one star than from the other.

In our discussion of the two quantities, *brightness* refers to the apparent brightness seen from Earth, and *luminosity* refers to the total amount of power emitted by a star. You may occasionally find the term *absolute luminosity*; this simply emphasizes that we are speaking of the radiation actually being emitted, not the radiation reaching Earth.

luminosity The rate at which electromagnetic energy is emitted.

Apparent Magnitude

One of the greatest astronomers of antiquity was Hipparchus, a Greek thinker who lived in the second century BC. Hipparchus compiled a catalog of some 850 stars, listing each star's location in the sky, along with a number that designated its brightness. To indicate the brightnesses of stars, he divided all visible stars into six groups, calling the brightest stars in the heavens *magnitude one* stars and the dimmest he could see *magnitude six* stars. Other stars fell in between, with differences between magnitudes representing equal differences in

FIGURE 12-1 Some of the stars of the constellation Orion appear bright because of their proximity, whereas others are inherently so luminous that they look bright from Earth even though they are very far away. Betelgeuse, the star that appears orange at the upper left, is a variable star, about 430 light-years away, with radius 660 times that of the Sun and luminosity 4500–15,000 times that of the Sun. Rigel, the star that appears blue/white at the lower right, is about 730 light-years away, with radius 60 times that of the Sun and luminosity about 39,000 times that of the Sun. The stars appear to be different sizes in the photo. These differences are not due to the stars' actual sizes, but occur because brighter stars expose larger areas of the light-recording device. The scale of the image is 20° by 28°.

FIGURE 12-2   The basic idea of photometry is that the light from a star is focused onto a photocell, which measures the amount of light. Often a filter is used to allow only light of a certain wavelength range to enter. More modern methods use CCD technology rather than photocells.

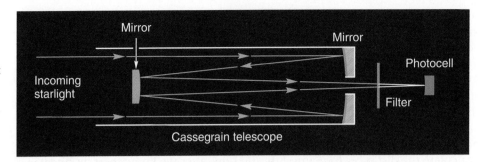

Incoming starlight

Mirror

Mirror

Photocell

Filter

Cassegrain telescope

The quantity described here is apparent magnitude rather than absolute magnitude, which refers to the luminosity of a star and is discussed later.

apparent magnitude A measure of the amount of light received from a celestial object.

Photometry and CCDs (charge-coupled devices) were discussed in Section 5-4.

Is our Sun the brightest star in the Milky Way galaxy?

FIGURE 12-3   Apparent magnitudes of some familiar objects.

brightness. The brightness difference between a third-magnitude star and a fourth-magnitude star on his scale thus was visually the same as that between a fifth- and a sixth-magnitude star. Although it may seem odd to assign the larger number to the dimmer star, we might appreciate his reasoning by thinking of the brighter star as a first-class star and the dimmest as a sixth-class star.

Today we use a slightly revised version of Hipparchus' *apparent magnitude* scale, for we measure the brightnesses of stars by photographic and electronic methods. Photographic techniques began to be used for such measurements in the mid-1800s. When these measurements were made, astronomers found that if two stars differ by one magnitude, we receive 2.5 times as much light from the brighter star as from the dimmer one. For example, a fifth-magnitude star is about 2.5 times brighter than a sixth-magnitude star. Moving up from fifth to fourth magnitude means that the brightness increases another 2.5 times, making a fourth-magnitude star 6.25 times brighter than a sixth-magnitude star (2.5 × 2.5 = 6.25). A magnitude change of five would mean a change in brightness of 2.5 raised to the fifth power, 2.5^5, or about 100. Astronomers recreated the scale accordingly, *defining* a difference of five magnitudes as corresponding to a factor of exactly 100 in the light reaching us. This means that stars differing in apparent magnitude by one unit have a brightness ratio equal to the fifth root of 100, or 2.512.

Before electronic devices became common, astronomers measured light intensity from a star (photometry) by measuring the size and density of its image on a photograph (Figure 12-1). During the last few decades, the process used in photometry is similar to the way your automatic camera determines the brightness of the subject you are photographing. FIGURE 12-2 illustrates the idea. Today, however, video imaging is becoming common, and CCDs are used to detect and measure light intensity.

Although the eye is able to distinguish only a few different classes of stars (Hipparchus distinguished six), photometric methods allow us to discern the difference in brightness between two stars that may appear identical to the eye. The ability to measure magnitudes accurately has changed Hipparchus' unit-step system into a continuous one so that the magnitudes of stars are now measured to fractional values, to an accuracy of 0.001 magnitude or better.

When the magnitude scale is defined as described, we find that some stars in Hipparchus' first-magnitude group are much brighter than others in that group. If a star were 2.5 times brighter than first magnitude, it would have to be assigned a magnitude of zero. A star 2.5 times brighter than this would have a magnitude of −1, a negative number. Sirius, the brightest star in the night sky, is about 10 times brighter than the average first-magnitude star and has an apparent magnitude of −1.43. FIGURE 12-3 shows the approximate apparent magnitude of several objects, and Appendices E and F contain

Apparent magnitude scale (Figure 12-3)

Apparent magnitude	Object
−25	Sun
−20	
−15	
−10	100-watt bulb (at 100 feet) / Full Moon
−5	Venus (at brightest)
0	Sirius / Saturn (at brightest)
5	Naked-eye limit
10	Binocular limit
15	Pluto
20	Large telescope (visual limit)
25	Large telescope (photographic limit)

BRIGHT / DIM

TABLE 12-1

Magnitude Difference versus Ratio of Brightness

Magnitude Difference*	Ratio of Light Received	
1.0	2.5	
2.0	6.3	(2.5 × 2.5)
3.0	16	(2.5^3)
4.0	40	(2.5^4)
5.0	100	(2.5^5)
10.0	10,000	(2.5^{10})

* If the difference in magnitude between two stars is 3, we receive 16 times more light from one star than from the other.

tables of the brightest and nearest stars, respectively, including magnitude values. As shown in Appendix E, in addition to our Sun and Sirius, two other stars in our night sky (Canopus and Arcturus) have negative apparent magnitudes.

TABLE 12-1 shows the ratio of light received for a given difference in apparent magnitude. The following example shows how to use the chart.

EXAMPLE

The star W Pegasi is within the square of the constellation Pegasus (**FIGURE 12-4**). This star varies in brightness, getting as bright as eighth magnitude. Fomalhaut (the brightest star in the constellation Pisces Austrinus, just south of Aquarius) is a first-magnitude star. Which star appears brighter, and how many times more light do we receive from that star than from the other?

SOLUTION   To answer the first question, remember that the star with the lesser magnitude is the one that appears brighter. Fomalhaut thus is the brighter star. The difference in magnitude provides the information needed to answer the second question. The difference is 7. Table 12-1 does not list a difference of 7, but the values in the table can be used to determine the light ratio. A difference in magnitude 5 corresponds to a light ratio of 100, and a difference of 2 corresponds to a ratio of 6.3. A difference of 7 thus means that the ratio of light received is 100 × 6.3, or 630. We receive 630 times as much light from Fomalhaut as from W Pegasi (when the latter is at its brightest). [The magnitudes of Fomalhaut and W Pegasi are 1.17 and 7.9, respectively. For simplicity, we have rounded them off.]

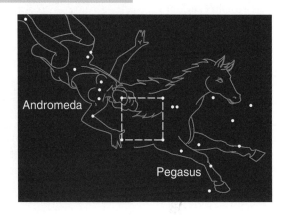

FIGURE 12-4   The great square of Pegasus is an obvious feature of the fall sky. The constellations Pegasus and Andromeda share one bright star, Alpheratz, shown as an eye of Andromeda.

TRY ONE YOURSELF
Barnard's star has an apparent magnitude of about 10. Suppose that on some particular night Mars is measured to have an apparent magnitude of 2. How many times more light do we receive from Mars than from Barnard's star?

As indicated in the example, telescopes allow us to see stars that are much dimmer than sixth magnitude. A telescope with a diameter of 12 cm (about 5 inches) might permit one to see, under perfect conditions, stars of about 13th magnitude. With a 5-meter telescope, one can photograph stars as dim as 25th magnitude.

It is valuable to know the brightnesses—the apparent magnitudes—of stars, but this is not a quantity that is inherent in the stars. Instead, it partially depends on their distances from Earth and therefore on the position of the Earth in the Galaxy. We would like to know the *absolute luminosity* of each star, for this would tell us something about the star itself, independent of the Earth's location. To do this, however, we must know the distance to the star.

parallax angle Half the maximum angle that a star appears to be displaced because of the Earth's motion around the Sun.

12-2 Measuring Distances to Stars

Recall from the earlier discussion of the heliocentric/geocentric debate (see Section 2-3) that the *absence* of observable stellar parallax was an argument against the heliocentric system. Copernicus, however, held that it was not observed simply because the stars are too far away. Not until 1838 was stellar parallax first observed, for the maximum angular displacement of the nearest star is only 1.52 arcseconds. Although the angle between lines from opposite sides of the Earth's orbit toward the star is 1.52 arcseconds, astronomers define the *parallax angle* as half of that, or 0.76 arcsecond in this case. (Parallax angle is defined in this way so that there is a straightforward application of the properties of right triangles; see **FIGURE 12-5**.)

Of course, the observation of stellar parallax is now an argument in favor of the heliocentric system.

The formula used to determine the distance to a star by parallax can be written as

$$\text{distance to star (light-years)} = \frac{3.26 \text{ light-years}}{\text{parallax angle in arcseconds}}.$$

Astronomers usually prefer to express parallax using a different distance unit, the **parsec**. With distance in parsecs, this equation simplifies to

$$\text{distance to star (parsecs)} = \frac{1}{\text{parallax angle in arcseconds}}.$$

In fact, this form of the distance equation *defines* the parsec: one parsec (abbreviated pc) is the distance from the Sun to a star that has a parallax angle of 1 arcsecond. The parsec is the unit astronomers normally use to express stellar distances. One parsec corresponds to about 3.26 light-years, or 206,265 astronomical units (AUs).

Only stars within about 120 parsecs (400 light-years) have parallax angles large enough (0.008 arcseconds) to permit accurate parallax measurements from Earth. In 1989 the European Space Agency launched *Hipparcos* (*HIgh Precision PARallax COllecting Satellite*), which measured the positions and parallaxes of some 120,000 stars to an accuracy of 0.001 arcseconds, and more than 2 million stars to an accuracy of 0.02–0.03 arcseconds, before its power failed. *Gaia*, the successor mission to *Hipparcos*, is named after the Greek Earth goddess, and will be launched in late 2011. Its mission is to give precise and detailed information about the billion brightest objects in the sky, thus giving us a large and precise three-dimensional map of the Milky Way. This map will allow astronomers to do an archaeological study of our Galaxy, uncovering its history and mysteries in the process. NASA is now planning *the Space Interferometry Mission* (*SIM*), scheduled for launch no earlier than 2016, which will be the first space mission with an optical interferometer as its primary instrument. *SIM* is designed to achieve an accuracy, over the whole sky, of 4 microarcseconds. This is 250 times better than the accuracy achieved by the *Hipparcos* mission. Such accuracy would revolutionize the field of astrometry—the precise measurement of the positions of stars—and allow *SIM* to probe nearby stars for Earth-sized planets.

Having more accurate stellar distances will be very valuable to astronomers for two reasons: first, because knowledge of the distance to a celestial object is often the key to determining other properties of the object, such as its luminosity, and second, because it will help us determine the distance scale of the universe more accurately. In the next section, we explain how knowing the distance to a star allows us to calculate the star's luminosity. In a later section we explain why the accuracy of our measurements of distances to very remote objects often depends on accurate knowledge of the distances to nearby stars.

Absolute Magnitude

In Section 11-2 we explained how we can calculate the luminosity of the Sun, using the distance from Earth to the Sun and the value of the solar power striking a square meter of the Earth. In a similar manner, we can calculate the luminosity of any light source if we know the distance to the light source and the brightness of the source. (The power striking a given area of Earth determines brightness.) We all make subconscious judgments similar to this when we look at a distant light at night—a street light, for example—and decide that the lamp is dim or bright. We subconsciously take into account how bright the light appears, as well as its distance from us, to decide (qualitatively) the wattage of the bulb. If we had the tools to measure brightness and distance quantitatively, we could calculate the lamp's power in watts.

FIGURE 12-6 emphasizes that if we know any two of the following three factors, we can calculate the third: apparent brightness (or, in astronomy, apparent magnitude),

parsec The distance from the Sun to an object that has a parallax angle of 1 arcsecond.

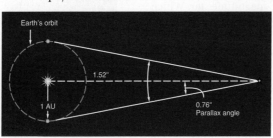

FIGURE 12-5 The nearest star has a displacement of about 1.5″. The parallax angle is defined as half of this, using one astronomical unit as the baseline. The drawing is far out of scale.

Earth's orbit

1.52″

1 AU

0.76″
Parallax angle

0.001 arcseconds corresponds to the angular size of a golf ball viewed from across the Atlantic Ocean. *Hipparcos'* mirror was so smooth that if it were scaled to the size of the Atlantic Ocean, its surface bumps would be less than 4 inches high.

Note the similarity of Figure 12-6 to the discussion in Section 6-1 of the parallax formula and the small-angle formula. Also note that distance is a key quantity in all of these relationships.

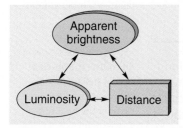

Apparent brightness

Luminosity

Distance

FIGURE 12-6 If any two of the three quantities are known, the other can be calculated.

TOOLS OF ASTRONOMY

Naming Stars

Most of the names of the constellations are Latin translations of the original Greek names. Most of the names of stars, however, are of Arabic origin. Ptolemy's *Almagest* (his summary of Greek astronomy, including a catalog of over 1000 stars) was preserved and passed on by Arab astronomers, who assigned names to numerous stars. The Arabic translation of the English article "the" is "al," which explains why so many names of stars begin with those letters—Alcor (in the constellation Ursa Major), Aldebaran (in Taurus), and Alpheratz (in Andromeda; Figure 12-4).

Only the brightest stars have popular names, however. Stars are assigned "official" names by several methods. The brightest stars in each constellation are given a Greek letter according to their brightness. Elnath, the second brightest star in Taurus, thus is β *Tauri*. The brightest star in Taurus is Aldebaran, but because Aldebaran is part of a binary system (to be discussed later in the chapter), it is named α *Tauri A*, and its dimmer companion is α *Tauri B*. Dimmer stars are given English letters followed by their constellation names (for example, *W Pegasi*; Figure 12-4).

We soon run out of letters in the Greek and English alphabets, of course, and thus, most stars are known only by a catalog number. Although the brightest star in Taurus is named *Aldebaran* and α *Tauri*, it is also known as *87 Tau*. Less distinguished stars don't have popular names and don't have letter designations. One near Aldebaran just goes by the tag *75 Tau*. Only the International Astronomical Union has the authority to assign designations to celestial bodies and any surface features on them; thus, do not be fooled by companies that let you pick a name for a celestial object for a fee.

luminosity, and distance. The connection between these three quantities will be used several times in the remaining chapters. Brightness is most easily measured, and it is always one of the two known quantities. When the other known quantity is distance, we can calculate the luminosity; when the luminosity is known, we can calculate the distance. (Interstellar absorption of light does affect the observed brightness, but it is generally small enough that it can be ignored.)

The triple connection between brightness, distance, and luminosity allows astronomers to calculate the luminosity of some 1000 nearby stars; that many stars are close enough for us to determine their distance fairly accurately by parallax. If we want to know a star's luminosity in watts, we can use the procedure discussed for the Sun in Section 11-2. Usually, however, astronomers use another method to state the intrinsic luminosity of a star. The ***absolute magnitude*** of a star is defined as the apparent magnitude that the star would have if it were located 10 parsecs from the Earth. The absolute magnitude therefore becomes a way to compare a star's actual brightness; it is a very fundamental property of a star (**FIGURE 12-7**). A method of calculating the absolute magnitude of a star is illustrated in the next Tools of Astronomy box.

absolute magnitude The apparent magnitude a star would have if it were at a distance of 10 parsecs.

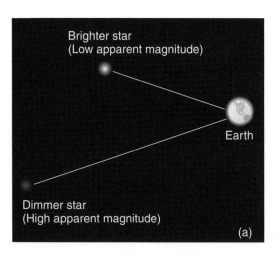

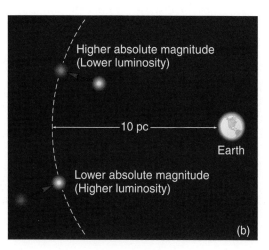

FIGURE 12-7 (a) As seen from Earth, one star may appear brighter than another star because it is closer, not because it is inherently brighter. (b) By calculating the magnitudes of stars at the same distance from Earth (10 parsecs), we can determine which stars have higher luminosity (lower absolute magnitude).

TOOLS OF ASTRONOMY

Calculating Absolute Magnitude

We can find the apparent magnitude of an object by simply observing it, but its absolute magnitude is not as easy to measure. After all, we cannot just move an object to a distance of 10 parsecs from Earth to observe it. Instead, we calculate the absolute magnitude. The relationship between an object's apparent magnitude (m), absolute magnitude (M), and distance (d, in parsecs) can be expressed as

$$m - M = 5 \times \log(d) - 5.$$

The difference ($m - M$) is called the **distance modulus**.

To illustrate the use of this equation, let's calculate the absolute magnitude of Sirius. Sirius has an apparent magnitude of -1.43 and is 2.6 parsecs from Earth. We substitute these values into the equation and solve for the absolute magnitude.

$$-1.43 - M = 5 \times \log(2.6) - 5$$
$$-1.43 - M = 5 \times 0.42 - 5$$
$$M = 1.47$$

What is the physical meaning of this equation? Let's use Sirius as an example. According to the inverse square law, if we "move" Sirius from its actual distance of 2.6 to 10 parsecs from us, it would appear $(2.6/10)^2$ or 1/15 times as bright. From Table 12-1 we see that this corresponds to a change in magnitude of about 2.9. Therefore, Sirius' magnitude at 10 parsecs would be about $(-1.43) + 2.9 = 1.47$, exactly the same value obtained previously.

12-3 Motions of Stars

You sometimes hear references to the "fixed stars." In fact, the stars are not fixed, but are moving relative to the Sun. This motion is obviously not visible to the naked eye, for the constellations have retained their shapes fairly well over the centuries; however, in 1718, Edmund Halley discovered that stars do move with respect to one another and therefore constellations do gradually change their shapes.

FIGURE 12-8 shows two photographs, taken 22 years apart, of a magnified portion of the sky. The arrows point to a particular star, Barnard's star, which has moved noticeably during that time. Barnard's star is the second closest star to the Sun and shows the greatest motion as observed from Earth. It moves at the rate of 10.3 arcseconds per year. Motion expressed as the angle through which a star moves each year (as seen from the Sun) is called the **proper motion** of the star. (You might speculate as to how a star could have improper motion, but the name comes from an old use of the word "proper" meaning "belonging to," for this is the motion that actually belongs to the star, as opposed to observed motion due to Earth's movement.) Only relatively nearby stars show proper motion and, thus, the background stars might still be called *fixed* stars, although they only appear to be fixed because their proper motion is too small to detect. FIGURE 12-9 shows how proper motion has changed the shape of the Big

proper motion The angular velocity of a star as measured from the Sun.

FIGURE 12-8   (a) The arrow indicates a star named Barnard's star. (b) This photo, taken 22 years later, shows that Barnard's star has moved noticeably during that time. (The width of each photo is about 1 degree, and thus, the full Moon would cover about half the width of the photo.)

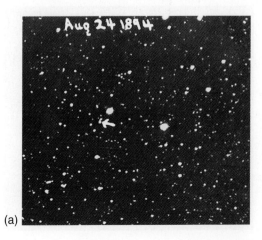

(a)

(b)

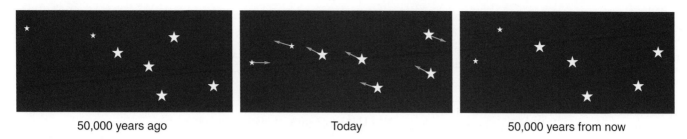

50,000 years ago   Today   50,000 years from now

FIGURE 12-9   The center drawing shows the Big Dipper as it is today; the arrows indicate the proper motions of its stars. From these motions, we can conclude that it once had the shape shown in the left drawing and will some day have the shape shown in the right one.

Dipper over the past 50,000 years and how its shape will continue to change in the future. We normally identify constellations by the stars that appear brightest, but because these stars are often the closest ones, they have the greatest proper motion. Constellations thus gradually change shape over the ages.

Proper motion does not tell us the actual velocity of a star in normal units of velocity, but we can calculate it by using the small-angle formula. (We used this formula to determine the Moon's size in Section 6-1. Instead of the width of the Moon, we calculate the distance the star moved.) Of course, one must know the distance to the star to make such a calculation (**FIGURE 12-10**). In doing the calculation, only the speed of the star *across* our line of sight—its *tangential velocity*—is being computed.

The velocity of a star toward or away from the Earth—its *radial velocity*—is easier to detect and measure than its tangential velocity. We discussed in Section 4-7 how this measurement is made and emphasized that the Doppler effect measures only the star's radial velocity.

A star's actual motion relative to the Sun—its **space velocity**—is a combination of its radial and tangential velocities. Because these two are at right angles to one another, we can use the Pythagorean theorem (**FIGURE 12-11**) to calculate the star's space velocity.

Naturally, the Earth's movement in its orbit affects the observed motions of stars and must be taken into account in calculating the velocity of a star. After this is done, we have the star's velocity relative to the Sun. We see later that there are methods for determining the Sun's movement relative to the distant, "fixed" stars. If this motion is taken into account, we can determine a star's velocity relative to the distant stars.

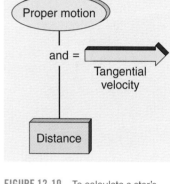

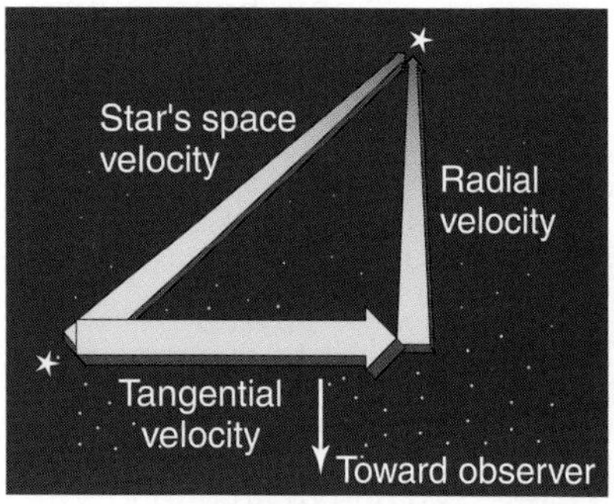

FIGURE 12-10   To calculate a star's tangential velocity, we must know both its proper motion and its distance.

space velocity The velocity of a star relative to the Sun.

FIGURE 12-11   Because radial velocity and tangential velocity are at right angles, we can use the Pythagorean theorem to add them.

(Tangential velocity)^2 + (Radial velocity)^2 = (Space velocity)^2

EXAMPLE

Barnard's star has a proper motion of 10.3 arcseconds per year and is about 1.82 parsecs away. Using the Doppler effect, we measure the star's radial velocity to be −111 kilometers/second. (The minus sign indicates that the star is moving closer to us.) Calculate the space velocity of Barnard's star.

SOLUTION   Using the small angle formula we find

$$\text{distance traveled across the sky per year} = \frac{10.3''/\text{yr}}{206,265} \times 1.82 \text{ pc}.$$

But 1 parsec is 206,265 AU, 1 AU is about 1.5×10^8 kilometers, and 1 year is about 3.15×10^7 seconds. Therefore, Barnard's tangential velocity is

$$\text{tangential velocity} = \frac{10.3''/\text{yr}}{206,265} \times 1.82 \text{ pc} \times 206,265 \frac{\text{AU}}{\text{pc}} \times \frac{1.5 \times 10^8 \text{ km/AU}}{3.15 \times 10^7 \text{ s/yr}} = 89 \text{ km/s}.$$

The star's space velocity is

$$\sqrt{89^2 + 111^2} = 142 \text{ km/s or about } 320,000 \text{ mi/h}.$$

12-4 Spectral Types

FIGURE 12-12 The colors of the stars are obvious in this photo of Orion. The photo was made from three exposures taken sequentially, so the stars are smeared in the image due to the Earth's rotation. Rigel is at lower right. The scale of the image is 19° by 21°.

Are all stars the same yellow-white color?

FIGURE 12-13 Annie J. Cannon (1863–1941), a member of the Harvard College Observatory for almost 50 years, was the founder of the spectral classification scheme in use today.

FIGURE 12-12 is a photo of Orion taken in an unusual way. During the 30-minute exposure, the camera was held steady, but its focus was changed in steps so that the stars became more and more out of focus as they drifted by toward the west (the right). This causes each star to form a fan-shaped image, and it reveals the colors of the stars. Most of the brightest stars appear as blue. The star that appears as red at upper left is Betelgeuse, and the red "star" at lower center is the Orion nebula.

As we explained in Section 4-4, the color of a star is determined by its temperature. In fact, the wavelength at which a star emits most of its energy provides an accurate way to measure the star's temperature. Refer back to the blackbody curve (intensity/wavelength graph) of the Sun in Figure 11-17, which shows that the Sun's energy output peaks near the center of the visible region and indicates a temperature of about 5800 K for the surface of the Sun.

Another method of analyzing the spectrum provides an independent measure of temperature. This method depends on the absorption of radiation at various wavelengths—the absorption spectrum. As we have seen, the absorption of certain wavelengths is what allows us to determine the chemical elements in a star's atmosphere; however, the absorption by an atom depends not only on what element it is but also on the state of its electrons (that is, whether some electrons have been moved to higher energy levels or stripped from the atom). In a gas at higher temperature, more atoms are at higher energy levels, and the transitions that occur from these atoms are different from the transitions that take place in atoms of a cool gas (whose atoms are at lower energy levels). This provides a method of determining temperature other than by the intensity/wavelength graph. The Tools of Astronomy box on page 349 explains this method in greater detail.

Annie Jump Cannon (**FIGURE 12-13**), an astronomer at Harvard College Observatory, devised a classification scheme for spectra and separated several hundred thousand stars according to the strength of the hydrogen lines in their spectra. The classes were first labeled alphabetically, with "A" having the strongest lines, then "B," and so forth. It was later understood, through the work of Cecelia Payne-Gaposchkin, that the classes represented stars of various temperatures. The classes were then rearranged in order of temperature, and some classes were dropped as redundant. The spectral classes used today are designated, from hottest to coolest, as O B A F G K M (**FIGURE 12-14**).

The hottest stars, the blue-violet O stars, range in temperature from about 30,000 to 60,000 K. The blue-white B stars have a temperature range of 10,000–30,000 K, whereas the white A stars range in temperature from 7500 to 10,000 K. The ranges for the yellow-white F stars, yellow G stars, and orange K stars are 6000–7500, 5000–6000, and 3500–5000 K, respectively. The coolest stars are the red M stars, with temperatures less than 3500 K.

Within each spectral class, stars are subdivided into 10 categories, called *spectral types*, and are indicated by attaching an integer from 0 to 9 to the original spectral class. For example, our Sun is listed as a G2 star. You need not be concerned about these subdivisions, but you should remember the classification scheme from hottest

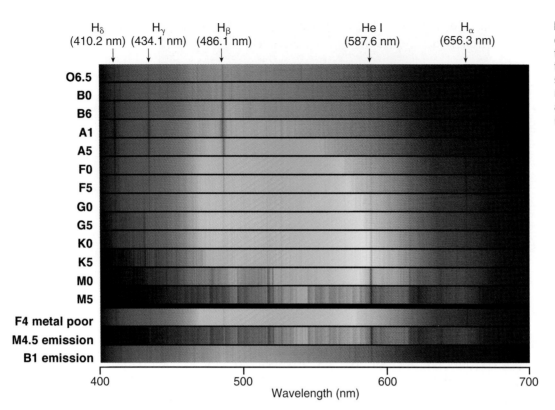

FIGURE 12-14 The spectra of stars of various spectral types, from O to M, with their classes and subclasses shown at left. A few of the major absorption lines are identified at the top (1 nm = 10 Å).

to coolest as being O B A F G K M. The traditional mnemonic is "Oh, Be A Fine Guy/Girl, Kiss Me," although you may want to devise your own memory aid.

The Hertzsprung-Russell Diagram

Suppose that an extraterrestrial being, unfamiliar with humans, somehow learned the age and height of each person in your neighborhood and then used these values to plot a graph, such as that shown in **FIGURE 12-15**. The first thing to notice in this graph is that a pattern exists: There is some relationship between height and age for humans. The graph also implies an important fact about the life cycle of humans, even though no one person had been analyzed through his or her entire life. The alien could logically hypothesize that an individual spends most of his or her life at about the same height, that height being the tallest achieved by the person. In our study of stars, we have a similar chart, called the *Hertzsprung-Russell diagram*.

Early in this century, Ejnar Hertzsprung, a Danish astronomer, and Henry Norris Russell, an American astronomer at Princeton University, independently developed the diagram now named for them and often called simply the H-R diagram. The diagram shows that a pattern exists when stars are plotted by two properties: their temperature (or spectral type) and their absolute magnitude (or luminosity). **FIGURE 12-16** is similar to the diagram Russell plotted in 1913. The stars are not evenly distributed over the entire chart, but seem to group along a diagonal line. If we include many more stars than those on Russell's first diagram, we obtain a plot as shown in **FIGURE 12-17**. (Notice the Sun's position on the diagram.)

> **Hertzsprung-Russell (H-R) diagram** A plot of absolute magnitude (or luminosity) versus temperature (or spectral type) for stars.

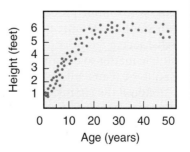

FIGURE 12-15 A plot of age versus height of people in a neighborhood might look like this. Each dot represents one person.

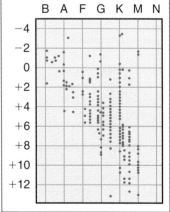

FIGURE 12-16 This graph is similar to the first H-R diagram, plotted by Henry Norris Russell in 1913. He plotted absolute magnitude versus spectral type and saw an obvious pattern in the distribution of stars. Notice that his category labeled N contained no stars.

FIGURE 12-17 A modern H-R diagram shows that stars fall into various categories, including main sequence stars, white dwarfs, giants, and supergiants. The temperature increases as you move to the left on the diagram. The purple lines show how we can calculate the absolute magnitude of a main sequence star from the H-R diagram; the star in this case is an A0 star, with a temperature of 10,000 K.

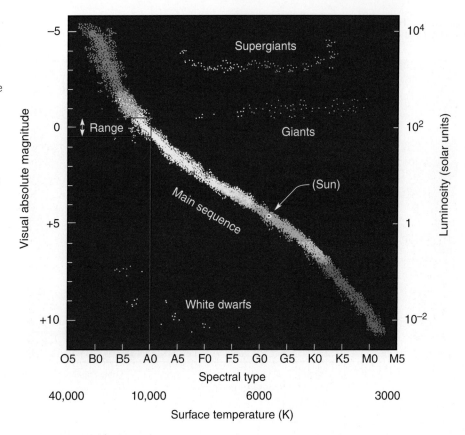

About 90% of all stars fall into a group whose properties are such that the plotted values form a band running diagonally across the H-R diagram. This band is called the ***main sequence***. Other groups are named as shown in the diagram. The significance of the groups is discussed later in this chapter and in the next few chapters, along with the life cycle of stars.

Spectroscopic Parallax

main sequence The part of the H-R diagram containing the great majority of stars; it forms a diagonal line across the diagram.

As we have seen, the distances to the nearest stars can be measured using stellar parallax. The H-R diagram provides another method of measuring such distances, one that is not confined to neighboring stars. Although this method does not involve parallax, it is called ***spectroscopic parallax***.

spectroscopic parallax The method of measuring the distance to a star by comparing its absolute magnitude with its apparent magnitude.

As we show later, it is possible to determine from the spectrum of a star whether that star is a main sequence star, a giant (or a supergiant), or a white dwarf. Suppose that we observe a particular star and determine that it is a main sequence star. As we have seen in our discussion of the blackbody curve and Wien's law in Section 4-4, it is a fairly routine procedure to ascertain the temperature of a star. Let's suppose that our star has a temperature of 10,000 K. The region of the main sequence where stars have such a temperature is marked on the H-R diagram of Figure 12-17. It is now a simple matter to determine that our chosen star has an absolute magnitude between about +0.5 and −0.5.

In practice, the temperature cannot be measured with absolute precision and some error thus is introduced in this manner also.

Recall the connection between absolute magnitude, apparent magnitude, and distance. If we know two of these, we can calculate the third. The use of the H-R diagram has allowed us to determine, within a small range, the absolute magnitude of the star. Because apparent magnitude is directly measurable, we can calculate the star's distance.

As indicated, we know the absolute magnitude only within a certain range, and therefore, our precision in determining distance is limited. Keep in mind, however, that this is *always* the case with a measurement. In using spectroscopic parallax, the source of the error is obvious, but as we have seen, the determination of distance by

TOOLS OF ASTRONOMY

Determining the Spectral Type of a Star

Before studying this Tools of Astronomy box, you may find it helpful to reread a related box in Chapter 4 entitled "The Balmer Series."

Hydrogen is the most common gas in the atmosphere of stars and, thus, it will serve as our example here. Each hydrogen atom has one electron in orbit around its nucleus. Suppose that the hydrogen gas near a star is cool enough that most of its electrons are in the ground state ($n = 1$ in **FIGURE B12-1**). When an atom with its electron in the ground state absorbs a photon, the photon's energy is such that the electron moves from the ground state to one of the other energy levels. These lines form the Lyman series, which is in the ultraviolet region of the spectrum; therefore, they do not

appear in the visible part of the spectrum and cannot be seen in Figure 12-14.

Now consider a hotter star, one with a significant number of atmospheric hydrogen atoms with electrons in the second energy level. For one of these atoms to absorb a photon, that photon must be of the right energy to move the electron from level 2 to higher levels. Figure B12-1 shows that the wavelengths of these photons are part of the Balmer series and are in the visible range. Look at the spectra of Figure 12-14. The hydrogen-α line is indicated at 656 nanometers, in the red part of the spectrum, far to the right of the other lines shown. The hydrogen-β line is at 486 nanometers, in the blue part of the spectrum. The other two Balmer lines shown in Figure B12-1 (hydrogen-γ, at 434 nanometers, and hydrogen-δ, at 410 nanometers) are in the violet part of the spectrum. The spectrum of the cool K5 star does not show a pronounced absorption line for the energy associated with hydrogen-β. The reason is that the star is not hot enough for many of its atmospheric hydrogen atoms to be in the level 2 state.

Look at the hydrogen-β absorption line in the hotter stars. As one moves to hotter and hotter stars, the absorption becomes more intense. In fact, the line is most pronounced in A-type stars. In the spectra of the hottest O- and B-type stars, the hydrogen-β absorption line is again less pronounced. The reason in this case is that in these very hot stars, most hydrogen atoms have so much energy that their electrons are at level 3 or the atoms are ionized. There thus are few electrons at the second level to absorb photons that correspond to the hydrogen-β absorption line.

Only one line of the Balmer series of hydrogen has been considered here. A complete analysis would require examining a number of lines of several different chemical elements. The spectral type of a star is determined by the relative intensities of these lines.

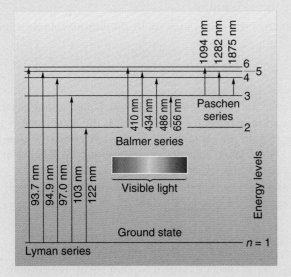

FIGURE B12-1 The energy levels of the hydrogen atom. If many atoms are in the $n = 2$ state, their absorption of photons will cause a dark line in the visible portion of the spectrum.

---

trigonometric (stellar) parallax is also limited in accuracy, both because of the difficulty in measuring the small angles involved and because other motions must be taken into account, such as the proper motion of the star being measured. All measurements contain error. The important thing in science is to recognize how great the likely error is.

Because of the limits in the accuracy of spectroscopic parallax, it is most useful when applied to groups of stars that are nearly the same distance from Earth. As we show in the next chapter, it is common for stars to be grouped with many other stars in large clusters. The stars of each cluster are very close to one another compared with their distance from Earth, and we use statistical methods to combine individual measurements of spectroscopic parallax to determine the distance to the cluster. The resulting value has much less error than the distance we calculate to any individual star.

Such averaging over a number of cases to improve accuracy is a valuable and common technique in science.

◼ EXAMPLE ◼

Spica, the brightest star in the constellation Virgo, is a B1-type main sequence star with a temperature of about 25,000 K. Its apparent magnitude is about 0.96. From these data, determine whether its distance from Earth is less than 10 parsecs, approximately 10 parsecs, greater than 10 parsecs, or much greater than 10 parsecs.

SOLUTION   Referring to Figure 12-17 and using the fact that Spica is a B1-type star, we determine that its absolute magnitude is between about −3 and −4.5 (although a safer range might be from −2.5 to −5.0).

Next, let's compare our absolute magnitude values with Spica's observed apparent magnitude. Spica's apparent magnitude is 0.96, and we can reason that because its apparent magnitude has a greater value than its absolute magnitude, Spica is farther than 10 parsecs away. (Remember that greater magnitude means a dimmer star and, thus, its apparent brightness is less than its brightness would be at 10 parsecs.) In fact, from the difference between +0.96 and −3 or −4, we can conclude that it would have to be moved fairly far to bring it as close as 10 parsecs. How far? The difference corresponds to about 4 to 5 magnitudes. According to Table 12-1, this means that we receive 40 to 100 times less light than if the star were at a distance of 10 parsecs. Using the inverse square law, we find that the star is about 6 to 10 times farther than 10 parsecs, or between 60 and 100 parsecs from us.

Actually, Spica is about 80 parsecs, or 260 light-years, from us. Its absolute magnitude is about −3.6.

TRY ONE YOURSELF
The star 40 Eridani is a main sequence star of spectral type K1, which means that its temperature is 4500 K. Its apparent magnitude is +4.4. (Actually, this is the magnitude of the brightest star of a triple-star system. In the example above, Spica is a binary system. Multiple star systems are discussed later.) What can you conclude about the distance to 40 Eridani?

> Maury worked at Harvard College Observatory with Annie Jump Cannon and Henrietta Leavitt, who are discussed elsewhere in this chapter. Maury learned astronomy as a student of Maria Mitchell (see Chapter 10).

> Do not confuse *luminosity class* with *spectral type* discussed earlier.

> **luminosity class** One of several groups into which stars can be classified according to characteristics of their spectra.

Luminosity Classes

Recall from Chapter 4 that a solid object emits a continuous spectrum because its atoms interact with one another and thereby distort their energy levels. This smears what would be separate, discrete wavelengths into a continuous range of wavelengths. Recall also that an absorption spectrum is produced when light passes through a star's atmosphere. In the 1880s, Antonia Maury discovered that absorption lines are also subject to a smearing effect and that, in general, they are not fine lines. This discovery has become very valuable in classifying stars.

The atmosphere of a main sequence star is fairly dense, and therefore, its atoms collide frequently, distorting their energy levels; as a result, what would have been a thin absorption line turns into a broader line. Red giant stars have thinner atmospheres and, thus, the broadening does not occur to the extent seen in a main sequence star. Their spectral lines are narrower (**FIGURE 12-18**). Through the examination of the extent of line broadening, as well as other subtle differences in spectra, stars are classified into various ***luminosity classes*** that are located on the H-R diagram as shown in **FIGURE 12-19**.

The example of spectroscopic parallax in the preceding section involved only main sequence stars. The classification of stars into luminosity classes allows spectroscopic parallax to be used with any star. If we know a star's luminosity class and temperature, we can read its absolute magnitude (and thus luminosity) from the appropriate location on the H-R diagram.

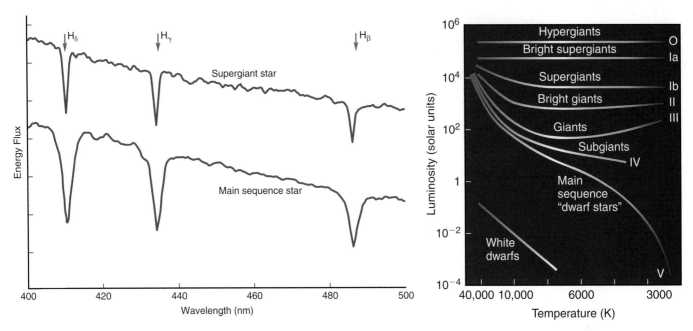

FIGURE 12-18 A comparison of the absorption spectrum of a 10,000 K supergiant star to that of a 10,000 K main sequence star. The denser atmosphere of the main sequence star results into broader hydrogen lines. The line strength and broadening depend on the number density of the atoms in the atmosphere.

FIGURE 12-19 Stars can be classified according to luminosity class, which is determined by analysis of their spectra.

Analyzing the Spectroscopic Parallax Procedure

Observe that in using the method of spectroscopic parallax to find a star's distance from us, we use two triple connections (**FIGURE 12-20**): By knowing the temperature of a star (that is, its spectral type) and its luminosity class, we can determine its absolute magnitude. Then, by knowing its absolute magnitude and its apparent magnitude, we can calculate its distance. This last triple connection is used in a different way than it was before. For nearby stars, distance and apparent magnitude were used to determine absolute magnitude.

Absolute magnitudes calculated for nearby stars enabled astronomers to draw the H-R diagram. Once the patterns of that diagram were known, they could be applied to stars that are too far away to permit distance measurement by parallax. We reasonably assume that these stars fit the same H-R diagram pattern as do nearby stars and, thus, we can then use the diagram to determine the absolute magnitudes of these faraway stars.

The basic procedure outlined here is often used in astronomy (and in other sciences). By observing familiar objects (nearby stars, in this case), we see patterns and formulate laws (or statements of relationships). We then assume that these patterns and laws hold for more distant objects of the same type. This allows us to use the same patterns to learn more about those distant objects. At times it appears that the whole system is a house of cards ready to fall down, but usually cross-checks are available; measurements can be taken in several ways, thereby permitting us to verify theories by independent measurements.

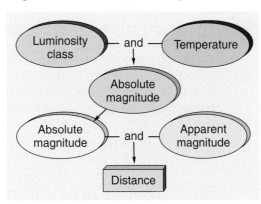

FIGURE 12-20 From the luminosity class and temperature of a star, its absolute magnitude can be determined. From this and its apparent magnitude, its distance can be determined.

Luminosity and the Sizes of Stars

Although the sizes of a few of the largest stars have now been directly detected by interferometry methods, in general a star is observed as only a point of light. How, then, can we determine the size of an object so distant that it appears to have no size?

Consider the group of stars in the lower left corner of the H-R diagram, the **white dwarfs**. These stars are hot, but their location in the lower part of the diagram indicates that they are intrinsically dim. One might think that a hot star would be bright, for this is not only the pattern seen in the main sequence, but it also follows from the intensity/wavelength graph: As an object becomes hotter, the wavelength at which the intensity of the emitted radiation is maximum becomes shorter, and the *total amount of radiation per unit area from the star increases.* How, then, can these stars be dim? The answer is that they are small. Being hot, each square meter of their surface emits more energy per second than a cooler star, but they simply have small total surface areas. (Recall our discussion of the blackbody curve and Stefan-Boltzmann law in Section 4-4.) The name *white dwarf* indicates their temperature (white-hot) and their size.

An everyday example might help here. You can adjust the burner of an electric stove to various temperatures. As the burner gets hotter, its color changes from dull red to orange-red (and perhaps to orange). At the same time, its overall brightness increases. Imagine, however, that you have a very small burner and you make it orange-hot. If it is small enough, it will emit less total radiation—it will be less bright—than a very large burner that is not as hot. The small, hot burner corresponds to a white dwarf in that its small size causes its overall luminosity to be less than average.

On the other hand, consider the **giants** and **supergiants**. How do we know that these are large stars? Because they are very bright in spite of their low temperatures. The giants at the upper right corner of the H-R diagram are often called *red giants,* paralleling the name given the white dwarfs. (A single stove burner set to "warm" does not yield much light, but if enough of these relatively cool burners are lit, they provide quite a lot of red light.)

In these examples of white dwarfs, giants, and supergiants, qualitative reasoning was used to learn something about their sizes; however, determination of stellar sizes is not limited to these classes of stars; nor is it limited to qualitative methods. Knowing the temperature of an object, we know the amount of energy emitted each second per square meter of its surface (that is, the energy flux).

Then if we know the total energy emitted by the object each second (that is, the luminosity, by knowing the absolute magnitude), we can calculate the area of its surface and therefore its diameter. In other words,

$$\text{Luminosity} = \text{area} \times \text{energy flux} \propto \text{radius}^2 \times \text{temperature}^4.$$

A star may be very luminous and large in size but have a relatively low temperature, or it may be very luminous with an average size if it has a high temperature.

white dwarf The burnt-out relic of a low-mass star. (Typical diameter is 0.01 that of the Sun—about the size of the Earth.)

giant star A star of great luminosity and large size (10 to 100 times the Sun's diameter).

supergiant A star of very great luminosity and size (100 to more than 1000 times the Sun's diameter.)

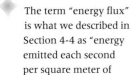 The term "energy flux" is what we described in Section 4-4 as "energy emitted each second per square meter of area."

EXAMPLE

We observe the spectrum of a star and find its luminosity class and spectral type. In addition, using Wien's law and the star's spectrum, we find that it has a temperature twice that of the Sun. From the temperature and luminosity class of the star, we find, using the H-R diagram, its luminosity, which is about that of the Sun. What is the radius of this star?

SOLUTION   A star's luminosity (L) is related to its radius (R) and temperature (T):

$$L_{\text{star}} \propto R^2_{\text{star}} \times T^4_{\text{star}}.$$

This expression also holds true for the Sun:

$$L_{\text{Sun}} \propto R^2_{\text{Sun}} \times T^4_{\text{Sun}}.$$

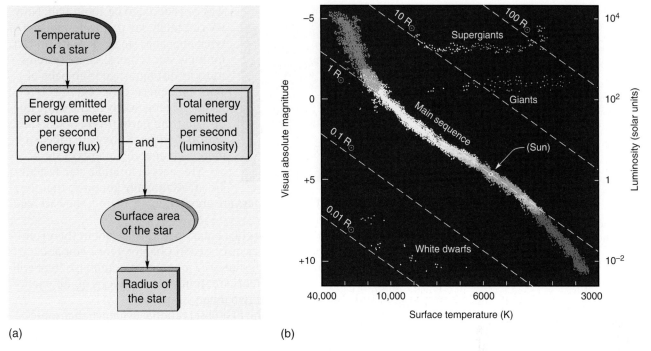

FIGURE 12-21 (a) If a star's temperature and the total power emitted by it are known, the triple connection shown here can be used to calculate its radius. (b) This H-R diagram (the same as in Figure 12-17) includes information about the radii of stars.

In this example, we have $T_{star} = 2 \times T_{Sun}$. Also, $L_{star} = L_{Sun}$, which allows us to set the right-hand sides of the above two expressions equal to each other. We get

$$R^2_{star} \times (2 \times T_{Sun})^4 = R^2_{Sun} \times T^4_{Sun} \text{ or } R^2_{star} \times 16 = R^2_{Sun}.$$

Therefore, the radius of this star is 1/4 that of the Sun.

TRY ONE YOURSELF
Find the radius of a star if its luminosity is 10,000 times that of the Sun and its temperature is about 3870 K. (Hint: The star's temperature is about two thirds that of the Sun.)

FIGURE 12-21a illustrates the triple connection in the chain of reasoning. We find that stars on the main sequence range from about 0.1 times the Sun's diameter up to as much as 20 times the Sun's diameter. Nonmain sequence stars, however, differ in size even more: White dwarfs may be as small as 0.01 of the Sun's size and supergiants may be 100 times larger in diameter than the Sun. The H-R diagram in Figure 12-21b is similar to that of Figure 12-17, but it also includes information about the radii of stars, calculated by using the Stefan-Boltzmann law.

It is always desirable to have a second, separate method of measuring a quantity. This allows us not only to check our theories regarding the object being measured, but also to check our measurement techniques. Fortunately, there is a second method for measuring the sizes of stars, although only a few stars can be measured by this technique. To see how it works, some discussion of multiple star systems is necessary.

12-5 Multiple Star Systems

A few decades ago, astronomy books reported that about one fourth of the objects that appear to be single stars really contain two or more stars in a close grouping. Books published a decade ago reported this as about one third. A few years ago, it

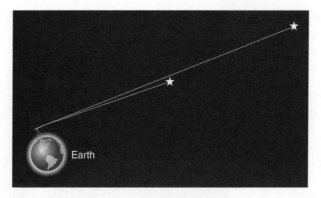

FIGURE 12-22 Two stars are said to form an optical double if they appear close together but have no actual relationship to one another.

optical double Two stars that have small angular separation as seen from Earth but are not gravitationally bound.

binary star system A system of two stars that are gravitationally bound so that they orbit one another.

visual binary An orbiting pair of stars that can be resolved (normally with a telescope) as two separate stars.

was considered to be about half. Now we can safely say that *more than half* of what appear as single stars are, in fact, multiple star systems. These systems must be distinguished from pairs of stars that appear close together as a result of being nearly in the same line of sight from Earth (**FIGURE 12-22**). These *optical doubles*, as they are called, are merely chance alignments of stars.

The stars in multiple star systems are gravitationally bound so that they revolve around one another. When two stars are gravitationally bound, they are said to be a *binary star system*.

FIGURE 12-23 illustrates how the two stars of a binary pair revolve about their common center of mass, in the same manner as discussed in Chapter 3 for a star and a planet. Although groups of more than two stars are common, our emphasis here will be on binary systems, as multiple star systems are simply combinations of, for example, a binary system and a single star or of two binary systems.

In 1802, William Herschel obtained the first observational evidence that some double stars orbit one another. (This was also the first direct evidence of gravitational force at work outside our solar system.) Herschel had taken an interest in double stars in hopes of using them for parallax measurements, but his discovery of binary star systems turned out to be much more important to astronomy, for it is by such systems that we are able to determine the mass of stars.

Binary systems are classified into several categories according to how they are detected. We discuss each category in turn and then explore what we can learn from binary stars.

Visual Binaries

A *visual binary* is a system in which the pair of stars can be resolved as two individual stars in a telescope. Using the largest telescopes, perhaps 10% of the stars in the sky are visual binaries. In a small telescope, only a small fraction of these can be resolved, but some are beautiful sights, particularly when the two stars are very different in color. **FIGURE 12-24a** is a photograph of Albireo, a visual binary in Cygnus, taken by an amateur astronomer. If you have access to a telescope, find Albireo during a clear summer night.

If two stars appear close together in the sky, how can astronomers tell if they are a binary system? The best way is simply to observe the pair over a period of time and look for signs of revolution. **FIGURE 12-25** shows a binary (at the right in the photograph) that reveals obvious orbital motion over a number of years; however, things

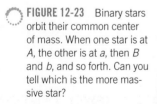

FIGURE 12-23 Binary stars orbit their common center of mass. When one star is at *A*, the other is at *a*, then *B* and *b*, and so forth. Can you tell which is the more massive star?

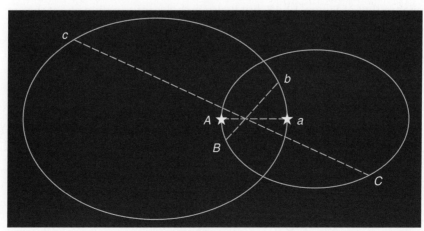

are not usually this easy, because in order for us to be able to resolve a binary pair, the stars must be either very close to Earth or very far apart. The Albireo pair, for example, is separated by some 4100 AU. As indicated by Kepler's laws, a great distance of separation indicates a long period of revolution, and the Albireo binary is thought to have a period of many thousands of years. Because no detectable orbital motion has occurred in the relatively short time over which we have photographic records, it is difficult to confirm that such a system is indeed a binary pair. We do know, however, that the two stars are about the same distance from us, and therefore we think that they are gravitationally bound.

A second method of determining whether a pair is indeed a binary system employs that powerful astronomical tool, the Doppler effect, which leads to the next type of binary system.

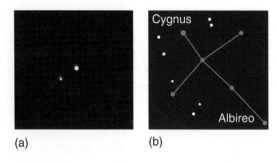

(a) (b)

FIGURE 12-24 (a) Albireo is a binary pair that shows an obvious color difference between the two stars. (b) Albireo is the bottom star of the cross of Cygnus, about 386 light-years from Earth.

Spectroscopic Binaries

FIGURE 12-26 shows two spectra of the star κ (the Greek letter *kappa*) Arietis, taken at different times. The upper spectrum contains more lines than the lower. Look closely and you will see that this is the result of each line in the lower spectrum having broken into two lines in the upper spectrum. Continuing observations of this star reveal that the spectrum repeatedly goes through a cycle in which each line gradually breaks into two, which spread until they reach a maximum separation and then come together again. The explanation for this is that we are seeing not one, but two stars. They are revolving around one another so that the Doppler effect causes us to see separate spectral lines when one star is moving toward us and the other away. Such a binary system is called a **spectroscopic binary**.

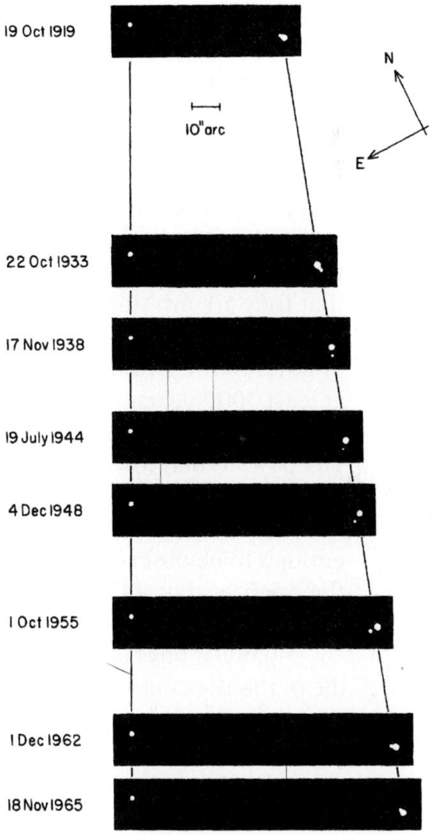

The Doppler effect shows only the radial components of the stars' motion; that is, the motion toward or away from us. If a binary pair is oriented so that its plane of revolution is perpendicular to the line of sight from Earth, no Doppler shift is observed in its spectral lines. On the other hand, if the plane of revolution is tilted directly toward the Earth, the Doppler effect allows measurement of the stars' actual velocities during the part of their orbits when one is moving directly toward us and the other directly away (**FIGURE 12-27**). At any

FIGURE 12-25 The visual binary Kruger 60 can be seen during its 45-year period. In addition, its proper motion away from the star at the left can be seen.

19 Oct 1919

22 Oct 1933

17 Nov 1938

19 July 1944

4 Dec 1948

1 Oct 1955

1 Dec 1962

18 Nov 1965

spectroscopic binary An orbiting pair of stars that can be distinguished as two due to the changing Doppler shifts in their spectra.

FIGURE 12-26 Two spectra of the spectroscopic binary κ Arietis are shown here. Lines that are single in the bottom spectrum are split in the upper one. When single lines appear, the two stars are moving at right angles to the line of sight. When the lines are double, one star is moving toward us and the other is moving away. Compare this figure with the next one, and mentally note the stars' positions on their orbits to get single or double lines.

FIGURE 12-27 When the binary stars whose spectra are shown in Figure 12-26 are moving perpendicular to the line of sight, no Doppler shift is observed. The Doppler effect is seen only when the stars have a component of motion toward or away from us. The spectra shown in Figure 12-26 are snapshots corresponding to specific positions of the stars on their orbits. As the stars orbit their center of mass, the lines continually shift between these two snapshots.

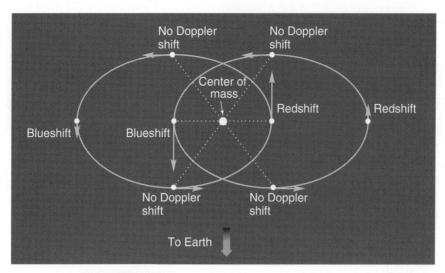

other position or any other orientation of the plane of revolution, we detect only the radial component of the stars' motions.

In the handle of the Big Dipper is a particularly interesting example of binary stars (**FIGURE 12-28**). As pointed out in Section 5-3, good eyesight reveals two stars at that location, the brighter one named Mizar and the dimmer one Alcor. Although it had been thought that Alcor and Mizar form an optical double rather than a gravitationally bound binary, more recent observations of their radial velocities indicate that they are orbiting one another. In any case, if you view Mizar and Alcor through a telescope, you will see that Mizar itself is two stars (Figure 5-10): Mizar A (the brighter of the two) and B are a widely separated visual binary and have a period of at least 3000 years.

In 1889, it was found that Mizar A is a spectroscopic binary with a period of only 104 days. When Mizar B was scrutinized, its spectral lines did not separate into two parts, but instead moved back and forth; first they redshifted and then blueshifted. Mizar B is indeed a binary star, but the spectrum of its companion is not bright enough to be observable. The shifting spectrum that is observed is the spectrum of the brighter star of the pair. We deduce the existence of the companion from that motion.

Finally, it has been found that Alcor is a spectroscopic binary. The dot in the handle of the Big Dipper thus in fact is six stars!

Mizar B is called a *single-line spectroscopic binary system* rather than a double-line system such as Alcor.

Eclipsing Binaries

light curve A graph of the numerical measure of the light received from a star versus time.

A special case of binary stars occurs when their plane of revolution is along a line from Earth so that one star moves in front of the other as they orbit. The star Algol, in the constellation Perseus, is a good example of an eclipsing binary. Its brightness changes periodically, and in 1783, John Goodricke, an amateur astronomer, explained the brightness change as being due to a dimmer companion passing in front of the brighter. **FIGURE 12-29a** shows a graph of the light received from Algol as time passes, called the system's *light curve*. Every 69 hours the apparent magnitude of Algol changes from 2.3 to 3.5. This happens when Algol A is partially eclipsed, as shown in Figure 12-29b. Midway between these dips in the light curve we see smaller dips, caused by the dimmer companion being eclipsed.

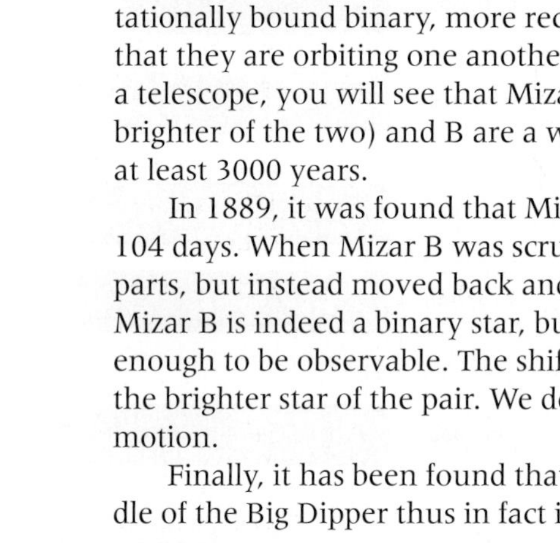

Since Goodricke's discovery that Algol is a binary star, its nature has been confirmed from

FIGURE 12-28 Even a small telescope reveals that Alcor and Mizar are three stars.

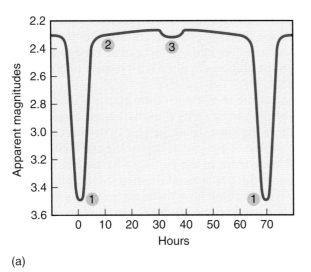

(a)

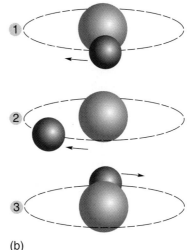

(b)

FIGURE 12-29 The light curve of Algol (a) is explained by the eclipsing of its components, one of which is much darker than the other. Positions 1, 2, and 3 in the light curve correspond to the numbers in part (b).

spectroscopic evidence. Like Mizar B, its companion is too dim to show its own spectrum, but the spectral lines of Algol A move back and forth in rhythm with its cycle of brightness changes. Like most eclipsing binaries, it can now be classified as both an eclipsing binary and a spectroscopic binary.

Other Binary Classifications

There are at least two other ways to detect binary stars. Sometimes a star is seen to shift back and forth in its position among the other stars, indicating that it is revolving around an unseen companion. Several such systems, known as ***astrometric binaries***, have been found.

If one star of a binary system is much hotter than the other, their spectra differ enough from each other that it is possible to ascertain that the spectrum we see is not from a single star, but is a composite of two spectra. Such a system is called a ***composite spectrum binary.*** This provides another method of detecting binary stars, but nothing can be learned of the motions of the stars in such a system.

astrometric binary An orbiting pair of stars in which the motion of one of the stars reveals the presence of the other.

composite spectrum binary A binary star system with stars having spectra different enough to distinguish them from one another.

12-6 Stellar Masses and Sizes From Binary Star Data

Binary stars are interesting in themselves. For example, astronomers speculate on the stability of a planetary system around a star that is part of a binary system and on how conditions would be different on such planets because of the extra star; however, the major importance of binary stars is that they allow us to measure stellar masses. Recall from Section 7-2 that we can calculate the mass of a planet from the orbit of one of its moons and that likewise we can calculate the mass of the Sun from the orbits of the objects circling it. The calculation involves Kepler's third law as revised by Newton. Stellar masses of binary systems are calculated in the same way.

To do such a calculation, we must know two things about the orbit of one (or both) of the stars: the size of the ellipse (its semimajor axis) and the period of revolution. The period is easy to ascertain in all cases except when it is extremely long. Determining the size of the ellipse is a little more complicated.

Figure 12-23 shows the orbits of a visual binary pair as seen straight on. The center of mass of the two stars can be determined from the fact that it must always lie along a line connecting them. Each of the orbits is an ellipse, and the center of mass of the two stars is located at one of the focal points of each ellipse; however, this is not what we usually observe. The explanation for this is that we are not seeing the ellipses straight on. A circle, when viewed at an angle, yields an elliptical shape (**FIGURE 12-30**), and an ellipse

FIGURE 12-30 A circular shape, such as the basketball rim, appears elliptical when viewed at an angle.

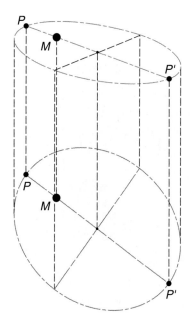

FIGURE 12-31 This figure shows the apparent path (above) and the true path (below) of a binary star. Although both shapes are ellipses, if you turn your book, you can see that point *M* is at a focus of the lower ellipse.

viewed at an angle also yields an ellipse, but one of a different shape than the original. The fact that the center of mass is not at one of the ellipses' foci can be used to determine how much the ellipses are tilted with respect to our line of sight (**FIGURE 12-31**).

After the true shape of the ellipse is determined, its size can be calculated using the small-angle formula. (We have to know the distance to the pair to do this calculation.) Knowledge of the size of one of the stars' ellipses, along with knowledge of the period of its motion, allows us to calculate the *total mass* of the two stars, using Kepler's third law. To determine how the mass is distributed between the two, we need only consider the ratio of the two stars' distances to the center of mass. This is analogous to the way that we can calculate the weight of each of the people on a seesaw if we know the total weight of the two people and know how far each is sitting from their center of mass if they are balanced (**FIGURE 12-32**). In Section 3-6 we worked out an example of a binary system in which we found the mass of each star by using this method.

In the case of a spectroscopic binary, if we are confident that the plane of the stars' revolution lies very close to our line of view, we can do a similar calculation. In this case, however, instead of calculating the size of the orbit with the small-angle formula, we calculate it from a knowledge of the maximum speed of the star in orbit and the period of the orbit. Because it is difficult to know the inclination of the orbit, we cannot calculate the masses; however, we can obtain valuable information about *average* masses of stars in a great number of spectroscopic systems by assuming an average inclination of the orbits. (Because any orbital inclination from 0° to 90° is equally likely, if enough systems are included in the analysis, the average inclination will be 45°.)

Eclipsing binaries that are also spectroscopic binaries provide us with a way of measuring not only the masses of the two stars but also their sizes. Because the fact that they eclipse one another means that the inclination of their orbit with our line of view is zero (or very nearly so), their Doppler shift tells us their velocities. Knowing their velocities and the time it takes to complete an eclipse, we can calculate the size and luminosity of each star. **FIGURE 12-33** shows a simplified case that illustrates this calculation. We can also determine the size of a star if we know its luminosity, distance, and apparent magnitude. Eclipsing binary systems give us an independent method of measuring star sizes.

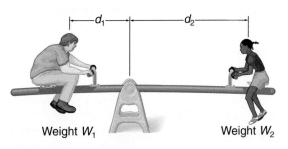

FIGURE 12-32 Starting from the simple relationship $W_1 d_1 = W_2 d_2$, it can be shown that, if W is the total weight of the two people,

$$W_1 = \frac{W \times d_2}{d_1 + d_2}.$$

This same method allows us to determine the mass of each star in a binary system if we know the total mass and the distance of each star from the center of mass.

FIGURE 12-33 Suppose Doppler effect data tell us that the relative velocity of these two stars as they pass one another is 8.0×10^2 km/s (which is 6.9×10^7 km/day). Assume that in the leftmost drawing of the stars, the small one is moving to the right. At point B on the light curve, it began to be hidden by the large star. At point D it starts to emerge. From B to D is about 0.8 days, and thus, it took the small star this long to cross the large one. The diameter of the larger star must then be (0.8 days) × (6.9×10^7 km/day), or 5.8×10^7 km. In a similar manner, using the time period between points B and C, we can calculate the diameter of the smaller star.

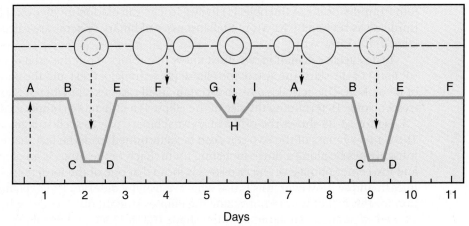

12-7   The Mass-Luminosity Relationship

Suppose we plot a graph of the masses of several stars compared with their luminosities. **FIGURE 12-34** shows such a graph, plotting mass in solar masses along the horizontal axis (*x*-axis) and luminosity in solar units along the vertical axis (*y*-axis). There is an obvious correspondence for main sequence stars: More massive stars are more luminous.

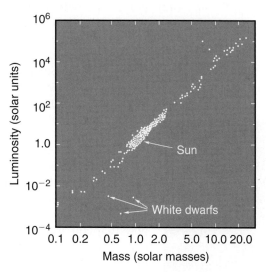

FIGURE 12-34 When the luminosities of stars are plotted against their masses, an obvious relationship is seen for most main sequence stars. The three at the bottom that do not fit here are white dwarfs. (The scale at the bottom is a logarithmic scale.)

This ***mass-luminosity diagram*** was produced from knowledge of binary stars that are close enough to Earth to yield the necessary data; however, it is reasonable to assume that more distant stars also follow this pattern. (Otherwise, we would be claiming that we live in some special place in the universe where stars behave differently.) The graph of Figure 12-34 is empirical; that is, it is plotted from values that result directly from measurements, rather than from a theory of how mass and luminosity should be related. Because the points that represent main sequence stars lie along an approximately straight line, we can determine the mathematical equation for that line. The equation is found to be

$$L = M^{3.5} \text{ (for main sequence stars)},$$

where both *L*, the luminosity of a star on the main sequence, and *M*, the star's mass, are given in solar units.

We can use this equation to calculate the luminosity of a star that has three times the mass of the Sun:

$$L = M^{3.5} = 3^{3.5} = 3 \times 3 \times 3 \times \sqrt{3} = 46.8.$$

This star's luminosity, thus, is about 47 times that of the Sun.

The mass-luminosity relationship is valuable to astronomers in investigating less accessible stars and in constructing and evaluating hypotheses concerning the life cycle of stars. This is discussed in the next chapter.

mass-luminosity diagram
A plot of the mass versus the luminosity of a number of stars.

12-8   Cepheid Variables as Distance Indicators

Eclipsing binary stars are sometimes called eclipsing *variables*, as their light intensity varies over time. Other types of variable stars are also found in the heavens. One particular type is of importance to us here because it provides a method of measuring distances.

In 1784, John Goodricke (the same astronomer who explained Algol's variations) discovered that the star δ Cephei varies in luminosity in a regular way but that its variations cannot be explained by an eclipse of a binary companion. This star changes its apparent magnitude between 4.4 and 3.5 every 5.4 days, with a light curve as shown in **FIGURE 12-35**. Soon thereafter, other stars exhibiting this

Goodricke made his discovery when he was 19 years old, just 2 years before his untimely death.

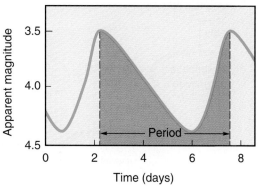

FIGURE 12-35 This is the light curve of δ Cephei, the prototype of a class of stars that have light curves with this shape and are therefore called Cepheid variables.

δ is the Greek letter "delta."

characteristic light curve—brightening quickly and then dimming more slowly—were seen. Doppler effect data show that such stars are actually pulsating—changing size—in rhythm with their changes in luminosity. It is fairly easy to identify this class of stars, now called *Cepheid variables*, or *Cepheids*.

Each Cepheid has a very constant period of variation; the periods range from about 1 day to about 3 months for different Cepheids. We discuss the reason for the pulsations of these stars in Section 14-6.

In 1908, a Harvard College astronomer, Henrietta S. Leavitt, published data on variable stars in the Small and Large Magellanic Clouds (**FIGURE 12-36**). She discovered that in the case of Cepheid variables, the brighter variables have the longer periods.

It may seem strange that period and *apparent* magnitude are related, for apparent magnitude is not a quantity intrinsic to the star; rather, it depends on our distance from it. The reason the relationship exists is that all of the stars of the Magellanic Clouds are about the same distance from us—at least, compared with the distance to the clouds. To see this, suppose that you are on a hill outside your town at night, observing the street lights of the town (each of which, we will assume, has the same absolute luminosity). The lights will *appear* to be many different brightnesses, depending on each light's distance from you. Now suppose that the hill is high enough so that you can use a small telescope to observe the lights of a town 200 miles away. You will find that the lights of this town all appear about equally bright, as each of them is 200 miles away, give or take only a few miles.

The stars of the Magellanic Clouds do not all have the same intrinsic (absolute) luminosity, but because they are all at about the same distance, their apparent magnitudes are related directly to their absolute magnitudes. For example, if the absolute magnitude of a certain star in the Large Magellanic Cloud is five magnitudes less than its apparent magnitude, the absolute magnitude of *each* star in the cloud will be five magnitudes less than its apparent magnitude.

Astronomers quickly realized the importance of the relationship between the magnitude and the period of Cepheids (**FIGURE 12-37**): The period (which is easy to determine) allows us to determine the absolute magnitude, and we can use this quantity along with apparent magnitude to determine the distance to the variable star (using the relationship between absolute magnitude, apparent magnitude, and distance that was explained in the Tools of Astronomy box on page 344). The problem was that only the *apparent* magnitudes of the Cepheids in the Magellanic Clouds were known. The absolute magnitudes of these stars could not be determined because the clouds are too far away for parallax to be observed. The apparent magnitudes of Cepheids in the clouds are around 15; they appear dim. However, it was obvious from Cepheid variables nearer the Earth that Cepheids, as a class, are very luminous stars. This provided qualitative evidence that the Magellanic Clouds are at a great distance from Earth.

All that was needed was to find one Cepheid variable near enough to the Earth that its distance could be measured by parallax, but there is none. Beginning in 1917, Harlow Shapley (to be discussed more fully when we study galaxies) worked out a complex statistical method to determine distance to Cepheids within our Galaxy. Using such methods, by the early 1950s, astronomers were confident that they had determined correct

FIGURE 12-36 The Large Magellanic Cloud (LMC, on the right) and the Small Magellanic Cloud (SMC, on the left) are irregular galaxies, satellites to our Galaxy. They appear to the naked eye as hazy cloudlike patches in the sky of the Southern Hemisphere. They were first reported to Europeans by Magellan after his voyage around the world.

Polaris is a Cepheid variable with a small magnitude variation of between 1.9 and 2.1 and a period of 3.97 days.

FIGURE 12-37 The distance modulus of a star (the difference between its apparent and absolute magnitudes) gives us its distance in parsecs. Because the period of a Cepheid star gives us its absolute magnitude and because its apparent magnitude is easily measured, we can find the distance to the star.

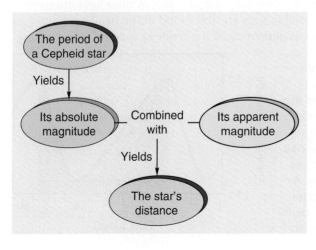

Henrietta Leavitt

Henrietta Swan Leavitt (**FIGURE B12-2**), one of seven children, was born on July 4, 1868. She graduated from Radcliffe College (then known as the Society for the Collegiate Instruction of Women) in 1892. In her senior year she took a course in astronomy, and after teaching and traveling for a few years, she joined the Harvard College Observatory in Cambridge, Massachusetts in 1895 as a student and volunteer research assistant. The quality of her work resulted in a permanent position on the staff in 1902. Soon she was named chief of the photographic photometry department.

Leavitt's primary work was with variable stars, and she discovered many of them. In 1908 she published a list of 1777 variable stars in the Magellanic Clouds; the list included a table of 16 Cepheid variables on which she had very precise data. Concerning these she inserted a comment: "It is worthy of notice that in Table VI the brighter variables

have the longer periods."[1] This brief comment went unnoticed by other astronomers. Later that year, because of ill health, she was forced to return to her family home in Wisconsin.

During her years of illness in Wisconsin, Leavitt continued her work (the observatory sent stellar photographs to her). When she returned to Cambridge in 1912, she published a report on 25 Cepheid variables. In it she stated, "A remarkable relation between the brightness of these variables and the length of their periods will be noticed."[2] She included a graph showing the relationship between period and brightness, but although she recognized the importance of the discovery, her duties at the observatory (and the nature of the work being done there) prevented her from following up on it.

Henrietta Leavitt was one of a group of some 40 women hired by Edward Pickering at the Harvard Observatory, starting in the 1880s, when it was unusual for women to work outside the home. Although women had made contributions to astronomy prior to this time, many still considered science an inappropriate field for them. Pickering was not really a progressive thinker in hiring the women; he did so because they would work for less money than men. Nonetheless, he was rewarded by the significant advances made by many members of the group, particularly Annie Jump Cannon, Antonia Maury, and Henrietta Leavitt.

Although her health was never good and her hearing was impaired from childhood, Leavitt worked at the Harvard Observatory until her death from cancer in 1921.

FIGURE B12-2   Henrietta Leavitt (1868–1921).

1. Henrietta Leavitt, "1777 Variables in the Magellanic Clouds," *Annals of the Harvard College Observatory* 60 (1908): 107.
2. Henrietta Leavitt, *Periods of 25 Variables in the Small Magellanic Cloud*, Harvard College Observatory Circular no. 173 (March 3, 1912).

distances to several Cepheids. This allowed them to plot a graph of period against *absolute* magnitude, as shown in **FIGURE 12-38**.

Brighter Cepheids have longer pulsation periods. One of the reasons that it took from 1917 to the early 1950s for astronomers to determine the correct relationship between Cepheids' periods and absolute magnitudes (the *period-luminosity relationship*) accurately is that there are actually two different types of Cepheid variables. Cepheid I stars are younger and richer in metals than Cepheid II stars. The two groups differ from each other in luminosity by about a factor of four, or 1.5 magnitudes, for the same period. We can now detect the difference between the two types, and the uncertainties surrounding the period-luminosity relationship have been greatly reduced with time.

To show how the period-luminosity chart can be used to determine distances, suppose that a Cepheid I variable has a period of 10 days and an apparent magnitude of 14. Figure 12-38 shows that a Cepheid I with a period of 10 days has an absolute magnitude of about 24. Now that both apparent and absolute magnitudes are known,

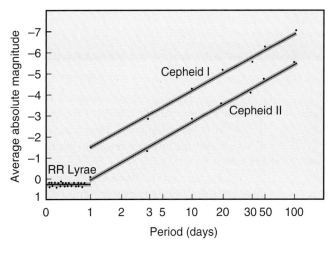

FIGURE 12-38   The period-luminosity diagram for Cepheid (and RR Lyrae) variables. Luminosity and absolute magnitude are intricately related because they are two different ways of representing the same quantity: the true energy output of a star.

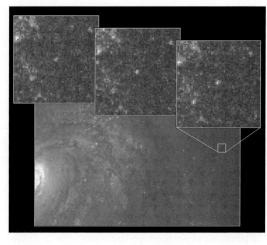

FIGURE 12-39   A Cepheid variable in the galaxy M100, seen by the *HST*. The measured period of 51.3 days indicates a distance from Earth of 56 million light-years.

Corresponding to normal metric usage, 1000 parsecs is one kiloparsec, abbreviated kpc, so the Large Magellanic Cloud is 48 kpc away.

Using light echoes, astronomers measured the distance to RS Pup, a Cepheid in the constellation of Puppis, to be 6500 ± 90 light-years!

we can calculate its distance. (Because the star's apparent magnitude is greater than its absolute magnitude by 18, you can see that the star must be extremely far away. In fact, using the distance modulus relationship on page 344, we find a distance of about 40,000 parsecs.) Use of Cepheid variables tells us that the Large and Small Magellanic Clouds are, in fact, 48,000 and 56,000 parsecs away, respectively.

In measuring magnitudes, however, uncertainties are introduced by the fact that the gas and dust in interstellar space absorbs and scatters starlight. These effects of the interstellar material on starlight have been well studied and can be corrected for; we examine them in Section 13-1.

A new way to accurately measure the distance to a Cepheid is based on the presence of echoes. When a Cepheid is embedded in a nebula, light from the star reaches us in two ways; first, directly from the star, and second, after reflecting from the dust of a part of the nebula. The luminosity of the star changes in a distinctive pattern (Figure 12-35). The reflected light from a blob in the nebula follows the exact same pattern but with a delay, because the star's light must first travel from the star to the nebula. This time delay, "light echo," when multiplied by the speed of light, gives us the distance between the star and the blob. Knowing this distance and the angular separation between the blob and the Cepheid we can find the distance to the star. This type of distance measurement is purely geometrical and does not depend on assumptions about the star's properties.

Figure 12-38 includes another type of variable stars, the RR Lyrae stars. They are also used to measure distances, but because they are less luminous than the Cepheid stars, they can only be seen to smaller distances. They have periods ranging from a few hours to one day, and their luminosity is approximately constant. (In Section 14-6, we examine the reason behind the variability of both Cepheid and RR Lyrae stars.)

Although astronomers have detected only a few thousand Cepheid variables, their importance greatly outweighs their number. Because they are among the very brightest of stars, we can see them not only in distant parts of our own Galaxy, but in other galaxies, giving us a method of measuring distances to these faraway worlds (**FIGURE 12-39**).

Conclusion

As we have shown in this chapter, once we know the distance to a star, many other properties become available to us. Knowing distance and proper motion, we can calculate a star's tangential velocity. The Doppler effect allows us to measure radial velocity, and from these two velocities, we can determine a star's actual motion relative to the Earth and to other stars. From distance and apparent magnitude, we can calculate a star's absolute magnitude. This knowledge can be combined with knowledge of the star's temperature to calculate its size.

On the other hand, the patterns in the H-R diagram allow us to determine a star's absolute magnitude without first knowing its distance so that we can use the distance versus absolute magnitude versus apparent magnitude relationship in reverse to determine stellar distances.

The discovery and analysis of binary stars have provided a second method of learning a star's size and also a method of measuring a very fundamental property of a star, its mass.

Finally, a particularly useful type of star, the Cepheid variable, provides a tool to measure even greater distances, opening up the entire field of galactic astronomy.

STUDY GUIDE

1. Suppose you observe a previously uninvestigated star and find its apparent magnitude. To determine its absolute magnitude, you need to know the star's
   A. distance.
   B. color.
   C. velocity.
   D. brightness as seen from Earth.
   E. Doppler shift.

2. The absolute magnitude of a star is
   A. the same as the apparent magnitude.
   B. equal to the greatest the apparent magnitude can be, in the case of a variable star.
   C. equal to the apparent magnitude if the star is 10 parsecs away.
   D. the size of a star from one side to the other.
   E. equal to the brightness of the star on the clearest night.

3. A magnitude change of 1 corresponds to a brightness change of about 2.5 times. A magnitude change of 2 corresponds to a brightness change of about
   A. 1.25 times.
   B. 2.5 times.
   C. 5 times.
   D. 6.3 times.
   E. 15.9 times.

   For the next three questions, refer to the following table:

Star	Apparent Magnitude	Absolute Magnitude	Distance (parsecs)
Sirius	−1.5	—	2.7
Vega	0.0	0.5	—
Antares	0.9	−5.1	160.0
Fomalhaut	1.15	—	6.9

4. Which star appears dimmest to an observer on Earth?
   A. Sirius.
   B. Vega.
   C. Antares.
   D. Fomalhaut.
   E. [The answer cannot be determined from the information given.]

5. Which of the following is about the correct distance for Vega?
   A. 4.2 parsecs.
   B. 8.1 parsecs.
   C. 11.7 parsecs.
   D. 26.7 parsecs.
   E. [The answer cannot be determined from the information given.]

6. Fomalhaut's absolute magnitude is about
   A. 0.75.
   B. 2.
   C. 12.
   D. 22.
   E. 20.75.

7. Suppose that Star X is twice as far away as Star Y. The parallax angle of Star X is about
   A. half that of Star Y.
   B. the same as that of Star Y.
   C. twice that of Star Y.
   D. four times that of Star Y.
   E. [The answer cannot be determined from the information given.]

8. The distances to nearby stars can be measured by
   A. comparing the apparent magnitudes of several stars.
   B. bouncing radar pulses from their surfaces.
   C. measuring the time it takes light to get here from them.
   D. measuring their shifting motion against background stars through the year.
   E. [Both B and C above.]

9. Star S and Star K have the same tangential velocity, but Star S is closer to Earth. Which has the larger proper motion?
   A. Star S.
   B. Star K.
   C. They both have the same proper motion.
   D. [The answer cannot be determined from the information given.]

10. Star S and Star K have the same tangential velocity, but Star S is closer to Earth. Which has the larger radial velocity?
   A. Star S.
   B. Star K.
   C. They both have the same radial velocity.
   D. [The answer cannot be determined from the information given.]

11. If a star is 100 light-years away, what is its approximate distance in parsecs?
   A. 3000 parsecs.
   B. 900 parsecs.
   C. 30 parsecs.
   D. 9 parsecs.
   E. 1/3 parsec.

12. To use spectroscopic parallax, we must know
   A. the star's diameter.
   B. the star's temperature.
   C. the distance from the Earth to the Sun.
   D. the star's luminosity class.
   E. [Both B and D above.]

13. If the temperature of a star increases without a change in the star's size, its point on the H-R diagram moves
   A. up and to the left.
   B. up and to the right.
   C. down and to the left.
   D. down and to the right.

14. Stars on the main sequence that have a small mass are
   A. bright and hot.
   B. dim and hot.
   C. dim and cool.
   D. bright and cool.
   E. [Any of the above; there is no regular relationship.]

15. To observe spectroscopic binaries, we rely on
   A. knowing the composition of the individual stars.
   B. our knowledge of the distance separating the stars.
   C. our knowledge of the distances from Earth to the stars.
   D. the change in light intensity as the stars orbit.
   E. the Doppler effect.

16. The stars of binary star systems
   A. revolve around a point midway between their centers.
   B. revolve around a point somewhere between their centers (but not necessarily midway).
   C. revolve such that one star remains still, and the other revolves around it.
   D. do not revolve around each other.

17. Binary star systems are especially important to us because they allow us to calculate the _____ of stars.
   A. compositions
   B. proper motions

   C. radial velocities
   D. temperatures
   E. masses

18. Which type of binary system provides the most information about its component stars' masses and sizes?
   A. Eclipsing binaries.
   B. Spectroscopic binaries.
   C. Visual binaries.
   D. Composite spectrum binaries.
   E. [All of the above about equally.]

19. We can determine the size of stars by
   A. measuring the size of their image in a telescope.
   B. a measurement using a spectroscope.
   C. calculations based on temperature and absolute magnitude.
   D. Doppler effect measurements.
   E. mass measurements.

20. Cepheid variables can be used to find distances because their
   A. luminosity is related to their period.
   B. radial velocity is related to their mass.
   C. distance is related to their mass.
   D. magnitude is related to their color.
   E. period is related to their radial velocity.

21. If the orbital plane of a certain binary star system is not parallel to our line of view, the maximum radial velocity of one of its stars measured by the Doppler effect will be
   A. less than its true velocity.
   B. greater than its true velocity.
   C. either greater or less than its true velocity, depending on other factors.

22. If a star has a parallax angle of 0.20 arcseconds, how far away is it?
   A. 0.20 parsec.
   B. 1.0 parsec.
   C. 2.0 parsecs.
   D. 5.0 parsecs.
   E. [The answer cannot be determined from the information given.]

23. When a star's spectrum is redshifted as a result of the Doppler effect, we know that the star is
   A. much cooler than average.
   B. slightly cooler than average.
   C. about average temperature.
   D. hotter than average.
   E. moving away from Earth.

24. Spectroscopic parallax allows us to measure
   A. the distances to stars using the Earth's orbital motion.
   B. the distances to stars using the Earth's rotational motion.
   C. the distances to stars using the H-R diagram.
   D. the temperatures of stars using their spectra.
   E. the radial speed of stars using their spectra.

25. Considering only stars on the main sequence, the most massive stars are the
   A. hottest and brightest.
   B. hottest and dimmest.
   C. coolest and brightest.
   D. coolest and dimmest.
   E. [No general statement can be made.]

26. To calculate (or judge) the velocity of an object moving in the sky, what quantities must we know?

27. Define and distinguish between radial velocity and tangential velocity.

28. Define and distinguish between apparent magnitude and absolute magnitude. To which is luminosity most closely related?

29. Describe the most direct method of measuring the distances to stars.

30. How is the radial velocity of a star measured?

31. What two quantities are plotted on an H-R diagram?

32. Sketch an H-R diagram, labeling the axes and showing the general location of the following: main sequence, red giants, supergiants, and white dwarfs.

33. Explain the method of spectroscopic parallax.

34. It would seem that a hot star should be a very luminous star. How can white dwarfs be dim stars?

35. Distinguish between optical doubles and binary stars. Why is one of these much more important to astronomers than the other?

36. Name and describe at least three methods of detecting binary stars.

37. Sometimes we can tell by the spectrum of a star that it is part of a binary system. Explain how this is done.

38. Describe the problem presented by the fact that the plane of revolution of a binary is unlikely to be parallel to our line of view.

39. Name two quantities that must be known to determine the absolute magnitude of a star (other than a Cepheid variable).

40. Explain how Cepheid variables are used to determine distances.

1. The apparent magnitude of the star Proxima Centauri (the closest nighttime star) is about 11, and the apparent magnitude of 61 Cygni B is about 6. Which star appears brighter? How many times more energy do we get from it than from the dimmer star?

2. Both "parsec" and "light-year" might sound—to non-astronomy students—to be units of time. Define each. Why do astronomers find "parsec" to be a useful unit?

3. The apparent magnitude of a star is fairly easy to determine. What else must be known to calculate absolute magnitude? Why is it important to know the absolute magnitude of a star?

4. Why do only nearby stars show measurable proper motion? In general, how does the tangential velocity of nearby stars compare to that of distant stars?

5. In Figure 12-25 (Kruger 60), the binary pair is getting farther from the single star. From this series of photographs it would be impossible to tell which has the proper motion. Still, the binary pair would be taken as the most likely candidate. Why?

6. Draw an H-R diagram, labeling the axes and including the names of at least three classes of stars.

7. It is possible for a binary star system to fall into two or more classifications (such as visual binary and eclipsing binary). Describe some situations in which this might be the case.

8. Explain why there is such a simple relationship between the apparent and absolute magnitudes of stars in the Magellanic Clouds.

9. Do you think that it is reasonable to use the H-R diagram, which is plotted from data on nearby stars, for stars much more distant from us? Why or why not?

10. In a group of stars that are gravitationally bound, we observe a red star and a blue star. Both seem to have the same radius. Which one will look brighter? Explain your answer.

11. A star's luminosity class plays an important role in determining its luminosity. What are we looking for in a star's spectrum to determine its luminosity class?

12. Why is spectroscopic parallax called "parallax" even though no angle measurement is involved?

13. Would a mass-luminosity diagram (Figure 12-34) show the same relationship if the scales were independent of the Sun instead of being multiples of the Sun's values?

14. We observe a binary system composed of two stars of the same radius and temperature. We observe their elliptical orbits edge on (so that we get full eclipses during each cycle). Sketch the light curve for this eclipsing binary. How would the light curve change if the two stars were of the same radius but one of them were blue and the other yellow?

15. Why was it important to find the distance to a Cepheid variable after Leavitt had plotted the data for Cepheids in the Magellanic Clouds?

1. What is the ratio of light received from stars that differ in magnitude by 15?

2. About how many times as much light reaches us from Antares, the brightest star in Scorpius (apparent magnitude +1.0), than from τ (the Greek letter "tau") Ceti (apparent magnitude +3.5)?

3. The star Ross 128 has a parallax angle of 0.30 arcseconds. How far away is it in parsecs? In light-years?

4. α (the Greek letter "alpha") Centauri is the nearest naked-eye star to the Sun. Its apparent magnitude is +0.01, and its distance is 1.35 parsecs. Is its absolute magnitude greater or less than 0?

5. Use data from the last question to calculate the parallax angle of α Centauri.

6. Altair, a star visible in the summer and fall, is on the main sequence, is spectral type A7 (8300 K), and has an apparent magnitude of +0.77. Is Altair closer than 10 parsecs, about 10 parsecs away, somewhat farther than 10 parsecs, or much farther than 10 parsecs?

7. Rigel, a bright supergiant, is a B8 Ia star with a temperature of about 10,300 K. Its apparent magnitude is +0.14. What can you conclude about its distance from us?

8. A certain main sequence star has a mass four times greater than the Sun. How does its luminosity compare with the Sun's?

9. A binary system, observed face on, has a period of 4 years and parallax 0.2 arcseconds. The length of the semimajor axis is 0.4 arcseconds. What is the total mass of the system? (Hint: You will need to first find the distance to the binary system. Then use the small-angle formula to find the length of the semimajor axis in AU. Finally, use Kepler's third law to find the total mass.)

10. Use the data given in the caption of Figure 12-33 to calculate the diameter of the smaller star.

1. "Origins of the Stellar Magnitude Scale," by J. B. Hearnshaw, in *Sky & Telescope* (November, 1992).

2. *Stars and Their Spectra*, by J. Kaler (Cambridge University Press, 1997).

3. "Reading the Colors of Stars," by C. Sneden, in *Astronomy* (April, 1989).

4. "Accretion Disks in Interacting Binary Stars," by J. K. Cannizzo and R. H. Kaitchuck, in *Scientific American* (January, 1992).

5. "Bonuses of the Microlensing Business," by M. Mateo, in *Sky & Telescope* (September 1997).

6. "Extragalactic Eclipsing Binaries: Astrophysical Laboratories," by I. Ribas, in *New Astronomy Reviews* (July 2004).

7. "The Stars that Revealed the Vastness of the Universe," by B. Dorminey, in *Astronomy* (September 2009).

Quest Ahead to Starlinks
http://physicalscience.jbpub.com/starlinks

Starlinks is this book's online learning center. It features **eLearning**, which contains chapter quizzes and other tools designed to help you study for your class. You can also find **online exercises**, view numerous relevant **animations**, follow a guide to **useful astronomy sites** on the Internet, or even check the latest **astronomy news** updates.

CALCULATIONS

EXPANDING THE QUEST

STARLINKS

Interstellar Matter and Star Formation

13

WHEN WE GAZE AT THE STARS, THEY APPEAR TO BE UNCHANGING point-like sources of light. This was at the core of ancient beliefs about the heavens. Not until Galileo observed sunspots on the surface of our star did we start thinking of stars as changeable. We now know that all stars change, some gradually enough that we do not notice the changes during our lifetimes but others rapidly and dramatically. We know that the evolution of stars is a cyclic process. Stars are born out of the gas and dust that exist in the space between them, called the *interstellar medium*. The stars evolve and finally die and in the process return some of their mass back to the interstellar medium through explosive events and stellar winds. This recycled material will eventually form a new generation of stars.

The true-color chapter-opening image of the giant galactic nebula NGC3603 shows the entire cycle of stellar evolution. A blue supergiant star (a star at the final stages of evolution) is shown at the upper left of center, along with its grayish blue ring of glowing gas. Two jets are returning processed material back to the interstellar medium (blobs to the upper right and lower left of the star). A cluster of stars appears near the center of the image, dominated by young, hot stars and early O-type stars. A large cavity around the cluster has been created as a result of the interaction between the ionizing radiation and winds emitted from these massive stars and the interstellar

This true-color image of the giant galactic nebula NGC3603 was taken in 1999 by the *Hubble Space Telescope*. It illustrates the entire life cycle of stars. First, the Bok globules and giant gaseous pillars, then disks around protostars, followed by the evolved massive stars in the young cluster, and finally, the blue supergiant with its ring and jets. (The text here describes the positions of these objects in the photograph.)

All cross references to chapters, sections, figures, and tables pertain to the main text, *In Quest of the Universe, Sixth Edition. In Quest of the Solar System* contains Chapters 1–11 and 19 of the main text. *In Quest of the Stars and Galaxies* contains Chapters 1–5 and 11–19 of the main text.

An everyday definition of *medium* is "an intervening substance through which something is transmitted." Light is transmitted through empty space but also through the interstellar medium.

molecular cloud A cold (10–50 K), dense (10^3 – 10^6 particles/cm^3), massive (from a few to a few million times the mass of the Sun) interstellar cloud; it consists mainly of molecular hydrogen, with traces of carbon monoxide and other molecules.

The Advancing the Model box on page 373 explains why this hypothesis for the dark areas was abandoned.

IRAS, a joint project of the United States, the United Kingdom, and The Netherlands, was launched in 1983. During its 10 months of operation, it detected about 350,000 infrared sources and extended infrared emission from a number of objects.

interstellar cirrus Faint, diffuse dust clouds found throughout interstellar space.

medium. This interaction is clearly shown in the form of the giant gaseous pillars to the right of the cluster. The dark clouds at the upper right (called Bok globules) are probably at an earlier stage of star formation. Finally, at the lower left of the cluster are two compact, tadpole-shaped emission nebulae. These structures may correspond to gas and dust evaporating from disks around protoplanets.

In this chapter, we discuss all but one of these objects. The image clearly shows that the interstellar medium plays a crucial role in the evolution of stars and galaxies. Stars form in **molecular clouds**, which are the densest regions of the medium and where molecular hydrogen can form; in turn, stars replenish the medium with matter and energy during their lifetimes through stellar winds, planetary nebulae, and supernova explosions (which we describe in the next two chapters). The result of this interplay affects the rate at which a galaxy depletes its gaseous content and thus its star-forming period.

In previous chapters, particularly Chapter 12, we discussed the tools and techniques by which astronomers measure the properties of stars and discern relationships among these properties. In the next few chapters, we show how these tools and techniques have enabled us to learn about the universe beyond the solar system. Astronomers study the dust and gas that exist in interstellar space and are learning how this material forms into new stars, how these stars live out their lives, and how they eventually "die."

Astronomers have amassed a tremendous amount of information about the life cycles of stars. Still, you may not be surprised to learn that many of our theories concerning stellar evolution are somewhat tentative. Many questions remain.

13-1 The Interstellar Medium

There are two seemingly contradictory truths about the space between the stars. First, a large amount of gas and dust exists in that space. Second, the space between stars is very nearly a perfect vacuum. The secret to reconciling these statements is to realize the vastness of interstellar space. Although the molecular density (the number of molecules per cubic centimeter) of the dust and gas in space is very small—smaller than the best vacuum in a laboratory on Earth—space is so vast that the total amount of material between stars is enough to affect the light that passes through it. In addition, as we explain later, this amount of material is enough to provide for the formation of new stars.

Interstellar Dust

Astronomers have been aware of dark areas in the sky (**FIGURE 13-1**) for ages. Although it was once thought that these dark patches are simply spaces between the stars that allow us to see into the darkness beyond, we now know that the dark areas are caused by giant clouds of interstellar dust that block light from stars behind them. In the 1930s, astronomers became aware that grains of dust exist not just in clouds, but also throughout interstellar space. The *Infrared Astronomical Satellite (IRAS)* and the *HST* have enabled astronomers to photograph the diffuse interstellar dust.

FIGURE 13-2a is an *IRAS* image of the constellation Chameleon. Figure 13-2b is an image of a region of Orion (*HST*) as it appears in near-infrared. The wispy appearance of the material in the images and its similarity to cirrus clouds on Earth lead to its name, **interstellar cirrus**. Cirrus clouds emit infrared radiation because they are warmed slightly by light that they absorb.

Wispy interstellar cirrus clouds span huge volumes of space, from parsecs to tens of parsecs across. To appreciate this size, recall that a parsec is somewhat greater than 3 light-years and that the solar system is only about 0.001 light-years across.

Interstellar dust grains make up only 1% of the mass of the interstellar medium, but dust has very definite effects on light. We have seen one of these effects, the reduction in the amount of light from distant stars—and sometimes the complete extinction of that light. Another effect results from the fact that although dust grains

(a)

(b)

(c)

(d)

FIGURE 13-1 Interstellar clouds of dust hide the stars behind them. (a) The swanlike nebula Barnard 163 is within an emission nebula complex (IC 1396) in the constellation Cepheus. (b) The dark molecular cloud Barnard 68 is 300 light-years away in the constellation Ophiuchus. Its mass is about 1.5 solar masses, at a temperature of only 10 K (one of the coldest objects in the universe). Its average diameter is about 12,000 AU. Even though molecular clouds are stellar nurseries, this one seems to be stable (except for a long-period pulsation, probably due to the influence of a shockwave from an exploding star). (c) The Horsehead nebula results from clouds of dust blocking light from a bright nebula behind (at the right center of the photo). It is in the constellation Orion, about 1500 light-years from us. The bright star at the left center is Alnitak, the easternmost star in the belt of Orion, 251 parsecs (820 light-years) from us. (d) A composite color image (in the visual part of the spectrum) of the Horsehead Nebula and its immediate surroundings. Red color is for the hydrogen (H-alpha) emission from the HII region, brown for the foreground obscuring dust, and blue-green for scattered starlight.

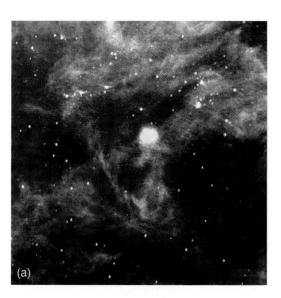

(a)

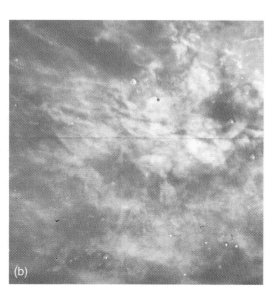

(b)

FIGURE 13-2 (a) A composite image, taken by *IRAS*, of the constellation Chameleon, at wavelengths of 12 micrometers (coded blue), 25 micrometers (coded green), and 60 micrometers (coded red). Stars are brighter at 12 micrometers and appear as blue dots in the picture. (b) The *HST* obtained this near-infrared image of the edge of the Orion Nebula. The image reveals the diffuse, patchy clouds of the nebula.

Is the space between the stars empty?

The effect by which starlight is blocked completely by interstellar material is called *interstellar extinction*. It is due to scattering and absorption. A similar phenomenon occurs when the headlights of oncoming cars appear dimmer while driving in fog.

Polycyclic refers to multiple loops of carbon atoms. *Aromatic* refers to the types of chemical bonds existing between the carbon atoms. *Hydrocarbon* refers to the composition of carbon and hydrogen atoms.

fluorescence The process of absorbing radiation of one frequency and re-emitting it at a lower frequency.

may absorb some light, they scatter much of it, and this scattering is more efficient for light of shorter wavelengths. As a result, blue light is scattered more than red light (FIGURE 13-3). The light from distant stars thus is *reddened* by the dust through which it passes.

The *interstellar reddening* caused by scattering should not be confused with the redshift caused by the Doppler effect. In the case of interstellar reddening, the positions of the spectral lines are not changed, but in the case of Doppler redshift, the spectral lines actually shift as illustrated in FIGURE 13-4. Also, some objects might be intrinsically red and, thus, their color is not related to interstellar reddening.

How large are the interstellar dust grains? We know that they must be smaller than the wavelengths corresponding to visible light, because otherwise they would not scatter blue light more than red. This means that the dust grains are the size of particles of cigarette smoke and smaller, typically a fraction of a micron across. (You may have noticed that a cloud of cigarette smoke has a blue tint.)

What is the dust made of? Spectral analysis indicates that it contains silicate grains and carbon in the form of graphite, similar to the carbon that is found in comets. An important family of organic molecules observed in the interstellar medium is the PAHs (*polycyclic aromatic hydrocarbons*). They are common to daily life, many of them are known (or suspected) carcinogens, and they form by incomplete burning of carbon-containing fuels (such as wood, coal, diesel, and tobacco). PAH molecules **fluoresce** (absorb starlight and re-emit it at lower energies) most noticeably in the infrared part of the spectrum; this allows the *Spitzer Space Telescope* to easily observe them and, as a result, the star-forming regions in which they reside.

The presence of PAHs in dark interstellar dust clouds suggests that they are produced by the same processes as the silicon (which is like sand on Earth). These organic molecules also provide a tool to understanding the abundances of chemicals related to life. In Chapter 15, we discuss the origin of the interstellar material and the related hypotheses about the composition of the dust.

Interstellar Gas

Gas in space accounts for 99% of the total mass of the interstellar material. In most cases the gas is extremely diffuse (so that for most purposes the volume it occupies may be considered a vacuum); however, in some

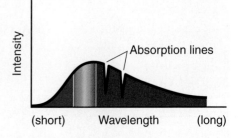

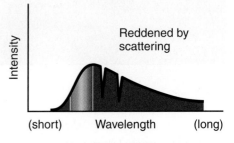

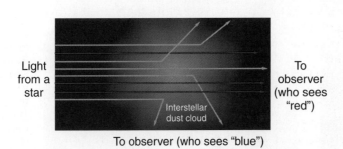

FIGURE 13-3 Grains of dust scatter blue light more efficiently than red light and, thus, a star seen through a cloud of dust appears redder than it would if the dust were not present. The amount of scattering and absorption depends on the number density of particles, the wavelength of light, and the cloud thickness.

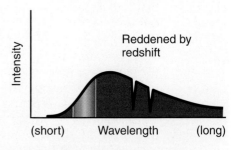

FIGURE 13-4 (top) A blackbody (intensity-wavelength) curve showing two (exaggerated) absorption lines. (middle) This is the same source with its light reddened by scattering. The absorption lines remain in the same place. (bottom) Again, the same source but with its light reddened by the Doppler shift. The entire curve is shifted to the right in this case.

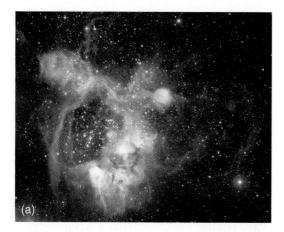

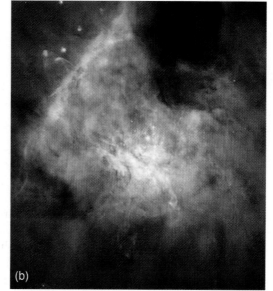

FIGURE 13-5 (a) This ESO/MPG image of the central region of the emission nebula N44 in the Large Magellanic Cloud is a composite of three images taken at the H-alpha (red color), singly-ionized (blue), and doubly-ionized (green) oxygen lines. The field size is 8.5' × 8.5'. (b) This is a mosaic of 45 images of the Orion nebula taken by the *HST*. Light emitted by oxygen is shown as blue. Hydrogen emission is shown as green, and nitrogen as red. The overall color balance is close to what an observer near the nebula would see. (c) The Tarantula nebula is the largest emission nebula in the sky, a bit less than the size of the full Moon. It is located in the Large Magellanic Cloud, about 170,000 light-years from us, and measures more than 1000 light-years across. This three-color VLT image was taken at 485 nm, 503 nm, and 657 nm. The predominantly red light comes from hydrogen atoms (the H-alpha line), and the green-blue light comes from hydrogen atoms (H-beta) and doubly ionized oxygen ions. At the heart of the nebula is a cluster of young, massive, hot stars (2 to 3 million years old) whose ultraviolet light excites the emission nebula. To its right, the reddish bubble is created by a massive star blowing material into the cloud. At the upper right is another cluster of bright, massive stars, about 20-million years old; some of the cluster's stars have already exploded as supernovae.

cases the gas clusters together into giant clouds. The gas between the stars reveals its presence in several ways.

- Clouds of gas can be seen in photographs of the sky. The red glow behind the Horsehead nebula in Figure 13-1c is a giant cloud of gas that is glowing due to ultraviolet light from hot stars within it. The glow is due to a *fluorescence* process; that is, the atoms of the gas absorb ultraviolet light from nearby hot stars, which causes the atoms to become energized, and then they lose their extra energy by emitting light at lower energies. (Recall our discussion of Kirchhoff's laws in Section 4-5.) Of particular importance is the Hydrogen-alpha line produced in the red part of the spectrum (see Figure B12-1). FIGURE 13-5 shows other examples of such bright ***emission nebulae***.

  Such nebulae are direct evidence of gas atoms in the interstellar medium. A typical emission nebula temperature is about 10,000 K, and a typical mass is between 100–10,000 solar masses. The term *nebula* (plural nebulae) has its origin in the adjective *nebulous*, which means "lacking definite form or limits." This is an apt description of an interstellar cloud, for such clouds have no definite boundaries. A cloud is called a nebula if it is dense or bright enough to show up in a photograph.

- Interstellar gas causes absorption lines in stellar spectra. Astronomers can distinguish these absorption lines from the absorption lines of a stellar atmosphere in three ways. First, absorption lines due to interstellar gas tend to be narrower

emission nebula
Interstellar gas that fluoresces due to ultraviolet light from a star near or within the nebula.

FIGURE 13-6 (a) Absorption lines in the spectrum of starlight passing through an interstellar cloud can result from both the star's atmosphere and the cloud. (b) This graph is a highly magnified portion of the intensity-wavelength graph of light from a star. It shows three absorption lines. The outer two are much finer than the central one, indicating that they were caused by interstellar gas.

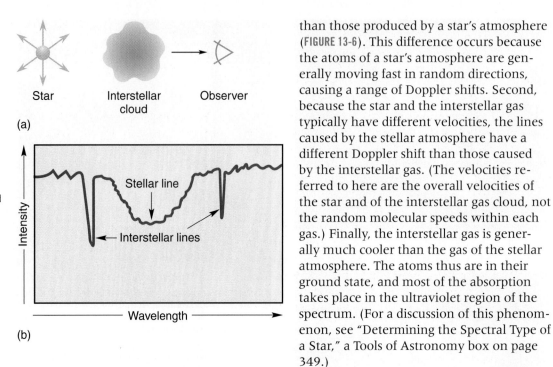

Star   Interstellar cloud   Observer

(a)

(b)

Intensity

Stellar line

Interstellar lines

Wavelength

than those produced by a star's atmosphere (**FIGURE 13-6**). This difference occurs because the atoms of a star's atmosphere are generally moving fast in random directions, causing a range of Doppler shifts. Second, because the star and the interstellar gas typically have different velocities, the lines caused by the stellar atmosphere have a different Doppler shift than those caused by the interstellar gas. (The velocities referred to here are the overall velocities of the star and of the interstellar gas cloud, not the random molecular speeds within each gas.) Finally, the interstellar gas is generally much cooler than the gas of the stellar atmosphere. The atoms thus are in their ground state, and most of the absorption takes place in the ultraviolet region of the spectrum. (For a discussion of this phenomenon, see "Determining the Spectral Type of a Star," a Tools of Astronomy box on page 349.)

- In 1951, two American astronomers, E. M. Purcell and H. I. Ewen, used a specially built radio telescope to detect radiation emitted by interstellar hydrogen. Scientists had predicted that hydrogen atoms should emit radiation with a wavelength of 21 centimeters, but Purcell and Ewen's telescope was the first to reveal it. Hydrogen is the most common element in stars, and astronomers had expected it would also be the most common element in interstellar space. The 21-centimeter radiation confirmed this expectation and allows us to spot interstellar hydrogen at great distances, for the interstellar medium absorbs very little of the 21-centimeter radiation.

- In addition, we are able to detect radio emission lines from interstellar gases—including water, carbon monoxide, ammonia, and formaldehyde—at other wavelengths. However, these gases are much less abundant than hydrogen. Hydrogen (in neutral, ionized, or molecular form) comprises about 70% of the mass of the interstellar medium, with helium making up most of the remaining mass and other elements (such as carbon and silicon) accounting for only a few percent of the total.

As we explain in Chapter 16, the 21-centimeter radiation is very valuable to us in our efforts to map the Milky Way Galaxy.

[The Pleiades] were filled with an entangling system of nebulous matter which seems to bind together the different stars with misty wreaths and streams of filmy light all of which is beyond the keenest vision and the most powerful telescopes.
Edward Emerson Barnard, 1887, examining a long-time exposure photograph.

reflection nebula
Interstellar dust that is visible due to reflected light from a nearby star.

Clouds and Nebulae

As we indicated earlier, the gas and dust in the interstellar material are not uniformly distributed; they tend to clump together into what we call *interstellar clouds*. The interstellar cirrus clouds are made up mostly of gas whose molecules are widely dispersed. At sea level on Earth, there are about 2×10^{19} molecules of air per cubic centimeter. The interstellar cirrus clouds contain fewer than 1000 molecules per cubic centimeter. It is difficult to determine the shape and size of these diffuse clouds, but most of them seem to be wispy sheets rather than spherical clumps. The sheets are probably less than one parsec thick, but many parsecs wide and long. The total mass of a cirrus cloud might equal the mass of a small-to-average star.

If a cloud is near a hot star, the ultraviolet radiation from the star causes the cloud to fluoresce (as in Figure 13-5), and the cloud is then known as an *emission nebula*. A diffuse cloud of dust that happens to be behind or beside a bright star may be visible to us because it refracts and reflects the star's light toward us. **FIGURE 13-7** is a photograph of such a *reflection nebula*. When we see a reflection nebula, we are seeing light that has been scattered in the cloud and, thus, the nebula appears blue. **FIGURE 13-8a** is a diagram of a star surrounded by an emission nebula with a reflec-

ADVANCING THE MODEL

Holes in the Heavens?

Sir William Herschel, the discoverer of Uranus, proposed that the dark patches we see in the sky are simply large spaces between the stars that allow us to see into the dark void beyond. How do we know that this is not the case?

Today, of course, infrared imaging allows us to detect the dust that makes up the "holes in space," but before we ever made infrared images of the dust, an argument based on geometry gave astronomers reason to doubt Herschel's idea. Stars are not all at the same distance from us; they are not on a "celestial sphere." Instead, they are spread out so that some are relatively nearby and others are very far away. For us to be able to see through gaps between the stars, the gaps thus would have to be similar to tunnels, and the tun-

nels would have to be perfectly aligned with Earth. Suppose you are deep in a forest. Even with open clearings among the trees, you might not be able to see beyond the forest. For you to see outside there would have to be straight, open trails. In the case of stars, it is extremely unlikely that so many open trails ("tunnels") would be perfectly aligned with Earth. As astronomer Arthur Cowper Ranyard put it in 1894, "The probabilities against such a radial arrangement with respect to the Earth's place in space [seem] to my mind to conclusively prove that the narrow dark spaces are due to streams of absorbing matter, rather than to holes or thin regions in bright nebulosity."

tion nebula nearby. Figure 13-8b is a photograph of a real example that is similar to the diagram. In the photograph, the pink emission nebula (the Trifid nebula, in Sagittarius) surrounds hot central stars that provide the energy to make the nebula fluoresce. Above it is a reflection nebula.

The Trifid nebula also contains dust lanes that absorb the light from the emission nebula, extinguishing this light and appearing dark. Dust clouds such as this may have densities ranging from as low as 1000 to perhaps 1,000,000 particles per cubic centimeter. (This is still not very dense compared with number densities on Earth, as

FIGURE 13-7 The reflection nebula surrounding the Pleiades (an open cluster about 436 light-years from us in the constellation Taurus). The dimmer stars are part of the cluster and at the same distance as the bright O- and B-type stars. The spikes from each star are caused by diffraction around secondary mirror supports in the telescope. The nebula is not a remnant of the cloud that formed the cluster, but a different interstellar cloud through which the cluster happens to be currently drifting. The cluster, often called the "Seven Sisters" (daughters of Pleione and Atlas), is the most famous galactic star cluster in the heavens since antiquity. According to Greek mythology, the "seven sisters" were half-sisters of the Hyades and were saved by Zeus from the pursuit of the giant Orion by being transformed into a group of celestial doves.

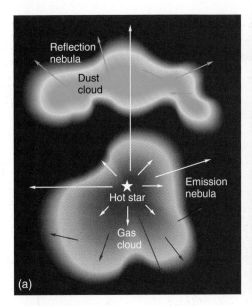

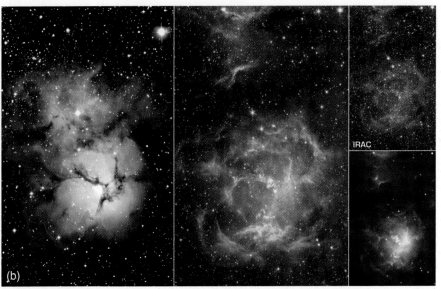

FIGURE 13-8 (a) A hot star inside a cloud of gas causes the cloud to glow, so it is known as an emission nebula. When the star's light passes through the dust cloud, some of the blue light is scattered. If the Earth happens to be located to the left of the page, this dust cloud appears as a blue reflection nebula. If the Earth is located at the top of the page, and if the cloud is very thick, the star's light may not be able to penetrate it, and the cloud appears as a dark nebula. In the case shown, however, some light gets through the cloud. It is reddened in the process, making the star appear redder than it really is. (b) The Trifid nebula, 5400 light-years away in Sagittarius, is a red (or pink) emission nebula that appears to be divided into three parts by dark nebulae that form lanes blocking the light from behind. Above the Trifid nebula is a blue reflection nebula that belongs to the same molecular cloud as the Trifid nebula. The *Spitzer* false color, infrared views show bright star-forming activity where the dark lanes of dust are in the visible image. The IRAC image is a three-color composite: 3.6 μm (blue), 4.5 μm (green), and 8 μm (red); the new stars uncovered by *Spitzer* are visible as yellow or red spots. The MIPS image was taken at 24 μm and shows cool material falling on the growing young stars in the nebula. The middle panel is a combination of *Spitzer* data from both instruments. Star formation is thought to have been triggered by the massive, type-O star seen as a white spot at the nebula's center.

dark nebula An interstellar molecular cloud whose dust blocks light from stars on the other side of it.

As we will discuss later, the interstellar material is constantly in a state of flux, for it is being used up as new stars are formed and being replenished as old stars die.

air at sea level is still 20 trillion times more dense than the densest interstellar cloud.) A dust cloud is able to extinguish light from a star behind it only if the cloud is extremely large, perhaps hundreds of parsecs across. The photographs in Figure 13-1 and **FIGURE 13-9** contain other examples of *dark nebulae*.

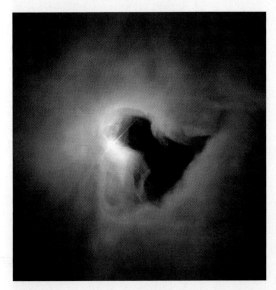

FIGURE 13-9 The reflection nebula NGC1999 in Orion. It is illuminated by the bright, young star to the left of center. The dark nebula, located right of center and in front of NGC1999, is a condensation of cold molecular gas and dust so thick that it blocks light.

Although it is the dust of these molecular clouds that is responsible for blocking starlight, most of the particles in such nebulae are atoms, about 75% hydrogen and 23% helium, with the remaining 2% being heavier elements. The high densities and low temperatures in dark nebulae lead to the formation of molecular hydrogen.

If we take an inventory of the dust, gas, and stars in our Galaxy, we find that interstellar matter contributes about 15% of the total visible mass. In Section 16-3 we discuss how the distribution of molecular clouds in our Galaxy allowed us to infer its spiral structure.

As we mentioned earlier, a number of elements related to life have been discovered in interstellar space. In addition to PAHs, UV observations of HD124314, a reddened star in the constellation Centaurus, showed the presence of molecular nitrogen, the most prevalent element in Earth's atmosphere. Radio observations of the star-forming area W3 (OH) in our Galaxy indicate the presence of fila-

Blue Skies and Red Sunsets

The same phenomenon that explains the reddening of starlight by interstellar dust also explains why the sky is blue and sunsets are red (or at least orange). As we pointed out in the discussion of interstellar reddening, dust scatters blue light more than it scatters red. A similar effect happens in the Earth's atmosphere, where the scattering of sunlight is what gives the daytime sky its color. The sky appears blue simply because particles in the air (and air molecules) scatter higher frequency (blue) light more than lower frequency light. It follows from this that light that reaches us directly from the Sun without scattering is somewhat deficient in these higher frequencies. Consequently, the Sun seen from Earth's surface appears redder than it appears when viewed from a vantage point in space. From Earth, the Sun appears somewhat yellow, but from space it is closer to white.

When we look at the Sun as it is setting, the sunlight that reaches our eyes has traveled even farther through the atmosphere than sunlight at midday. Thus, even more of the high-frequency light is scattered away, and the Sun appears red. This effect is especially noticeable if the atmosphere contains significant amounts of dust. The dust that remained in the air from major volcanic eruptions in 1991 caused beautiful red sunsets for the next few years.

ments of methyl alcohol, up to 3000 AU long, wrapped around a stellar nursery. The amount of methyl formate (a product of alcohol and formaldehyde commonly used as an insecticide) in a typical dust cloud in our Galaxy is about 10^{27} gallons! Finally, *Spitzer* (infrared) observations showed the presence of significant amounts of icy organic materials (including water, methanol, and carbon dioxide) in planet-forming disks around young stars in the constellation Taurus.

13-2 A Brief Woodland Visit

Imagine that you are a visitor from another star system who has landed in a forest on Earth for a 2-day visit. What could you learn about the life cycles of the woodlands during your short visit? You would see small and large trees, but would you be able to determine that trees progress from small to large? In 2 days you would have no chance to see any growth. You would see decayed material on the forest floor, but it might not be obvious that this is what remains of once-standing trees. Perhaps you might be lucky enough to see a change take place. You might see a limb fall from a tree. Would you conclude that trees get smaller by dropping their branches? Or perhaps you might see an entire tree fall. What would that tell you?

Like an extraterrestrial forest visitor, we humans are brief tourists in the universe. During our short lifetimes, we see very little change in the heavens. As we measure time, the stars change slowly. We do have an advantage over the forest visitor in that previous generations have handed down their observations to us so that our learning cycle is longer than one lifetime. However, the earliest reports of astronomical observations are only a few thousand years old; this is a minute time period compared with the total life of the heavens.

It is possible, however, to learn about the life cycle of stars. We are fortunate that a few stellar events happen quickly enough for us to observe them over human lifetimes or (in some cases) over much shorter times. Although astronomers cannot experiment directly with their subject matter, they can observe tremendous numbers of stars in various stages of development. This, along with earthbound experiments on the material of which stars are made, allows scientists to develop theories of stellar life cycles in which we can have reasonable confidence.

13-3 Star Birth

Like other scientific theories, today's theories of star birth and star death developed slowly. Successful theories do not spring fully developed into a scientist's mind. Theories in astronomy result from long struggles with data from the heavens, often accompanied by experimentation here on Earth.

The story of today's understanding of the life cycles of stars starts early in the 20th century, with astrophysicist Henry Norris Russell playing the major role. The H-R diagram is the key to understanding the lives of stars, but it is not easy to read. Russell's early theories held that stars begin their lives as red giants, become O- and B-type main sequence stars, and then move down the main sequence, gradually dimming as they live out their lives. When new evidence arrived, particularly from H-R diagrams of clusters and from new knowledge of nuclear energy processes, Russell realized that stars live most of their lives on the main sequence with very little change in their positions and that the red giant stage is not the beginning of a star's life but is near the end. We start at the beginning.

The Collapse of Interstellar Clouds

Astronomers estimate that there are 5000 giant molecular clouds in our Galaxy.

Stars are born in the coldest places in the Galaxy, the giant molecular clouds (GMCs) of interstellar space. These clouds have a temperature of about 20 K (−420°F) and consist mainly of molecular hydrogen. The low temperature of a GMC implies that particles in it have low speeds, and therefore, the corresponding gas pressure is small. As a result, gravity can bring interstellar material together by overwhelming the pressure pushing this material apart. All that is needed is a relatively dense region so that atoms and dust particles are close enough for gravity to take over. A typical cloud may be 50 parsecs across and may contain as much as a million solar masses of material. The densest regions in a GMC form dark nebulae where new stars are born.

FIGURE 13-10b is a picture of a very small portion of M-16, the Eagle nebula (shown in Figure 13-10a). The dark pillar-like structures are part of a GMC and, thus, they are made up of gas and dust. Intense ultraviolet radiation from hot, massive newborn stars that lie at the upper right, beyond the photo, causes the columns. FIGURE 13-11 illustrates the process. Prior to the sequence shown in the figure, a molecular cloud was fairly evenly spread throughout this region of space. In part (a) of the figure, ultraviolet radiation from the upper right is blowing the cloud back and evaporating gas outward from the cloud's surface. At the same time, the radiation illuminates the surface of the cloud, causing it to glow.

The name EGG also fits the globules because they are embryonic stars.

Certain parts of the cloud are denser than average. In part (b), a dense globule of gas is about to be uncovered. Globules like this are called EGGs, for "Evaporating Gaseous Globules." Radiation pushes the surrounding gas away from the EGG in much the same way that a strong wind on a beach blows sand along, uncovering shells and leaving trails of sand behind the shells. Part (c) of Figure 13-11 shows most of the nebula blown away, but a snake of gas and dust remains that has been protected from radiation by the EGG. Later, the nebula will blow still farther away, leaving behind a teardrop shape, shown in part (d). Even later, all of the nebula will disappear, leaving behind only the EGG.

What is an EGG? At least in some cases, it is the beginning of a star. Within the globule, gravitational forces between individual molecules and dust particles are sufficient to draw the particles together. The globule was in the process of this contraction when the ultraviolet light began blowing away the surrounding nebula. The contraction continues within the globule. When the center of the EGG gets hot enough to give off its own radiation, it will itself blow away any remaining parts of the nebula that have not fallen inward and a star will remain. Stars born in this manner probably do not have planets around them, for the radiation that causes the pillars probably disperses the material that would have ended up as planets.

Some stars form in more isolated conditions, without nearby hot stars blowing their raw material away. In these cases, parts of the nebula that remain near the

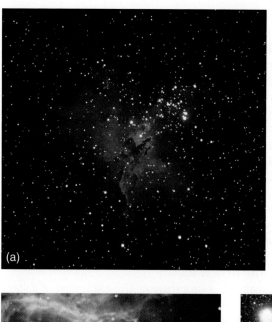

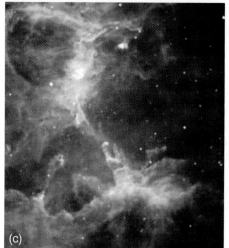

FIGURE 13-10 (a) The Eagle nebula in the constellation Serpens, about 7000 light-years distant and 20 light-years across. (Can you see the nebula as a landing eagle?) (b) This 1995 *HST* image shows material from a very small part of the Eagle nebula being swept away from dense areas where new stars are forming. The pillar to the left is about 1 light-year long; each pillar is a bit wider than our solar system. (c) A composite infrared image of Eagle's fiery heart, taken by ESA's *Infrared Space Observatory* (7.7 microns in blue; 14.5 microns in red). The temperature of the dust in the Eagle, seen as a "bluish fog," is about 170 K. (d) This 1999 infrared *HST* image of the Eagle nebula shows that only a small fraction of the areas that are opaque in visible, as shown in (b), remain so in the infrared. Instead of the substantial columns of material implied by the optical images, the infrared images show only isolated clumps of dust and gas.

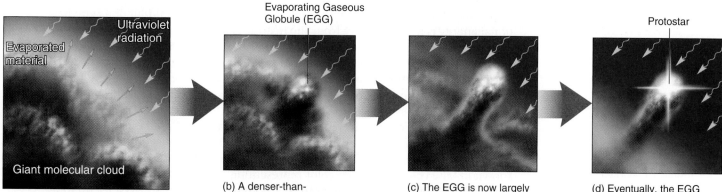

(a) The surface of a molecular cloud is illuminated by intense ultraviolet radiation from nearby hot stars. The radiation evaporates material from the cloud's surface.

(b) A denser-than-average globule of gas (an "EGG") begins to be uncovered. Because it is denser than its surroundings, it is not evaporated as quickly and is left behind. Young stellar objects begin to form within some EGGs.

(c) The EGG is now largely uncovered. The EGG protects a column of gas behind it, giving it a finger-like appearance.

(d) Eventually, the EGG separates from the cloud in which it formed. As the EGG itself slowly evaporates, the star within it becomes visible.

FIGURE 13-11 Pillars like those in Figure 13-10 are the result of strong ultraviolet radiation striking a giant molecular cloud.

embryonic star will continue to fall into it, perhaps resulting in a very large star or revolving around the star to become a planetary system. Such isolated massive stars and compact ionized hydrogen clouds have been observed in a number of galaxies in the Virgo cluster, at the boundary between a galaxy's halo and the intercluster space, and high up in the halo of our own Milky Way.

One question remains before we consider how an EGG becomes a star. What triggers the collapse of parts of a GMC to form globules? The answer may involve collisions between GMCs. Along the boundary of such a collision, molecules would be forced closer together, perhaps close enough that gravitation would take over and continue the compression. Another cause for collapse may be interstellar shock waves. As such a wave strikes the GMC, it forces parts of the cloud together enough for gravity to become dominant.

There are at least four possible sources of interstellar shock waves. We have seen one—radiation from hot, newly forming stars. Second, as we explain in the next section, very massive stars undergo a period during which they release enormous quantities of material at great speeds. These bursts are a source of shock waves in the surrounding GMC that can trigger the birth of new stars.

A third source of interstellar shock waves is a supernova—the explosion of a star as it nears the end of its life. We discuss supernovae in Chapters 14 and 15. Finally, as we discuss in Chapter 16, we know that tremendous shock waves move around the entire Galaxy, forming its galactic arms. These waves may be the most common trigger of star formation.

Protostars

Whatever causes the material within the core of a giant molecular cloud to start to collapse, once the collapse begins, the increased gravitational force causes the molecules of the core to pick up speed as they fall inward. The condition of fast-moving molecules is synonymous with high temperature and, thus, material near the center of the collapse becomes hot. Soon the higher pressure caused by the increase in temperature prevents further collapse of the central core. Material farther from the core continues falling inward, crashing down on the central core and causing a further increase in temperature.

The hot core is now called a ***protostar***. It receives its energy from the infall of material, so the ultimate source of its energy is gravitational. This is similar to the way a hammer that is repeatedly lifted and dropped onto a piece of metal heats both itself and the piece of metal. Each molecule that falls onto the core converts gravitational energy into thermal energy.

As the cloud contracts and its temperature increases, about half of its gravitational energy is radiated away from the heating center. As the center gets hotter and hotter, it radiates more and more energy. In only a few thousand years, the center is as bright as the Sun; however, we cannot see the center directly because the outer portion, which continues to fall relatively slowly toward the center, blocks most of the radiation. This ***cocoon***, or ***cocoon nebula***, absorbs the radiation from the center, becoming warmer as it does so.

A warm object emits infrared radiation, and it is the infrared radiation emitted from the cocoon that we see from Earth and that gives evidence for the existence of protostars. FIGURE 13-12 includes two infrared photographs, which therefore are shown in false color, with each color representing a different intensity of infrared radiation. Such false-color photographs of nonvisible sources are very useful tools in astronomy, but do not be fooled into thinking that the objects have colors like the photographs.

Evolution Toward the Main Sequence

FIGURE 13-13 shows the progress of a protostar of one solar mass on an expanded H-R diagram. (The main sequence is now compressed into the left side of the figure.) The diagram had to be expanded beyond the coolest stars on the main sequence because a protostar begins its life much cooler than any main sequence star. The details of

The material flowing from stars is commonly called the *stellar wind*. The *solar wind* that we detect (and which results in auroras on Earth) is far weaker than the stellar wind bursts described here.

protostar An object in the process of becoming a star, before it reaches the main sequence.

cocoon nebula The dust and gas that surround a protostar and block much of its radiation.

In speaking of evolutionary tracks and stellar evolution, astronomers are using the term *evolution* in a very different way than biologists do. One star is said to *evolve* as it lives its life. We are not speaking of an entire species evolving through generations.

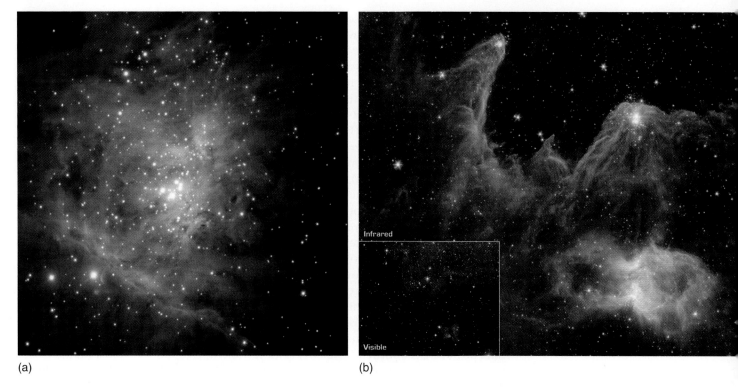

(a)   (b)

FIGURE 13-12 (a) A VLT color composite of 81 images of the central part of the Orion nebula. The image shows the Trapezium stars (near the center) and the associated cluster of about 1000 stars, about 1-million years old. This near-infrared image (1–2 microns) shows thermal radiation from the area's dust, which is heated by the intense UV and visible light from the stars. It also shows the warm, dense dust cocoons around the very young stars. (b) This is a comparison between the visible-light image (inset) and the *Spitzer* infrared image of an edge in the W5 region in the Cassiopeia constellation, 7000 light-years away. Infrared light from young stars (shown in white/yellow) inside the dusty pillars can escape and be detected, and the pillars themselves have been warmed by starlight and glow in the infrared.

this part of the star's **evolutionary track** are not known, but the protostar begins as a cool, dim object, warms up by gravitational contraction, and moves toward the main sequence. Finally, as the cocoon continues to contract, its smaller size causes it to appear dimmer and, thus, the star begins to move downward on the diagram as it nears the main sequence.

Motion on the H-R diagram does not mean the star is actually moving in space, of course. Recall from Figure 12-15, the diagram of people's height versus age, that as a person ages, his or her dot moves on this diagram. This is analogous to the motion of a star on the H-R diagram. (However, a star spends most of its life on the main sequence and, thus, its "motion" on the H-R diagram is very slow for most of its life.)

As time passes, the center of the protostar continues to shrink and become hotter, emitting more radiation all of the time. This radiation gradually blows away the outer portions of the cocoon. Some of the cocoon may not be blown away but may condense to form planets, as discussed in Chapter 7. In some stars, the cocoon is apparently blown away by sudden bursts of stellar winds. Certain stars, named **T Tauri** stars after the variable star T in the constellation Taurus, appear to be young stars undergoing some sort of instability that causes enormous flares. These flares are

evolutionary track The path on the H-R diagram taken by a star as its luminosity and color change.

T Tauri stars A class of stars that show rapid and erratic changes in brightness.

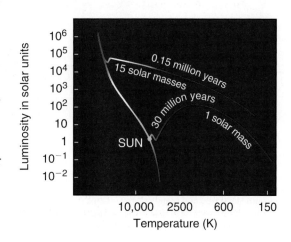

FIGURE 13-13 This H-R diagram shows temperatures much lower than on previous diagrams to include the early, cool stages of a star's formation. A more massive star ends up at a higher location on the main sequence because it is hotter and brighter. Because gravitational forces are greater on the more massive star, it spends much less time in its protostar stage.

The Horsehead nebula of Figure 13-1d is a smaller part of this same Orion nebula. New stars are thought to be forming in the dark horse head.

thought to play a part in blowing away the cocoon of newly forming stars, particularly K- and M-type stars, which, being at the bottom of the main sequence, are of low mass.

Computer simulations and observations of massive star-forming regions support the idea that massive stars form in a way similar to that of less massive stars. That is, they form through gravitational collapse and mass accretion with disks. The most massive stars, the O- and B-type stars, follow a track on the H-R diagram toward the top of the main sequence (Figure 13-13). These stars are much fewer in number than stars of lesser mass; however, they are so energetic that they signal their presence by emitting ultraviolet light, which stimulates the hydrogen of their cocoon (and the nebula beyond the cocoon) to emit its own light, forming an emission nebula. **FIGURE 13-14a** is a photograph of the Orion nebula, an emission nebula lit by energy from new stars within it. The very hot blue star in Figure 13-14b is experiencing a continuous mass loss from its outer layers. The two symmetric shells of material (emission nebulae) on either side of the star were formed during past vigorous outbursts.

The infalling particles of massive stars experience much greater gravitational force than do particles in low-mass stars, and massive stars reach the main sequence much faster. M-type stars remain protostars for hundreds of millions of years. The Sun, with more mass, spent about 30 million years in this phase; however, the most massive stars remain protostars for only tens of thousands of years. Stars of low mass may undergo a period of instability just before joining the main sequence. Massive stars do likewise, but their instability is more violent; O and B stars blow off material at supersonic speeds during this time, creating a shock wave in the surrounding material. This shock wave may be one of the triggers that starts the collapse of other portions of the interstellar cloud to form more stars.

The great amount of radiation from massive stars is what limits how massive a star can be. Astronomers calculate that a star with a mass greater than about 150 solar masses emits radiation so intense that it prevents more material from falling into the star, thereby limiting the star's size. This upper limit is a conservative estimate based on *HST* observations of hundreds of massive stars in the Archer cluster, one of the densest clusters in our Galaxy, 25,000 light-years away. It is also consistent with

(a)   (b)

FIGURE 13-14 (a) This is a composite *HST* and *Spitzer* image of the Orion nebula, an emission nebula that is our closest massive star formation region. The Orion nebula is easily visible in small telescopes or even binoculars. The Trapezium, the collection of four massive stars, is at the cloud's center. Green reveals hydrogen and sulfur gas heated by the UV light of the Trapezium stars. Red and orange reveal organic molecules (PAHs). (b) The blue star HD148937 (the brightest in a triple-star system in the constellation Norma) is experiencing a continuous mass loss from its outer layers. The emission nebulae (NGC6164, NGC6165 on either side of the star) were formed during past vigorous outbursts.

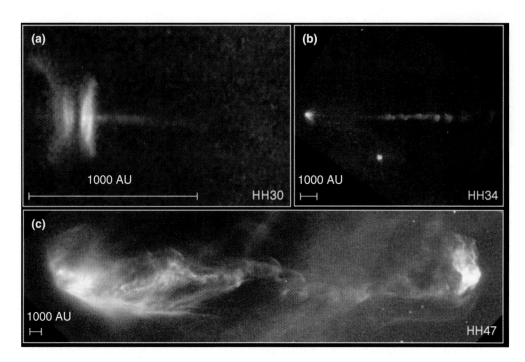

FIGURE 13-15 Composite *HST* images, in visible light, of jets emanating from three different protostars. "HH" stands for Herbig-Haro objects, the bright knots of hot, ionized material that appear to be moving away from their corresponding protostars. (a) An edge-on disk of dust around a protostar associated with HH-30, about 450 light-years away in the constellation Taurus. The protostar is hidden behind the densest and opaque part of the disk but illuminates the disk's top and bottom surfaces. (b) The jet in HH-34 shows a beaded structure, suggesting an episodic ejection of dense parcels of material. It is about 1500 light-years away, close to the Orion nebula. (c) The protostar in HH-47 is hidden inside a dust cloud (near the left edge of the image) while its light is reflected by the white filaments on the left. The jet structure suggests that the protostar might be wobbling. It is about 1500 light-years away, at the edge of the Gum nebula.

statistical studies of clusters of smaller mass stars in our Galaxy and with observations of a massive star cluster (R136) in the Large Magellanic Cloud.

On the other hand, protostars with masses less than about 0.08 solar masses do not develop the necessary internal pressure and temperature to start hydrogen fusion. They heat up because of gravitational contraction but never become main sequence stars. Eventually, they contract as far as they can and then begin to cool, becoming planet-like objects—cold cinders in space—called brown dwarfs, which we study in the next chapter. Recall from Chapter 9 that Jupiter emits about twice as much energy as it receives from the Sun, and this comes from gravitational energy that remains from its formation.

Infrared observations have revealed that it is very common for protostars to be surrounded by disks of gas and dust. These observations fit well with the theory of the formation of the solar system and lead us to believe that planetary systems are common around stars. In addition, streams of material (jets) have been observed flowing from the poles of many protostars. **FIGURE 13-15** clearly shows the presence of disks and outflows in protostars. Even though the *HST* images are of three different objects, when compared together they give us a general view of how jets from protostars behave as they propagate in space. A jet emanates from a disk surrounding a young star. The jet remains narrow and shows a beaded structure and then creates a shock wave when it interacts with the surrounding interstellar medium.

accretion The process by which an object gradually accumulates matter, usually because of the action of gravity.

On one hand, a protostar's mass may increase by the process of *accretion*, as particles in the disk surrounding the protostar lose energy and spiral closer to it as a result of collisions within the disk. On the other hand, a protostar's mass may decrease as a result of the ejection of mass along the jets (**FIGURE 13-16**). (In some cases the outflow occurs along only one jet, whereas in other cases, the outflow is along two oppositely directed jets.) Astronomers hypothesize that these outflows play an important part in reducing the angular

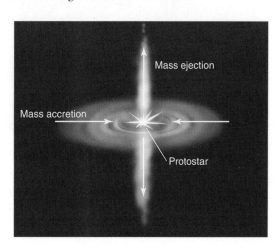

Mass ejection

Mass accretion

Protostar

FIGURE 13-16 A typical protostar strikes a balance between mass accretion from a surrounding disk of dust and gas and mass ejection in powerful jets of material.

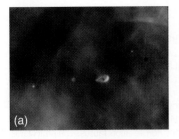 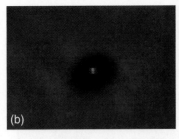

FIGURE 13-17 (a) Close-up of protoplanetary disks in the Orion nebula. The field of view is about 9000 AU across. (b) A very young star (between 300,000 and 1 million years old) surrounded by its protoplanetary disk. There is enough material in the disk to form at least seven planets like Earth. The disk is 600 AU across or 7.5 times the diameter of Pluto's orbit in our solar system.

The problems of star formation are wonderfully complex. In a place like Orion you can see the whole region disintegrating. Supernovas, stellar winds, shocks are slamming into the molecular cloud, sweeping it back, igniting a conflagration. Things are happening wholesale. You have to be careful in applying simple models.

Patrick Thaddeus, radio astronomer

momentum of stars and allowing them to grow by accretion; particles can fall inward (toward the star) after they shed their excess angular momentum. As we discussed in Section 7-6 about the formation of the solar system, theoretical models predict that the Sun should be spinning faster than it is. If, early in the Sun's formation, material were ejected from its poles, that ejection would have resulted in reducing the Sun's rotational speed to what is observed. Much remains to be learned about the part played by the ejection of matter from the poles of protostars.

The presence of disks of gas and dust surrounding newly formed stars is also shown in **FIGURE 13-17a**. This *HST* image of a small portion of the Orion nebula shows five stars, four of which are surrounded by disks of gas and dust. These protoplanetary disks will probably evolve on to form planets. Figure 13-17b is an image of a very young star surrounded by a protoplanetary disk, seen in silhouette against the background of the Orion nebula. Not all stars form such disks; for example, it is likely that massive stars or some stars in binary systems do not form protoplanetary disks. As recent observations suggest, however, these disks are found around most young, low-mass stars. The duration of the different phases that a young star goes through as it evolves depends a lot on its environment. If a young star has nearby companions, their UV radiation and stellar winds may disrupt the young star's accretion disk. The longevity of the star's jets will be affected by whether or not planets start forming in the disk, clearing away gas that feeds these outflows. **FIGURE 13-18** shows two examples of young stars making their presence felt in their surrounding medium.

Many details of our theories of the life cycles of stars come from computer modeling. For example, the image in **FIGURE 13-19** shows the inner region of a 50-solar masses gas ball 266,000 years after it started to collapse under its own gravity. New stars are shown in white, and a dense gas disk can be seen at lower center.

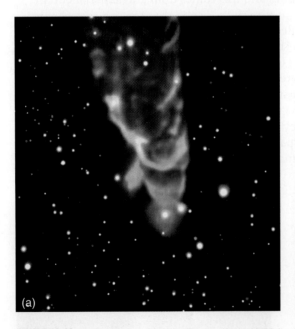 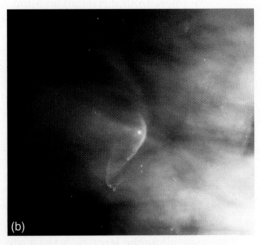

FIGURE 13-18 (a) This *Spitzer* false color image of HH 49/50 (in the Chamaeleon I star-forming region) shows a tornado-like feature, a shock front created by a jet emanating from a still-forming star off the upper edge of the image. The jet slams into interstellar dust clouds, heating the dust and causing it to glow in the infrared. The shape of the feature could result from magnetic fields or eddies. It is not clear if the star at the tip of the tornado is physically related to it. (b) A bow shock around the very young star LL Ori, in Orion's Great Nebula. Such shocks can be created in space when two streams of gas collide. The star emits a strong wind, which interacts with the surrounding medium.

Such computer simulations help us understand better the processes involved in star and planetary system formation.

Star Clusters

The Pleiades (Figure 13-7) is a small cluster of some 500 stars in the constellation Taurus. The brighter stars of the cluster are visible to the naked eye and are a beautiful sight even in a small telescope. **FIGURE 13-20** shows two other clusters of the same type, called *galactic clusters*, or *open clusters*.

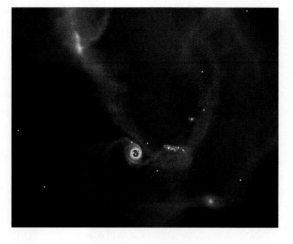

FIGURE 13-19 A frame from a computer simulation of star formation. The original 50-solar mass ball of gas is 1.2 light-years wide. The frame shows the inner 0.08 light-years wide core.

Galactic clusters are found primarily in the disk of the Galaxy. **FIGURE 13-21** shows another type of star cluster, called a *globular cluster*. Globular clusters contain hundreds of thousands of stars. They are not confined to the Galactic disk, but orbit the center of the Galaxy in all directions so that they spend most of their time outside the disk. **FIGURE 13-22** shows the location of the two types of clusters relative to our Galaxy.

Clusters are important to astronomers for two reasons. First, all of the stars in a given cluster are at about the same distance from us. This means that their apparent magnitude is a direct indication of their absolute magnitude—those that look the brightest really are the brightest. Second, the stars within a cluster began their formation at essentially the same time, when their interstellar cloud began its collapse. The evidence for this conjecture is based on regularities we see in H-R diagrams of clusters such as the one for a very young cluster shown in **FIGURE 13-23**. This means that they formed out of the same giant molecular cloud, and therefore, it is probably safe to assume that all of them have about the same chemical composition.

Much of what we know about star formation has come from examination of clusters. The fact that the stars in a cluster began forming at about the same time allows us to learn how stars of different mass have progressed at different rates. Figure 13-23 indicates that protostars of low mass have not yet reached the main sequence. H-R diagrams of clusters provide evidence that stars of low mass spend much more time in the protostar stage than do more massive stars. As we explain in the next chapter, H-R diagrams of older galactic clusters can be used in a similar

galactic (or open) cluster A group of stars that share a common origin and are located relatively close to one another.

globular cluster A spherical group of up to hundreds of thousands of stars, found primarily in the halo of the Galaxy. Globular is pronounced "glob" (as in "mud") rather than "globe."

The history of host galaxies leaves an imprint on the properties of its globular clusters. It is possible then that, due to collisions between galaxies, not all stars in a globular cluster have the same origin.

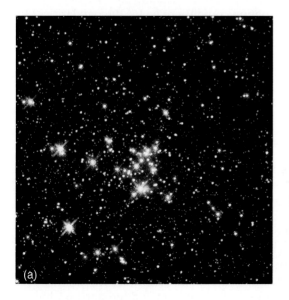

FIGURE 13-20 Many clusters are named by their number in the New General Catalog (NGC), which was compiled between 1864 and 1908. (a) The cluster NGC457, in Cassiopeia, is called the Owl cluster and contains about 80 stars. The bright star is φ (phi) Cassiopeiae. (b) A pair of open clusters (NGC869 and NGC884) in the constellation Perseus.

FIGURE 13-21 This is a globular cluster (M80). Globular clusters are gravitationally bound groups of hundreds of thousands of stars.

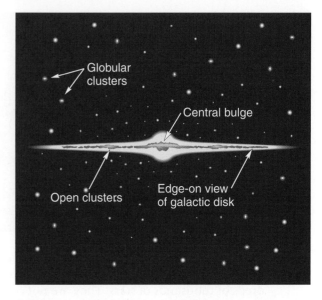

FIGURE 13-22 Galactic (open) clusters are found within the disk of the Galaxy, and globular clusters are found primarily in the halo surrounding the disk.

FIGURE 13-23 An H-R diagram of a very young cluster shows that low-mass protostars have not yet reached the main sequence. As time goes on, the H-R diagram of this cluster will look like the ones in Figure 14-6.

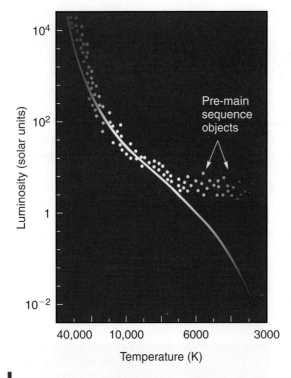

manner to reveal when stars end the main sequence part of their lives (see Figure 14-6).

Thus far, we have tracked the evolutionary path of stars through the protostar stage. When the core of a protostar becomes hot and dense enough, nuclear fusion begins. The new star becomes stable and begins the main sequence portion of its life. We discuss this stage of a star's life in Chapter 14.

As we observed earlier, the regularities we see in H-R diagrams such as Figure 13-23 prove the validity of the assumption that stars within a cluster began their lives at about the same time, for what other explanation could there be for the regularities? In the next chapter, we continue to examine the H-R diagrams of star clusters, for they are very valuable tools in our quest to understand stellar life cycles.

Conclusion

Interstellar space is so barren of matter that for many purposes it can be considered a vacuum; however, vast quantities of dust and gas do exist between the stars, and this material is important to astronomers for two main reasons. First, radiation from stars passes through the interstellar medium on its journey to Earth, and if astronomers are to understand the information carried by starlight, they must understand how the interstellar medium affects radiation. Second, interstellar matter is the stuff of which stars are made. If astronomers want to understand the life cycles of stars, they must first understand the substance from which the stars form. They must also use the entire spectrum to uncover details of the interstellar medium and the star-forming

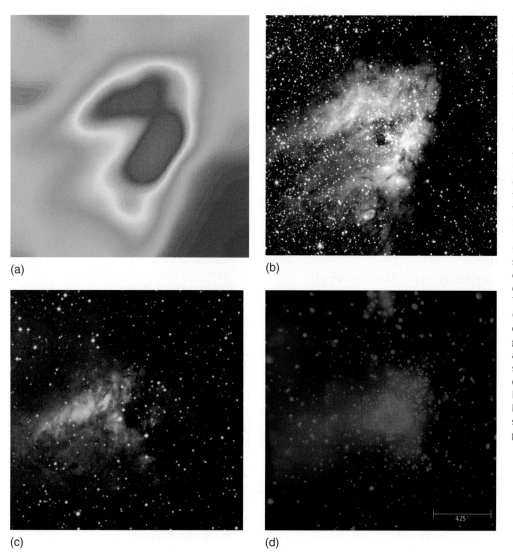

(a)

(b)

(c)

(d)

FIGURE 13-24 M17 is one of the most luminous and active star-forming regions in our Galaxy. The stars in the cloud are only a million years old. Each image is 17′ per side. (a) The VLA radio image shows the distribution of ionized hydrogen in the region, signifying active star formation. (b) The 2MASS infrared image reveals the dark molecular cloud; although it is too cold to emit visible or near-infrared light, its densest regions can still be seen as dark structures. (c) The DSS optical image shows dark regions (due to dust grains) superimposed on bright gas regions. (d) The *Chandra* X-ray image (processed to emphasize diffuse emission) shows hot gas (1.5–7 million K) flowing away from massive young stars; areas of low/high X-ray energy are shown in red/ blue. The hot gas appears to have carved the horseshoe-shaped cavity in the cool gas seen in the infrared.

regions. **FIGURE 13-24** shows the central region of the Horseshoe nebula (M17) in the radio, infrared, optical, and X-ray part of the spectrum.

Clouds of dust and gas in interstellar space differ in density, composition, and size. Interstellar cirrus is extremely diffuse, but some dust clouds are dense and large enough to extinguish light from stars on the other side of them. In some clouds, the processes of fluorescence and reflection provide beautiful images that can be appreciated by astronomers and nonastronomers alike.

Technological developments during the past few decades—particularly in microwave and radio telescopes—have provided many answers to questions about the interstellar medium, but, as is usual in science, they have also created additional questions. The quest to understand how stars are formed from the interstellar medium is of special interest to astronomers, and many mysteries remain, particularly about the initial steps in the formation of a star.

Whatever serves as the trigger to begin the collapse of portions of a giant molecular cloud, mutual gravitational forces between the particles continue and accelerate the compression. As the core of the cloud collapses, gravitational energy is transformed into thermal energy, which raises the temperature of the protostar. Finally, the inner portion of the collapsing matter becomes so hot and dense that nuclear fusion starts. At that time, the cloud begins its life on the main sequence of the H-R diagram. Examination of star clusters has shown the importance of a star's mass in determining how the star develops, for more massive stars become hotter and more luminous than less massive stars. In the next chapter, we show that mass is not only important in determining the early development of a star, but that it is the single most important property in determining the fate of stars.

STUDY GUIDE

1. In the densest interstellar cloud, the particle density is _____ the particle density of air at sea level on Earth.
   A. much less than
   B. about the same as
   C. much greater than

2. The reddening of starlight as it passes through interstellar dust clouds is due to
   A. the Doppler effect.
   B. fluorescence.
   C. scattering.

3. A nebula that glows due to fluorescence is called
   A. a fluorescence nebula.
   B. a reflection nebula.
   C. an emission nebula.

4. Absorption lines due to interstellar gas can be distinguished from spectral absorption lines caused by a star's atmosphere because
   A. their Doppler shifts are different.
   B. different chemical elements are found in the two different places.
   C. [The question is misleading; interstellar gas does not cause absorption lines.]

5. A protostar is
   A. a newly forming star.
   B. a main sequence star.
   C. a star nearing the red giant stage.
   D. a black dwarf.

6. H-R diagrams of young star clusters show
   A. fewer than about 10 stars.
   B. very few of the most luminous stars.
   C. that massive stars have not reached the main sequence.
   D. that low-mass stars have not yet reached the main sequence.

7. What is the primary difference between a star and a planet as each forms?
   A. Age.
   B. Mass.
   C. Volume.
   D. Chemical composition.
   E. Velocity.

8. Which of the following statements is true?
   A. Less massive protostars spend a longer time in the protostar stage than do more massive protostars.
   B. Protostars are surrounded by cocoons of gas and dust.
   C. Protostars radiate mainly in the infrared.
   D. [All of the above.]

9. Prior to reaching the main sequence, a star's energy comes from
   A. gravitation.
   B. nuclear fusion.
   C. nuclear fission.
   D. hydrogen conversion to helium.
   E. [Both B and D above.]

10. Although a star is on the main sequence, its energy comes primarily from
    A. gravitational shrinking.
    B. nuclear fusion.
    C. nuclear fission.
    D. proton instability.
    E. chemical reactions.

11. As a star is forming by the condensation of gases, the gases
    A. cool as they fall.
    B. heat up as they fall.
    C. stay about the same temperature.
    D. [Any of the above, depending upon the mass involved.]

12. A star is considered to begin its main sequence life when
    A. it starts to collapse.
    B. its protostar life begins.
    C. nuclear reactions start.
    D. it begins to move off the main sequence.
    E. its planetary system has formed.

13. The contraction of an interstellar cloud to become a star is caused by
    A. magnetic forces.
    B. electric forces.
    C. nuclear forces.
    D. gravitational forces.
    E. [Both A and B above, for they are closely related.]

14. An evolutionary track on an H-R diagram reveals
    A. the changes that occur in a star's life.
    B. the changes that occur as one star evolves into another.
    C. the motion of a star relative to others in its cluster.
    D. the motion of a star relative to others in the entire galaxy.
    E. [Both C and D above.]

15. Which of the following choices is *not* considered a possible trigger to begin the collapse of an interstellar gas cloud?
    A. A shock wave from a supernova.
    B. A shock wave occurring during the formation of very massive stars.

C. A shock wave resulting from radiation from nearby emission nebula.

D. A shock wave passing around the galaxy.

E. [All of the above are considered likely triggers.]

16. Clusters help us to learn about the evolution of stars because stars in a cluster
    A. are all about the same temperature.
    B. are all about the same mass.
    C. interact with one another and affect each other's evolution.
    D. are all about the same age.
    E. can be observed more easily than individual stars.

17. The H-R diagram for a single star cluster is different from the usual H-R diagram because all stars in the cluster have the same
    A. mass.
    B. temperature.

C. diameter.
D. rotational speed.
E. age.

18. What is a protostar?

19. List four possible events that may be responsible for triggering the collapse of interstellar clouds.

20. Which stars take longest to complete their protostar stage? Why?

21. What is the source of the energy that powers protostars?

22. Why is there a lower limit to the possible mass of a star? An upper limit?

23. Explain how galactic clusters provide observational evidence for theories of stellar evolution.

---

1. Starlight that has passed through a cloud of dust is reddened. So is starlight from a star that is moving away from us. In each case, how can we tell that the light from a star has been reddened? Perhaps what we see is the actual color of the star.

2. If light is passing through space, why does the universe look black except where we see dust, gas, or an object?

3. Describe the processes that produce each of the following types of nebulae: dark nebulae (like the Horsehead nebula), pink nebulae (like the N44 nebula), blue nebulae (like the one that appears near the Trifid nebula).

4. Explain how an examination of the H-R diagram for a large group of stars can tell us the relative lengths of the various stages of a star's life.

5. Why do astronomers feel that there must be a triggering mechanism for the beginning of an interstellar cloud's collapse? Why couldn't it happen on its own?

6. Explain the formation of "pillars" like those in Figure 13-10b.

7. Why do we think that most stars form from a *rotating* disk of gas? Couldn't it just as well be nonrotating?

8. Explain why the mass of a star determines the amount of time that the star remains a protostar.

9. The primary difference between protostars and main sequence stars is their energy source. What is the source of energy of each type of star?

10. Explain the value of star clusters to astronomers in their attempt to understand stellar evolution.

---

1. Along the line of sight to a particular star cluster, interstellar extinction allows only 20% of the light from the stars to pass through 1000 light-years of the interstellar medium. If this cluster is located 2000 light-years from us, what percentage of its light do we receive?

2. During our Sun's early stages of evolution, it had a surface temperature of 1000 K and luminosity 1000 times greater than it has now. What was our Sun's radius at this stage? Express this radius in astronomical units and as a multiple of the Sun's current radius. (Hint: 1000 K is about one sixth of the Sun's current surface temperature. Consider the Stefan-Boltzmann law.)

3. A star in Orion is located 0.1 degrees from the center of the nebula and is moving away from the center at 100 kilometers/second. How long ago did it get ejected from the center? (Hint: The distance to Orion is about 1500 light-years. Consider the small-angle formula.)

4. The temperature of dust grains in most interstellar clouds is about 300 K. Using Wien's law, calculate the wavelength at which most of the radiation is emitted by the grains. Is there a reason why astronomers are interested in using the infrared part of the electromagnetic spectrum to study these clouds?

Deep Sky Objects with a Small Telescope

Many beginning telescope users limit themselves to viewing the Moon and planets, but some interesting objects outside the solar system are accessible with a small telescope or even with binoculars. The star maps in this book can help you find the objects listed here. At the end of each listing is the time of year when the object is highest in the sky in the evening, but each may be seen in roughly the same place later at night, earlier in the year.

Orion nebula—A cloud of dust and gas 1500 light-years away. Its diameter is some 15 light-years. Find the Trapezium, four stars in a small dark area of the nebula (sometimes called the Fish Mouth). (January through March)

Alcor and Mizar—A naked-eye double star in Ursa Major. Your telescope will reveal that Mizar is itself a double star. (March through July)

ACTIVITIES

Sagittarius—Such a rich area of the sky for interesting objects that if you slowly scan it with your telescope, you will find several clusters and nebulae. (July and August)

Albireo—A binary pair with obviously different colors (Figure 12-24). It is the "beak" star of the swan Cygnus, or the bottom star of the Northern Cross. (August through October)

Andromeda galaxy—A spiral galaxy similar to the Milky Way. Can you detect an oval shape? (October through December)

Pleiades—An open cluster visible to the naked eye. It contains a few hundred stars (although only six or seven are visible with the naked eye) and is about 400 light-years away. See whether you can detect the nebula around the stars. (November through February)

h and χ (the Greek letter "chi") Persei—A pair of open clusters visible to the naked eye (though not as easily as the Pleiades). In the sword handle of Perseus, they are an excellent view in binoculars. (November through February)

To enjoy your telescope (or binoculars) fully, a good star atlas is almost a necessity. The following are recommended:

The Monthly Sky Guide (8th ed.). Ian Ridpath and Wil Tirion. Cambridge University Press, 2006.

Whitney's Star Finder (5th ed.). C. Whitney. New York: Knopf, 1989.

A Field Guide to Stars and Planets (4th ed.). J. Pasachoff. Boston: Houghton Mifflin, 1999.

To keep up with developments in astronomy, you might contact the nonprofit Astronomical Society of the Pacific, whose aim is to share the excitement of astronomy with a wider public. Its members include scientists, teachers, hobbyists, and thousands of people from around the world who enjoy reading about the universe.

EXPANDING THE QUEST

1. "The Stuff Between the Stars," by G. Knapp, in *Sky & Telescope* (May, 1995).

2. "The Early Life of Stars," by S. W. Stahler, in *Scientific American* (July, 1991).

3. "Interstellar Molecules," by G. L. Verschuur, in *Sky & Telescope* (April, 1992).

4. "Spying on Stellar Nurseries," by R. Jayawardhana, in *Astronomy* (November, 1998).

5. "Deciphering the Mysteries of Stellar Origins," by C. Lada, in *Sky & Telescope* (May, 1993).

6. "The Secrets of Stardust," by M. Greenberg, in *Scientific American* (December 2000).

7. "The Gas Between the Stars," by R. J. Reynolds, in *Scientific American* (January 2002).

8. "The Emptiest Places," by E. Scannapieco, P. Petitjean, and T. Broadhurst, in *Scientific American* (October 2002).

9. "Fountains of Youth: Early Days in the Life of a Star," by T. P. Ray, in *Scientific American* Special Edition (September 2004).

10. "Companion to Young Stars," by A. P. Boss, in *Scientific American* Special Edition (September 2004).

11. "Stellar Archaeology," by F. Reddy, in *Astronomy* (May 2005).

12. "Stellar Origins: From the Cold Depths of Space," by D. Shepherd, in *Sky & Telescope* (June 2007).

STARLINKS

Quest Ahead to Starlinks
http://physicalscience.jbpub.com/starlinks

Starlinks is this book's online learning center. It features **eLearning**, which contains chapter quizzes and other tools designed to help you study for your class. You can also find **online exercises**, view numerous relevant **animations**, follow a guide to **useful astronomy sites** on the Internet, or even check the latest **astronomy news** updates.

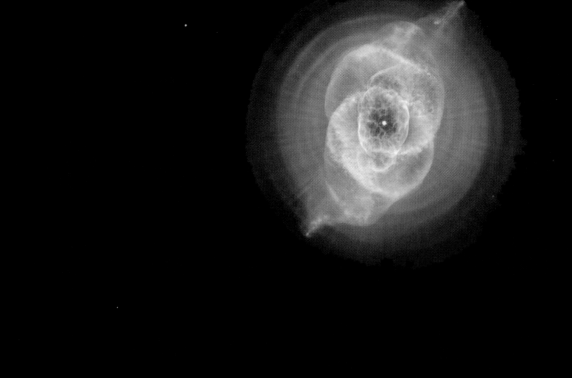

The Lives and Deaths of Low-Mass Stars

14

The "Cat's Eye nebula," officially named NGC6543, as seen by the *Hubble Space Telescope* (*HST*).

THIS BEAUTIFUL *HST* IMAGE SHOWS THE "CAT'S EYE NEBULA." The image reveals that the nebula contains at least 11 concentric gas shells (each being the edge of a bubble seen projected on the sky), jets of high-speed gas, and shock-induced knots of gas. Because of its age (estimated to be 1000 years), the nebula is a visual "fossil record" of the late stages of evolution of a dying star, which is expected to collapse into a white dwarf in a few million years.

The observations suggest that the dying star at the center has been ejecting its outer layers in a series of pulses at 1500-year intervals. The resulting dust shells create the observed structure of layered rings around the star. It is not yet clear what creates this pattern of mass loss every 1500 years, but it seems that about 1000 years ago the pattern changed; as a result, the nebula started forming inside the dusty shells and has been expanding ever since.

Such images reveal the beauty of distant dying stars, but we hope that as we study the lives and deaths of stars, you will also admire the beauty of the theory that underlies these events.

Stars form from the material of the interstellar medium. As a portion of a giant molecular cloud collapses, the matter in its core becomes hotter and denser until

All cross references to chapters, sections, figures, and tables pertain to the main text, *In Quest of the Universe, Sixth Edition. In Quest of the Solar System* contains Chapters 1–11 and 19 of the main text. *In Quest of the Stars and Galaxies* contains Chapters 1–5 and 11–19 of the main text.

nuclear fusion begins. When this happens, the object joins the family of stars, for it is then on the main sequence of the H-R diagram, where it will spend most of its life.

In this chapter, we describe stars of relatively low mass, those toward the bottom of the main sequence (including our Sun). We show that although all stars live their main sequence lives in much the same manner, they differ greatly in the way they die. Stars with very low masses seem to fade away, "giving up the ghost" gently. Others expand to become red giants and then puff away their outer layers, thereby revealing their previously hidden white-hot cores.

As we study the amazing events in the lives of stars, keep in mind that we are describing the results of various models of stellar interiors. We cannot examine the inside of a star directly. Instead, we observe the properties of the surfaces of stars and make mathematical models of what must be occurring inside them to produce our data. Such models have improved dramatically in the last 20 years as computer capabilities have increased; the calculations that produce today's models are numerous and complex.

Theoretical models cannot be divorced from reality, of course. They must correspond to observations. For example, the major supernova of 1987 (which we study in Section 15-3) served as a case history for testing models of supernovae. The supernova excited astronomers not only because it provided another opportunity to test their theories, but—as we will see—because it confirmed most of their predictions.

14-1 Brown Dwarfs

As we explained in the previous chapter, the mass of a collapsing object determines how long the object remains in the protostar stage. After a star has completed its protostar stage and reached the main sequence, its mass continues to be the dominant property determining the steps it passes through during its life; therefore, before discussing stellar life cycles, we should examine the question of what limits the mass of a star.

The maximum mass that a star can have is probably about 150 solar masses. If a protostar has more mass than this, it collapses quickly and develops tremendous amounts of energy and internal pressure. This pressure is so large that it overwhelms gravity and blows the star apart before it can establish equilibrium on the main sequence.

On the other hand, suppose that a very small portion of an interstellar cloud becomes compressed enough for the gravitational force to take over and continue the collapse. Calculations show that if this protostar has a mass greater than about 8% of the Sun's mass, the object's core becomes hot and dense enough for nuclear fusion to begin. In that case, the object joins the main sequence as a full-fledged star. But what if the protostar has less than enough mass for fusion to begin? In this case, the object still heats up due to gravitational contraction. FIGURE 14-1a shows hypothetical evolutionary trails of five protostars on an H-R diagram. The two with the most mass become stars on the main sequence. Two other protostars on the diagram—those with masses of 0.07 and 0.02 times the mass of the Sun—heat up so that they glow with a dull red color, but nuclear fusion never starts in their cores. Their dull color gives them their name: *brown dwarfs*. After a brief initial phase where some deuterium is burned, brown dwarfs continue to cool, releasing the heat left over from their birth, and become progressively fainter. The fifth object has the mass of the planet Jupiter, which may once have emitted a small amount of visible light, but now is too cool to do so.

The question of how small an object can be and still be classified as a brown dwarf is open for discussion. In order to separate large planets from brown dwarfs, we set the lower limit for the mass of a brown dwarf to be 13.6 times the mass of Jupiter; more massive objects undergo deuterium burning, whereas those with mass greater than 65 times the mass of Jupiter undergo both deuterium and lithium burning. Brown dwarf mass limits thus are between 0.013 and 0.08 solar masses.

Brown dwarfs have been predicted since 1963, but until one was found, astronomers were not able to confirm that it is possible for such a small portion of an

brown dwarf A star-like object that has insufficient mass to start nuclear reactions in its core and thus become self-luminous.

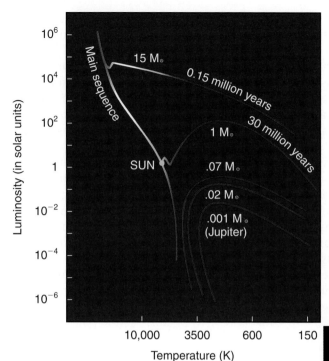

(a)

FIGURE 14-1 (a) Protostars with masses less than about 0.08 solar masses never reach the main sequence. Of the five stars whose evolutionary paths are shown here, two become main-sequence stars, two become brown dwarfs, and one becomes a planet. (b) The small object just right of center is GL229B, a brown dwarf, photographed by the *HST* in far red light. The dwarf is a companion to a much brighter star, Gliese 229, that is off the left edge of the image. The two objects are about 18 light-years away from us in the constellation Lepus and are separated from one another by at least 4 billion miles. (c) This VLT near-infrared composite image shows the brown dwarf (2M1207a) and its companion planet (2M1207b) in the 8-million-year-old TW Hydrae association, located about 230 light-years away.

(b)

(c)

interstellar cloud to collapse. Until 1994, the only objects whose existence was confirmed that were near the required size were the planets of our solar system. The first brown dwarf was seen in October 1994, using adaptive optics with the 60-inch telescope on Mount Palomar. It appears very close to another star, a red dwarf, and a year after the discovery, astronomers confirmed that the brown dwarf is actually gravitationally bound to the nearby star. Figure 14-1b is an *HST* photograph of the brown dwarf, called GL229B; it has about 20 to 50 times the mass of Jupiter and about the same diameter as Jupiter.

Brown dwarfs bridge the gap between stars and planets. Their properties reveal new insights into how stars and planets form. Even though brown dwarfs have been discovered orbiting stars, many of them seem to be isolated. How do brown dwarfs form? Two main scenarios have been proposed for isolated brown dwarfs. According to the first scenario, brown dwarfs form just like stars, by contraction in an interstellar cloud of dust and gas. In the second scenario, when a very young star in a multiple star system gets ejected from the system, it loses access to the surrounding accretion disk; this effectively stops its further growth by accretion, and it ends up as a brown dwarf. Computer simulations that follow the collapse of an interstellar gas cloud support this scenario. Such simulations show that the process of star formation is chaotic; in new-born stellar groups the more massive objects gather more gas than lower mass objects, the latter being ejected from the groups so quickly that they do not manage to accrete enough gas to become stars.

The preponderance of evidence supports the idea that brown dwarfs form like stars, however VLT and *Spitzer* observations support the presence of dusty disks around young brown dwarfs by detecting their infrared emission. Just like young stars, the majority of brown dwarfs are surrounded by such disks at an age of about a million years, and disks are also present around brown dwarfs as old as 10 million

A red dwarf is a small-mass star with low temperature and, as a result, it appears red.

Current predictions suggest that there should be twice as many brown dwarfs as main sequence stars. As of 2009, there are about 620 known brown dwarfs.

Jets are ubiquitous in the universe. They are observed over an enormous range of masses (from young stellar objects to active nuclei of galaxies) and lengths (from a few AUs to millions of light-years).

years. Other spectroscopic observations show that the accretion of material from the disks around brown dwarfs occurs at a slower pace but is otherwise similar to that from disks around young stars. Finally, brown dwarfs have been observed in pairs, and in 2005, astronomers confirmed the presence of a 5 Jupiter-masses planet orbiting a young, 25 Jupiter-masses brown dwarf (Figure 14-1c); VLT observations in 2007 show the presence of jets emanating from this brown dwarf, a behavior similar to that of young stars. *Spitzer* has also observed the beginnings of what might become planets around brown dwarfs by detecting clumps of tiny dust grains and crystals orbiting five brown dwarfs.

After the discovery of brown dwarfs, two new spectral types were introduced below the long-known cool M dwarfs. The L and T types, loosely defined at present, cover the temperature range from 3500 to about 1000 K. Objects in this range are almost invisible at optical wavelengths, with most of their emission coming out in the infrared. As the surface of a brown dwarf cools below 1500 K, something dramatic happens that changes its appearance; large amounts of methane form, making the T-type brown dwarfs the coolest objects detected so far.

Notice that we have avoided referring to a brown dwarf as a star. The reason for this is that the word *star* is reserved for an object massive enough to sustain nuclear fusion in its core, that is, to have reached stellar maturity on the main sequence.

14-2 Stellar Maturity

Do stars simply ignite and start fusion reactions immediately?

When the center of a collapsing star becomes hot and dense enough, hydrogen fusion begins. Gravitational contraction serves as the energy source for protostars, heating them and causing the emission of radiation, but main sequence stars—including our Sun—have as their energy source the fusion of hydrogen into helium.

Stellar Nuclear Fusion

In Section 11-2, we described the fusion reactions that occur in the Sun. That series of reactions, known as the proton–proton chain, is the predominant reaction that provides the energy for stars of low mass, up to about 1.5 solar masses. For the sake of completeness, it is repeated in FIGURE 14-2.

Stars that have masses greater than about 1.5 solar masses have higher temperatures in their cores, and a different chain of nuclear reactions dominates. This series of reactions involves carbon, nitrogen, and oxygen, and is called the **carbon cycle** or the **CNO cycle**. Its steps are shown in FIGURE 14-3. If you examine the steps of the CNO cycle, you will see that its overall effect is to

carbon (or **CNO**) **cycle**
A series of nuclear reactions that results in the fusion of hydrogen into helium, using carbon-12 in the process.

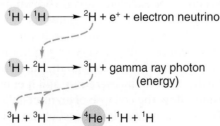

FIGURE 14-2 In the proton–proton chain, four hydrogen nuclei are changed to one helium nucleus, releasing energy in the process. The symbol e^+ denotes a positive electron, a positron. Dashed arrows indicate a nucleus taking part in the next reaction. In the third reaction, the second hydrogen-3 nucleus comes from a second occurrence of the two previous steps.

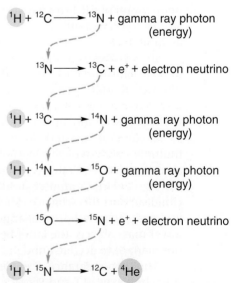

FIGURE 14-3 The CNO cycle changes four hydrogen nuclei into one helium nucleus, with an attendant release of energy.

transform four hydrogen nuclei into one helium nucleus, just like the proton–proton chain. Carbon-12 is one of the nuclei that participates in the reaction, but for each carbon-12 nucleus that enters the reaction, one is produced at the end; therefore, there is no net change in the carbon. It participates only as a facilitator. (This is similar to the role played by catalysts in some chemical reactions. The catalyst stimulates the reaction but is not used up in it.)

The Stellar Thermostat

Thermostats on the walls in our homes have two features: a thermometer to read the temperature and a switch to make the furnace or air conditioner turn on or off. If the thermometer indicates that the temperature is below some preset level, the switch turns on the furnace until the thermometer indicates that the temperature is high enough, whereupon the switch turns off the furnace. The core of a main sequence star has an analogous regulating mechanism, which controls the rate of consumption of the hydrogen fuel.

Suppose that somehow the rate of hydrogen fusion in a star begins to increase. The extra energy produced causes the temperature of the core to increase; however, when the temperature of a gas increases, the gas expands. This expansion serves as a switch to decrease the rate of fusion. This occurs because in the expanded gas, the average distance between hydrogen nuclei has increased, and the increase in distance means fewer collisions and therefore fewer fusions. The result is that the expansion causes the rate of energy production to decrease, which causes the expansion to stop. In this way equilibrium is reached, and fusion reactions occur at a uniform rate.

On the other hand, if the rate of fusion in a star's core somehow decreases, the core contracts, decreasing the distance between nuclei. This causes the fusion rate to increase. Another effect that occurs when the core contracts is the conversion of gravitational energy into heat, just as occurred when the original interstellar cloud collapsed; therefore, energy comes both from an increase in fusion and from gravitational energy, and both of these energies contribute to the thermostat that brings the star once again to equilibrium. The overall effect of the stellar thermostat is that nuclear fusion proceeds at a rate that is just enough to balance the force of gravity that tends to compress the star.

Main Sequence Life of Stars

The stellar thermostat is important during the main sequence life of a star. In a main sequence star, hydrogen in the core of the star is continually being converted to helium as the nuclear reaction occurs. This causes the number of particles in the core to decrease gradually (as *four* hydrogen nuclei are converted to *one* helium nucleus), and this results in the core shrinking slightly. In turn, the contraction causes the temperature to rise within the core, and this increases the rate of fusion; therefore, the core releases more energy, and as this energy flows outward, it causes the outer portion of the star to expand somewhat. These steps are outlined in **FIGURE 14-4**. We see the effects of these changes on the H-R diagram (**FIGURE 14-5**), for the increase in energy from the star means that it has a greater luminosity, and thus, it moves upward on the diagram. In addition, the expansion of the star causes its outer layers to cool a little,

The analogy of a thermostat is an alternative description of hydrostatic equilibrium, which we discussed in Section 11-3.

When you pucker your lips and blow on your hand, your breath feels cold because it expands as it moves from inside your mouth to the outside air. Similarly, gases in a bottle of carbonated drink expand and cool when you open the bottle, creating a small cloud in the bottle's neck.

When you pump air into a bicycle tire with a hand pump, the compressed air gets warm, making the pump warm, too.

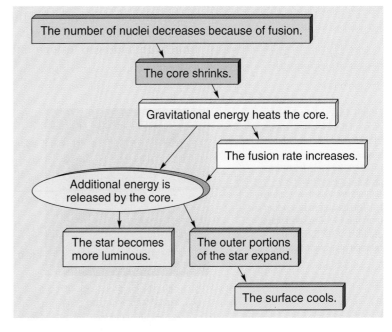

FIGURE 14-4 This diagram shows why a star becomes more luminous and cooler as it ages on the main sequence. Each arrow indicates that one event causes the next.

FIGURE 14-5 As stars age on the main sequence, they change slightly in temperature and luminosity. (The Tools of Astronomy box on page 396 shows how we calculate main sequence lifetimes.)

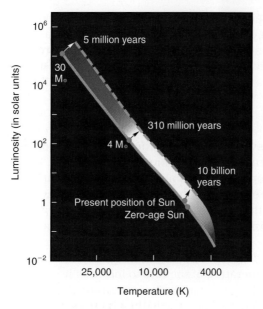

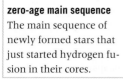

zero-age main sequence
The main sequence of newly formed stars that just started hydrogen fusion in their cores.

and this causes the star to move to the right on the diagram, toward lower temperatures. These changes are the reason the main sequence has a perceptible width. Stars start their lives as ***zero-age main sequence*** stars on the left side of the strip and then move up and to the right as they age.

The changes that the Sun is undergoing will have drastic effects on the Earth. The Sun began its main sequence life about 5 billion years ago, at which time its chemical composition was about 74% hydrogen, 25% helium, and the remaining 1% other heavier elements. As a result of the nuclear reactions, the Sun's core has contracted a bit. The Sun's luminosity has increased by about 40%, and its core now contains more helium than hydrogen (about 64% helium and 35% hydrogen). There is still enough hydrogen in its core for our Sun to continue on the main sequence for another 5 billion years, at which time it will be twice as luminous as it is now. This will raise the average temperature of the Earth by about 20°C (36°F), enough to melt the polar caps and drastically change the Earth's climate. Because the Sun was dimmer in the past, liquid water on the Earth's surface would have been frozen. This contradicts geological observations of sedimentary rocks, whose presence requires flowing liquid water. In order to resolve this contradiction, called the *faint young Sun paradox*, scientists have proposed that strong volcanic outgassing of greenhouse gases during Earth's early history kept the temperature high enough for liquid water to exist. It is possible, however, that the Earth experienced a number of periods when the oceans froze over completely.

Stars that are more massive than the Sun have a much greater fusion rate because their cores have greater pressures and higher temperatures—the thermostat is set higher in massive stars. This means they are more luminous. (Recall from our discussion in Section 12-7 the mass–luminosity relationship for main sequence stars: the more massive, the more luminous.) The greater fusion rate also causes these massive stars to use up their core hydrogen in a much shorter time. The most massive stars fuse hydrogen so quickly that their cores run out of hydrogen in only a few million years. Calculations for the least massive stars, on the other hand, show that they continue with hydrogen fusion on the main sequence for hundreds of billions of years.

FIGURE 14-6 (a) An H-R diagram of the Pleiades reveals that its most massive stars have started leaving the main sequence. The arrows indicate the evolutionary paths they must have taken. (b) The cluster M11 is older than the Pleiades, and stars are turning off the main sequence at a lower point.

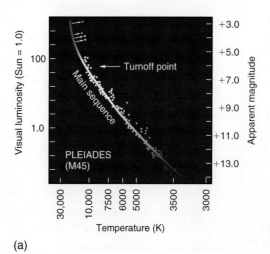

(a)

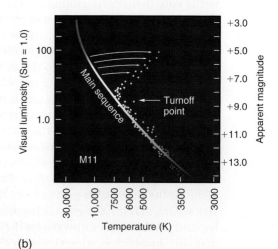

(b)

Observational evidence for the shorter lifetimes of massive stars comes from galactic clusters. **FIGURE 14-6a** is the H-R diagram of the Pleiades. Although this is considered a young cluster, its most massive stars are already leaving the main sequence. Arrows on the figure indicate their evolutionary paths after they end their main sequence lives. Figure 14-6b is a similar diagram for the cluster M11, a cluster older than the Pleiades. The most massive stars have moved even farther to the right whereas stars of less mass are now leaving the main sequence. Assuming that the stars in a cluster were all formed at approximately the same time, H-R diagrams such as those shown in Figure 14-6 allow us to find the age of a cluster. Stars at the **turnoff point**, which corresponds to the top of the group of stars still on the main sequence, are just now exhausting their core hydrogen and starting to move off the main sequence; therefore, the main-sequence lifetime of the stars at the turnoff point is equal to the age of the cluster.

14-3 Star Death

Until the end of their lives on the main sequence, the primary difference between the evolution of stars of various masses is in the amount of time they spend as protostars and as main sequence stars. For example, a one-solar-mass protostar takes about 30 million years to become a main-sequence star, whereas a 15-solar-masses protostar takes only 150,000 years. The main-sequence lifetime of a one-solar-mass star is about 10 billion years, whereas that of a 15-solar-masses star is only about 10 million years. From this point on, however, the mass of a star determines which of several very different paths its life will take. We discuss each of these paths in turn, beginning with stars of very low mass. Because mass classifications are fairly arbitrary and astronomers have not agreed on official names for the different classes, we group them in the following categories: very low mass (less than 0.4 M$_\odot$), moderately low mass (0.4–4 M$_\odot$), moderately massive (4–8 M$_\odot$), and very massive stars (greater than 8 M$_\odot$), where M$_\odot$ is the mass of the Sun.

14-4 Very Low Mass Stars ($<$0.4 M$_\odot$)

Recall from the discussion in Section 11-3 of energy transport within the Sun that little convection takes place in the Sun except in the outer layers. That is why after hydrogen is used up in the core of a star like the Sun, the core is not replenished with fresh hydrogen from outside. This causes the star to move from the main sequence.

In the least massive stars (those with a mass of less than about 0.4 solar masses), however, convection occurs throughout most or all of the star's volume (**FIGURE 14-7**); therefore, hydrogen from throughout the star is cycled through the core, and the entire star runs low on hydrogen at the same time. When this happens, the rate of fusion in the core decreases. As described earlier, this causes the core to contract and its temperature to increase as gravitational energy is converted into **thermal energy**. In a more massive star, this thermal energy is carried outward by radiation, but a very low mass star distributes its thermal energy by convection. The result is that the entire star contracts and heats up. On the H-R diagram, the star moves toward the lower left (**FIGURE 14-8**) and becomes a **white dwarf**.

Computer models indicate that a star with very low mass requires more than 20 billion years to complete the nuclear burning of its hydrogen fuel and end its main sequence life. Because this is greater than the age of the universe, the white dwarfs that astronomers see could not have originated in this manner. For this reason, we discuss white dwarfs again later in this chapter.

The name M11 means that this cluster is number 11 in the Messier Catalog of Nebulae and Star Clusters, originally developed in 1781 by the French astronomer Charles Messier (1730–1817).

turnoff point The point on an H-R diagram of a cluster of stars where the stars are just leaving the main sequence.

thermal energy Energy that is due to the random motions of molecules and atoms of a substance.

white dwarf The burnt-out relic of a low mass star.

In everyday language, the word "heat" is often used when "thermal energy" would be more correct. Strictly speaking, heat is the exchange of thermal energy.

TOOLS OF ASTRONOMY

Lifetimes on the Main Sequence

The amount of time that a star spends on the main sequence depends on two things: the amount of hydrogen in its core and its rate of hydrogen consumption. The relationship can be expressed as a proportionality:

$$t \propto \frac{\text{amount of hydrogen}}{\text{rate of hydrogen consumption}},$$

where t represents the star's lifetime and the symbol "$\propto$" means "proportional to." Because all stars have roughly the same proportion of core hydrogen at equivalent points in their lives, the amount of hydrogen in a star is proportional to its mass M. The rate at which the hydrogen is consumed depends on the star's luminosity. Therefore,

$$t \propto \frac{M}{L}.$$

In Section 12-7, we explained that a star's luminosity (L) is proportional to its mass (M) raised to the power of 3.5. Thus,

$$t \propto \frac{M}{M^{3.5}} \propto \frac{1}{M^{2.5}}.$$

Using the Sun as a standard so that t_{Sun} is the lifetime of the Sun on the main sequence and M is measured in units of the Sun's mass, the previous expression can be written as an equation:

$$t = \frac{t_{\text{Sun}}}{M^{2.5}}.$$

Then the main sequence lifetime of a star with a mass of 3 solar masses can be calculated as follows:

$$t = \frac{t_{\text{Sun}}}{M^{2.5}} = \frac{t_{\text{Sun}}}{3 \times 3 \times \sqrt{3}} \approx 0.064 \times t_{\text{Sun}}.$$

A star with three times the Sun's mass thus will live on the main sequence only about 6% as long as the Sun. Because the Sun's main sequence life (t_{Sun}) is about 10 billion years, this star's life will be about 640 million years. Similarly, a star with a third of the mass of the Sun will remain on the main sequence for 16 times longer than the Sun, about 160 billion years.

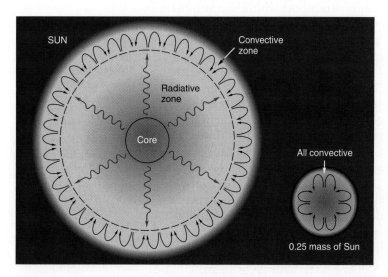

FIGURE 14-7 The Sun (at left) contains a large radiative zone. A star of very low mass, on the other hand, consists of one convective zone, meaning that all of its material mixes.

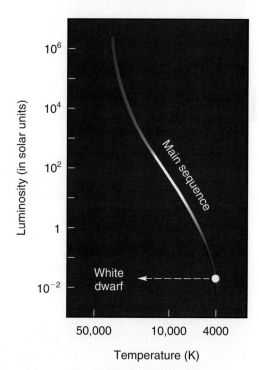

FIGURE 14-8 Very low mass stars spend their main sequence lifetimes at the very bottom of the main sequence. Although the universe is not old enough for any to have completed that stage of their life, when they do, they will slowly become hotter and join the category of white dwarfs.

14-5 Beyond the Very Low Mass Stars: The Red Giant Stage

About 90% of the stars in the sky are on the main sequence. This tells us that a typical star spends 90% of its luminous lifetime there. In all cases except for stars of very low mass (those with mass less than about 0.4 solar masses), the next step for stars leaving the main sequence is essentially the same. They become red giants.

The process of becoming a red giant starts when the core begins to run low on hydrogen fuel. Until then, the nearly constant production of fusion energy kept the core from collapsing. Now, lacking a source of energy to fight gravitational collapse, the core starts to shrink dramatically; however, just as in the case of a protostar, the contraction converts gravitational energy into thermal energy and radiation. In fact, the energy produced in the core actually increases over what it was when the energy source was fusion. The resulting increase of radiation from the core causes the shell of material around the core to heat up enough that hydrogen fusion begins there (FIGURE 14-9).

Now gravitational energy is being converted to thermal energy in the core and fusion is producing energy in a shell surrounding the core. These two sources of energy in the star's center result in a large increase in radiation from the center and cause the outer part of the star to expand. When a gas expands, it cools, and the outer portion of the star does just that. The star's surface temperature thus decreases, and the star moves to the right on the H-R diagram. At the same time, it moves upward, toward the red giant region, because its total luminosity is increasing because of the additional energy being produced in the core.

It may seem strange that a star can decrease its surface temperature and at the same time increase its luminosity, for when an object cools, it emits *less* radiation per square meter of its surface. The key to understanding this apparent contradiction is to realize just how large the red giant becomes. The Sun, at this stage in its life as a giant, will have expanded so that it encompasses the orbit of Mercury! It is true that each square meter of the Sun's cooler surface will emit less radiation, but the surface will have become so tremendously large that the total radiation emitted will be greater than before. FIGURE 14-10 illustrates how the Sun will change its position on the H-R diagram. As the Sun expands, it will move from point *A*, its turn-off point from the main sequence, toward point *B*. At this point as a red giant star, our Sun will be about 1000 times brighter than it is today, even though its surface temperature will be only about 3500 K. The outer planets will lose their atmospheres because of the

The cooling of a gas as it expands is what causes the gas escaping from an aerosol spray can to feel so cold.

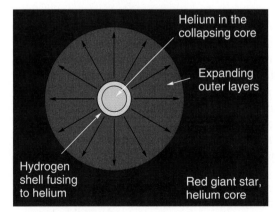

FIGURE 14-9 As the core of a star shrinks, it heats up, causing additional hydrogen fusion in a shell surrounding the core. (The core and hydrogen-fusing shell are actually a much smaller fraction of the star than is shown here.)

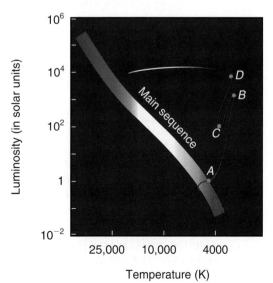

FIGURE 14-10 The Sun will enter the first stage of its death when it leaves the main sequence at point *A*. The helium flash (which we discuss in the next section) will occur when it reaches the luminosity and temperature of point *B*, and this will cause a quick decrease in its luminosity until it reaches point *C*. Helium fusion around the Sun's inner carbon core will then move it up to point *D*, where it will reach its greatest luminosity.

TABLE 14-1

A Typical Red Giant

Absolute magnitude	-1
Luminosity (solar units)	500
Mass (solar units)	1
Diameter (solar units)	200
Average density	10^{-7} g/cm³ (Sun's average density = 1.4 g/cm³)
Surface temperature	3600 K

increase in brightness and some of the inner planets will be vaporized. **TABLE 14-1** lists some properties of a typical red giant.

14-6 Moderately Low Mass Stars (0.4–4 M☉)

All main sequence stars become red giants (or supergiants, which we study in the next chapter) when they leave the main sequence. How stars change during the red giant stage, however, depends on how massive they are. We first consider the moderately low mass stars on the main sequence. This group includes stars from about 0.4 solar masses up to about 4 solar masses, and therefore, it includes our Sun. Even among members of this group, there are differences in the conditions that develop in their cores as they proceed to become red giants and in the way that the next cycle of nuclear fusion (helium fusion) begins in their cores. Because of these differences, we first consider stars with mass less than about 2 solar masses.

Electron Degeneracy and the Helium Flash

The core of a star at the end of its main sequence life consists of helium nuclei intermingled with electrons, a mixture that has properties similar to a regular gas. In this case, a simple relationship exists between pressure, temperature, and volume. For example, an increase in pressure exerted on a normal gas causes its volume to decrease and its temperature to increase. That is why gravitational pressure causes a star's core to contract and heat up when hydrogen fusion shuts down at the end of main sequence life.

As a moderately low mass star becomes a red giant, some unusual conditions start to develop in its core. Under the increasing temperatures and pressures, the free electrons in the core are forced to get very close to each other. After the density of the core reaches a certain value, the electrons are so closely packed together that any additional compression violates a very important natural law. According to this law, two electrons cannot have completely identical parameters describing their state. In a simple analogy, if we are playing a game of musical chairs, with each chair corresponding to a specific state for the electrons, only one electron is allowed to occupy each chair. As a result, the electrons produce a large pressure that resists any attempt to compress them further. At this point, the core stops contracting. The matter of the core is now said to be **degenerate**, and the pressure that supports it against gravity is called *electron degeneracy pressure*.

When a normal gas is heated, its pressure increases and the gas expands. In degenerate matter, pressure does not depend on temperature; it depends only on density. The properties of degenerate matter lead to a strange phenomenon. Refer to the graph in **FIGURE 14-11a**, which shows the relationship between the radius and the mass of a main sequence star. As you would expect, the more massive the star, the larger is its radius; however, the degenerate core of a red giant does not act this way, as shown in Figure 14-11b. The more massive the core of a red giant, the smaller is its radius!

electron degeneracy The state of a gas in which its electrons are packed as densely as nature permits.

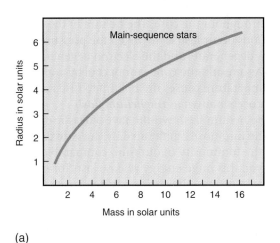

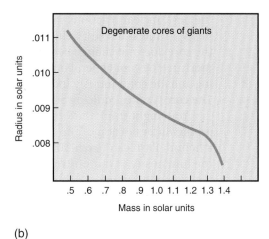

(a)   (b)

FIGURE 14-11 (a) In the case of normal, main sequence stars, the more massive the star, the larger is its radius. (b) In the case of the degenerate core of a red giant, however, the more massive the core is, the *smaller* is its radius.

This is very different from our experience with nondegenerate—regular—matter, and it has a profound influence on the next stage of the star's evolution.

As the hot, degenerate core of a red giant radiates its thermal energy outward, hydrogen continues to fuse to helium in a shell around the core. The hydrogen-fusing shell gradually works its way outward, all the time dumping its helium "ashes" onto the core. Because of the strange relationship between the mass and the radius of a degenerate core, the increased mass falling onto the core causes the core to shrink further and further. This, in turn, causes the core's temperature to continue to rise. Finally, the temperature of the red giant's core reaches a critical value, calculated to be about 100,000,000 K. At this point, shown as point *B* in Figure 14-10, helium nuclei begin to combine, forming carbon. The first steps in helium fusion reactions that produce carbon and oxygen from helium are shown in more detail in **TABLE 14-2** and **FIGURE 14-12**. Helium fusion does not occur at lower temperatures because helium nuclei repel each other with a much stronger force than hydrogen nuclei do, as each helium nucleus contains two protons, compared with only one proton for a hydrogen nucleus. At higher temperatures, helium nuclei move faster and therefore can get close enough for fusion to occur.

A helium nucleus is also known as an *alpha particle*, and the conversion of three helium nuclei into a carbon nucleus is called the *triple-alpha process*.

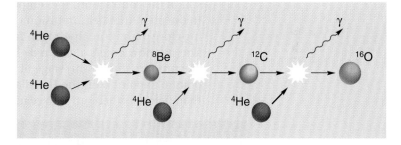

FIGURE 14-12 This diagram shows schematically the steps that occur as helium fuses into carbon in the core of a red giant.

Helium fusion does not proceed smoothly, however. Degenerate matter is a good conductor of thermal energy and, thus, the additional energy produced by

---

TABLE 14-2

Helium Fusion Reactions that Occur in Red Giants

Reaction	Explanation
$^4_2H + ^4_2H \rightarrow ^8_4Be + \gamma$ (energy)	Two helium nuclei combine to produce a beryllium-8 nucleus and a gamma ray (γ).
$^4_2H + ^8_4Be \rightarrow ^{12}_6C + \gamma$ (energy)	Another helium nucleus combines with the beryllium nucleus to form a carbon-12 nucleus and another gamma ray.
After sufficient carbon has been generated by the triple-alpha process, carbon nuclei can capture alpha particles (helium nuclei) to produce oxygen.	
$^4_2H + ^{12}_6C \rightarrow ^{16}_8O + \gamma$ (energy)	Carbon in the core fuses with helium and produces oxygen and another gamma ray (and more energy).

the new nuclear reactions spreads rapidly throughout the core. The helium in the core heats up and as a result helium fusion reactions occur faster; however, the pressure in the core does not change because the core is degenerate, and the electron degeneracy pressure does not depend on temperature. As a result, the core cannot cool by expanding and the continuously increasing core temperature causes the helium to fuse faster. The core "ignites" quickly and violently, a process known as the **helium flash**.

You might think that the violence of the helium flash would have a drastic effect on the outer portion of the star. After all, the luminosity generated by the core during a helium flash can reach 100 billion times the luminosity of our Sun, comparable to the luminosity of our entire Galaxy; however, this huge energy release lasts for only a few seconds, and most of the energy never makes it to the surface. Instead, computer models show that the helium flash changes the core drastically. Most of the energy released during a helium flash goes into heating up the core and destroying the electron degeneracy, thus restoring the stellar thermostat. The energy that escapes the core is absorbed by the layers overlying the core, which are opaque, possibly causing some mass from the surface of the star to be lost.

After the helium flash occurs, signaling the start of helium fusion in the core and providing the star with a new energy source, the star contracts slightly and its surface temperature increases. You might find this to be a very counterintuitive result. Because the core is not degenerate anymore, it expands like a normal gas. As a result, the temperature around the core decreases, and this cooling slows down the hydrogen fusion reactions in the shell around the core. Because this shell provides most of the red giant's luminosity, the resulting decrease in energy output allows the star's outer layers to contract and heat up. This stage in the evolution of the star is shown in Figure 14-10 as the movement from point *B* to point *C*.

The by-products of helium fusion reactions are carbon and oxygen nuclei. Our Sun will take about 100 million years to convert all the helium in its core to carbon and oxygen. At this point, because there is no energy source in the core, the core contracts until the electron degeneracy pressure stops the contraction again. As a result of the contraction, the temperature in a thin shell surrounding the core increases to the point that helium fusion reactions begin in this shell. While all this is happening close to the core, hydrogen fusion continues in a shell around the helium shell. The star now enters a new red giant phase. FIGURE 14-13 shows the structure of a star like our Sun during this stage of its evolution. During this time, the star evolves from point *C* to point *D* on the H-R diagram of Figure 14-10. Its size is now even larger than before. When our Sun reaches this stage, about 5 billion years from now, it will have expanded to enfold Mercury, Venus, and the Earth (FIGURE 14-14). Its luminosity will reach 10,000 times its current luminosity.

What will happen to the Earth? As the outer layers of the red giant approach Earth, our planet will find itself speeding through the hot gas of the Sun's atmosphere. Earth's atmosphere will be ripped away. The planet will be slowed by friction and will begin falling toward the Sun's center. Meanwhile the Earth's crust will melt and then vaporize. Our

helium flash The runaway helium fusion reactions that occur in the core during the evolution of a red giant.

FIGURE 14-13 When a red giant reaches the stage where it has a carbon core, the heat from that shrinking core ignites helium fusion in a shell around it. At the same time, hydrogen is fusing in a second shell beyond the first. This activity occupies very little of the volume of the star.

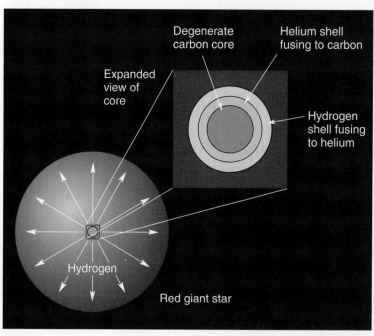

planet will become one with the Sun. Mars and the Jovian planets will mostly evaporate away.

In this section, we concentrated on stars with mass less than about 2 solar masses. Computer simulations indicate that stars more massive than about 2 solar masses do not experience a helium flash. The cores of these stars get hot enough toward the end of hydrogen burning that they make a smooth transition to helium burning without becoming degenerate. When their supply of helium is exhausted, however, their carbon core shrinks and becomes degenerate just as it does for lower mass, Sun-like stars; therefore, in both cases, the interior of the new red giant has the structure of Figure 14-13.

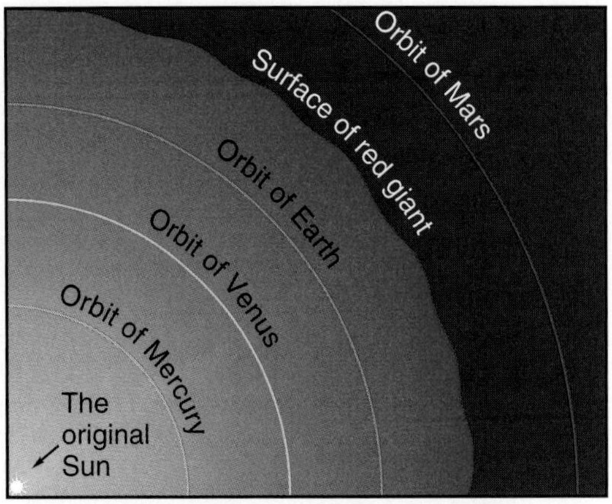

FIGURE 14-14 When the Sun becomes a red giant, it will expand until its surface is somewhere between Earth and Mars. Compare this with its present size.

Stellar Pulsations

During its evolution, a star continuously tries to remain in equilibrium. As its outer layers contract and expand, responding to the changes occurring in the core, the star's luminosity changes drastically, and these changes are often periodic. Astronomers have been able to deduce this behavior by using spectroscopic observations and the Doppler effect, and they have catalogued more than 20,000 pulsating stars, with periods that range from a few seconds to hundreds of days. Two types of pulsating stars are especially important to astronomers because they can be used as distance indicators. In Section 12-8 we discussed one of these groups, the Cepheid variables, which are pulsating stars with periods between 1 and 100 days. In this section, we explain the reason for the pulsations for both the Cepheids and another important group of distance indicators, the RR Lyrae variables.

As for any periodic phenomenon, a particular mechanism is responsible for the oscillations of a star's outer layers. When helium gets compressed at the outer layers of a Cepheid, the compression ionizes the helium atoms instead of increasing their temperature. Because the resulting gas of ionized helium is opaque, the layers trap heat from inside the star. This increases the pressure under the layers, forcing them to move outward. As a result of the expansion, the outer layers cool. The helium atoms are not ionized any more. The helium gas in the layers becomes transparent, and the trapped energy is released. Finally, the outer layers fall inward, compressing the helium and starting the cycle again. The conditions necessary for such pulsations to occur exist in a narrow strip on the H-R diagram, called the *instability strip* (FIGURE 14-15).

After helium fusion reactions begin in the core, a *massive* star moves across the middle of the H-R diagram. As the star crisscrosses the red-giant region of the H-R diagram, it can become unstable and pulsate while inside the instability strip, becoming a Cepheid variable. After the helium flash stage, *low mass stars* also crisscross the instability strip, but through its lower end. Some of these low mass stars become variable stars with periods shorter than a day and approximately constant luminosity (100 times that of our Sun). These low mass variables are called **RR Lyrae variables**. The relationship between the period and luminosity of RR Lyrae stars allows astronomers to find distances to globular clusters, in the same way that Cepheids are used to find distances to other galaxies.

In some cases, the pulsations of a star can be so large that the speed with which the outer layers are moving outward is larger than the escape speed from the star. As a result, the star loses its outer layers to outer space, contributing to the recycling of interstellar material through the processes of star formation and stellar evolution.

instability strip A region of the H-R diagram where pulsating stars are found.

RR Lyrae variables Pulsating stars with periods shorter than a day.

Mass Loss From Red Giants

During its main sequence life, our Sun continually sheds material as the solar wind. Orbiting X-ray and ultraviolet telescopes such as the *High Energy Astrophysical Observatories* and the *International Ultraviolet Explorer* have detected emission spectra from around other main sequence stars, indicating that stars commonly have hot chromospheres and coronas just as our Sun does. It seems safe to assume that they

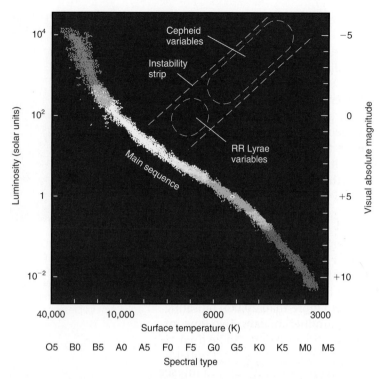

FIGURE 14-15 A section of the instability strip above the main sequence. Only two types of variable stars are shown. A star passing through this strip as it evolves becomes unstable and pulsates.

FIGURE 14-16 Star HD65750, a red supergiant, is losing material at a high rate. It is surrounded by a reflection nebula.

also have stellar winds like the Sun. The solar wind carries away about 10^{-14} of the Sun's mass each year. This is an extremely small amount, for at this rate, the Sun will lose only 0.01% of its mass in 10 billion years. Mass loss is much more important for red giants.

We are able to detect a slight Doppler shift in emission lines from gas near red giants. Based on the amount of the Doppler shift, astronomers conclude that the gas is being blown off red giants at speeds between 10 and 20 km/s, and that a typical red giant loses about 10^{-7} solar masses per year. **FIGURE 14-16** shows mass loss from HD65750, a giant star surrounded by a reflection nebula.

One reason for the greater mass loss in red giants is that these stars are so much larger than Sun-like stars that they exert less gravitational force on material near their surfaces. In addition, the helium-burning process that provides much of the energy for red giants is unstable and varies greatly in response to small changes in temperature. As a result, a red giant undergoes instabilities and pulsations as helium fusion increases and decreases. In fact, during a portion of a red giant's life, it is common for the star to pulsate with periods of less than one day, as we discussed in the previous section.

Planetary Nebulae

planetary nebula The shell of gas that is expelled by a low mass red giant near the end of its life.

The chapter-opening figure and **FIGURE 14-17** show photographs of objects called ***planetary nebulae***. Planetary nebulae were given that name decades ago because, when viewed in a small telescope, many of them have a blue-green color that resembles the color of Uranus and Neptune viewed through similar telescopes. In fact, planetary nebulae have nothing to do with planets. Spectroscopic analysis reveals—again by the Doppler effect—that these nebulae are made up of material moving outward from a very hot, low mass central star. Many planetary nebulae appear to be doughnut shaped, but this is because we are looking through a shell of material when we view them, as explained in **FIGURE 14-18**. The material glows because ultraviolet radiation from the central hot star causes it to fluoresce.

The fluorescence produces an emission spectrum, and the green color is due primarily to ionized oxygen. (Recall the fluorescence of a comet's coma and tail.)

At least two models attempt to explain planetary nebulae:

• One model holds that the pulsations in the core of a red giant continue to increase in intensity until the outer layers of the star also become unstable. With each pulse, part of the star's envelope is blown away. After perhaps 1000 years—

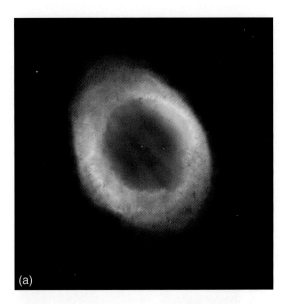

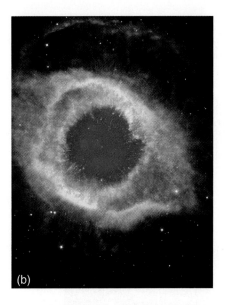

FIGURE 14-17 (a) This *HST* image of the Ring nebula (M57), in the constellation Lyra, is in approximately true colors: blue for very hot helium; green for ionized oxygen; and red for ionized nitrogen. It seems as if we are looking down a barrel of gas ejected from a dying star (with surface temperature of about 120,000 K) thousands of years ago. The nebula is about 2000 light-years away and 1 light-year in diameter. (b) The Helix nebula (NGC7293), 690 light-years away, in the constellation Aquarius. The rings are about 1.5 light-years across. In this composite *HST*/Cerro Tololo/*Spitzer* image, infrared data are shown in red and visible-light data in blue (OIII filter) and green (Hydrogen-alpha filter). The observed structure is created by a central binary system, with two gaseous disks nearly perpendicular to each other. Thousands of blobs resembling comets are seen; each holds about an Earth-mass in hydrogen and other gases and is twice the size of our solar system.

a short time by stellar standards—the entire outer portion of the star will have been ejected, forming a shell around the original core.

- A second model, now gaining in favor, holds that the stellar winds emitted from a dying star occur in definite stages. At the earliest stage, the outer layers of the star are blown off as a slow, dense wind at speeds of perhaps 90,000 kilometers/hour. This finally leaves behind a hot, dense star that emits high-energy radiation. Now the character of the wind changes so that it becomes a low-density but very fast wind at speeds of about 5 million kilometers/hour. When this second wind catches the first, it pushes the first forward, causing a region where there is a very dense wave. (To picture this, imagine that a group of walkers is being overtaken and run into by a group of runners. Where the two groups are colliding, the people are bunched very close together as the runners push the walkers into one another.) In addition, the intense radiation causes that dense area to glow, and this is the shell that we see as a planetary nebula. In the next chapter, when we discuss supernovae, we see that some hourglass-shaped planetary nebulae have been discovered that can be explained using this model.

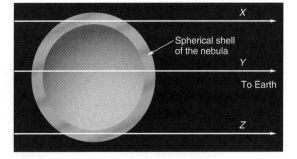

FIGURE 14-18 Three lines of sight through the spherical shell of a nebula. Lines *X* and *Z* pass through much more of the glowing nebula than does line *Y*; thus, the nebula appears brighter here.

Whichever model is correct (or perhaps a combination of the two will be found to be the best explanation for the phenomenon), not all planetary nebulae appear as rings. The Dumbbell nebula of **FIGURE 14-19a** shows just a hint of a spherical structure. The *HST* image of a planetary nebula in part (b) reveals that the initial ejections from the star were periodic and fairly uniform (as evidenced by the concentric spheres). The bright inner regions of the nebula must have been produced by irregular ejections of material, and dense clouds of dust condensed from the ejected material. The "twin jet" nebula of **FIGURE 14-20a** is a clear example of a bipolar planetary nebula. The

In addition, a 2004 study of the central objects in eleven planetary nebulae has found that 10 of them show clear evidence for radial velocity oscillations. This indicates the gravitational influence of a companion star and provides a simple explanation for the nonspherical structures and outflows observed in many planetary nebulae.

DATA PAGE

Planetary Nebulae

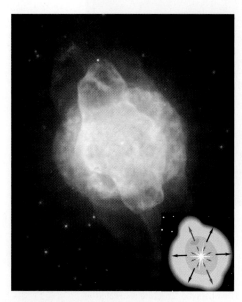

NGC3918 features an inner balloon of gas starting to break through the spherical outer envelope in two places.

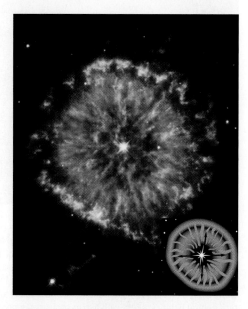

NGC6751 is 6500 light-years away. Hot blue gas forms a rough ring around the central star, pushing the cooler orange gas into an outer ring with streamers toward the central star.

Hubble 5 is a butterfly or bipolar nebula. The lobes are fast winds of gas extending beyond the disk formed early in the star's death.

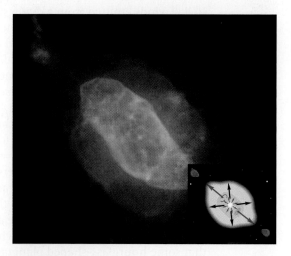

NGC7009 has a barrel-shaped envelope of material. Regions of low-density gas have shot out through each end of the barrel, glowing faintly.

QUICK FACTS

- Planetary nebulae have nothing to do with planets, but are clouds of glowing gas and dust thrown off from moderately massive stars changing from red giant stage to white dwarfs. The shapes we see depend on how the outflows interact with each other and with the star's magnetic field.

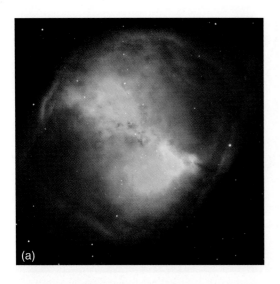

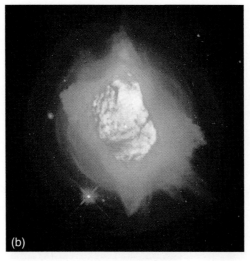

FIGURE 14-19 (a) A three-colored composite image of the Dumbbell Nebula (M27) taken by ESO's VLT. The nebula is about 1200 light-years away. (b) This *HST* image of NGC7027 is a composite of images in the visible and infrared. The central part is not solid, as it may appear in the false-color image. The dense clouds are shown in yellow.

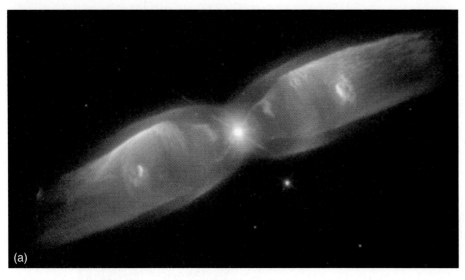

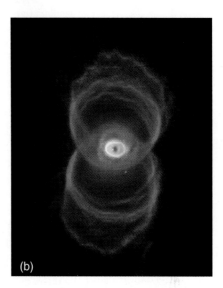

FIGURE 14-20 (a) M2-9 is an example of a "butterfly" or bipolar planetary nebula. The central star is known to be in a very tight binary system. A disk that surrounds both stars is seen in some *HST* images and has a size 10 times the diameter of Pluto's orbit. The nebula is 2100 light-years away in the constellation Ophiuchus. In this *HST* image, red corresponds to neutral oxygen, green to singly ionized nitrogen, and blue to double-ionized oxygen. (b) This is a composite of three *HST* images of the Hourglass nebula (MyCn18). Red, green, and blue represent ionized nitrogen, hydrogen, and doubly ionized oxygen, respectively. The nebula is 8000 light-years away.

measured speed of the ejected gas is more than 720,000 miles/hour. Observations show that the nebula's size increases with time, suggesting that it formed just 1200 years ago. The shape of the Hourglass nebula shown in Figure 14-20b is thought to result from the expansion of a stellar wind that moves fast inside a slowly expanding cloud (which is denser close to its equator than near its poles).

Only a few thousand planetary nebulae have been observed. Although this may seem a large number, it is very small compared with the number of stars, and this fact indicates that planetary nebulae do not last long; however, it is possible to make direct observations of cosmic evolution. FIGURE 14-21 shows two examples of *proto-planetary nebulae*, which represent the transition phase between the last stages of a red giant star's life and the early stages of a planetary nebula after the star has ejected most of its mass. The proto-planetary nebula phase is thought to last between a few hundred and 1000 years. The images in Figure 14-21 show that the ejected material forms complex shapes and symmetries, implying that some powerful processes are at work. It is thought that magnetic fields probably play a crucial role in creating these shapes and symmetries; support for this was provided by observations in late 2002,

proto-planetary nebula An object in transition between the last stages of a red giant star's life and its planetary nebula phase.

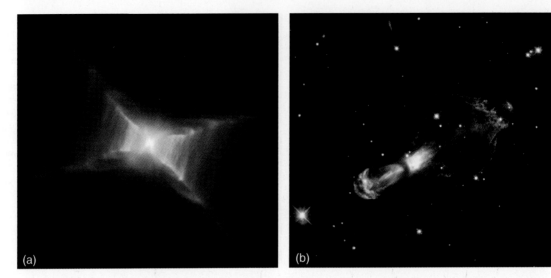

FIGURE 14-21 (a) In this *HST* image of the Red Rectangle, a nebula about 2300 light-years away in the constellation Monoceros, the straight features (appearing like rungs on a ladder) imply episodes of mass ejection from the central object (a close binary system) that began about 14,000 years ago and repeat every few hundred years. In a few thousand years, as the central star becomes smaller and hotter, a flood of ultraviolet light will make the nebula fluoresce, producing a planetary nebula. The surrounding dust particles are seen now because they reflect starlight from the star. Molecules mixed in with the dust emit light in the red part of the spectrum and hydrocarbons are probably forming in the cool outflow. (b) The Calabash proto-planetary nebula. It is also known as the Rotten Egg nebula because of the large amounts of sulphur compounds present, which would produce an unpleasant smell if we could smell in space. The gas (shown in yellow) has a speed of up to 1.5 million km/hour, and when it rams into the surrounding medium it forms shock-fronts on impact, which heat the surrounding gas. Light from hydrogen and ionized nitrogen arising from these shocks is shown in blue.

which showed that the magnetic field close to a number of aging stars is 10 to 100 times stronger than our Sun's.

Observations by ESA's *Infrared Space Observatory (ISO)* have found evidence that complex organic molecules form rapidly during the proto-planetary phase. It is likely that proto-planetary nebulae are huge factories of organic molecules. *ISO* also found that the dusty torus surrounding a very hot star at the center of one of the brightest planetary nebulae known, the Bug nebula (**FIGURE 14-22**), contains hydrocarbons, carbonates, water ice, and iron. The presence of carbonates (such as calcite) is very interesting because in our environment they form when carbon dioxide dissolves in liquid water and forms sediments; the observations suggest that carbonates can form through other processes in space.

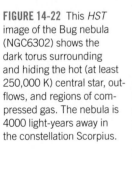

FIGURE 14-22 This *HST* image of the Bug nebula (NGC6302) shows the dark torus surrounding and hiding the hot (at least 250,000 K) central star, outflows, and regions of compressed gas. The nebula is 4000 light-years away in the constellation Scorpius.

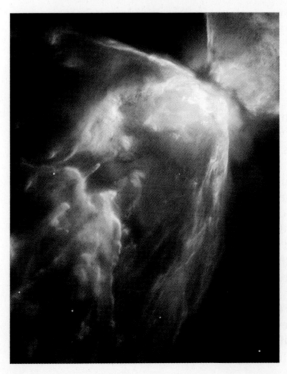

FIGURE 14-23 provides two more examples of the different shapes found among planetary nebulae. The material that we see around the central star quickly dissipates into space and becomes part of the interstellar medium. As the nebula dissipates, the hot, bright core of the star begins to peek through. This core may have a temperature of 100,000 K, and its appearance causes the star to shift its position on the H-R diagram. To include this stage on an H-R diagram, the diagram must be extended toward higher temperatures than have been included in previous diagrams. **FIGURE 14-24** shows that the star moves far to the left. The core remains at its very lumi-

FIGURE 14-23 (a) An *HST* image of the Boomerang nebula. It is a young nebula, in the constellation Centaurus, about 5000 light-years from us. The bow-tie shape of the nebula is created by a 500,000 kilometer/hour wind that results in a solar mass of material being lost every 1000 years (10 to 100 times more than other similar objects). The rapid expansion enables the nebula to be the coldest known region in the universe, just one degree warmer than absolute zero; as such, this nebula is the only object found so far that has a temperature lower than the background radiation. (b) The Ant nebula. This *HST* image shows the "ant's" body as two lobes protruding from the dying star.

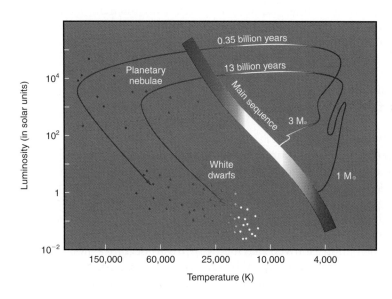

FIGURE 14-24 This H-R diagram has been expanded to the left—to higher temperatures—in order to include stars that are passing through the planetary nebula stage. The evolutionary tracks of a one-solar-mass star and a three-solar-mass star are shown.

nous stage for a relatively short time; then it quickly moves down the H-R diagram and collapses to become a white dwarf.

14-7 White Dwarfs

White dwarfs were discovered long before astronomers predicted their existence. In 1844, the German astronomer Friedrich Bessel (1784–1846) noticed that the star Sirius (in the constellation Canis Major) wobbled back and forth slightly. He hypothesized that an invisible companion orbiting Sirius causes the wobble. Bessel's calculations showed that the unseen companion (called Sirius B) has a mass about the same as Sirius (now called Sirius A). Sixteen years after Bessel had died, the American telescope maker Alvan G. Clark (1804–1877) discovered the companion of Sirius and, in doing so, became the first person to see a white dwarf. **FIGURE 14-25a** shows Sirius B as it appears in a much larger telescope. Figure 14-25b shows seven white dwarfs in M4, a compact cluster of stars.

A binary system in which the motion of one star reveals the presence of the other is called an *astrometric binary*. Bessel discovered, also in 1844, that Procyon, the brightest star in Canis Minor, is an astrometric binary.

FIGURE 14-25 (a) Sirius is revealed to be two stars in this telescopic view. The arrow points to Sirius B. The spikes that radiate from Sirius A are caused by diffraction around mirror supports within the telescope. (b) The image at left is a cluster of stars 7000 light-years away. The *HST* image at right is a small part of this cluster, only 0.63 light-years across. It contains seven white dwarfs (inside the blue circles) among the other much brighter stars.

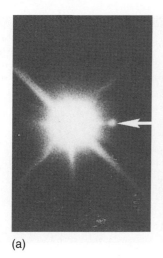

(a)

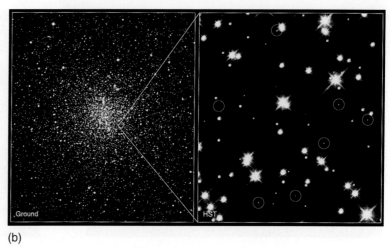

(b)

More than 1000 stars have been identified as white dwarfs, but none is bright enough to be seen by the unaided eye. Astronomers estimate that about 10% of all stars are white dwarfs, but this is a rough estimate, since white dwarfs are difficult to see.

White dwarfs are the cores of red giants that remain after the outer layers of the original stars have blown away. White dwarfs that formed from low mass stars are made up of helium and carbon. White dwarfs formed from more massive stars contain oxygen, neon, and sodium, and may contain atomic nuclei as massive as iron. Because white dwarfs consist of degenerate matter, the more massive a white dwarf, the smaller it is (as shown in Figure 14-11b). As a white dwarf gets older, its temperature and luminosity decline without a significant change in its size, for electron degeneracy prevents it from contracting further.

The Chandrasekhar Limit

Even though electron degeneracy supports the white dwarf against collapsing completely, there is a limit to the amount of pressure that degenerate electrons can withstand. In 1930, a 19-year-old student on his way from his native India to graduate school in England calculated this limit, showing it to be 1.4 solar masses. Subrahmanyan Chandrasekhar was awarded a share of the 1983 Nobel Prize in Physics for his discovery, and the limit is known as the ***Chandrasekhar limit***.

Chandrasekhar's calculations showed that if a white dwarf becomes more massive than 1.4 solar masses, the pressure caused by gravity at its surface is greater than the maximum pressure a degenerate gas can support. This means that white dwarfs must have masses less than 1.4 times the Sun's mass. Main sequence stars with masses of up to about 4 solar masses (which we have classified as moderately low mass stars) can end up as white dwarfs only because they lose mass during the red giant and planetary nebula stages. Stars more massive than these stars retain cores whose mass exceeds the Chandrasekhar limit and cannot form white dwarfs. Such moderately massive and massive stars end their lives in a dramatically different way than do less massive stars. We discuss the more massive stars in the next chapter.

Characteristics of White Dwarfs

White dwarfs have been observed with surface temperatures from 4000 to 85,000 K, but computer models predict that it is possible for them to have even higher temperatures. The masses of white dwarfs range from perhaps 0.02 solar masses up to 1.4 solar masses. Because the typical white dwarf is comparable in size to the Earth, the density of these stars is very high, about 10^6 grams per cubic centimeter. A teaspoon (about 5 cm³) of white dwarf material would weigh 5 tons! **TABLE 14-3** lists the properties of a "typical" white dwarf.

When the Sun becomes a white dwarf, its mass will be about 0.6 of its present mass. The remainder of its material will have been puffed away during its red giant

---

> Young people have made many major breakthroughs in science; Newton (age 23 in 1665) and Einstein (age 26 in 1905) are famous examples.

Chandrasekhar limit The limit to the mass of a white dwarf star, above which it cannot be supported by electron degeneracy and cannot exist as a white dwarf.

TABLE 14-3

A Typical White Dwarf

Absolute magnitude	+11
Mass	0.8 solar masses
Diameter	10,000 km (3/4 Earth diameter)
Density	10^6 g/cm^3
Surface temperature	15,000 K
Surface gravity	400,000 times Earth's

Earth

White dwarf

Surface of the Sun

FIGURE 14-26 When the Sun becomes a white dwarf, it will be slightly larger than today's Earth and therefore much smaller than the present Sun. Its mass, however, will be 0.6 of what it is now.

stage and blown away during its planetary nebula stage. It will be nearly as small as the Earth (**FIGURE 14-26**), with luminosity about a 10th of its current luminosity. As time passes, our Sun's luminosity will keep decreasing, and it will simply fade away.

To get to the white dwarf stage, Sun-like stars must go through the following stages: protostar, main sequence star, red giant, and planetary nebula. Protostars produce energy from gravitational contraction. Main sequence stars produce energy from nuclear fusion. Red giants produce energy from gravitational contraction of their cores (until the cores become degenerate) and from fusion within shells around the cores. White dwarfs, however, do not produce energy. They are hot because of leftover energy, and as they radiate this energy away, they get cooler. After billions of years, a white dwarf will have cooled enough that it no longer radiates in the visible region of the spectrum. It will appear on the H-R diagram at a position below and to the right of the bottom of the main sequence. As billions more years pass, it will cool further to become a ***black dwarf***, the burned-out cinder of a once-proud star. It is unlikely that the universe is old enough for many, if any, black dwarfs to have formed.

FIGURE 14-27 reviews the evolutionary steps taken by very low mass and moderately low mass stars. All begin as protostars and end up as black dwarfs. The end state of more massive stars—the moderately massive and very massive categories—is even more exotic, and we discuss it in the next chapter.

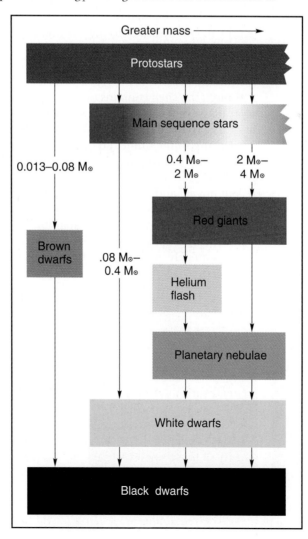

Greater mass ⟶

Protostars

Main sequence stars

0.013–0.08 M☉

0.4 M☉– 2 M☉

2 M☉– 4 M☉

Brown dwarfs

.08 M☉– 0.4 M☉

Red giants

Helium flash

Planetary nebulae

White dwarfs

Black dwarfs

FIGURE 14-27 This diagram shows the evolutionary steps taken by the stars discussed in this chapter. Unless we include planets in the category, it is probable that no black dwarfs have ever formed, and thus, their box could have been left off the chart. The boxes for the top two categories are open at the right to indicate that more massive stars also fall into these categories. Keep in mind that the limiting masses in each case are known only approximately.

black dwarf The theoretical final state of a star with a main sequence mass less than about 4 solar masses, in which all of its energy sources have been depleted so that it emits no radiation.

Novae

In 1054 AD, Chinese scholars recorded what they called a "guest star," a new star never seen before. In the next chapter we discuss this star, which was bright enough that it could be seen in the daytime.

Ancient astronomers believed that the sky never changed; however, in Chapter 12 we showed that stars move and that the constellations gradually change shape. Other changes also occur in the heavens. In 1572, Tycho Brahe observed a "new star." His excitement at the discovery is revealed in his words in a nearby Historical Note box. We know now that the phenomenon Tycho observed and similar phenomena observed by others since his time are not actually new stars, but are a phenomenon associated with white dwarfs.

Consider a binary star system in which one star is more massive than the other. The more massive star will end its main sequence life before its smaller companion. The massive star will then become a red giant, and if it is in our moderately low mass category, it will shed its outer portion and become a white dwarf. This will leave a white dwarf in orbit with a main sequence star, which then, in its turn, ends its fusion-burning life and begins to grow into a red giant.

At some point, the material of the outer portion of the new red giant is attracted to the white dwarf with more force than is exerted from its own core. This material is pulled away from the giant star. Although some might fall directly onto the white dwarf, analysis shows that most of the material pulled from the red giant will go into orbit around the white dwarf, forming an *accretion disk* (FIGURE 14-28).

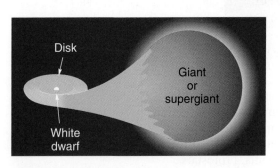

FIGURE 14-28 In a binary system composed of a white dwarf and a red giant, material from the growing giant will fall toward the dwarf and go into orbit around it.

accretion disk A rotating disk of gas orbiting a star, formed by material falling toward the star.

In the accretion disk, matter swirling around the white dwarf is heated as it falls inward, another example of gravitational energy being changed into thermal energy. Collisions within the disk cause its material—mostly hydrogen from the outer portions of the red giant—to fall inward onto the surface of the white dwarf. The hydrogen builds up there, becoming denser and hotter. When the temperature reaches about 10 million kelvin, hydrogen ignites in an explosive fusion reaction. The explosion, which occurs *only* on the surface of the white dwarf, blows off the outer layers of the white dwarf. Although the shell contains only a relatively tiny amount of mass (perhaps 0.0001 solar masses), it can cause the white dwarf to become 10,000 to 100 million times brighter (10 to 20 magnitudes) within a few days.

nova (plural **novae**) A star that suddenly and temporarily brightens, thought to be due to new material being deposited on the surface of a white dwarf.

During Brahe's time, people thought such a newly brightened star was a new star and called it a *nova*, the Latin word for new. In actuality, the star is not new, but is simply the sudden brightening of an old star. FIGURE 14-29 shows one of the brightest of recent novae, and TABLE 14-4 contains data for a typical nova.

Because so little material is blown off during a nova, the explosion does not disrupt the binary system. The companion star soon resumes transferring matter to the white dwarf. Depending on how fast the hydrogen is transferred, fusion can be reignited as quickly as a few months later, or 10,000 to 100,000 years may be required for a recurrence of the nova. Binary systems including a white dwarf and (usually) a red dwarf star (like our Sun but less massive) are called *cataclysmic variables*. There are various types of cataclysmic variables, including *classical novae* (which undergo huge explosions) and *dwarf novae* (which erupt in smaller explosions), but these differences are not important here. They have orbital periods from 1 to 15 hours, and their sizes are

FIGURE 14-29 These photos of Nova Cygni 1975 show its dimming from magnitude 2 at maximum light to magnitude 15.

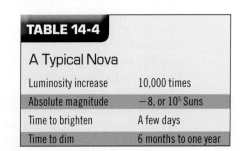

TABLE 14-4	
A Typical Nova	
Luminosity increase	10,000 times
Absolute magnitude	−8, or 10^5 Suns
Time to brighten	A few days
Time to dim	6 months to one year

HISTORICAL NOTE

Tycho Brahe's Nova

Tycho Brahe had been taught that the heavens were perfect and unchanging; therefore, when he saw a "new star"—a nova—he was very excited. (We know now that the "nova" observed by Tycho was actually a supernova.) His writings reveal his feelings:

In the evening, after sunset, when, according to my habit, I was contemplating the stars in a clear sky, I noticed that a new and unusual star, surpassing all the other stars in brilliancy, was shining almost directly above my head; and since I had, almost from boyhood, known all the stars of the heavens perfectly (there is no great difficulty in attaining

that knowledge), it was quite evident to me that there had never before been any star in that place in the sky, even the smallest, to say nothing of a star so conspicuously bright as this. I was so astonished at this sight that I was not ashamed to doubt the trustworthiness of my own eyes. But when I observed that others, too, on having the place pointed out to them, could see that there really was a star there, I had no further doubts. A miracle indeed, either the greatest of all that have occurred in the whole range of nature since the beginning of the world, or one certainly that is to be classed with those attested by the Holy Oracles.

about the Earth–Moon distance. It is estimated that there are about a million such variable stars in our Galaxy.

14-8 Type I Supernovae

White dwarfs consist mostly of degenerate carbon (but they probably contain heavier elements as well). Imagine a white dwarf in a binary system accreting material from its companion. When the accretion brings the mass of the white dwarf above the Chandrasekhar limit, electron degeneracy can no longer support the star, and it collapses. The collapse raises the core temperature enough that carbon fusion suddenly begins. This is somewhat similar to the helium flash during the red giant stage, except that now carbon rather than helium is the fuel. Runaway fusion reactions result in carbon fusing to elements as massive as iron. The fusion reactions increase the temperature in the interior, but the pressure remains the same, as the core matter is degenerate. As a result, the white dwarf cannot expand and cool. As the temperature increases, so does the rate of the fusion reactions, in a catastrophic fashion. At some point, the temperature becomes so high that the electrons are no longer degenerate; they start behaving like an ideal gas, resulting in an explosive expansion of the white dwarf. Whereas a red giant undergoing a helium flash has a surrounding envelope of gases to control its explosion, the white dwarf doesn't. The star explodes completely.

The destruction of a white dwarf in this manner is called a **supernova**. The name is unfortunate, for a supernova is not just a "super nova." A nova might reach an absolute magnitude of -8 (about 100,000 Suns), but a supernova attains a magnitude of -19 (10 *billion* Suns). When a supernova is observed in another galaxy, it may shine brighter than all the rest of the galaxy combined.

Astronomers classify supernovae into two categories, *Type I* and *Type II*, depending on their spectra. The spectrum of a Type II supernova contains prominent hydrogen lines, but the spectrum of a Type I supernova does not contain any hydrogen lines. Furthermore, Type I supernovae are divided into three subclasses: Ia, Ib, and Ic, again depending on their spectra. Astronomers think that *massive* stars that have lost different proportions of their outer layers before exploding cause Types Ib and Ic. These supernovae are found only near regions of recent star formation; however, Type Ia supernovae are found even in regions where there is no star formation. Because white dwarfs contain no hydrogen, they are thought to be responsible for Type Ia supernovae. Type II supernovae result from the explosion of a single star; we discuss them in the next chapter.

supernova The catastrophic explosion of a star, during which the star becomes billions of times brighter.

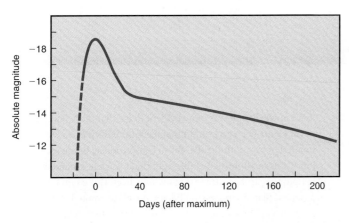

FIGURE 14-30 A Type Ia supernova, thought to be produced by the destructive explosion of a white dwarf. The light curves of brighter supernovae diminish slower that those of less bright supernovae.

A Type Ia supernova reaches maximum brightness after a few days. It fades quickly for about a month and then declines in brightness more gradually until it dissipates in about a year. FIGURE 14-30 shows the light curve of a Type Ia supernova. Theoretical models indicate that its energy (after the initial explosion) comes from radioactive decay of nuclei produced in the explosion. The light curves and spectra of all Type Ia supernovae are very similar. This is surprising given the varied range of stars from which they start. It is likely that the slow process of accretion that brings these white dwarfs to the point of destruction erases most of the initial differences between the original stars. This similarity among the light curves of Type Ia supernovae is a powerful tool that astronomers use to trace the history of the expansion of the universe, as we discuss in Chapter 18.

A Type Ia supernova remnant is shown in FIGURE 14-31. The X-ray spectrum of the remnant DEM L71 shows a high concentration of iron atoms relative to oxygen and silicon. The mass of the ejected material was found to be comparable to the mass of the Sun. These observations show that we are looking at the remnants of an exploded white dwarf.

Even though astronomers agree on the basics of Type Ia explosions, there are still some very important details that are not well understood. We need to understand better the nature of the explosion and how symmetric it is. We also need to understand better the methods by which white dwarfs gain enough mass to exceed the Chandrasekhar limit. Direct spectroscopic evidence for the material surrounding a star before exploding into a Type Ia supernova (SN2006X) was found in 2007 using VLT, VLA and *HST* data; this supports the widely accepted model of a white dwarf orbiting, and feeding from a companion star. Also, the observation of a strong X-ray source before the explosion, and the lack of such a source after, would strongly support the model of a binary system of a white dwarf and companion star; after all, as material from the companion hits the surface of the white dwarf, it will heat the star and produce strong X-rays before the explosion, whereas after the star explodes, it will be undetectable in X-rays.

Extreme helium stars contain almost no hydrogen but are instead dominated by helium, with significant amounts of carbon, nitrogen, and oxygen.

However, in addition to accretion of mass from a companion giant star, merging white dwarfs could also produce supernovae. The mass of the merged object is only briefly above the Chandrasekhar limit, before the object explodes. In 2006 astronomers announced that spectroscopic observations of a rare group of stars, called "extreme helium stars," support theoretical predictions of the composition of a star formed by the merger of a helium-rich white dwarf and a more massive carbon–oxygen white dwarf. If the mass of the star is above the Chandrasekhar limit, the star will become a Type Ia supernova; if it is less, the new merged star will become an extreme helium supergiant star. SN2006gz (and possibly SN2003fg) is an example of a Type Ia supernova that resulted from the merging of two white dwarfs; it showed a very strong spectral signature of unburned carbon and evidence for compressed layers of silicon (which formed during the explosion), both supported by computer models of merging white dwarfs.

Finally, there is observational evidence for a new class of super-

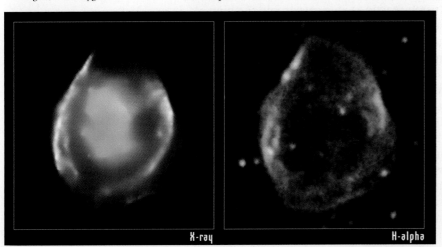

FIGURE 14-31 The left panel shows X-ray data of the supernova remnant DEM L71 taken by the *Chandra Observatory*. A hot, 10 million K inner cloud of glowing silicon and iron is shown in aqua, surrounded by an outer blast wave. The right panel shows this blast wave at optical wavelengths. After a white dwarf explodes, the expanding material drives a shock wave in front of it and into the surrounding interstellar medium (the bright outer rim). The pressure behind this shock wave drives another shock wave inward that heats the expanding material (the cloud shown in aqua). The size and temperature of this remnant indicate that it is several thousand years old.

novae, called Prompt Type Ia. It includes a population of massive stars that evolve very quickly in a very dense environment. Accreting material from a companion star in such an environment at an early stage, the star will explode in only about 100 million years, much shorter than other Type Ia supernovae. Examples of such objects include Kepler's supernova in our galaxy (so named because Kepler observed it more than 400 years ago), DEM L238/L249 in the Large Magellanic Cloud, and others at greater distances. More data and better models are needed to understand Type Ia events, which play such an important role as yardsticks of cosmic evolution.

What happens to the remains of a supernova? They disperse into space. **FIGURE 14-32** is a photograph of the Vela nebula, the remains of a Type II supernova that exploded some 11,000 years ago. To calculate its age, astronomers determined the Doppler shift of the part of the remnants that is moving toward us and calculated when the explosion had to have occurred to produce the present speed. In Chapter 15, we discuss Type II supernovae and describe examples of past and recent supernovae.

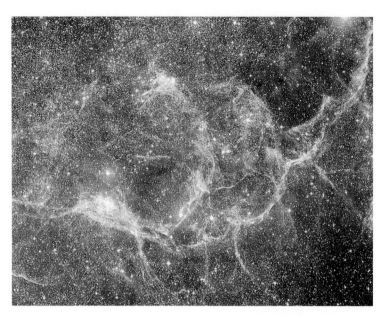

FIGURE 14-32 The Vela nebula is a Type II supernova remnant. (Photos of other supernova remnants appear in Chapter 15.)

Conclusion

Mass is the single most important factor that determines a star's properties and how the star evolves. The least massive protostars never get hot and dense enough to join the main sequence and become hydrogen-fusing stars. These so-called brown dwarfs heat up from gravitational contraction during their protostar stage; however, after that they have no heat source and, thus, they cool to become dark, planet-size cinders. More massive protostars become stars on the main sequence, where they fuse hydrogen to helium (and, in more massive stars, to heavier elements). During their main sequence lives, stars become slightly more luminous and cooler, resulting in a small change in their position on the H-R diagram. The more massive a star is, the less time it takes to exhaust its fuel and end its main sequence life.

In stars massive enough to support fusion but with masses less than about 0.4 solar masses, their entire supply of hydrogen mixes as fusion takes place. Although the universe is not old enough for any of these stars to have had time to complete their main sequence lives, computer models show that when they run low on hydrogen, they will shrink and heat up until electron degeneracy sets a lower limit to their size. They will then live out their lives as white dwarfs until they, too, cool to become burned-out cinders.

Moderately low mass stars fuse hydrogen in their cores without any intermixing with the outer portions. When these stars run low on fuel, they become brighter, larger, and cooler; that is, they become red giants. Their electron-degenerate cores heat up until they become hot enough to begin helium fusion. When the Sun and other stars of similar mass reach this point, a helium flash occurs; the stars then consume their helium quickly, getting hotter and dimmer for a time.

During the red giant stage, stars lose great amounts of matter from their outer layers because of intense stellar winds. At the end of the red giant stage, Sun-like stars blow away their outer shells and become beautiful planetary nebulae. Left behind when the nebula has dispersed into space is the star's white-hot degenerate core—a white dwarf. White dwarfs are dim and hard to see, but a white dwarf that is part of a binary system will advertise its presence as a nova if its companion, as it swells to become a red giant, leaks material onto the hot surface of the white dwarf.

When the mass of a white dwarf that is accreting matter from a companion or merges with another white dwarf becomes greater than the Chandrasekhar limit, the star explodes in a brilliant supernova. This completely destroys the star, dispersing its radioactive remnants through space.

Stars with masses greater than about 4 solar masses end the red giant portion of their lives in a different way than stars of lower mass. In the next chapter we describe the behavior of these stars, including the supernovae that they produce.

STUDY GUIDE

1. The length of a star's main sequence lifetime is determined by the star's
   A. carbon content.
   B. distance from the center of its host galaxy.
   C. surface temperature.
   D. mass.
   E. spectral type.

2. Why do stars of greater mass live longer on the main sequence than stars of lesser mass?
   A. The massive stars have more hydrogen fuel.
   B. The massive stars use their fuel more slowly.
   C. The massive stars go through many stages of fusion.
   D. [More than one of the above.]
   E. [None of the above; the statement in the question is false.]

3. If the rate of hydrogen fusion within the Sun were somehow to increase, the core would
   A. collapse and the Sun would grow cool.
   B. collapse and heat up further.
   C. expand and therefore tend to slow the fusion.
   D. expand and, therefore, increase in temperature.
   E. stay the same size but become hotter.

4. Why does hydrogen fusion occur only in a star's center?
   A. Only near the center is there enough heat and pressure.
   B. Only near the center is there enough hydrogen that is not mixed with other elements.
   C. Only near the center is the speed of light favorable for the reaction to occur.
   D. Heat is transferred down to the center during the main sequence life of the star.
   E. [The statement is false; fusion occurs throughout the star's volume. This is what causes the surface to be bright.]

5. Why do massive stars run out of hydrogen in their cores faster than less massive stars?
   A. Their hydrogen fuses faster because of greater pressure.
   B. There is less hydrogen in their cores.
   C. The cores of less massive stars contain a greater percentage of helium, which slows hydrogen fusion.
   D. The cores of less massive stars contain a lesser percentage of helium, which slows hydrogen fusion.
   E. [The statement is false; more massive stars do not run out of hydrogen faster than stars of less mass.]

6. The Sun will at some time in the future become
   A. a red giant.
   B. a white dwarf.
   C. a black dwarf.
   D. [All of the above.]
   E. [None of the above.]

7. Which choice lists the stages in a star's life in correct order?
   A. Main sequence, protostar, white dwarf, red giant.
   B. Protostar, red giant, main sequence, white dwarf.
   C. Protostar, main sequence, white dwarf, red giant.
   D. White dwarf, protostar, main sequence, red giant.
   E. Protostar, main sequence, red giant, white dwarf.

8. Red giants are more luminous than white dwarfs because
   A. red giants are hotter.
   B. red giants are closer.
   C. red giants have greater radii.
   D. [All of the above.]
   E. [None of the above; the statement is not true.]

9. Why are all known white dwarfs relatively close to the Sun?
   A. White dwarfs are formed only in our neighborhood of the Galaxy.
   B. Light from distant white dwarfs has not yet reached the Earth.
   C. No white dwarfs are bright enough to be seen at great distances.
   D. Light from distant white dwarfs is too redshifted to be seen.
   E. [The statement is false; white dwarfs are seen at all distances from the Sun.]

10. What is the difference between the Sun and a one-solar-mass white dwarf?
    A. The Sun has a greater radius.
    B. The Sun has more hydrogen.
    C. They have different energy sources.
    D. [All of the above.]
    E. [None of the above.]

11. One of the causes for the phenomenon called a nova is
    A. the fusion of iron in the core of a massive star.
    B. the infall of material onto a neutron star from a white dwarf.
    C. the transfer of material onto a white dwarf in a double star system.
    D. the collapse of a protostar.
    E. the death of a massive star and the formation of a black hole.

12. The nuclear reactions in a star's core are kept under control so long as
    A. the star's luminosity depends on its mass.
    B. the pressure of the gas in the core depends on its temperature.
    C. the star's density depends on its mass.
    D. the star's mass depends on its temperature.

13. A planetary nebula is
    A. the vastly expanded shell of a dying star.
    B. a cloud of gas out of which stars form.
    C. a cloud of cold dust in space.
    D. the same as a white dwarf.
    E. a circular ring around a black hole.

14. Compared with a young star of the same mass, an older star contains
    A. more hydrogen.
    B. more elements heavier than hydrogen.
    C. about an equal amount of each element.
    D. more of every element.
    E. [No general statement can be made.]

15. The mass of a white dwarf is
    A. always greater than 4 solar masses.
    B. always less than 0.5 solar masses.
    C. less than the mass of its original main sequence star.
    D. greater than the mass of its original main sequence star.
    E. [Two of the above.]

16. A brown dwarf is
    A. the final fate of all stars.
    B. the final fate of stars like the Sun, but not of less massive stars.
    C. a stage of a star's life prior to the white dwarf stage.
    D. a stage of a star's life after the white dwarf stage.
    E. a warm starlike object that has too little mass to support fusion in its core.

17. When more material falls onto a white dwarf, its size
    A. increases.
    B. remains the same.
    C. decreases.
    D. [Any of the above, depending on the nature of the white dwarf.]

18. When the core of a star shrinks after hydrogen fusion stops,
    A. the core cools and the star expands.
    B. the core cools and the star contracts.
    C. the core heats and the star expands.
    D. the core heats and the star contracts.

19. The Chandrasekhar limit is a limit to a star's
    A. size.
    B. volume.
    C. density.
    D. mass.
    E. [Both A and B above.]

20. Some supernovae occur when the Chandrasekhar limit is exceeded in a white dwarf. In this case, the energy producing the explosion is
    A. chemical in nature.
    B. nuclear fusion.
    C. nuclear fission.

    D. degenerate electrons.
    E. gravity.

21. The primary source of energy that powers a main sequence star is
    A. gravity.
    B. chemical energy.
    C. nuclear fission.
    D. nuclear fusion.
    E. fossil fuel.

22. A star is considered to begin its main sequence life when
    A. it starts to collapse.
    B. nuclear reactions start.
    C. its protostar life begins.
    D. it begins to move off the main sequence.
    E. its planetary system has formed.

23. When a star "moves off" the main sequence to become a red giant,
    A. it moves across the sky toward regions where stars are hotter.
    B. it moves across the sky toward regions where stars are cooler.
    C. it stays in the same place in the sky, but becomes hotter.
    D. it stays in the same place in the sky, but becomes cooler.
    E. [Either A or B above, depending on the mass of the star]

24. When a star ends its protostar stage, what determines where it ends up on the main sequence?
    A. Its location in the sky.
    B. Its radial velocity.
    C. Its transverse velocity.
    D. Its mass.
    E. [Both B and C above.]

25. Which stars have longer lifetimes on the main sequence?
    A. The most massive stars.
    B. The least massive stars.
    C. Stars in binary systems.
    D. Stars in star clusters.

26. A planetary nebula occurs
    A. in a newly forming planetary system.
    B. in an old, dying planetary system.
    C. in a white dwarf near the end of its life.
    D. at the end of the life of a red giant.
    E. in the core of a white dwarf.

27. A white dwarf (compared to the Sun) has
    A. a very high average density and a high surface temperature.
    B. a very low average density and a high surface temperature.
    C. a very high average density and a low surface temperature.
    D. a very low average density and a low surface temperature.
    E. [No general statement can be made.]

28. What property of a star is most important in determining the stages of its evolution?

29. Why do stars of the lowest mass not become red giants?

30. What causes the core of a star to heat up after its hydrogen fusion ceases?

31. List the stages in the life of a one-solar-mass star.

32. What will happen to the Earth when the Sun becomes a red giant? How large will the Sun become?

33. Starting at the center and going outward, list the various layers of a red giant that contains a carbon core.

34. Describe two hypotheses designed to explain what causes a planetary nebula. Describe the appearance of a planetary nebula. What causes it to glow?

35. Describe a white dwarf with regard to size, mass, luminosity, and temperature.

36. What supports a white dwarf from further collapse?

37. What happens to the size of a main sequence star when mass is added to it? What happens to the size of a white dwarf when mass is added to it?

38. Why does a recurrent nova continue to flare up time after time?

1. What is a brown dwarf? What determines if a protostar becomes a brown dwarf or a main sequence star?

2. Gravity causes protostars to collapse. What stops the collapse?

3. Massive stars have much more hydrogen in their cores than do less massive stars. Why, then, do they run out of hydrogen faster than stars of low mass?

4. We see relatively few white dwarfs compared to main sequence stars, but astronomers are confident that white dwarfs are very common. Explain this discrepancy.

5. What determines whether the carbon cycle will occur in a particular star?

6. Based on how stars evolve during their main sequence lifetime, explain why the main sequence on an H-R diagram should not be a thin line, but should instead have a thickness.

7. What causes a star's radius to increase as it becomes a red giant? What causes its surface to cool?

8. Explain why a Sun-like star, after it becomes a red giant, undergoes a helium flash, but a star a few times more massive than the Sun begins helium fusion gently.

9. Explain what causes the pulsations of RR Lyrae and Cepheid variables.

10. Explain how star clusters provide evidence that supports our theories of what happens to stars after their main sequence life ends.

11. What do planetary nebulae have to do with planets? What are they and how did they get the name "planetary nebulae"?

12. Compare parts (a) and (b) in Figure 14-6. Although these are generally true to the actual data, they were not drawn carefully from the data. Look at the plotted stars and explain what tips you off that the diagrams do not actually fit real data.

13. Explain the meaning of the two graphs of Figure 14-11, and explain what happens to make the two types of stars act differently.

14. Why are novae and Type Ia supernovae not included in Figure 14-27, which shows the evolutionary steps of normal stars?

1. The Sun's life on the main sequence is expected to be 10 billion years. What is the life expectancy of a star with a mass half that of the Sun? What is the life expectancy of a star with a mass 10 times that of the Sun? (Hint: See the Tools of Astronomy box on page 396.)

2. Assume that you observe a circular nebula on the celestial sphere. Its angular size is 1.2 arcminutes, and it is located about 3000 light-years away. Material in the nebula is expanding at the rate of 20 km/s. How long ago did the central star lose its outer layers? (Hint: You will first need to use the small-angle formula to find the radius of the circular nebula.)

3. The star at the center of a planetary nebula has luminosity of 1000 times the Sun's and temperature of 100,000 K. What is the radius of the white dwarf in solar units? (Hint: Use the Stefan-Boltzmann law.)

4. We mentioned in the text that during a nova explosion, a white dwarf becomes 10,000 to 100 million times brighter. Show that this corresponds to a change in magnitude by 10 to 20.

5. The star Zeta Geminorum is a Cepheid variable with a period of about 10 days. Interferometric observations show that the star's angular diameter changes by about 4.7×10^{-8} of a degree during its cycle. Doppler measurements show that the change in the star's diameter is about 8.55 million kilometers. Show that the star is about 338 parsecs away.

1. "Planetary Nebulae," by N. Soker, in *Scientific American* (May, 1992).

2. "The Lives of Stars: From Birth to Death and Beyond," by I. Iben and A. V. Tutukov, in *Sky & Telescope* (December, 1997; January, 1998).

3. "Giants in the Sky: The Fate of the Sun," by J. Kaler, in *Mercury* (March/April, 1993).

4. "Stellar Metamorphosis," by S. Kwok, in *Sky & Telescope* (October, 1998).

5. "White Dwarfs: Fossil Stars," by S. Kawaler and D. Winget, in *Sky & Telescope* (August, 1987).

6. "The Mystery of Brown Dwarfs Origins," by M. Subhanjoy and R. Jayawardhana, in *Scientific American* (January 2006).

7. "How to Blow Up a Star," by W. Hillebrandt, H.-T. Janka, and E. Müller, in *Scientific American* (October 2006).

8. "Supersoft X-ray Stars," by P. Kahabka, E. P. J. van den Heuvel, and S. A. Rappaport, in *Scientific American* Special Edition (September 2004).

9. "The Extraordinary Deaths of Ordinary Stars," by B. Balick and A. Frank, in *Scientific American* (July 2004).

10. "The Discovery of Brown Dwarfs," by G. Basri, in *Scientific American* Special Edition (September 2004).

Quest Ahead to Starlinks
http://physicalscience.jbpub.com/starlinks

Starlinks is this book's online learning center. It features **eLearning**, which contains chapter quizzes and other tools designed to help you study for your class. You can also find **online exercises**, view numerous relevant **animations**, follow a guide to **useful astronomy sites** on the Internet, or even check the latest **astronomy news** updates.

EXPANDING THE QUEST

STARLINKS

The Deaths of Massive Stars

15

After and before (arrow) images of the region around the great supernova of 1987. SN1987A was discovered near the Tarantula nebula in the Large Magellanic Cloud, a companion galaxy to our Milky Way.

IT WAS WINDY AND THE ROOF WAS BROKEN THAT FEBRUARY night in 1987. Ian Shelton, resident astronomer at the Las Campanas Observatory in Chile, had to open the roof by hand in order to use the 10-inch telescope he had repaired. Ian was responsible for maintaining the larger 24-inch telescope, but seldom got to use it himself, because visiting astronomers kept it pretty busy. Shivering in the mountain air, he climbed a ladder and pushed the corrugated metal aside.

This night had started like most others. Winds up to 40 miles/hour pounded the cinderblock telescope shack, but once Ian had opened the roof, he was able to begin the work he liked: photographing the sky. This night he decided to photograph the Large Magellanic Cloud, which is not a cloud at all, but the nearest galaxy to the Milky Way. It and its companion, the Small Magellanic Cloud, are visible only from the Southern Hemisphere. To get a good bright photo, he decided to take a 3-hour time exposure. (Long exposures are the norm for astronomers.) His telescope, however, was not like more modern telescopes that use motors and computers to guide them so that they automatically follow the stars. Instead, he had to look through the eyepiece constantly and guide the telescope manually.

All cross references to chapters, sections, figures, and tables pertain to the main text, *In Quest of the Universe, Sixth Edition. In Quest of the Solar System* contains Chapters 1–11 and 19 of the main text. *In Quest of the Stars and Galaxies* contains Chapters 1–5 and 11–19 of the main text.

At 3 o'clock the sky suddenly went dark; the wind had blown the roof shut and jarred the telescope. What a night! Thinking that his photograph had been ruined, he considered going to bed, but then decided to develop the photographic plate and see what he had.

In the dim light of the darkroom, he examined the result. The Magellanic Cloud was apparent, but when things start going wrong, nothing stops the slide—there was a flaw on the plate. A bright blotch appeared near the edge of the galaxy's image. Of course, a bright star looks just like a blotch, but Ian knew the sky well enough to know that there is no bright star where the spot appeared. He had been using photographic plates of the same type for some time, however, and he knew that a flaw of this size would be extremely unusual. What was the alternative? A new star—a supernova? None bright enough to be seen with the naked eye had appeared since 1604. If this were a supernova, it certainly was a bright one.

The graduate school dropout argued with himself for 20 minutes before going outside to check the sky with his own eyes. There it was, the first naked-eye supernova since 1604, an object that would be studied by astronomers for years and one that has already caused us to change our theories about the death of stars. The unpromising night had yielded the astronomical event of the century, and Ian Shelton was the first to observe it.

Within hours, nearly all of the radio and optical telescopes in the Southern Hemisphere were focused on SN1987A, as it was named. The *International Ultraviolet Explorer* satellite was recording radiation from the supernova by the next day; the Japanese X-ray satellite *Ginga* was observing the area within weeks. The neutrino detection teams in Cleveland, Ohio and Kamioka, Japan ran to check their records of the past few days. High-altitude planes and balloons were launched to observe the region without interference from Earth's atmosphere. Said one astronomer, "It's so exciting. It's hard to sleep."

In Chapter 14 we described how stars with masses less than about four solar masses evolve from protostars to white dwarfs. In general, we might say that such stars end their lives in a fairly uneventful manner. Although all but the very low mass stars become red giants and some put on beautiful displays as planetary nebulae, all-in-all, low mass stars die not with a bang, but with a whimper; however, if a white dwarf happens to be part of a binary system, things may get explosive. The resulting novae and supernovae give a preview of what happens when massive stars die.

You might say that the smallest stars don't die—they just fade away.

A massive star's death is truly a major event. Instead of gently puffing away their outer layers, some stars blow away their outer portions in cataclysmic supernova explosions, such as the one described in the chapter opening. Many end up as superdense neutron stars, sending out powerful lighthouse beacons across space. The most massive stars, however, end their lives as something even more exotic—as black holes.

15-1 Moderately Massive and Very Massive Stars (>4 M☉)

Although the final destinies of moderately massive stars (perhaps 4 to 8 solar masses) and very massive stars (greater than about 8 solar masses) are far different, they behave similarly during most of their lives and, thus, we consider them together for now. Recall from Figure 14-7 that the internal structure of main sequence stars depends on their mass. In a star of mass greater than about 4 solar masses, the core temperature and pressure are greater than in stars of low mass and so is the temperature difference between the core and outer layers. As a result, the energy created in the core by fusion reactions

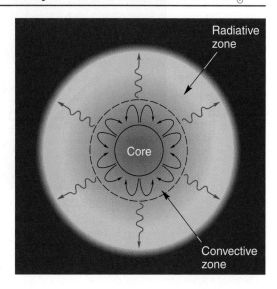

FIGURE 15-1 The internal structure of stars with mass greater than 4 solar masses. Energy transport from the core occurs by convection in the inner part of the star and by radiation in the outer part.

FIGURE 15-2 This *HST* image of Betelgeuse is the first direct image of a star (other than the Sun). It reveals a huge atmosphere with a mysterious hot spot on the star's surface. In the constellation Orion (see Figure 1-8), Betelgeuse is the bright star in the hunter's right shoulder.

Size of Star

Size of Earth's Orbit

Size of Jupiter's Orbit

supergiant The evolutionary stage of a massive star after it leaves the main sequence.

is transported outward by convection, whereas the low density in the outer layers allows energy transport by radiation (**FIGURE 15-1**).

The stars that have the most mass live very short main sequence lives. A star 15 times as massive as the Sun takes only about 10 million years to burn up its hydrogen. Like a less massive star, it then expands to become a red giant. During its red giant stage, the 15-solar-mass star changes gradually from a hydrogen-fusing central core to one that fuses helium. Because of its greater mass and corresponding greater core temperature, it becomes brighter than the standard red giant, and we call it a **supergiant**. The most prominent supergiant is Betelgeuse, the bright star in Orion's right shoulder (**FIGURE 15-2**). A supergiant may be a million times brighter than the Sun and have an absolute magnitude of −10. Figure 12-17 shows the position of red supergiants on the H-R diagram, and **TABLE 15-1** lists some typical data for a supergiant.

As the core pressure and temperature increase inside a supergiant, the products of the previous cycle of fusion reactions become the fuel for the next cycle of nuclear reactions. Heavier elements are continuously produced, including neon, silicon, and even iron. As a result, successive cycles proceed faster. New shells of material are generated around the core, and new cycles of fusion reactions occur at the outer layers. The star continuously enters new red-giant phases and repeatedly moves back and forth on the H-R diagram (**FIGURE 15-3a**). The various reactions and their approximate longevity are shown in **TABLE 15-2** for a 15-solar-mass star. The core of the red supergiant now contains several layers, as illustrated in Figure 15-3b.

FIGURE 15-3 (a) This H-R diagram shows the evolution of two massive stars. Massive stars do not experience a helium flash; instead, when core helium fusion reactions start, the evolutionary track turns sharply downward in the red-giant region. (b) Most of the volume of a 15-solar-mass red giant is hydrogen, but the central part is multilayered. The hypothesized mass of each layer is shown.

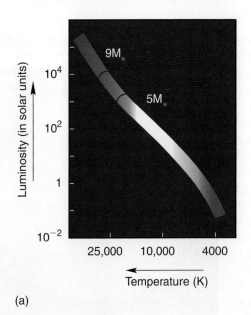

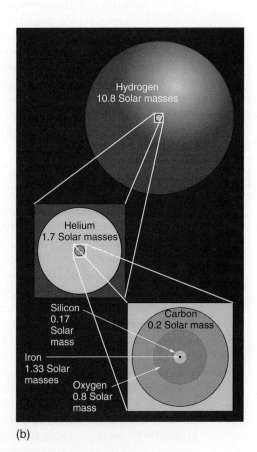

(a)

(b)

TABLE 15-1

A Typical Supergiant

Absolute magnitude	−5
Luminosity	10,000
Mass (solar units)	12
Diameter (solar units)	1200
Average density	10^{-8} g/cm^3
(Sun's average density = 1.4 g/cm^3)	
Surface temperature	3600 K

TABLE 15-2

The Evolution of a 15-Solar-Mass Star

Element Fused	Fusion Products	Time	Temperature
Hydrogen	Helium	10,000,000 years	4,000,000 K
Helium	Carbon	>1,000,000 years	100,000,000 K
Carbon	Oxygen, neon, magnesium	1000 years	600,000,000 K
Neon	Oxygen, magnesium	A few years	1,000,000,000 K
Oxygen	Silicon, sulfur	1 year	2,000,000,000 K
Silicon	Iron	A few days	3,000,000,000 K

15-2 Type II Supernovae

We classify supernovae into two types, based on their spectra. A Type I supernova has no hydrogen lines in its spectrum, and most supernovae of this type (Ia) are thought to be produced by white dwarfs in binary systems, as we described in Chapter 14. Type Ia supernovae are powered by nuclear energy released by runaway nuclear reactions as the white dwarf completely disintegrates. The spectra of Type II supernovae, on the other hand, reveal prominent hydrogen lines, indicating that these supernovae result from stars that still contain hydrogen in their outer layers. Type II supernovae (along with Ib and Ic) are powered by gravitational energy that is released as gravity continuously collapses the core. Differences between the two types of supernovae are outlined in **TABLE 15-3**; typical light curves are shown in **FIGURE 15-4**.

The process by which Type II supernovae occur is not well understood, but various models have been proposed. According to the leading model, after the core of a supergiant has reached the stage where silicon has formed and begins fusing to iron, things start to happen quickly. Computer modeling indicates that the silicon layer fuses to iron in only a few days. Considering that stellar lifetimes are measured in millions and billions of years, a few days is a remarkably short time. But after this, things change even faster!

Silicon fuses to iron in the center of the star. Previously, when a new central core (first of helium, then carbon, then oxygen, and finally silicon) was formed, the material of that core began to fuse to heavier elements when it got hot enough. Iron is different. All of the fusion reactions in Table 15-2 produce energy when they occur. When silicon nuclei fuse to form iron, for example, energy is produced; however, if iron were to fuse together to form even more massive nuclei, energy would be *absorbed* instead of released. In other words, if iron is to fuse, it must have a supply of energy; therefore, as the iron core shrinks and heats up, it does not fuse to something more massive. Instead, once its mass reaches the Chandrasekhar limit, the core collapses violently. In much less than a second, the tremendous pressure generated by gravity in the

TABLE 15-3

Typical Supernovae

	Type Ia	Type II
Spectrum	No hydrogen lines	Prominent hydrogen lines
Magnitude at peak	−19	−17
Light curve	Sharp peak	Broader peak
Expansion rate	10,000 km/s	5000 km/s
Mass ejected	0.5 solar masses	5 solar masses

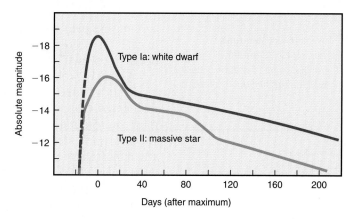

FIGURE 15-4 Typical light curves from the two types of supernovae are shown. Type Ia supernovae get brighter and have a sharper peak. The curve is dashed at the beginning, during the time that the luminosity of the star is usually not observed.

ADVANCING THE MODEL

Nucleosynthesis

Only hydrogen, helium, and lithium, the very lightest elements, were present in the very early universe. All of the other heavier elements that we now see around us (up to iron) were produced inside stars through nuclear fusion reactions; however, the fusion process cannot produce elements heavier than iron. Instead, these elements can be produced in a process where neutrons are added to the nuclei of atoms. This addition creates different, heavier isotopes and can take place because neutrons are neutral particles; thus, they do not feel any electrical repulsion from the charged nuclei.

There are two different stellar environments in which the process of neutron capture can happen. The first environment is inside very massive stars when they explode as supernovae. In such an event, the build-up proceeds via the "rapid-process." The explosion creates an enormous blast of neutrons that pulverizes atomic nuclei, which have no chance of decaying by emitting an electron and turning into a stable atom of the next element in the Periodic Table. As a result, nuclei that are extremely neutron-rich are formed, which finally decay very fast.

A second environment is inside normal stars during the end of their lives when they burn their helium. Here the addition of neutrons is done rather gently via the "slow-process." About half of all the elements heavier than iron are believed to be formed by this process, which takes place during a specific stage of stellar evolution, just before an old star expels its outer layers and dies as a white dwarf. During the slow-process, neutrons enter the nucleus of an atom, turning it into an isotope. This neutron capture continues until there are too many neutrons inside the nucleus for the isotope to remain stable. Then the isotope emits an electron and becomes a stable atom of the next element on the Periodic Table.

Computer-based stellar models predict that the slow-process is very efficient in stars whose metal content is low (metal-poor stars). Such stars were born at an early stage in our Galaxy and are therefore quite old. In these stars, the slow-process is expected to produce nuclei all the way up to the most heavy and stable ones, those with atomic numbers around 82 (the lead region). The discovery in late 2001 of metal-poor stars with high abundances of lead showed an excellent agreement between the predicted and observed abundances, reinforcing our current understanding of how the slow-process operates in the interiors of stars.

How does the slow-process operate? When a carbon-13 nucleus (which includes 6 protons and 7 neutrons) is hit by a helium-4 nucleus (which has 2 protons and 2 neutrons), they fuse to form oxygen-16 (which has 8 protons and 8 neutrons) while also releasing a neutron. A neutron is also released in a similar fusion reaction that involves a neon-22 nucleus and a magnesium-25 nucleus. It is these neutrons that become the building blocks for making the heavier elements. The carbon-13 isotope is in turn produced by fusion of a normal carbon nucleus (carbon-12) and a proton (a hydrogen nucleus). In order for this process to work, however, there must be a region in the star where sufficient amounts of carbon and hydrogen coexist. This is problematic because most hydrogen nuclei have already fused to heavier nuclei, including carbon. Current models of stellar interiors and the discovery of "lead" stars suggest that there must be a moderate mixing between the outer hydrogen-rich stellar layers and the inner carbon-rich regions. It is not yet clear how this mixing process works.

We described neutrinos in Chapter 11, during the discussion of fusion reactions in the Sun.

core pushes pairs of electrons and protons together to form a neutron and a neutrino in each case. Most of these neutrinos escape from the core, carrying a large amount of energy, which causes the core to become cooler and therefore smaller and denser.

As the core collapses, its density increases and very quickly reaches the density found inside nuclei, which is about 4×10^{14} g/cm^3. It is very difficult to compress the core beyond this density. Computer simulations indicate that when the core reaches this stage, its innermost part rebounds—it stops contracting and expands outward—sending a strong pressure wave outward. Meanwhile, the outer layers of the core are falling inward to fill the gap left when the iron core shrank. The rebounding surface collides violently with the infalling material and produces two effects: first, the collision is energetic enough to cause iron to fuse into heavier elements. (This requires an *input* of energy. The necessary energy comes from the collision.) Second, the collision sends shock waves outward that throw off the outer layers of the supergiant. Some astronomers question whether the shock wave is energetic enough to cause the star to explode. One theory holds that as the shock wave begins to lose its energy, it is heated by great numbers of neutrinos escaping the core. The heating action takes less than a second, and at the end of it, the newly energized shock wave has enough energy to blow away the outer portion of the star.

We know that a major fraction of the energy of a supernova appears in neutrinos. If this energy is included, Type II supernovae are by far more luminous than Type Ia.

(a)

(b)

FIGURE 15-5 Supernova 1993J was discovered in a spiral arm of the galaxy M81 on March 28, 1993. (a) It is identified by an arrow in the left photo, taken March 30, when it had an apparent magnitude between 10 and 11. (b) Nothing appears at that spot in the right photo, which was taken earlier. The other dots in the images are stars in our Galaxy and are not associated with M81.

Some Type II supernovae are more luminous than others, but a typical one reaches a peak absolute magnitude of -17. This is nearly a billion times brighter than the Sun! **FIGURE 15-5** shows a supernova that occurred in the galaxy M81 in March 1993. The other dots in the photograph are not part of M81, but are stars in our Galaxy. Except for the supernova, stars in M81 show up only as a haze.

Physicists have long recognized the special place that iron holds in the list of chemical elements. When lighter nuclei are fused to form heavier nuclei, energy is produced, but only so long as the newly formed nucleus is not more massive than iron. Elements heavier than iron cannot be produced without some source of energy and, thus, they are not formed spontaneously in nature. What, then, is the source of the heavy elements that we find here on Earth? Supernovae are a part of the answer. Supernova explosions in the distant past produced some of the heavy elements and blasted them away into space to become part of the interstellar material from which the Earth was formed. In the Advancing the Model box on "Nucleosynthesis," we describe in more detail the processes involved in nucleosynthesis, the building up of heavier elements from lighter ones.

Detecting Supernovae

Three supernovae in our Galaxy have been seen with the naked eye; they occurred in the years 1054, 1572, and 1604. The most spectacular on record occurred in the constellation Taurus on July 4, 1054, and was observed by Chinese astronomers, who reported that it was bright enough to be seen in daylight and to read by at night. It remained bright for a few weeks, and then gradually faded until it disappeared from view after about two years. Invention of the telescope was centuries away, of course, so the Chinese of the 11th century had no way to continue observing their "guest star" (as they called the object). In 1731, an amateur astronomer reported a small nebula in Taurus, and 200 years later Edwin Hubble discovered that the nebula consists of material expanding at a rate such that it must have begun its expansion at the time the Chinese reported the guest star. **FIGURE 15-6** is a photograph of the supernova remnant,

Pulsar

FIGURE 15-6 The Crab nebula is the remnant of a supernova that was seen in the year 1054. The arrow indicates the position of the Crab pulsar.

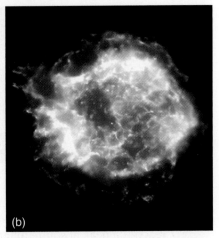

FIGURE 15-7 (a) The Veil nebula, shown here, is part of the Cygnus loop, a supernova remnant 2500 light-years away. (b) This *Chandra* X-ray image of Cassiopeia A shows the regions of greatest intensity in low-, medium-, and high-energy X-rays in red, green, and blue, respectively. The red material on the left outer edge is rich in iron. The bright greenish white region on the lower left is rich in silicon and sulphur. In the blue region on the right edge, low- and medium-energy X-rays have been filtered out by a cloud of dust and gas in the remnant.

Edwin Hubble (1889–1953), for whom the space telescope is named, was an important American astronomer whose name will appear often in the next two chapters.

TABLE 15-4

Some Supernova Remnants

Remnant	Distance (Light-Years)	Diameter (Light-Years)	Age (Years)
Cassiopeia A	10,000	15	330
Crab nebula	6000	4.4	955**
Cygnus loop	2500	100	20,000
Vela nebula	300*	2300	11,000
Tycho's supernova	9800	11	435
Kepler's supernova	20,000	14	405

*This is the distance to the nearest part of the nebula.

**Because the Crab nebula is 6000 light-years from us, the supernova actually occurred about 7000 years ago (6000 + about 955 years since 1054), but we speak of it as if it happened about 955 years ago.

called the Crab nebula because of its shape. Telescopic observations taken over the last half-century show its growth; they reveal that its outer portions are still expanding outward at about 1400 km/s and that it is now about 4.4 light-years in diameter.

We have also been able to find remnants of the supernova seen by Tycho Brahe in 1572 (which provided evidence to him that the heavens were not unchanging) and the one seen by Johannes Kepler in 1604. Kepler's supernova occurred 4 years before the invention of the telescope; since then there have been no supernovae in our Galaxy seen with the naked eye.

TABLE 15-4 lists some major supernova remnants. Images of two of these—the Cygnus loop and Cassiopeia A—appear in **FIGURE 15-7**.

15-3 SN1987A

When a new astronomical object is observed, a telephone call or telegram is sent to the International Astronomical Union's Central Bureau for Astronomical Telegrams, to establish priority of discovery. Telegram number 4316, received on February 24, 1987, read as follows:

> W. Kunkel and B. Madore, Las Campanas Observatory, report the discovery by Ian Shelton, University of Toronto, of a mag 5 object, ostensibly a supernova, in the Large Magellanic Cloud. . .

This supernova (shown in this chapter's opening image) is officially known as SN1987A ("A" because it was the first discovered that year). One of the first things

ADVANCING THE MODEL

Supernovae from Moderately Low Mass Stars?

A white dwarf accreting material from a binary companion continues to do so until its mass reaches the Chandrasekhar limit. It then collapses and suddenly initiates carbon fusion. A Type Ia supernova is the result. Some stellar models indicate that the same process might be responsible for Type II supernovae in red giants of close to 4 solar masses. Consider the point in its evolution when such a star is a red giant with a carbon core, surrounded by a shell of helium that is fusing together into carbon and dropping the carbon onto the core. If this model is correct, a 4-solar-mass star would form enough carbon that the mass of its core could reach the Chandrasekhar limit. When this occurs, the carbon core collapses, heats up, and undergoes violent carbon fusion.

According to this model, the sudden detonation of carbon is powerful enough to blast the remainder of the star away in a supernova explosion. Because the outer portions of the red giant are made up primarily of hydrogen, the spectrum of the hot expanding shell of gas will contain hydrogen lines and will have the characteristics of a Type II supernova.

The mass of the star discussed here is only about 4 solar masses, which puts the star in the category of moderately low mass stars, a group we described in the last chapter as evolving into planetary nebulae and then ending up as white dwarfs. If the model presented here is correct, some may die in a much more explosive way.

that astronomers did after the discovery of SN1987A was to examine recent photographs of the region where it occurred, to determine which star was its source. It is fortunate that the supernova occurred in a region of the Large Magellanic Cloud where new stars were known to be forming, for this meant that astronomers already had an interest in the region and many photographs of it could be found. Two very luminous blue stars were found very close together at the location where the supernova occurred. This caused confusion because it was thought that red giants, not blue stars, explode as supernovae. The explanation appeared when a shell of gas was detected (by the *International Ultraviolet Explorer* satellite) about a light-year from the central star. It appears that the shell consists of material that was shed from what was originally a 20-solar-mass red supergiant. The star ejected this material about 20,000 years ago, leaving behind its hotter, blue surface. Because 20,000 years is little more than an instant in a star's life, we can say that the red supergiant ejected its surface material shortly before it exploded as a supernova.

Supernova theory had predicted that a burst of neutrinos would be emitted by the explosion. As we discussed in Section 11-3, neutrinos are nuclear particles produced within stars and are the subject of research at various neutrino detectors around the world. Such devices make a record when they detect neutrinos. Our theory of supernovae was confirmed when neutrino researchers in the United States and Japan checked their records and reported that the number of neutrinos had increased a full three hours before the supernova was seen. The delay was due to the very low tendency of neutrinos to interact with matter. After they were formed in the core, most of these neutrinos traveled the remaining volume of the star and escaped into space; however, the shock produced when the rebounding core of the star collided with its infalling outer layers took about 3 hours to reach the star's surface. Only when this shock reaches the layers closest to the surface can we see the increase in the star's luminosity. By this time, the neutrinos were 3 hours ahead of the light and remained so for the rest of their travel between the star and Earth, a distance of about 169,000 light-years. Using the number of neutrinos detected and other factors (such as the sensitivity of the detector and the inverse square law of radiation discussed in Chapter 4, which applies to both light and neutrinos), astronomers were able to learn a great deal about this violent explosion. Over a period of about 10 seconds, SN1987A emitted a total of about 10^{58} neutrinos, which carried off an amount of energy equivalent to 100 times the energy that our Sun has emitted in its entire lifetime. As we mentioned earlier, if we take into account the energy carried off by neutrinos, Type II supernovae are by far more luminous than Type Ia.

Supernova theory is still young, and advances will have to be made before it can be considered well founded.

If neutrinos had no mass, they should have arrived at almost the same moment. The observed delay in the arrival times of the different neutrinos allowed astronomers to set an upper limit to the mass that they have.

FIGURE 15-8 (a) The light echo of SN1987A. Some light from the explosion was deflected by two dust sheets (about 470 and 1300 light-years from the supernova). Because it covers a longer path to reach us, it is seen after the star has faded away. This image was made by subtracting images taken before and after the explosion. (b) This *HST* image was taken in light emitted by nitrogen gas (658 nm) for the central part and using a visible filter (550 nm) for the debris. It shows the structure of the explosion debris of SN1987A, which is expanding at about 6 million miles per hour. The debris is resolved into two blobs moving in opposite directions along the short axis of the inner ring surrounding the supernova. The hourglass-shaped shell made by the three rings is probably the result of the interaction between slow/fast moving winds produced by the star before/after it became a blue supergiant. (c) The shock wave from the explosion hits the edge of the cavity that was carved by the high-speed wind from the star and produces a dramatic increase in X-ray radiation. This is shown in the *Chandra* image (on the left), revealing multimillion-degree gas at the location of the optical hot-spots seen in the *HST* image (on the right).

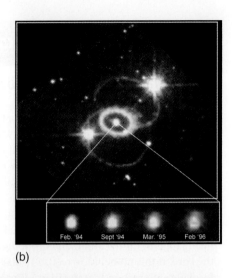

(a)   (b)

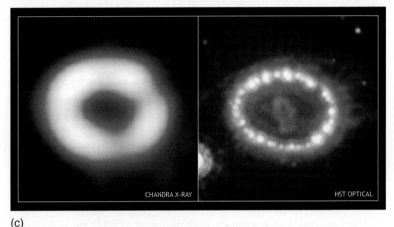

(c)

Besides its mass, other factors, such as the star's angular momentum, play a role in determining what the star leaves behind.

We continue to observe and learn from SN1987A. **FIGURE 15-8a** shows its light echo, the deflection of light by two dust-sheets near the supernova. In Figure 15-8b we see the three-ring structure of SN1987A and its evolution between 1994 and 1996. Figure 15-8c shows the glowing gas ring around SN1987A, heated by the collision of a shock wave from the explosion. Observations of SN1987A and other supernovae provide valuable information about these self-destructing stars and the influence they have on their surroundings, but there are still many mysteries about supernovae.

Even in the case of SN1987A, a proposed competing theory for the origin of the nebula holds that the progenitor star was an unstable, massive blue supergiant; such stars, called *luminous blue variables*, eject material in volcano-type eruptions a few times before finally going supernova. A few examples similar to SN1987A have been observed, supporting this competing view that blue supergiants can explode even before becoming red supergiants. Probably the best known example of a luminous blue variable is Eta Carinae (**FIGURE 15-9**), a bright and massive star in our

FIGURE 15-9 A composite *Chandra*/*HST* image of Eta Carinae. Dust and gas emission is shown in blue. X-ray emission produced as the stellar outflow rams into the interstellar medium is shown in orange and yellow.

Galaxy (between 100 and 150 solar masses) at only 7,500 light-years away; it is consuming its fuel at an incredible rate and it is very unstable. When it finally goes supernova, it will rival the Moon in brilliance.

Whatever their details, supernova explosions are singular in the lives of stars. In some cases, it appears that the entire star is blown apart, including the core, but in other cases, the core is left behind as a tiny remnant of the once-mighty star. The nature of this leftover core depends on whether the star was originally in our moderately massive or our very massive class. In either case, a unique, peculiar object is formed.

15-4 Neutron Stars

The neutron was discovered in 1932 by James Chadwick, although its existence had been suspected for some years before that. Only 2 years later, in 1934, the first suggestion was made that some stars might collapse into a strange object consisting of nothing but neutrons.

Theory: Collapse of a Massive Star

In the previous few chapters we have described much of the theory concerning the lives of stars, but have only occasionally cited evidence for it. (As we explained in Chapter 14, much of the evidence of premain and postmain sequence life comes from examining star clusters.) In addition, we have not detailed the many steps that accompanied the development of the theory. There is an interesting story, however, that illustrates a portion of the theory of stellar life cycles. It not only shows how the theory was confirmed, but also how the confirmation provided further knowledge—a common occurrence in science.

A hypothesis worked out in the 1930s predicted that after the mass of a star's core increases beyond the Chandrasekhar limit, the star will collapse further and its electrons and protons will combine to form neutrons, resulting in a *neutron star*. Just as electron degeneracy supports a white dwarf against collapsing under gravity, neutron degeneracy does the same for a neutron star. The hypothesis predicted that the remains of a moderately massive star's collapsed core would become a neutron star—a tremendously compressed star with a mass between 1.4 and about 3 solar masses. **TABLE 15-5** shows the properties of such a star. To try to imagine it, picture the entire mass of a star larger than the Sun compressed into a ball the size of a small city, about 20 kilometers across. The particles in a neutron star are packed so tightly together that a cubic centimeter (about the size of a sugar cube) of the star would have a mass of about a trillion kilograms; also, the gravitational force at the star's surface is so powerful that a marshmallow hitting the surface would release the energy of a Hiroshima-type bomb.

Astronomers had little hope of finding such a star because, despite its high temperature, its extremely small size results in it being very dim. The idea thus was put on astronomy's back burner as an interesting hypothesis but one that seemed beyond our ability to confirm or deny. In 1967, an accidental discovery changed this.

The diameter of a typical neutron star is only 0.2% of the diameter of a white dwarf, yet the neutron star is a billion times denser.

neutron star A star that has collapsed to the point at which it is supported by neutron degeneracy.

Observation: The Discovery of Pulsars

In 1967, Jocelyn Bell, a graduate student in astronomy at Cambridge University in England, was working with Antony Hewish and a group of researchers who were searching for quasars, energetic stellar sources that we discuss in Chapter 17. The radio telescope she was using for her research did not look at all like the giant radio telescope dishes normally associated with radio astronomy. Instead, it looked like a field of clotheslines covering a total of 4.5 acres (**FIGURE 15-10a**). It was designed to detect faint radio sources and to see quick changes

TABLE 15-5	
A Typical Neutron Star	
Mass	1.5 solar masses
Diameter	20 km (width of a small city)
Density	10^{15} g/cm^3
Temperature	10,000,000 K

FIGURE 15-10 (a) Part of the radio telescope that first detected a pulsar. (b) The chart of pulses from the Crab pulsar indicates their regularity. The difference between pulse duration and pulse period is illustrated. The pulsar's period is 33 milliseconds.

(a)

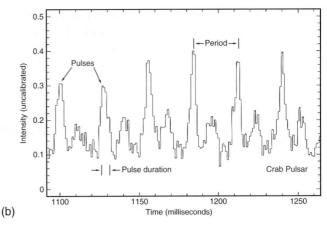
(b)

Six or eight weeks after starting the survey, I became aware that on occasion there was a bit of "scruff" on the records, which did not look exactly like a scintillating source, and yet did not look exactly like man-made interference either.

Jocelyn Bell, on first noticing a pulsar signal

pulsar A pulsating radio source with a regular period, between a millisecond and a few seconds, believed to be associated with a rapidly rotating neutron star.

in their energy. In the course of research for her dissertation, Bell found a new, unexpected, and unexplained source of radio waves. The signal from it pulsed rapidly, about once every 1.3 seconds. This was a much more rapid pulsation than had ever been observed from a stellar energy source.

The researchers' first thought was that the received radio waves were of terrestrial rather than celestial origin. A check of local radio transmitters, however, failed to indicate a source. In addition, the signal was detected about 4 minutes earlier each night than the previous night. Recall that a given star sets four minutes earlier each night as a result of the Earth's moving around the Sun. The researchers concluded therefore that the source was in the sky and was not of human origin.

Their next thought was that a signal from an extraterrestrial race had been detected. In fact, the source was referred to for a short while as an LGM (the initials for "Little Green Men," a reference to science-fiction-type extraterrestrials). This speculation was abandoned for a couple of reasons. First, the pulsations continued in a very regular fashion, instead of changing as they would if they contained a message. More convincing, however, was the discovery by Jocelyn Bell of three more such sources in other directions in the sky, each with its own characteristic rate of pulsation. It was highly unlikely that several different civilizations were sending such signals toward us at the same time, and thus, the sources had to be natural ones. They were renamed **_pulsars_**.

The first pulsar detected had a period of 1.3373011 seconds. Such great precision is possible in measuring the rate of a regularly repeating phenomenon because we can measure over a great number of cycles and then divide by that number, obtaining the time for a single cycle. Figure 15-10b shows a record of pulses from the Crab pulsar, whose period is 33 milliseconds; although the pulses are extremely regular, they do vary in intensity.

All of the pulsars that were found had a pulse with a duration of about 0.001 seconds. This immediately revealed to the astronomers an upper limit to the size of the object emitting the signals. The objects could be no greater than about 0.001 light-seconds in diameter, which corresponds to 0.001 seconds × 300,000 kilometers/second = 300 kilometers. To see how astronomers could reach such a conclusion, even before the nature of the pulsation was known, consider what we would observe if the Sun were to brighten instantaneously. We would not see this instantaneous brightening as being instantaneous at all, for light from the part of the Sun closest to us would reach us about two seconds before light from its limb (**FIGURE 15-11**). Because the Sun is about 2 light-seconds in radius, we would see the intensity of light build up over 2 seconds. Likewise, if the Sun suddenly shut off, the dimming would appear to take 2 seconds, rather than appearing to happen all at once.

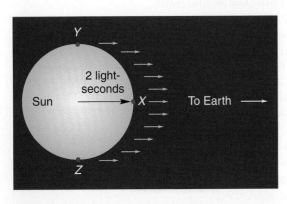

FIGURE 15-11 The arrows indicate light that left the Sun at the same time. Because point *X* is about two light-seconds closer to Earth than points *Y* and *Z*, its light reaches us 2 seconds sooner. Thus, if the Sun were to increase in brightness instantaneously, we would see the light increase gradually over 2 seconds.

The smallest stars known in 1967 were the white dwarfs, but these are Earth-size objects, not small enough to emit pulsations that last only 0.001 seconds; at least, not if the pulsation is caused by a change in the light emitted from the entire object. The pulsar thus must be even smaller than a white dwarf. How could a star be so small? Enter the hypothetical neutron star.

15-5 The Lighthouse Model of Neutron Stars/Pulsars

Today, nearly 1800 pulsars have been detected in our Galaxy, most with pulses between 0.1 and 4 seconds. As with any scientific theory, a model that explains pulsars must work for all these various cases.

Theory: The Emission of Radiation Pulses

Let us consider how an object might emit pulses of radiation. One way would be for its surface to vibrate up and down. (Some kinds of variable stellar objects are known to do this, including Cepheid variables, as we discussed in Chapter 14.) Not only did the short duration of flashes from pulsars seem to indicate that they were not white dwarfs, but when astronomers considered the nature of the material of a white dwarf and the gravitational force on its surface, calculations showed that the surface of a white dwarf could not vibrate as quickly as once per second. Neutron stars are much denser than white dwarfs and have a much greater gravitational force on their surface and, thus, their surface could beat up and down more quickly. In fact, calculations showed that they should be unable to oscillate as *slowly* as once per second. That is, vibrations of white dwarfs were eliminated as a possibility because they have periods that are too long to explain the observed faster pulses, whereas vibrations of neutron stars were eliminated because they have periods that fall below the range of the slower observed pulses.

A second possible way for an object to emit radiation in pulses is by an eclipsing binary process, with a bright object orbiting a dark one; however, the radiation curve of pulsars is different in nature from one that can be explained by eclipses. In addition, Kepler's third law (as modified by Newton) for an eclipsing binary with a period of 1 second or less suggests that the average distance between the two stars must be less than a few thousand kilometers. This distance is even smaller than the diameter of a white dwarf. Now, neutron stars are small enough that two of them could orbit each other with a period in agreement with those observed for pulsars; however, this scenario actually leads to the neutron stars spiraling closer together (a result of Einstein's general theory of relativity, which we discuss later in this chapter). This would decrease the neutron stars' orbital period, which contradicts the observed increase in the pulsars' periods. Therefore, eclipsing binaries must be ruled out.

A final mechanism for producing pulses is by radiation that comes from a small part of the surface of a rotating object. This is the mechanism that causes sailors at sea to observe pulses of light from a lighthouse. On a foggy night, they might see the lighthouse beam sweeping through the fog, but on a clear night the sailors would see the light only when it shines directly at them. This makes it appear to blink on and off. Could a star rotate with a period as short as that observed for pulsars? A star the size of the Sun would be torn apart by such fast rotation, but white dwarfs or neutron stars would have two advantages in this regard: their smaller size would mean that less force would be needed to retain their surfaces under fast rotation, and their small size and great mass would result in a much greater gravitational force on their surfaces than is experienced on the surface of the Sun. Calculations showed that a white dwarf might be able to withstand the forces involved in rotating with a period of 1 second, and perhaps with a period of 0.25 seconds, but certainly no faster than that. A neutron star, on the other hand, would have no difficulty rotating with a period of a fraction of a second.

The observed stretching out of any sudden change in the Sun would not be due to our distance from it, but rather to its size. Be sure to understand this idea; it is a useful tool in astronomy that tells us the maximum size of some objects.

Compare Figure 15-10b to a light curve for an eclipsing binary—Figure 12-33, for example.

As we discussed in Section 7-6, the lack of a shorter rotation period for the Sun posed problems for astronomers in understanding the formation of the solar system from a nebula.

It is easy to see what would cause either a white dwarf or a neutron star to rotate so fast. As a spinning object decreases in size, its rotation rate increases. White dwarfs, and especially neutron stars, are so small that they would be expected to be spinning very fast.

Logic seemed to be pointing more and more to the neutron star as the explanation for pulsars. In analogy with a sailor's lighthouse, the model developed to explain how neutron stars create pulses is called the *lighthouse model*.

Recall from Chapter 11 that the Sun has a magnetic field. When the theory of the existence of neutron stars was developed back in the 1930s, it was suggested that such a star might have an extremely strong magnetic field, as the star is the compacted core of a main sequence star that presumably would have had a magnetic field. During its main sequence period, the star's magnetic field was spread out over its entire surface and was "frozen-into" the star's ionized gases. As the star evolved and collapsed into a neutron star, its surface area was reduced by many orders of magnitude, and thus, its magnetic field was concentrated on a much smaller area. As a result, the magnetic field of a neutron star is very strong. This strong magnetic field is a necessary part of the lighthouse model. Even though a neutron star is mostly made up of neutrons, the presence of a magnetic field requires the existence inside the star of protons and electrons. It is thought that inside a neutron star, neutrons flow without any friction and protons move around without any resistance. The solid crust of a neutron star is a few kilometers thick and consists of a lattice of atomic nuclei and electrons. The high electrical conductivity of the crust allows strong currents to flow in it and thus anchors the magnetic field.

Neutron stars with extremely strong magnetic fields (10^{14} to 10^{15} times that of Earth's) are called *magnetars*. Satellite observations confirm their existence and we now know of about a dozen such objects. The strong magnetic field of a magnetar stresses its solid crust to the point of triggering starquakes that snap the field lines; this produces violent and recurring X-ray bursts. We do not know the evolutionary relationship between magnetars (thought to be about 10,000 years old) and pulsars (thought to be much older). It is possible that magnetars do not constitute a rare class of pulsars and that at least some pulsars, during the early stages of their life cycles, go through a magnetar phase.

As we discussed when describing the solar system, it is common for the magnetic poles of a planet to be out of alignment with the axis of the planet's rotation. Refer back to Figure 6-23 to see the location of the Earth's magnetic pole in the Northern Hemisphere. Also, the magnetic field of the Earth traps charged particles, and these particles result in the auroras seen near the Earth's magnetic poles. In the case of a fast rotating neutron star, the rapidly changing strong magnetic field gives rise to an electric field that continuously rips charged particles from the star's surface. The charged particles are quickly accelerated near the neutron star's magnetic poles. These charged particles spiraling around magnetic fields produce a narrow cone of radiation in the forward direction. The overall picture is that of two oppositely directed beams of radiation emanating from the star. The energy associated with these beams is extremely intense, and each beam (of radio waves and other radiation) is emitted from each magnetic pole. If the star's magnetic poles were located off the rotation axis, these beams would sweep through space as the star spins (FIGURE 15-12). Then, if the Earth were located in the path of a beam, we would see a pulse of radio waves each time the beam sweeps by us. In other words, according to this model, the

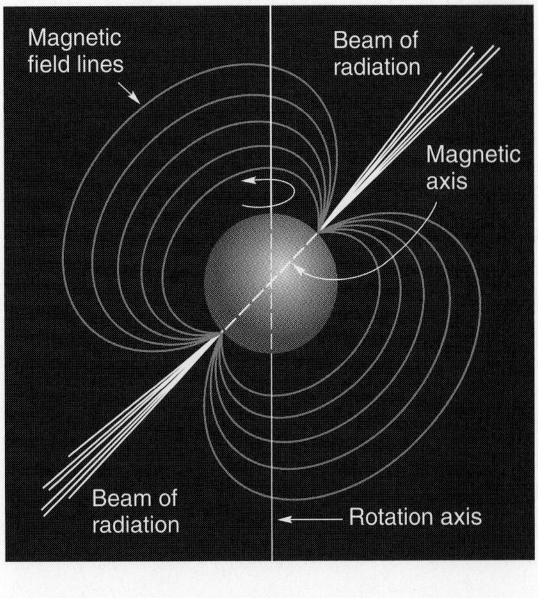

lighthouse model The theory that explains pulsar behavior as being due to a spinning neutron star whose beam of radiation we see as it sweeps by.

FIGURE 15-12 A beam of radiation is emitted near each magnetic pole of the pulsar. As the star rotates, the beam sweeps through space.

observed pulses do not correspond to a "pulsing" source. Instead, energy is continuously being emitted by the pulsar, and we simply observe a pulse every time the star's beam sweeps by Earth.

The lighthouse model also predicts the existence of many pulsars we could not observe from Earth, for we would see only those whose lighthouse beam happens to sweep by us. Because the duration of every flash from a pulsar is very short compared with its period, we can conclude that the angular size of the beam is small (**FIGURE 15-13**). This means that from Earth we would see only a small percentage of the pulsars that exist. It also means that we may never see a pulsar in SN1987A, depending on how the leftover rotating neutron star aligns with respect to Earth.

In 1967, the lighthouse model seemed to be a logical explanation for the pulsar observations, but to test it, more pulsars would have to be found. Good evidence for the model would be finding a pulsar related to the remnants of a supernova, where neutron stars are theorized to be.

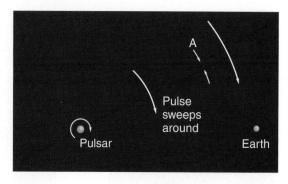

FIGURE 15-13 The angular width (A) of the beam from a pulsar determines how long we see its pulse. No pulsar has a long-duration signal, which indicates that pulsars' angular beam widths are narrow and therefore that we are seeing only a small fraction of the pulsars that exist. The sizes of the pulsar and Earth are not in scale.

Observation: The Crab Pulsar and Others

It was only a matter of months after the discovery of the first pulsar that one was found in the Crab nebula (shown in Figure 15-6). It might have been found sooner, but its rate of pulsation is much faster than was expected; it has a period of 0.033 seconds, so that it blinks 30 times per second. This short period finally ruled out completely the possibility of a spinning white dwarf as the source of pulses; a white dwarf would tear apart if it rotated 30 times per second.

Not only was the Crab pulsar flashing more frequently than others yet discovered, but it was emitting great amounts of energy in the radio region of the spectrum—the radio luminosity of the Crab's pulse is about 0.0025 of the total energy emitted by the Sun. Astronomers at the University of Arizona then observed that it also pulses in visible light, with the same pulse rate as it does in radio waves. Since that time, astronomers have found that the Crab pulsar emits radiation in all regions of the spectrum and that the total energy emitted from the Crab nebula is more than 100,000 times the energy from the Sun. **FIGURE 15-14a** is a *Chandra* X-ray image of the Crab pulsar and its neighborhood. We clearly see tilted rings of high-energy particles to a distance of a light-year from the pulsar, and jet-like features emanating from the pulsar. The *HST* image shown in Figure 15-14b indicates the presence of a knot on one side of the pulsar and a ring-like "halo" on the opposite side. The knot, pulsar, and center of the halo line up with the direction of a jet of X-ray emission.

Astronomers wondered what the source of the Crab pulsar's energy was. This question had been asked even before its pulsar had been discovered. Astronomers had long been puzzled by the luminosity of the nebula itself; it emits more energy than was thought possible. Both of these energy problems were solved when it was discovered that the Crab pulsar is slowing down; its pulses were found to be growing slightly less frequent, corresponding to an increase in its period by about 15 microseconds a year! It was hypothesized that the source of the energy that powers the nebula is the rotational energy of the pulsar. As the pulsar spins, its magnetic field propels electrons out into the nebula. These electrons are the cause of the nebula's great luminosity, but in being swept from the pulsar, the electrons, in turn, slow down the pulsar. If this hypothesis is correct, the amount of rotational energy lost as the object slows its spinning should correspond to the amount of energy emitted by the nebula. Calculations showed that the two amounts of energy did indeed correspond. The theory was confirmed quantitatively.

Direct observational evidence for the existence of a braking torque slowing down a pulsar was found in 2006. Observations of the pulsar PSR B1931+24 show that it periodically turns "off" (that is, there is no radio emission from the pulsar) for periods of 25 to 35 days and then turns "on" again for 5 to 10 days; during the "on" phases, its rotation slows down about twice as fast as during its "off" phases. This suggests

The pulse rate for a pulsar is sometimes observed to increase suddenly (this is called a *glitch*). This occurs because as the neutron star slows its spinning, forces on its surface change. It is possible that these changes cause "starquakes" and sudden changes in the pulsar's shape. It has also been suggested that perhaps the glitches result from the interaction between the pulsar's crust and the layer below it.

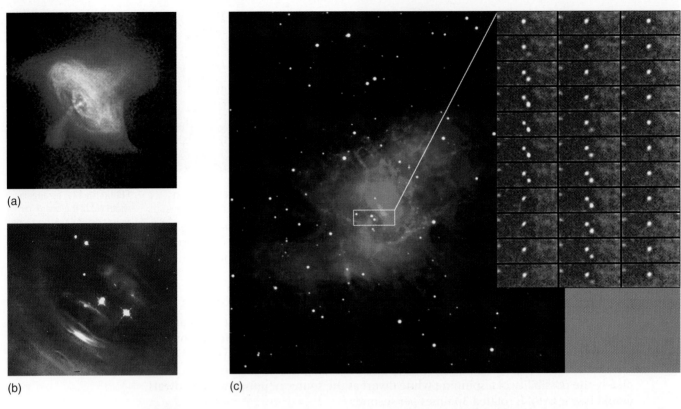

(a)

(b)

(c)

FIGURE 15-14 (a) This X-ray image clearly shows the Crab pulsar, the rings that surround it, and the jet-like structures. The X-rays from the Crab nebula are produced by fast-moving particles spiraling around magnetic fields in the nebula. The diameter of the inner ring is larger than 200 times the diameter of our solar system. (b) The Crab pulsar and other features seen in this *HST* image. The knot may be a "shock" in the jet, while the ring may be the boundary between the polar wind and jet and an equatorial wind that powers a larger torus of emission that surrounds the pulsar. (c) A time sequence for the Crab pulsar is shown in context against an image taken using an optical filter. The mosaic of 33 time slices is ordered from top to bottom and from left to right, with each slice representing about 1 millisecond in the period of the pulsar. The brighter, primary pulse is visible in the first column: the weaker, broader interpulse can be seen in the second column.

Three-dimensional simulations show that the spin of pulsars with periods in the range of 15 to 300 milliseconds is determined by the asymmetric shock wave created when the star's iron core collapses, and not by the spin of the original star.

that for this pulsar, about half of the slowing-down rate is due to the wind of fast-moving charged particles flowing out into space and the currents creating the pulsed radio emission; the other half is due to emission of radiation because of its rotating magnetic field.

The pulsar in the Crab nebula is spinning faster than most others because the supernova in which it had its start was so recent; it occurred only 955 years ago. As time passes, this pulsar will gradually slow down. As it loses its rotational energy, the intensity of its pulses will also decrease, and it will no longer emit X-rays. Finally, in tens of thousands of years, it will be just another radio pulsar with its nebula spread so far that it can no longer be seen. The final fate, however, of a pulsar depends on its neighborhood. Consider a pulsar in a binary system, with its companion being a low-mass star. As the pulsar slows down with time, while the companion star evolves and expands toward becoming a red giant, the pulsar might accrete material from the outer layers of the companion. As this material strikes the pulsar's surface at an angle and at high speed, it will cause the pulsar to spin faster. Astronomers believe that a small group of about 100 pulsars (with periods of 1 to 10 milliseconds) has been "spun up" by such transfer of matter. The companion in such a binary system might not survive the strong radiation of the pulsar and might eventually be destroyed. An example of such a millisecond pulsar system is shown in **FIGURE 15-15**. The observations show that the companion is not the expected white dwarf (the end stage of a low mass star), but a bloated red star. There is also a large amount of gas in the system, suggesting that we are seeing a millisecond pulsar just after it has been "spun up" by its companion star.

Neutron stars in binary systems give rise to two other exotic phenomena. As the neutron star captures material from the companion star, its strong magnetic field fun-nels this material onto the magnetic poles of the neutron star. The resulting high-speed impact creates extremely hot regions that emit large amounts of X-rays. As the neutron star rotates, the rotating beams of X-rays sweep the sky and may be observed as pulses of X-rays. On the other hand, if the magnetic field of the neutron star is not strong enough, the infalling material may be distributed more evenly over the surface of the neutron star. Because most of this material is hydrogen, the large pressures and temperatures in this layer result in fusion reactions that create a layer of helium under a layer of infalling hydrogen. When the conditions are such that helium fusion reactions begin, the surface of the neutron star gets very hot, and a sudden burst of X-rays is emitted that lasts for a few seconds. As a result of this emission, the star cools, but as new hydrogen

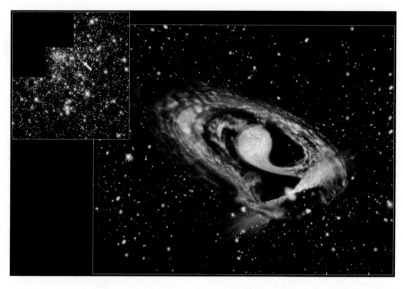

FIGURE 15-15 The millisecond pulsar system in the globular cluster NGC6397 in Ara (the Altar). In the *HST* image inset, the arrow marks the companion star. The artist's impression shows the pulsar (in blue with two radiation beams) and its bloated red companion star. After the pulsar has been spun up, it can no longer accrete gas from its companion.

flows onto the neutron star, this process repeats itself every few hours or days. This is an explosive thermonuclear process similar to a nova (where the degenerate star was a white dwarf, as we discussed in Chapter 14).

Normally, no nebula is found surrounding pulsars, for the nebula has long since dispersed. To confirm further that pulsars are indeed the neutron stars predicted to be left behind in supernovae, astronomers looked for more instances of pulsars associated with expanding nebulae. Since the discovery of the Crab pulsar, another has been found in the Vela nebula (Figure 14-32). The bizarre image of the Vela pulsar, shown in **FIGURE 15-16a**, suggests that after the star exploded, jets with unequal thrust along the neutron star's poles accelerated it like a rocket. Observed anisotropies in supernova explosions also lead to the resulting neutron stars moving at high speeds, up

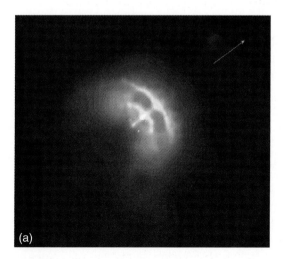

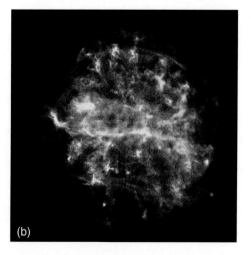

FIGURE 15-16 (a) A *Chandra* image of the compact nebula around the Vela pulsar. The two bows are thought to be the near edges of rings of X-ray emission from energetic particles produced by the pulsar at the center. The rings represent shock waves caused by matter moving away from the pulsar. The jets emanating from the pulsar are perpendicular to the bows and are along the direction of the pulsar's motion, as indicated by the arrow. (b) A *Chandra* image of G292.0+1.8, a young pulsar surrounded by a rapidly expanding shell of gas, 36 light-years across, that contains large amounts of oxygen, neon, magnesium, silicon, and sulfur. The estimated distance is about 20,000 light-years and the age about 1600 years.

ADVANCING THE MODEL

The Pulsar in SN1987A?

After the discovery of SN1987A in February 1987, astronomers began to look for a pulsar at its center. They were not particularly surprised that the pulsar had not been visible immediately because debris from the explosion would block our view of it.

On January 18, 1989, nearly 2 years after the supernova was discovered, an international team of astronomers observed the supernova from Cerro Tololo Inter-American Observatory in Chile for nearly 7 hours, using a detector that is sensitive to visible and near-infrared light. The signals contained pulses! To check their procedure and their instruments, the experimenters turned the telescope to another object that they knew did not pulse. Indeed, no pulses were detected, and measurements of that object's brightness corresponded to previous brightness measurements. They announced that the pulsar in SN1987A had been discovered!

Astronomers were excited about the new pulsar for at least three reasons: First, the discovery confirmed the theory that links pulsars and supernovae. Second, the pulsar was pulsing 1968.629 times per second, much faster than any other known pulsar. Although astronomers expected that a new pulsar would spin faster than an old one, this great rotation rate was difficult to explain, and it indicated that more theoretical work was necessary. Third, the pulses varied slightly in frequency over the 7-hour observation. This also could not be explained using present supernova theories.

Unfortunately, when the astronomers repeated their observations a week later in an attempt to confirm the discovery, they could find no pulses in the signal. Continued observation of the supernova through the remainder of 1989 failed to reveal pulsations. The only explanation seemed to be that, during the January 18 observation, light from the pulsar had reached Earth through a temporary gap in the debris cloud around the pulsar.

As 1989 passed, theoreticians worked to adapt pulsar models to the new findings, and experimental astronomers continued to look for the pulsar's reappearance, while concern was rising that perhaps the initial observation was flawed. The data and observational methods were examined over and over, but no flaw could be found. Then, at the February 1990 meeting of the American Association for the Advancement of Science, the solution to the problem was reported. An old video camera in the observatory was discovered; if the camera was left on during an observation, it sometimes caused an electronic pulse with a frequency of 1968.629 cycles per second—exactly that found for the pulsar. Apparently the camera had been on the night of the pulsar's "discovery," and conditions had been just right for its signals to join the signals received from the sky.

There are lessons to be learned from the would-be discovery of the SN1987A pulsar. Although skeptics might point to this as a failure of science, it is actually an example of a successful use of the scientific method, for it illustrates how science corrects itself. Although the team showed courage in reporting the mistake, their embarrassment would have been greater if someone else had found it. The supposed discovery prompted re-analysis of our theories of pulsars, and we should be better prepared for whatever we find when—and if—the pulsar of SN1987A is finally discovered.

to 1100 kilometers per second! These speeds are greater than those of ordinary stars in our Galaxy and, therefore, many neutron stars can escape its gravitational pull. Figure 15-16b is another spectacular image of a young, oxygen-rich supernova remnant with a pulsar at its center.

Astronomers are now confident that they have found the neutron stars that theory predicted 75 years ago; however, locating the pulsar associated with a given supernova remnant is not easy. When the star explodes, any asymmetry in the explosion might kick the resulting pulsar away from the site of the explosion. In other cases, the pulsar's beams might not sweep past the direction of the Earth and, thus, it is never seen. It is also possible, as is suspected for SN1987A, that the stellar core might collapse into a black hole. The Advancing the Model box on "The Pulsar in SN1987A?" in this section describes what was thought to be its discovery in 1989.

15-6 Moderately Massive Stars—Conclusion

FIGURE 15-17 reviews the steps a moderately massive star takes from protostar to pulsar/ neutron star; however, only a small fraction of all stars are in this category, for two reasons. First, the lifetimes of such stars are short compared with the lifetimes of stars with lower mass. Even if they are being formed at the same rate, not as many thus will be in existence at any given time. Second, only stars in a small range of mass end up with a core too massive to be supported by electron degeneracy and of low enough mass to be supported by neutron degeneracy. We know that neutron stars are more massive than the Chandrasekhar limit of 1.4 solar masses, but we are not sure of the upper limit to their mass. (We certainly do not have any neutron-degenerate matter on Earth with which to experiment.) Theory indicates, however, that the limit falls somewhere around 3 solar masses. Only stars that end up with masses greater than—but not much greater than—1.4 solar masses thus can ever become neutron stars and send out their characteristic lighthouse beam of radiation. When a star collapses to form a neutron star, most of its metal-rich material from its interior is returned to its surrounding region.

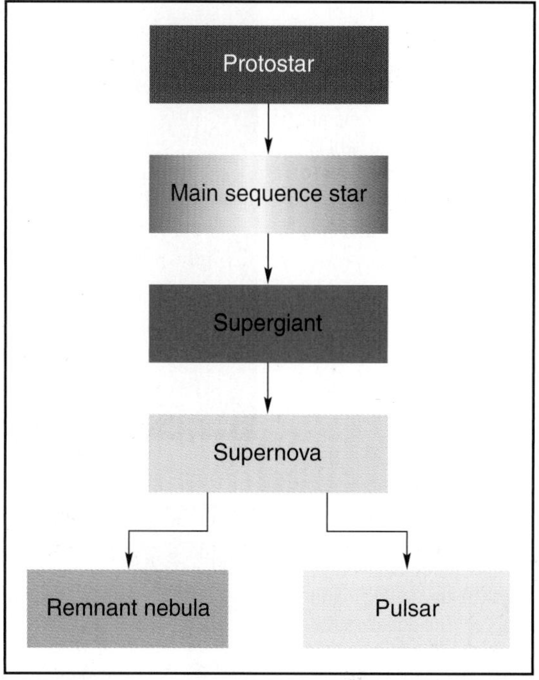

15-7 General Relativity

FIGURE 15-17 These are the steps a moderately massive star takes from protostar to its final stages.

Albert Einstein stated that in his youth he wondered whether, if he were moving at the speed of light, he could see himself in a mirror. Such questions led him in 1905 to develop the **special theory of relativity**, which we discussed in Chapter 3. This theory makes the perception of electromagnetic waves independent of the motion of the observer. It allows us to answer Einstein's mirror question with a "yes," but it reveals a link between the nature of space and time that does not appear in our everyday perception of the universe. Special relativity has some interesting effects, but it is Einstein's expansion of the theory to the **general theory of relativity** (or **general relativity**) that must be used to describe the fate of massive stars.

In Section 3-8 we described how the general theory of relativity provided a different way of explaining what we normally call gravitation. We began the discussion with a "thought experiment" involving a woman in a spaceship far from Earth. We reached the conclusion that if the spaceship were accelerating upward with an acceleration equal to the acceleration due to gravity on Earth, the woman could not distinguish her situation in the accelerating spaceship from one in a stationary spaceship on Earth. That is, an object would seem to fall in the accelerating spaceship just as if the ship were in a gravitational field. General relativity is based on the **principle of equivalence**, which states that there is no way to distinguish between an accelerating reference frame in a location where there is no gravitational field and a nonaccelerating reference frame that is in a gravitational field.

Now we extend the thought experiment to include electromagnetic radiation. Suppose that the woman is in a spacecraft on the surface of the Earth, not accelerating. The spacecraft has a window on one side, and through the window a beam of light enters parallel to the floor. The principle of equivalence leads to a prediction that the beam will bend downward as it crosses the spacecraft. To see why this is so, imagine that the spacecraft is not on Earth but is in deep space, with an acceleration upward equal to the acceleration due to gravity. A quick burst of light enters the craft, as shown in **FIGURE 15-18**. As the craft accelerates upward, gaining speed all the time, the light takes the path shown in Figures 15.18b and c. From our point of view, in the unaccelerated frame of the page in the book, the light continues in a straight line. From the point of view of the woman in the accelerating craft, however, the light will have bent downward, as shown in **FIGURE 15-19**.

The principle of equivalence says that the same bending should happen when the spacecraft is stationary on the surface of the Earth. In this case, however, the bending will appear to be due to gravity.

special theory of relativity A theory developed by Einstein that predicts the observed behavior of matter due to its speed relative to the person who makes the observation.

general theory of relativity A theory developed by Einstein that expands special relativity and presents an alternative way of explaining the phenomenon of gravitation.

principle of equivalence The statement that effects of acceleration are indistinguishable from gravitational effects.

If gravity bends light, shouldn't we see it curve here on Earth?

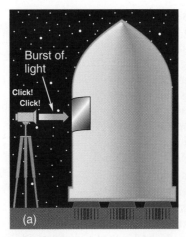  

FIGURE 15-18 A burst of light entering the accelerating spaceship continues in a straight line (a), but because the spaceship is accelerating upward (b), the light is getting closer to the floor (c).

FIGURE 15-19 As seen by a person in the spacecraft (a), the light beam bends downward (b) as if falling to the floor (c). The acceleration of the beam would be measured to be exactly equal to the acceleration of the ship.

The drawings that show the bending of light exaggerate the amount of bending, of course, for in the time required for a beam of light to cross a room-sized spacecraft, its acceleration would not result in enough curvature of the light's path to make it measurable, but if the woman's craft were in an extremely strong gravitational field, where the acceleration caused by gravity is 10 billion times greater than on Earth, the bending would be appreciable.

If the principle of equivalence is valid, light should bend in the presence of a massive object. This prediction of the gravitational bending of light was confirmed by measurements taken during a solar eclipse in 1919, as we discussed in Section 3-9, and other eclipses since then. The general theory of relativity thus presents a different way of looking at the phenomenon we call gravitation. In Chapter 3 we described how the theory explains gravitation not as a force but as the result of the curvature of space. The particular result of interest to us here is that the theory predicts that electromagnetic radiation responds to this curvature in a way that makes it seem to be responding to the force of gravity. This discussion will continue to speak of the force of gravity in Newtonian terms, but we will accept the Einsteinian prediction that electromagnetic radiation, including visible light, appears to respond to this force.

General relativity has survived every test to which it has been put (some of which were described in Chapter 3), and it is a well-accepted theory—not a hypothesis, as is sometimes suggested by the media.

A Binary Pulsar

In 1974, Joe Taylor, a professor at the University of Massachusetts, and Russell Hulse, a graduate student working with him, were using the giant radio telescope at Arecibo to study pulsars. They discovered a very unusual pulsar, whose pulse period of about

59 milliseconds seemed to be changing in a regular way. Sometimes the pulses were received a bit later than expected, sometimes a bit earlier, and these variations repeated with a period of about 7.75 hours. After nearly a month studying the object, they deduced that the changes were due to the Doppler effect as this pulsar revolved around a companion in a binary system. The companion, however, could not be detected. As we explained in previous chapters, binary star systems are useful in determining the masses of stars, but when the spectrum of only one star is observed, the object's mass cannot be calculated; however, astronomers were very interested in determining the masses of the objects in this case, for it would provide the first measurement of the mass of a pulsar and thereby serve as a check on the neutron star/pulsar theory.

Fortunately, the short period of revolution of this binary system (about 7.75 hours) allowed Hulse and Taylor to observe many orbits over a few weeks. This permitted them to determine that the pulsar's orbit was precessing at the rate of about 4.2 degrees/year, a precession much greater than Mercury's because of the great mass of the pulsar and its nearby companion.

The amount of precession predicted by general relativity depends only on the masses of the two objects in orbit and their distance of separation. If the masses of the objects had been known, the binary pulsar would have served as one more experimental check on general relativity. On the other hand, if the applicability of the theory is assumed, general relativity can be used to calculate the masses of the two objects. By 1978, observational data suggested that both stars have masses very close to 1.4 solar masses (the Chandrasekhar limit), strengthening the conclusion that the companion star was also a neutron star. After 3 decades of observing the system, the orbits of the two objects are known very precisely, allowing a precise calculation of their masses. The pulsar has a mass of 1.441 solar masses and its companion's mass is 1.387 solar masses. The system has been perfect for studying the predictions of general relativity: It was the first instance of the use of general relativity to calculate a stellar property and the first precise determination of the mass of a neutron star.

What makes this binary system special, however, is that it indirectly confirms another of general relativity's predictions. As we discussed before, according to the general theory of relativity, matter tells space how to curve and space tells matter how to move. If the distribution of matter in a volume of space changes (for example, because of the motion of two stars in a binary system), the resulting changes ("ripples") in the curvature of the surrounding space may propagate outward as a ***gravitational wave***, carrying away energy and angular momentum. As a result, general relativity predicts that the two stars will spiral closer together. As of 2004, the observed decay rate of the period is in excellent agreement (at the 0.13% level) with the value of 75.8 microseconds per year predicted by the general theory of relativity. The separation of the two stars becomes smaller by about 3 millimeters/orbit, and the two stars will come together in about 300 million years. Even though we still have not directly observed gravitational waves, this binary system has made it very hard to doubt their existence.

15-8 The Fate of Very Massive Stars

Now that we have reviewed general relativity, let's return to the examination of the evolution and death of very massive stars. A very massive star proceeds through its life in basically the same manner as a moderately massive star, although each stage occurs more quickly. Very massive stars differ from moderately massive ones primarily in what happens to them when their core is compressed to a greater density than electron degeneracy can support. When this occurs in a moderately massive star, the resulting supernova leaves a neutron star at its center. In a very massive star, an even more spectacular event happens: The core becomes a black hole. The general theory of relativity plays a very important part in understanding black holes.

Joe Taylor and Russell Hulse won the 1993 Nobel Prize in Physics for this work.

We discussed the precession of Mercury's orbit in Section 3-9.

gravitational wave
Ripples in the curvature of space produced by changes in the distribution of matter.

The *L*aser *I*nterferometer *G*ravitational-wave *O*bservatory (LIGO) began full operations in 2005. It should be able to detect gravitational waves from collapsing cores of supernovae or colliding neutron stars up to 70 million light-years away.

TOOLS OF ASTRONOMY

The Distance/Dispersion Relationship

In Chapter 12 we described various methods to measure distances to stellar objects. Pulsars provide another method, which relies on two phenomena: each burst of radiation from a pulsar contains an entire spectrum of wavelengths, and different wavelengths of electromagnetic radiation travel through space at slightly different group speeds. For many purposes, it can be assumed that all wavelengths travel at exactly the same speed in space—the *speed of light* in vacuum—but in practice, because space is not a perfect vacuum, longer wavelengths travel at slightly lower group speeds than shorter wavelengths. Recall from Section 5-2 that this difference in speed—called *dispersion*—is the same property that causes chromatic aberration in lenses. The interstellar medium acts as tenuous plasma, and the number of charged particles along the line of sight between the pulsar and the observer affects the arrival times between the different wavelengths. Differences in the composition of the medium thus result in different dispersions.

When we stated in the text that the pulses from a pulsar are typically only 0.001 seconds in length, we were referring to pulses of a single wavelength. If we consider, for example, a pulsar that emits visible light as well as radio waves, we find that the visible-light portion of each pulse reaches us before the radio portion, although for any one wavelength of visible light, or any one wavelength of radio energy, the pulse length is the same—perhaps 0.001 seconds. The relative amount of dispersion might be stated as the time that elapses between the detection of a given wavelength of visible light and the detection of a given wavelength of radio energy in the same pulse.

Two factors determine the amount of dispersion: the distance the pulse travels through the interstellar medium and the dispersion properties of that medium. We have here another triple connection (**FIGURE B15-1**). The amount of dispersion can be measured directly. To the degree that one of the other two quantities is known, the final one thus can be calculated. If the dispersion properties of the interstellar matter between us and a pulsar are known, the distance to the pulsar can be determined. On the other hand, if distances can be determined by another method, this provides a means of learning more about the interstellar material.

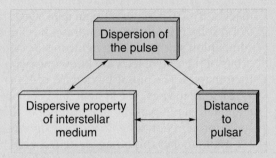

FIGURE B15-1 If any two of the three quantities are known, the third can be calculated.

Black Holes

Neutron degeneracy cannot support a neutron star whose mass is greater than about three solar masses. Such a star will collapse. How far it will collapse is still an open question, but there is no known force capable of keeping the force of gravity from collapsing a star to zero size. This is an unimaginable situation, for we cannot envision matter having no size whatsoever, especially an object several times more massive than the Sun. Fortunately, we do not have to answer the question of how far the star collapses to predict how it will appear from outside.

Shortly after Einstein introduced general relativity, Karl Schwarzschild calculated that when a star collapses to a dimension equal to or less than what is now called the **Schwarzschild radius**, light is unable to escape the object. As we discussed in Section 7-5, the escape velocity of a projectile at a distance r from the center of a star of mass M and radius R is given by

$$v_{esc} = \sqrt{\frac{2 \cdot G \cdot M}{r}},$$

Schwarzschild radius
The radius of the sphere around a black hole from within which no light can escape.

where G is the gravitational constant. For a projectile trying to escape from the star's surface, we have $r = R$ and, thus, the escape velocity increases as the star decreases in size. When the star's radius becomes so small that its escape velocity is greater than the velocity of light, the star has reached the Schwarzschild radius (R_S) (**FIGURE 15-20a**).

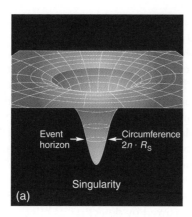
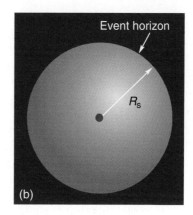

FIGURE 15-20 (a) The sphere at 1 R_s is the event horizon of a black hole. (b) The event horizon in (a) after a star has shrunk to inside the Schwarzschild radius; light can no longer escape its surface.

The size of the Schwarzschild radius depends on the mass of the star, for it is this mass that determines the force of gravity. The radius is given by

$$R_S = \frac{2 \cdot G \cdot M}{c^2} \text{ or (if } R_S \text{ is in km and } M \text{ in solar masses) } R_S \approx 3 \cdot M.$$

A star whose radius is less than the Schwarzschild radius is called a **black hole**; it is "black" because no light escapes from it, and "hole" because matter—or light—that falls into it can never be retrieved. A black hole exists if the star's radius is equal to or less than the Schwarzschild radius, but the radius of the black hole is considered to be the Schwarzschild radius.

The previous equation indicates that if matter falls into a black hole, its radius becomes larger. This is reasonable because as its mass increases it exerts more gravitational force and thus the escape velocity increases at any given distance. A star with a mass five times greater than the Sun's will have a Schwarzschild radius equal to about 15 kilometers. If we had the power to squeeze the mass of our Sun into a sphere of radius $R = 3$ kilometers, then it would become a black hole. This in no way would affect the Earth's orbit, because at 1 AU our planet is safely far away and Newtonian physics would still be all we need to describe its motion. The only problem, of course, would be the lack of sunlight and its negative effects on life on Earth, as we discuss in the Advancing the Model box on black holes.

A name given to the surface of the sphere formed at the Schwarzschild radius illustrates an important and interesting feature of black holes. The surface is called the **event horizon**. Just as we cannot see beyond the Earth's horizon, we cannot see inside the event horizon. Moreover, there is no way we can know about an event inside that sphere. Nothing that happens there is accessible to us and may just as well be in another universe, as we can have no experience of it. As far as we know, after a star collapses inside its event horizon, no force in the universe can stop the star from collapsing to a single point of infinite density. This point, at the center of the black hole, is called the **singularity**. Here the strength of gravity—and thus the curvature of space and time—is infinite. Time and space do not exist as separate entities at this point. Random things can happen here but do not affect the world outside the event horizon. It is the unpredictability of events and our lack of understanding as to what happens at a singularity that has forced astronomers to accept the proposition that singularities not surrounded (or shielded) by an event horizon cannot exist in nature. The structure of a black hole that is *not rotating* thus is easily described by the singularity and the event horizon.

Properties of Black Holes

Although the theory of black holes involves general relativity and therefore might be considered complicated, black holes themselves are very simple. A black hole can be described completely by only three numbers: one for its mass, one for its electric charge, and one for its angular momentum. A full description of any other object in

black hole An object whose escape velocity exceeds the speed of light.

Light can only exist at the speed of light in vacuum. The closer a light source is to a black hole, the more redshifted its light gets as it tries to escape.

event horizon A space-time boundary around a black hole from within which nothing can escape. Its radius is the Schwarzschild radius.

singularity The center of a black hole, where density, gravity, and the curvature of space and time are infinite.

Describing the structure of a *rotating* black hole requires the idea of an additional region surrounding the event horizon. In this region, the dragging of space and time is so severe that nothing can remain stationary; however, objects could travel through it without being sucked into the black hole.

In my talk, I argued that we should consider the possibility that the center of a [supernova remnant] is a gravitationally completely collapsed object. I remarked that one couldn't keep saying "gravitationally completely collapsed object" over and over. One needed a shorter descriptive phrase. 'How about black hole?' asked someone in the audience...Suddenly this name seemed exactly right.
John Archibald Wheeler, usually credited with coining the term "black hole"

the universe would involve a long list of properties, including size, color, chemical composition, texture, and temperature (in addition to the three properties named above). For a black hole, these other properties have no meaning. For example, chemical composition is meaningless because it does not matter what original material condensed to form the black hole. The matter no longer exists as a chemical element in the black hole. Whatever the properties of the material that formed the black hole, that information is forever removed from the universe.

The mass of a black hole, in principle, is easy to measure. It would be measured the same way as the mass of any other celestial object. That is, we would observe an object that is in orbit around the black hole and apply Kepler's third law to calculate the total mass of the black hole and the object.

A black hole might have an electric charge, either positive or negative. In principle, this charge could be measured by detecting its effect on charged objects near the black hole. In reality, a black hole is not expected to have an electric charge because if it had a positive charge, for example, the black hole would draw in negative charges until it became neutral. For this reason, electric charge is not normally considered in discussing black holes.

The final property that a black hole may have is angular momentum. As an object shrinks in size, it increases its rotation rate and, therefore, a black hole would be expected to spin rapidly. Results of the theory of general relativity show that this rapid rotation would change space-time around the black hole. In effect, the spinning black hole would drag nearby space-time around with it. Because of this, light passing near a rotating black hole on one side behaves differently than light passing by the other side. On one side the light is moving along with space-time, and on the other side, it is moving against the motion of space-time. This, of course, is not easy to imagine, for motions of space-time are not part of our everyday experience. If instead of light we used two spaceships orbiting the black hole in opposite directions, they would have two different orbital periods. We thus could measure the black hole's angular momentum by comparing these orbital periods. In 2004, scientists announced that they observed variations in the orbits of satellites around the Earth consistent with the expected distortions of space-time due to the Earth's rotation.

Because a black hole can be described completely by just three properties, astronomers say that "a black hole has no hair." Any properties of objects that fall into it are forever gone from the universe. They leave behind only the properties of mass, electric charge, and angular momentum. A black hole is a simple, "hairless" thing.

Detecting Black Holes

Actually, Pierre Simon Laplace proposed in 1798 that a very massive star might have such a strong gravitational force that light could not escape it. This was 100 years before Einstein's theory linked gravitation with the travel of light!

Astronomers predicted the existence of black holes in the 1930s when they realized that a star's mass may cause it to collapse beyond neutron degeneracy. A prediction of something as unusual as a black hole certainly calls for observational verification.

There is, of course, no hope of seeing a black hole directly, for we cannot see something from which no light escapes. If, however, matter were falling into a black hole, we should expect some of that matter to orbit the black hole in a manner similar to the way matter falling onto a white dwarf orbits it (causing a nova in the process). Because the gravitational field near a black hole is so strong, the orbital speed of nearby matter would be extremely great. As collisions among particles turned the regular orbital motion to random thermal motion, the matter would reach temperatures of hundreds of millions of degrees. FIGURE 15-21 illustrates material being pulled into orbit around a black hole from a companion binary star. Such hot material would radiate great amounts of energy, and because it is not yet inside the event horizon, we should be able to detect it. We can predict the characteristics of the radiation, and we know that the object should appear as an X-ray source.

Numerous X-ray sources have been found in the heavens by orbiting X-ray observatories. Are all of these black holes? Probably not. Only if one of these sources is found to be associated with a particularly massive star can we hope that it is a black hole. When we wish to know the mass of a star, we search for a binary system. Then if we find a binary system in which one of the stars is invisible with a mass greater than 4 or 5 solar masses, we can conclude that the star must be collapsed (or otherwise it would be visible). Finally, if the star emits X-rays characteristic of those predicted for a black hole, we would have good evidence for claiming to have found one.

In the 1960s, astronomers discovered Cygnus X-1, the first X-ray source in the constellation Cygnus. Emission from this source is highly variable and the time scale for the flickering can be as small as 0.01 seconds. Such a time scale sets an upper limit to the size of the object emitting the signals. In this case, the size of the emitting region cannot be larger than 0.01 light-seconds or about 3000 kilometers, which is smaller than the size of the Earth. Then, in 1971, they discovered that the location of Cygnus X-1 corresponds with a ninth-magnitude star named HDE226868, a blue supergiant of spectral type B0. A periodic Doppler shift of the spectrum of the supergiant indicates that it is part of a binary system with a period of 5.6 days, but its companion is invisible. A B0-type supergiant is expected to be a very massive star. Calculations reveal that if HDE226868 has the expected mass of about 30 solar masses, its companion must have a mass greater than 3 solar masses. If so, the companion is probably a black hole.

Cygnus X-1 was the first candidate for a black hole. Since its discovery, other black hole candidates have been found. A few are even part of eclipsing binary systems, thus providing us with further information about them.

In 1989, the star V404 Cygni, a G- or K-type star in our own Galaxy, called attention to itself by erupting with a powerful X-ray flare. In 1992, three European astronomers reported that Doppler shift analysis of V404 Cygni shows that it orbits an unseen companion with a period of 6.47 days (**FIGURE 15-22**). In this case, the mass of the visible star does not present a problem, for even if the calculation is based on its least possible mass, the dark companion must have a mass of at least 6.3 solar masses, and probably 8 to 12. V404 Cygni is now considered almost surely to be a black hole. It further convinces astronomers that black holes are a reality.

Another black-hole candidate is the flickering X-ray source A0620-00, which is also a member of a spectroscopic binary system. In this case, the visible companion (a K5 main sequence star) is relatively faint and orbits the X-ray source every 7.75 hours. Because we can observe the shifting spectral lines of both stars, the minimum mass of the X-ray source is about 3.6 solar masses and thus almost certainly a black hole. In Chapter 17 we discuss black hole candidates that astronomers are finding in the centers of some galaxies. The discovery of black holes not only confirms the theory of the death of the most

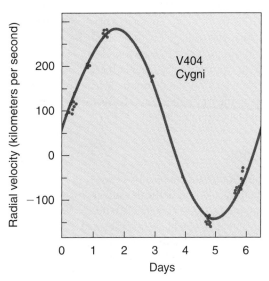

FIGURE 15-21 If a black hole and red giant or supergiant form a binary system, material will be pulled from the giant (left) and will swirl around the black hole, causing X-rays to be released from the heated material in the disk.

FIGURE 15-22 Applying Doppler shift analysis to the spectrum of V404 Cygni shows that its radial velocity changes with a period of 6.47 days. The dots indicate measured velocities (by the Doppler shift) at different times, and the curve is drawn to fit the data.

ADVANCING THE MODEL

Black Holes in Science, Science Fiction, and Nonsense

Black holes are fantastic objects and are fruitful subjects for science fiction as well as for a lot of nonsense. We will try to separate the science from the nonsense and the science fiction.

Nonsense

There is a common misconception that a black hole is a giant vacuum cleaner, sweeping up matter across wide portions of space. In fact, the gravitational force of a black hole is unusually great only near the black hole. To make this clear, suppose that the Sun could magically become a black hole without losing any mass. If it did so, the gravitational force it exerts on the Earth would not change. The force exerted on the Earth would remain the same, and the Earth would continue in its same orbit. The only difference to us on Earth would be that no radiation would arrive from the Sun. Newton's law of gravity states that the force of gravitational attraction between two objects depends only on the masses of the two objects and the distance between them. Although predictions from Einstein's theory differ from those of Newton's, these differences show up only where the forces are extremely great and, thus, Newton's theory can still be used to discuss Earth's orbit. The strength of Newton's gravitational force does not depend on the Sun's size, only its mass.

Where is the increased gravitational field, then? To understand this, note that with the Sun in its present state, the closest you can get to it (and still be outside) is its surface—some 700,000 kilometers from the center. If you could go down inside the Sun, the force of gravity on you would become *smaller*, for there would then be a gravitational force back toward the matter near the surface (**FIGURE B15-2a**). In fact, at the center of the Sun, you would be weightless, for you would be attracted equally in all directions.

The difference in the case of the solar-mass black hole is that now you can get closer to the center of the star while remaining outside its surface (Figure B15-2b). For such a black hole (only 6 kilometers across), you could get within a few kilometers of the center, and—as predicted by Newton's law of gravity—the force of gravity would be very great at these small distances. (Actually, Newton's law would not make accurate predictions this close to the black hole, but it correctly predicts that the force would be extremely large.)

Predictions from Science

Stephen Hawking made some interesting predictions concerning black holes. Based on calculations of the density of the universe at its beginning (discussed in Chapter 18), he predicted that it may be possible that *mini* black holes formed at the time. These might have been the size of a pinhead and

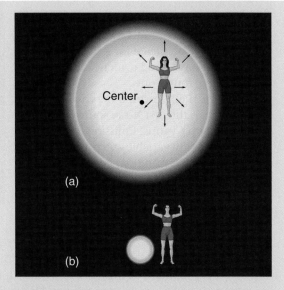

FIGURE B15-2 (a) If a person could exist inside the Sun, she would weigh less than she did on the surface, for gravitational forces would be exerted on her in all directions by parts of the Sun. (b) If the entire Sun could be shrunk to a small enough ball, a person could be the same distance from its center as in part (a) and still be outside its surface. She would then weigh much more than in part (a).

have had the mass of an asteroid. Alas, later theorizing by Hawking showed that if such mini black holes ever existed, they would have long since evaporated by radiation.

Unusual phenomena—by Earthly standards—occur as an object falls into a black hole. For one thing, as an object nears the black hole, it is pulled apart by tidal forces. Tidal forces result because of the difference between gravitational forces on one side of an object and the other. An object approaching a black hole would feel a much stronger gravitational force pulling on its side nearer the hole than on the other side. The force difference would pull the object apart. It would be impossible for a person to fall into a black hole and remain intact.

Even stranger would be the observation of something falling into a black hole. Forget for now the destruction caused by tidal forces. Einstein's theories of relativity tell us that if we could watch the object fall, we would never see it reach the event horizon. We would see it getting closer and closer to the event horizon and getting redder and redder (Doppler shift-like), but because of the distortion of time that is predicted by the theory of relativity, it would take forever—according to our observation—for the object to reach the event horizon. As time is reckoned on the object, however, it would fall into the hole very quickly.

ADVANCING THE MODEL

Black Holes in Science, Science Fiction, and Nonsense *(Cont'd)*

Science Fiction

Where does an object go when it falls into a black hole? Writers have speculated that it may appear at another place or another time. Such travel through "hyperspace" or through time has lent itself to numerous science fiction plots. If it indeed occurs, we should see "white holes" where matter and energy are appearing out of nowhere. (In the language used, the matter and energy enter a black hole, pass through a "worm hole," and emerge from a white hole.) No such phenomenon has been observed.

Even more speculative is the idea that the matter may come out in another universe—not in another galaxy, but in a parallel universe. Because by definition we have no contact with such a universe, we have no way of verifying such speculation. It thus is not in the realm of science at all. (A hypothesis must be verifiable to be classified as scientific.) While the hypothesis of white holes in our universe might perhaps qualify as a scientific hypothesis, the speculation of a parallel universe must remain pure science fiction.

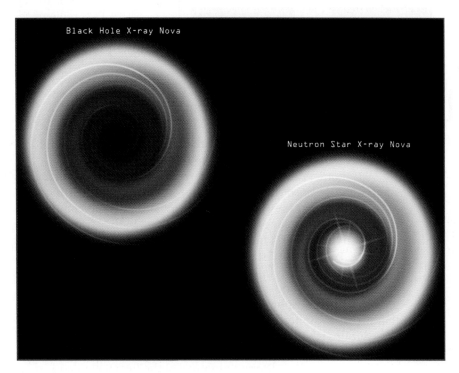

FIGURE 15-23 Gas from the companion star is drawn by gravity onto the compact object (black hole or neutron star) following a swirling pattern. As it nears the event horizon of the black hole (or the surface of the neutron star), a strong gravitational redshift makes the gas appear redder and dimmer. In the case of the black hole, when the gas finally crosses the event horizon, it disappears from view and the central region is black. In the case of the neutron star, when the gas strikes the star's solid surface, it glows brightly.

massive stars but also serves as another confirmation of the theory of general relativity. Although the general public may still think that black holes are on the fringes of science, these objects are in fact firmly entrenched in astronomical theory.

Black-hole candidates have been observed to have masses that span the entire spectrum, from just barely above the minimum 3 solar masses or so that separates the neutron stars from black holes to billions of solar masses at the cores of galaxies. Observations and numerical simulations show a direct path for forming massive black holes through successive mergers of smaller ones; this is especially true for the massive black holes at the centers of globular star clusters and the supermassive black holes at the cores of galaxies. Radiation from matter in the accretion disk around a black hole and jets of matter and energy emanating from such a system are but two of the many ways that black holes sculpt the region around them. Since the early 1990s, orbiting X-ray observatories have been providing important data that allow astronomers to distinguish between black hole candidates and neutron stars. When a neutron star accretes matter, energy is released when the infalling matter

FIGURE 15-24 The steps in the life of a very massive star.

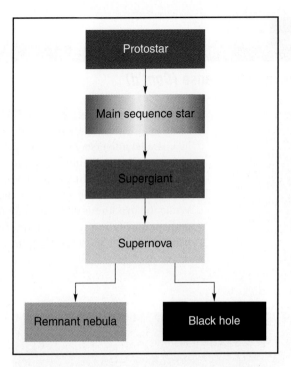

Black holes can also be formed by accreting white dwarfs and/or neutron stars if they accrete enough material. Two dead stars in a binary system coming together to form a single object may also give rise to a black hole.

hits the solid surface of the star; however, when a black hole accretes matter, there is no surface for the matter to hit. Instead, both the matter and the energy are lost after they cross the event horizon. A small amount of energy can escape just before the matter crosses the event horizon, but not as much as in the case of a neutron star. Observations made by X-ray satellites (*ROSAT*, *RXTE* and *Chandra*) support the idea of the event horizon around a black hole by showing that the X-ray emissions from neutron stars and black holes are different. **FIGURE 15-23** illustrates these differences based on observations of X-ray novae, so named because they occasionally erupt as brilliant X-ray sources and then settle into decades of dormancy.

FIGURE 15-24 reviews the steps taken by very massive stars as they progress from protostars to black holes.

15-9 Our Relatives—The Stars

The idea that the stars are our relatives might be surprising, but in fact, astronomers can show that humans and stars are related—albeit distantly.

Astronomers divide stars into three classes, called *Population I* stars, *Population II* stars, and *Population III* stars. The three populations are distinguished by the amount of heavy elements they contain. The Population III class has been theorized to include the very first generation of stars formed in the cosmos. These stars formed using only hydrogen and helium, the only elements available in the early history of the universe. Population II stars contain very little material in their atmospheres other than hydrogen and helium. Heavier elements are produced in the cores of stars as fusion takes place and, thus, heavy elements do exist in their cores; however, little is seen in their atmospheres. The spectra of Population I stars, on the other hand, reveal that their atmospheres contain heavier elements. The separation of stars into these three groups is somewhat arbitrary, because a continuum actually exists in the chemical composition of stars, but the distinction is convenient.

The existence of different amounts of the heavy elements in different stars can be easily explained. Stars are formed from interstellar clouds of gas and dust, and most stars end their lives by blowing (or exploding) much of their mass back out into space; therefore, the material they expel into space contains some of the heavier elements produced within the star. Population II stars thus are those old stars that were formed from interstellar material long ago in the history of the universe, before the interstellar material became enriched in heavy elements. Population I stars, on the other hand, are young stars formed from material that contained the remains of previous generations of stars.

Theoretical calculations for Population III stars suggest that these stars lived for only about a million years before extinguishing themselves and showering the metals they had created in their cores into space. How then could we confirm their existence? We could be able to glimpse some signs of their existence by looking in the most distant realms of space. The proposed *James Webb Space Telescope*, the successor to the *HST*, could see these stars exploding as supernovae. *Integral*, a gamma-ray observatory currently in orbit around Earth, could also provide clues about these stars by observing some of the most violent events known in the universe: the gamma-ray

ADVANCING THE MODEL

Gamma-Ray Bursts (GRBs)

GRBs were discovered by accident in the late 1960s by satellites designed to observe clandestine nuclear detonations in space. By the mid-1980s, most astronomers thought that the bursts originated on nearby neutron stars in our Galaxy. This was based on spectroscopic observations that suggested the presence of intense magnetic fields. In 1991, observations by the *Compton Gamma Ray Observatory* showed that the GRBs were distributed isotropically, with roughly the same number of bursts in any direction in the sky.

A burst may last from a hundredth of a second to tens of minutes, and a few bursts are observed each day from somewhere in the universe. For a short time period, some bursts become the brightest objects in the gamma-ray sky (**FIGURE B15-3**). The distances to these energetic events were not known for over 20 years, until early 1997 when the Italian-Dutch satellite *Beppo-SAX* provided X-ray measurements that allowed astronomers to pinpoint the position of a burst to within a few arcminutes in the Orion constellation. This enabled them to take additional data using optical and other telescopes. These observations confirmed that gamma-ray bursts are very distant and, therefore, that extremely energetic explosions must cause them. Two years later, astronomers tracked the visible glow of another GRB while it was still emitting gamma rays. The intensity of the eruption was measured to be equal to that of millions of galaxies.

An analysis of thousands of bursts supports the idea that long bursts (those lasting more than 2 seconds) result from explosions of very massive stars (more than 30 solar masses); the star's core collapses to form a rapidly rotating black hole surrounded by a disk, while the star's outer layers are ejected to form a supernova. A short time later, the black hole-disk system produces energetic jets of high-speed particles that shoot out (**FIGURE B15-4**). The jets create a traffic jam close to the explosion, which quickly heats up and explodes. A shock wave is generated from this fireball and travels at a speed very close to the speed of light. If the jet and shock wave are aimed toward

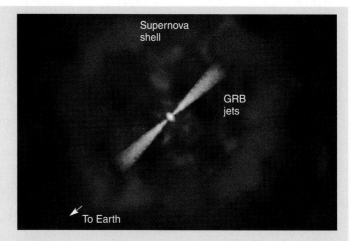

FIGURE B15-4 An illustration that shows the connection between a GRB and a supernova explosion.

Earth, we observe the emitted gamma and X-rays as a burst. When the jet interacts with the expanding supernova shell, it produces an afterglow in X-rays and other frequencies, which can last for days or even months. Support for this scenario was provided by a 21-hour *Chandra* observation in August 2002 of the afterglow of GRB020813, which revealed the presence of elements characteristically ejected by the supernova explosion of a massive star. Since then, additional support has been provided by similar observations of other GRB afterglows. A competing explanation is gaining support from simultaneous multi-wavelength observations made by *Swift*; namely, the burst corresponds to the release of huge amounts of energy carried by the strong magnetic field in the jets.

A similar analysis supports the idea that short bursts (those lasting less than 2 seconds) result from mergers between a neutron star and a black hole, or a pair of neutron stars or black holes; the end product of such a merger is a black hole. Observations of short bursts, which are dimmer than long bursts, show that they originate from galaxies that contain very old stars and thus are different from the host galaxies of the long bursts. These observations are consistent with the idea of mergers between compact objects. The presence of flares and longer periods of activity following the initial burst distinguish mergers between black holes and neutron stars from mergers between neutron stars. A short burst (GRB050724) observed in 2005 provided the first evidence of a black hole's tidal effects stretching a neutron star into a crescent before swallowing it. Computer simulations of mergers between stars and black holes support such disruptions during the mergers of these compact objects. It is possible that some short GRBs are caused by soft gamma-ray repeaters and not by mergers; support for this was provided by LIGO, which could not measure any gravitational waves in the aftermath of a 2007 event (GRB070201).

FIGURE B15-3 An artist's impression of a GRB illustrating how it can flare dramatically over a short time period. There is no way to predict when or where a GRB will occur.

bursts, described in the previous Advancing the Model box. Some astronomers suspect that some of these bursts are created by the death of these first-generation stars.

In which population does our Sun belong? We can answer this question without knowing anything about the spectrum of the Sun. The solar system was formed from material that did not fall all the way into the Sun as the interstellar cloud collapsed; the planets were formed from the same material that made the Sun (see Section 7-6). The fact that Earth (and the other planets) contains heavy elements means that the cloud from which the planets formed contained those materials. The Sun is a Population I star (although it contains less heavy material than some other stars).

Fusion in the cores of stars continues to release energy as heavier and heavier elements are formed until iron is produced. In order for iron nuclei to fuse with other nuclei to form heavier elements, there must be an *input* of energy. The reaction that forms heavier elements from iron does not release energy; it absorbs it. Although most of the matter of which the Earth is made is less massive than iron, many elements are more massive than iron. Such matter could not have been formed by fusion within the core of a star. It was instead formed during a supernova explosion, when tremendous amounts of energy were available. Most of the energy of a supernova is used up in releasing radiation (especially neutrinos) and in blasting away the outer parts of the star, but a small fraction is absorbed by the material of the star, where light elements are fused into heavy elements.

In addition to carbon, nitrogen, oxygen, and other heavy elements that are necessary ingredients for life, dust is also produced by supernova explosions. This process takes less than 10 million years. (Dust can also be produced by old, sun-like stars at the end of their lives. This process takes billions of years. As the dying star expels its outer layers, the gas expands and cools, allowing some matter in it to condense into dust grains. Spectroscopic observations show the presence of silicate particles in the expelled layers of oxygen-rich giant stars and silicon carbide particles in the expelled layers of carbon-rich giant stars.)

Radioactive elements are also produced by supernova explosions. *Integral* has confirmed that radioactive aluminum (Al-26) and iron-60 (Fe-60) are produced in such events in star-forming regions throughout our Galaxy. The relatively short half-life of Al-26 (only 720,000 years) provides a direct proof that nucleosynthesis is currently occurring in young stars. The presence of decay traces of this element in meteorites connects the element's radioactivity to the early solar system; this suggests that the solar nebula could have been near an exploding star.

This discussion has implications for us humans. The material that makes up our bodies is from the Earth. Where was it before it was part of the Earth? In an interstellar cloud. And before that? In a star!

Harlow Shapley, former director of the Harvard College Observatory, whose work we discuss later, listed cosmic evolution as 1 of the 10 revelations that have most affected modern humans' life and thought. He said, "Nothing seems to be more important philosophically than the revelation that the evolutionary drive, which has in recent years swept over the whole field of biology, also includes in its sweep the evolution of galaxies, and stars, and comets, and atoms, and indeed all things material."*

The half-life of Fe-60 is 1.5 million years. The ratio of Fe-60 to Al-26 is about 15%, in good agreement with theoretical estimates.

*Harlow Shapley, *Beyond the Observatory*. New York: Scribner, 1967, pp. 15–16.

Conclusion

All stars get their starts when shock waves compress parts of cold interstellar dust clouds. Most differences between stars—from the various stages through which each star will progress as it lives and dies, to how long each of those stages lasts—result entirely from differences in mass. The universe itself has not existed long enough for the least massive stars to have ended their lives, but the most massive stars have short lives; at the end they become the most astonishing things in nature: black holes. **FIGURE 15-25** summarizes the life cycles of all of the stars: very low mass, moderately low mass, moderately massive, and very massive stars.

We are star stuff.
Astronomer Carl Sagan

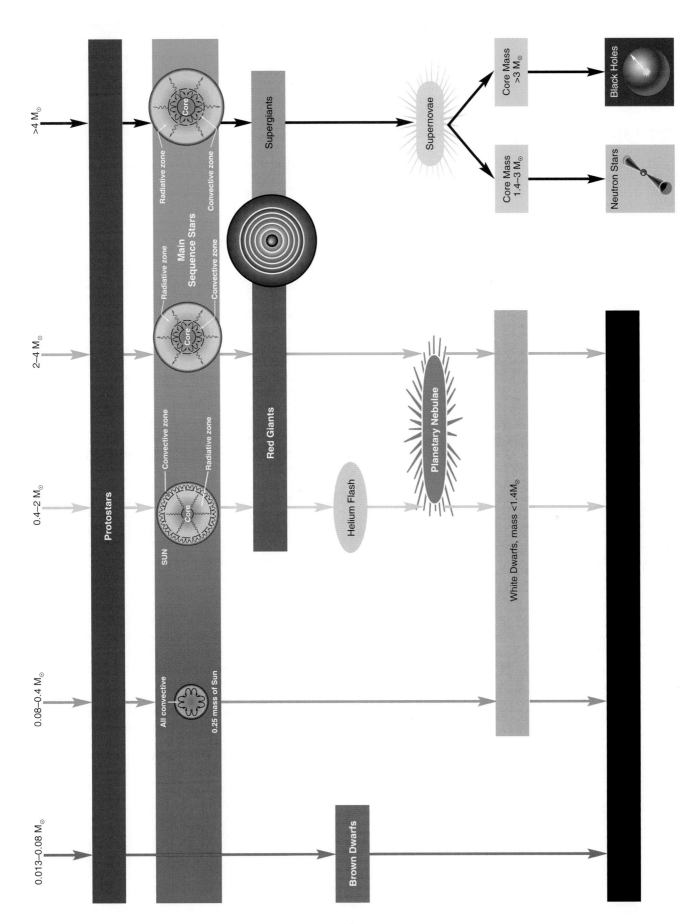

FIGURE 15-25 A summary of the steps in the life cycles of stars of various masses. The limits in mass are not well known.

With this chapter, we complete our examination of the stars. Just as our horizons expanded many times over when our focus changed from the solar system to the stars, they will expand again in the next chapter, where we begin to study the largest class of objects in the universe: the giant galaxies of stars.

STUDY GUIDE

1. The core of a mature supergiant
   A. is layered, with heavier elements in the center.
   B. is layered, with lighter elements in the center.
   C. is uniformly composed of hydrogen and helium.
   D. is uniformly composed of helium.
   E. is composed primarily of iron and carbon.

2. Which type of supernovae have hydrogen lines in their spectra?
   A. Type I (from binary systems).
   B. Type II (from single stars).
   C. [Neither type.]

3. The most massive element that can be formed by nuclear fusion with the liberation of energy is
   A. helium.
   B. carbon.
   C. oxygen.
   D. iron.
   E. lead.

4. The most recent naked-eye supernova was observed
   A. in 1054 by the Chinese.
   B. in the 1500s by Tycho Brahe.
   C. around 1600 by Kepler.
   D. in 1929 by Hertzsprung and Russell.
   E. in 1987 by Ian Shelton.

5. SN1987A was the explosion of a
   A. red giant.
   B. blue supergiant.
   C. white dwarf.
   D. pair of stars.

6. The pulse rate of pulsars is slowed because
   A. they convert rotational energy into radiation.
   B. they drag companion stars around.
   C. of friction with the interstellar medium.
   D. of the conservation of angular momentum.

7. The first candidate for a black hole was
   A. Cygnus X-1.
   B. Algol (the "Demon Star").
   C. Sigma Xi.
   D. Centaurus.
   E. Xi Ursa Majoris.

8. Theory predicts that a neutron star should spin fast because
   A. it was given increased speed by the supernova explosion.
   B. it was given increased speed by a companion star.
   C. it conserved mass as it collapsed.
   D. it conserved angular momentum as it collapsed.
   E. [The statement is false; neutron stars are not predicted to spin fast.]

9. Which of the following objects is a neutron star?
   A. The Crab nebula.
   B. A Cepheid variable.
   C. A dark nebula.
   D. A pulsar.
   E. [None of the above.]

10. Pulsars are
    A. larger and more massive than neutron stars.
    B. larger but less massive than neutron stars.
    C. smaller but more massive than neutron stars.
    D. smaller and less massive than neutron stars.
    E. the same as neutron stars.

11. Pulsars emit sharp bursts of pulses, each lasting for less than 1 second. From this we can conclude that pulsars
    A. have a great amount of energy.
    B. have little energy.
    C. are small in size.
    D. are moving rapidly away from us.
    E. are far away from Earth.

12. Which is the correct sequential order for the life cycle of a star?
    A. red giant, white dwarf, main sequence, protostar.
    B. protostar, main sequence, red giant, white dwarf.
    C. white dwarf, protostar, main sequence, red giant.
    D. protostar, main sequence, white dwarf, red giant.
    E. white dwarf, red giant, main sequence, protostar.

13. The first pulsar was discovered by
    A. a professional astronomer using a radio telescope.
    B. a professional astronomer using a visible-light telescope.
    C. a professional astronomer using an X-ray telescope.
    D. a professional astronomer using a CCD camera.
    E. a British graduate student using a radio telescope.

14. According to present theory, the pulses of radiation from a pulsar are due to
    A. pulsations of the surface of the star.
    B. pulsations from within the core of the star.
    C. eclipses of the star by a binary companion.
    D. rotation of the star.
    E. [Any of the above, depending upon the particular pulsar.]

15. What remains after a supernova?
    A. A main sequence star.
    B. A white dwarf.
    C. A neutron star.
    D. A black hole.
    E. [Either C or D above, depending on the mass of the star.]

16. Which of the following choices lists the stages in the life of *very massive stars* after they leave the main sequence?
    A. Supergiant, supernova, black hole.
    B. Supergiant, supernova, neutron star.
    C. Supergiant, white dwarf, black dwarf.
    D. Supergiant, planetary nebula, white dwarf.
    E. Planetary nebula, supergiant, black hole.

17. When more material falls into a black hole, the diameter of its event horizon
    A. increases.
    B. remains the same.
    C. decreases.
    D. [Any of the above, depending on the nature of the black hole.]

18. In which of the following categories are the objects the least dense?
    A. Main sequence stars.
    B. Nebulae.
    C. Pulsars.
    D. Red giants.
    E. White dwarfs.

19. Which objects have the greatest negative absolute magnitude when they are at their brightest?
    A. Black holes.
    B. Protostars.
    C. Pulsars.
    D. Supernovae.
    E. White dwarfs.

20. Which objects were first detected by Jocelyn Bell, a graduate student at Cambridge University in England?
    A. Black holes.
    B. Protostars.
    C. Pulsars.
    D. Supernovae.
    E. White dwarfs.

21. What do Cygnus X-1 and V404 Cygni have in common (other than that they are in the same constellation)?
    A. Both are neutron stars.
    B. Both are planetary nebulae.
    C. Both are white dwarfs.
    D. Both are probably black holes.
    E. Both are supergiants.

22. Which of the following statements is true?
    A. Black holes have been detected based on their gravitational effect on another star.
    B. Black holes have been detected based on radiation emitted from material falling into them.
    C. Black holes cannot be detected.
    D. [Both A and B above.]

23. The principal factor that determines the evolution of a star is
    A. its location in the sky.
    B. its radial velocity.
    C. its transverse velocity.
    D. its mass.
    E. [Both B and C above.]

(The following question is from an Advancing the Model box):

24. If the Sun could magically and suddenly become a black hole (of the same mass), the Earth would
    A. continue in its same orbit.
    B. be pulled closer, but not necessarily into the black hole.
    C. be pulled into the black hole.
    D. fly off into space.

25. Compare the luminosity of a supernova with that of a massive main sequence star.

26. What happens to the size of a main sequence star when mass is added to it? A white dwarf? A neutron star? A black hole?

27. Outline the stages in the lives of moderately massive and very massive stars. What causes them to end their lives differently?

28. What leads us to believe that the object seen by the Chinese in 1054 was a supernova?

29. Describe how pulsars were discovered. What led astronomers to conclude that they were neutron stars?

30. When astronomers were searching for the nature of pulsars, why, if one assumes that the pulses from a pulsar are caused by vibrations in their surfaces, did the observed rates of pulsation seem to rule out both white dwarfs and neutron stars?

31. Describe the lighthouse model of pulsars. What observation(s) led astronomers to select this model from among other suggested models?

32. What is a black hole? How can we expect to observe one?

33. Define *Schwarzschild radius* and *event horizon*.

34. What general statement can be made about the escape velocity from a black hole?

35. If the Sun (magically) became a black hole, what would be the effect on the Earth? Explain.

36. What is the observational evidence that black holes exist?

37. List in order the steps in the evolution of the most massive stars.

1. The fusion of lightweight nuclei is said to produce energy, yet fusion cannot occur until a high temperature is reached. Doesn't the high temperature indicate that energy is being supplied to the reaction rather than being produced by it? Explain.

2. Explain why, if a star could instantaneously change its luminosity, we would not see it as an instantaneous change. Explain how the time observed for the change allows us to calculate something about the star's size. Does this method yield the minimum or maximum size of the star?

QUESTIONS TO PONDER

3. When astronomers were looking for pulsars in super-nova remnants, they automatically searched the Crab nebula, for the result of a recent supernova. If the light-house model is correct, why was it unlikely that we would find a pulsar there (and therefore lucky that one was found)?

4. The Schwarzschild radius is not the radius of the matter that makes up the black hole. Of what is it the radius?

5. Explain why main sequence stars, white dwarfs, neu-tron stars, and black holes respond differently with regard to changes in their sizes when matter is added to them.

6. The two parts of Figure 15-5 were not taken with ex-actly the same orientation and are not printed with the same magnification. Match the stars in the two figures.

7. Explain why the observation that radiation from the first pulsar was appearing four minutes earlier each night led astronomers to conclude that the pulses were not of earthly origin (or, if they were, that astronomers had produced them).

8. The rotation of the Crab pulsar slowed due to the elec-trons sweeping away from it. Compare this with the problem of the rotational speed of the Sun that was dis-cussed in Section 7-6.

9. To calculate how fast a white dwarf (or neutron star) would pulse by vibrating its surface, we must know the strength of the gravitational field at its surface. What other property of the object must be known?

10. It is predicted that supernovae explode at the rate of one per second over the entire universe. Why don't we see more of them?

CALCULATIONS

1. What is the value of the Schwarzschild radius of a star of 7 solar masses?

2. If the mass of a black hole is doubled, how does the size of its event horizon change?

3. Betelgeuse is a red giant 428 light-years away. Tycho's supernova was 9800 light-years away. If Betelgeuse were to become a supernova, how many times brighter than Tycho's supernova would it appear?

4. In Section 15-5 we stated that Kepler's third law (as modified by Newton) for an eclipsing binary with a period of 1 second or less suggests that the average distance between the two stars must be less than a few thousand kilometers. Show that this is a correct state-ment. (Hint: What is the maximum mass that each star could possibly have?)

5. In Section 15-5 we stated that the strength of the magnetic field of a neutron star is orders of magnitude greater than that of its progenitor. Show that this state-ment is correct. (Hint: Would the radius of the progeni-tor star be greater than the radius of our Sun? How big is a neutron star? Compare the corresponding surface areas.)

6. Show how we obtain the relationship R_s (in km) = $3 \cdot M$ (in solar masses) from the expression of the Schwarzschild radius in Section 15-8.

EXPANDING THE QUEST

Information on the search for gravitational waves can be found at the following sites: http://www.ligo.caltech.edu (for the LIGO project), http://www.esa.int/science/lisa (for the LISA mission). Check also the American Museum of Natural History Project on Gravitational Waves.

Information on GRBs can be found at the site of the SWIFT mission: ttp://swift.gsfc.nasa.gov.

1. "The Great Supernova of 1987," by S. Woosley and T. Weaver, in *Scientific American* (August, 1989).

2. "Pulsars Today," by F. Graham-Smith, in *Sky & Telescope* (September, 1990).

3. "Do Black Holes Exist?" by J. McClintock, in *Sky & Telescope* (January, 1988).

4. "Wormholes and Time Machines," by P. Davies, in *Sky & Telescope* (January, 1992).

5. "Black Holes and the Information Paradox," by L. Susskind, in *Scientific American* Special Edition (May 2003).

6. "Eta Carinae's Year of Glory," by J. Roth, in *Sky & Telescope* (July 2003).

7. "Black Holes in the Middle," by S. Nadis, in *Astronomy* (March 2004).

8. "Catching Gamma-ray Bursts on the Wing," by G. Schilling, in *Sky & Telescope* (March 2004).

9. "Magnetars," by C. Kouveliotou, R. C. Duncan, and C. Thompson, in *Scientific American* Special Edition (September 2004).

10. "Binary Neutron Stars," by T. Piran, in *Scientific American* Special Edition (September 2004).

11. "The Weirdest Star in the Sky," by R. Villard, in *Astronomy* (March 2005).

12. "How to Blow Up a Star," by W. Hillebrandt, H.-T. Janka, and E. Müller, in *Scientific American* (October 2006).

13. "The Brightest Explosions in the Universe," by N. Gehrels, L. Piro, and P. J. T. Leonard, in *Scientific American* Special Edition (January 2007).

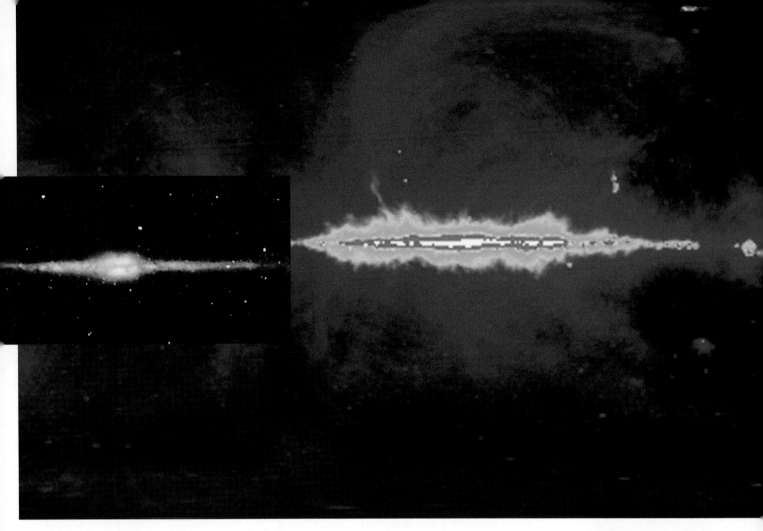

The Milky Way Galaxy

16

SERENDIPITY CAN BE DEFINED AS "MAKING A FORTUNATE and unexpected discovery by accident." An example occurred in 1933, when Karl Jansky of Bell Telephone Laboratories was using an antenna he had built to study radio static, hoping to improve radio telephone service to Europe. Jansky was studying the direction of arrival of static from thunderstorms. He detected radio waves he could not account for, and after much work, he determined that they were coming from the center of the Galaxy. Years later, in a letter to a colleague, Jansky wrote this.

> As is quite obvious, the actual discovery, that is, the first recording made of galactic radio noise, was purely accidental and no doubt would have been made sooner or later by others. If there is any credit due me, it is probably for a stubborn curiosity that demanded an explanation for the unknown interference and led me to the long series of recordings necessary for the determination of the actual direction of arrival.*

*Quoted by W. T. Sullivan III in *Serendipitous Discoveries in Radio Astronomy*, ed. K. Kellerman and B. Sheets. Green Bank, WV: National Radio Astronomy Observatory, 1983.

All cross references to chapters, sections, figures, and tables pertain to the main text, *In Quest of the Universe, Sixth Edition. In Quest of the Solar System* contains Chapters 1–11 and 19 of the main text. *In Quest of the Stars and Galaxies* contains Chapters 1–5 and 11–19 of the main text.

The *Cosmic Background Explorer (COBE)* satellite provided this false-color image (left) of the inner part of the Galaxy; the wide-angle view shows the nuclear bulge and the dense central plane near the bulge. The Milky Way is seen in radio wavelengths (right). Images of the Milky Way at different wavelengths were shown on page 96.

Serendipitous discoveries are fairly common in science, and analysis has shown that almost all of them have certain features in common. First, as Jansky wrote in his letter, the discoverer has a "stubborn curiosity that demand(s) an explanation for the unknown. . . ." The data that led to Jansky's discovery were only a series of tiny blips on a chart. Jansky could easily have put the chart aside and ascribed the unexpected blips to equipment problems or to any number of possible "unexplainable" sources. Instead, his stubborn curiosity led him to investigate.

A second feature common to serendipitous discoveries is that the discoverer is extremely knowledgeable about the subject that he or she is investigating. The fact that the discoveries are unexpected does not mean that just anyone can make them, for important discoveries are not made by the unprepared. Instead, they are made by competent observers engaged in serious scientific investigation.

Throughout this text we have seen how our perception of the universe has grown from a small universe centered on our immediate environment to a larger universe whose "center" was thought to be located farther and farther away. As we show in the next chapter, it now appears that the universe has no center. Hand in hand with this receding center has come a tremendous expansion of our realization of the universe's size. Galileo caused controversy not only because he placed Earth off-center, but also because the Earth played such a small part in his universe.

Yet Galileo did not realize that each of the stars he saw in the sky is another sun. When astronomers shift their focus from the solar system to the stars, their entire scale of thinking must expand. For example, as we have seen, the unit of measure used in the solar system, the astronomical unit (AU), is very inconvenient for stellar distances; therefore, astronomers use the light-year and parsec to describe such distances.

Another mental leap in our thinking about distances and sizes is required when we consider stars not as individual objects, but in galaxies. The leap is so great that it may be impossible to truly understand the distances involved. We'll see that our Galaxy contains about a trillion solar masses. Give some thought to what such a number means.

When we study objects at greater and greater distances from us, our knowledge of their nature and properties is more recent and less certain. In this chapter we discuss our Galaxy, of which the Sun is such an insignificant part, and describe how astronomers measure galactic properties. We then compare two different theories that attempt to explain the structure of galaxies like ours. Finally, we describe how our Galaxy formed and evolved to its present state. We'll see that in the case of galaxies, measurements are much less precise than we might be comfortable with.

16-1   Our Galaxy

▸ Galileo's contemporary, Giordano Bruno, did propose that the stars were suns, and he paid for his belief with his life; in 1600, he was burned at the stake for heresy.

FIGURE 16-1 shows part of the faint band of light that stretches around the sky, encircling the Earth at an angle of about 63° with respect to the celestial equator. The ancient Greeks named this hazy band *galaxies kuklos*, meaning "the milky circle." The Romans called it *via lactae*, "milky way." You are encouraged to find a good clear night sky away from city lights and look at the beautiful Milky Way, a sight many of us in the modern world never get a chance to experience. As shown on the star charts in your book, the Milky Way completely encircles the Earth, passing through the constellations Sagittarius, Aquila, Cygnus, Cassiopeia, and Auriga and between Gemini and Orion on the northern hemisphere of the celestial sphere. Then it passes through Monoceros, near Canis Major, and through Vela and Crux on the southern hemisphere.

Do some of the stars we see at night belong to other galaxies far, far away?

Milky Way Galaxy The galaxy of which the Sun is a part. From Earth, it appears as a band of light around the sky.

We know today that the haze of the Milky Way results from the many stars in the disk of our Galaxy. The Sun is one of about 200 billion stars that make up what we call the **Milky Way Galaxy**, or simply the Galaxy. Most stars in the Galaxy are arranged in a wheel-shaped disk that circles around a bulging center (FIGURE 16-2). The diameter of the Galaxy is about 50,000 parsecs (160,000 light-years), and the Sun is

FIGURE 16-1 This photo shows the Milky Way in Sagittarius. Compare it with Figure 16-3.

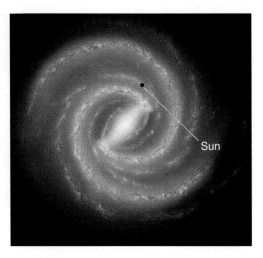

FIGURE 16-2 An artist's view of our Galaxy (seen face-on). The Sun is located about one third of the way out along the disk of the Galaxy. (See the Advancing the Model box "The Milky Way: A Barred Warped Galaxy.")

about a third of the way out from center (8000 parsecs or 26,000 light-years). The Galactic center is in the direction of Sagittarius (FIGURE 16-3) in our sky.

With only a few exceptions, every naked-eye object in our sky is part of the Galaxy. The Magellanic Clouds (see Figure 12-36) are exceptions that are visible from the Southern Hemisphere. These small galaxies are close to our Galaxy but are not part of it. An exception that those of us in the Northern Hemisphere can view in the fall sky is the Andromeda galaxy (FIGURE 16-4). It is the oval patch on the September sky chart at the end of this book.

The discovery that we live in a Galaxy of hundreds of billions of stars was made fairly recently. It began, though, 400 years ago when Galileo turned his telescope to the Milky Way and found that it is not made up of haze, but of stars far too numerous to count. The Milky Way appears misty because so many of the stars are so far away that the naked eye cannot distinguish individual stars and sees only the overall illumination from them. When Galileo looked through his telescope, he saw more haze behind the many stars his telescope revealed, and he concluded that it was caused by even more stars too faint to see individually.

William Herschel, discoverer of Uranus, wrote that through a telescope,

> We find [the stars] crowded beyond imagination along the extent of [the Milky Way]; . . . so that, in fact, its whole light is composed of nothing but stars of every magnitude from such as are visible to the naked eye down to the smallest points of light perceptible with the best telescope.

The telescopic view of the Milky Way led astronomers to conclude

In a spiral galaxy, are all the stars located in the spiral arms?

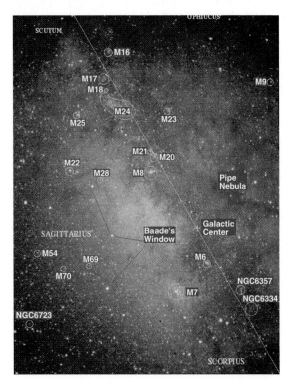

FIGURE 16-3 This image shows the region around the Galactic center near the border between Sagittarius and Scorpius. You can see the absorbing dust lanes that block the view of the Galactic center and some old bulge stars (yellow). The labels mark 17 Messier objects—including the nebulae Eagle (M16), Omega (M17), Lagoon (M8), and Trifid (M20)—globular clusters (M25, M22, M28, M54, M69, and NGC6723), open clusters (M6, M7, M23, M21, and M18), and the star cloud M24. Also marked are the dark Pipe nebula and Baade's Window (the region with the clearest view into the central bulge of our Galaxy).

FIGURE 16-4 The Andromeda galaxy (M31) is a spiral galaxy about 2.5 million light-years from our Galaxy. This galaxy can be seen with the naked eye as a fuzzy spot in the constellation Andromeda.

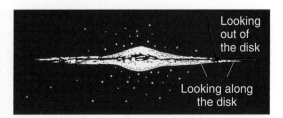

FIGURE 16-5 An edge-on view of our Galaxy. When we see the Milky Way in the sky, we are looking along the disk of the Galaxy. Otherwise, we are looking out of the disk.

that we live in a disk of stars. When we view the Milky Way, we are looking along that disk (**FIGURE 16-5**), and when we are looking in other directions, we are looking out of the disk.

Based on naked eye observations and despite some local variations, the Milky Way seems to be about as bright in one direction as in another. This suggests that we may be at its center. In the 1780s, to determine where the Sun lies relative to the disk, William Herschel and his sister Caroline made star counts in nearly 700 selected regions distributed around the sky. They reasoned that if more stars were found in one direction than in another, that direction could be assumed to be toward the center of the disk. Their conclusions not only confirmed the disk-like shape of the Galaxy but also indicated that the Sun is indeed at the center, for they saw about the same number density of stars in all directions in the Milky Way. **FIGURE 16-6** shows the shape they arrived at for the Galaxy. The Sun is nearly at its center.

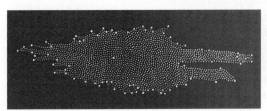

FIGURE 16-6 The Herschels' counting of stars led them to conclude that the Galaxy is shaped like this. The Sun is located at the bright spot within the Galaxy.

In the early part of the 20th century, Jacobus C. Kapteyn sought to find the Sun's location by analyzing the density of stars in various directions from the Sun. He did this by measuring not only the number of stars in each direction but also their distances from us. He found that the density of stars decreases in every direction from the Sun and, like William and Caroline Herschel, it was logical for him to conclude that the Sun is at the center of the disk.

The conclusion that the Galaxy centers on the Sun was viewed with skepticism, as you might expect. After all, we once thought that the Earth was the center of the universe, only to find that it circles the Sun. Were we now discovering that the Sun is the center? The evidence from the Herschels and from Kapteyn pointed to an affirmative answer, but the finding seemed to be contrary to the trend that began before written history and continued through Copernicus and Galileo.

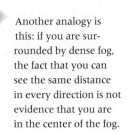

Jacobus Kapteyn (1851–1922) held a position at the University of Groningen in The Netherlands from the age of 27 until he retired at 70.

Today we know that the Sun is not at the center. To understand why the Herschels and Kapteyn obtained their erroneous results, imagine yourself standing in a large forest. Suppose you try to decide whether you are at the center of the forest by counting trees in all directions. Unless you are very near an edge, you will see the same number of trees in all directions even though you are nowhere near the center. The reason is that the trees themselves prevent you from seeing beyond a certain distance. If you cannot see to the edge of the forest, this method of determining your location is not valid.

The situation in the case of the stars is somewhat different, for the stars do not fill our view as do the trees in a forest. When we look out among the stars, however, interstellar dust and gas place a limit on how far we can see. The Herschels were unaware of the existence of this material. They assumed that they could see all the way to the edge of the group of stars within which our Sun lies. Because they could only see a limited distance and because the Sun is not near an edge, they concluded that it was at the center.

Another analogy is this: if you are surrounded by dense fog, the fact that you can see the same distance in every direction is not evidence that you are in the center of the fog.

Likewise, interstellar dust kept Kapteyn from counting the stars correctly. The density of stars at greater and greater distances from the Sun appeared to decrease because when Kapteyn counted stars at great distances, the interstellar dust prohibited him from seeing them all. In Figure 16-1, the dark areas stretching along the Milky Way are dust and gas clouds.

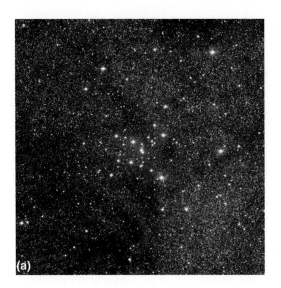

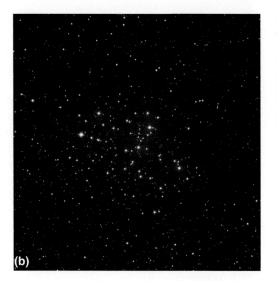

FIGURE 16-7 (a) The open cluster M7, about 1000 light-years away, in Scorpius; the Milky Way serves as the backdrop. The cluster, known to Ptolemy, includes about 80 stars. (b) The open cluster M6 (or Butterfly), in Scorpius.

These investigators reached erroneous conclusions simply because one of their assumptions—that there is nothing in space to block the view of distant stars—was wrong. Incorrect assumptions cause trouble not only in science but also in everyday life. Often we are not even aware of what our assumptions are, and this prevents us from even accepting the possibility that our conclusion may be in error.

Globular Clusters

As we discussed in Chapter 13, some stars begin their lives in galactic clusters. These clusters are called "galactic" because they are found within the disk of the Galaxy. The Pleiades (see Figure 13-7) is the most prominent example of this type of cluster, which may typically contain hundreds of stars. FIGURE 16-7 shows two other galactic or open clusters. FIGURE 16-8 shows two examples of a much larger

See also Figures 13-20 and 13-21 for more images of galactic and globular clusters.

FIGURE 16-8 (a) An *HST* image of M15, the most tightly packed globular cluster in our Galaxy, 33,600 light-years away in the constellation Pegasus. Observations hint at the presence of a 4000-solar-mass black hole at its core. The horizontal scale is about 9 light-years. (b) A VLT image of 47 Tucanae, 16,000 light-years away in the constellation Tucana. It can be seen by the unaided eye from the Southern Hemisphere. Its mass is about 1 million solar-masses, and it is 120 light-years across; it appears on the sky as big as the full moon. It contains at least 20 millisecond pulsars.

globular cluster A spherical group of up to hundreds of thousands of stars, found primarily in the halo of the galaxy.

type of cluster, a *globular cluster*. These beautiful, symmetrical clusters may look as though they have solid centers, but they are actually groups of hundreds of thousands of stars. The stars are so densely packed in the center of the cluster that we simply see a white area, not the individual stars. The average separation of stars near the center of a globular cluster is about 0.5 light-years. (In the Sun's region of space, stars are separated by an average distance of about 4 to 5 light-years.) Globular clusters are not confined to the disk of the Galaxy but are seen outside the disk.

While Kapteyn was seeking to determine the Sun's location in the Milky Way by studying star locations, Harlow Shapley was trying to do the same using globular clusters; however, he had a problem determining the distances to globular clusters, for they are much farther away than the stars Kapteyn was analyzing. Just a few years earlier, Henrietta Leavitt had discovered the relationship between the periods and the apparent magnitudes of Cepheid variables in the Magellanic Clouds. Shapley observed Population II Cepheids and RR Lyrae variable stars, which he considered to be like Cepheids but with shorter periods. Population II Cepheids are about 1.5 magnitudes dimmer than the classical (Population I) Cepheids of the same period observed by Henrietta Leavitt (see Section 12-8) and thus are less luminous by a factor of 4. RR Lyrae variables are commonly found in globular clusters; their average luminosity is about 100 times that of the Sun, and their periods are less than 1 day. Because these stars are easily identified from their periodic changes in luminosity, it is relatively easy to find their distances using their period-luminosity relation and the distance modulus (**FIGURE 16-9**). The distances to these variable stars correspond to the globular cluster distances.

FIGURE 16-9 After Shapley had determined the relationship between the periods and absolute luminosities of RR Lyrae variable stars, he could use a sequence like the one shown here to determine the distance to any RR Lyrae variable.

THE PERIOD OF THE CEPHEID/ RR LYRAE STAR

Yields → (Via the period-luminosity relationship)

ITS ABSOLUTE LUMINOSITY ← Combined with → ITS APPARENT LUMINOSITY

Yields

THE CEPHEID/ RR LYRAE'S DISTANCE

I did not doubt that the center of the entire galactic system must coincide with that of the globular clusters, and that it must have a mass very much greater than that of the Kapteyn system in order to prevent the high-velocity stars and the clusters from escaping altogether.
Jan Oort

In 1917, Shapley published results of his survey of distances and directions to the then-known 93 globular clusters. He showed that they are not distributed evenly around the sky, but tend to be located more on one side, centered about the constellation Sagittarius. In fact, they seemed to be distributed in a sphere centered on a point about 50,000 light-years away from the Sun. **FIGURE 16-10** shows the approximate distribution of globular clusters compared with the Herschels' model of the Galaxy. Shapley assumed that the clusters revolve around the center of the Galaxy and therefore concluded that this center lies at the middle of the group of globular clusters. This meant that the Galaxy is much larger than indicated by the Herschels' model, and that it is not centered about the solar system.

In the 1920s, further evidence indicated that the Sun is not at a unique position in the Galaxy. Jan Oort (who proposed the comet cloud that bears his name) and Bertil Lindblad studied the motions of great numbers of stars near the Sun. They

FIGURE 16-10 (a) The distribution of globular clusters as determined by Shapley is shown relative to the Sun and to the Herschels' model of the Galaxy. (b) The distribution of globular clusters and the nuclear bulge of the Galaxy shown through the interstellar dust. The lines mark the area blocked from view by the dust.

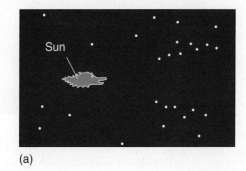

(a)

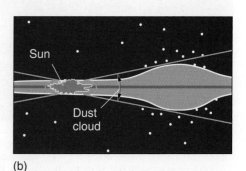

(b)

found that there is a pattern in their velocities, depending on their directions from the Sun. Kepler's third law, when applied to stars revolving around the center of the Galaxy, predicts that stars closer to the center should move faster and those farther from the center should move slower. This is the reason for the pattern of velocities shown in the Oort-Lindblad analysis. They concluded, as had Shapley, that the Galaxy's center was thousands of light-years away in the direction of Sagittarius.

It should be pointed out that Oort and Lindblad saw a pattern only after analyzing very great numbers of stars, for stars have random motions along with their pattern of motion around the Galactic center. This is similar to the way that cars on a multilane freeway have a pattern of motion in one direction, though at any given time certain cars may be changing lanes or otherwise deviating from the pattern. Oort and Lindblad ignored individual stellar motions and concentrated on patterns of average motions.

Finally, in 1930, the interstellar dust was discovered. This resolved the conflict between the conclusions of Herschel and Kapteyn on the one hand and of Shapley, Oort, and Lindblad on the other, for the interstellar dust had prevented Herschel and Kapteyn from seeing to the edge of the Galactic disk. It is interesting that interstellar extinction was the reason why both Kapteyn and Shapley were wrong. Because Kapteyn observed regions mostly in the Galactic disk (where extinction is strongest), he could not observe the far regions in our Galaxy, thus underestimating its size (to about 10,000 parsecs in diameter). A simple analogy is the case of a person standing in a dense fog, with limited visibility, trying to see the surrounding area. In Shapley's case, he observed globular clusters that are mostly above or below the plane of our Galaxy, in directions where extinction is not very strong; however, Shapley calibrated the period-luminosity relation by using variable stars near the Galactic plane, where extinction caused by dust in the disk dimmed the starlight, thus increasing the apparent magnitude of the stars. Shapley incorrectly concluded that the dimness of these stars implied large distances and thus calculated a diameter of about 100,000 parsecs. Although Shapley's original values for the size of the Galaxy had to be revised downward by a factor of about 2 when it was discovered that there are two types of Cepheid variables, his basic deductions were correct. He had shown that the Galaxy was much larger than previously thought. The boundaries of the universe were again pushed back.

16-2 Components of the Galaxy

We can describe four components of the Galaxy: the disk (which contains the Sun), the nuclear bulge, the halo, and the Galactic corona (**FIGURE 16-11**). We describe each of these components here and then discuss some of them in more detail in later sections.

The **disk** is the large, flat part of the Galaxy that rotates in a plane around its center. The disk contains individual stars, clusters of stars—particularly open clusters—and almost all of the gas and dust found in the Galaxy. Most of the stars in the disk are young, metal-rich, Population I stars. The disk appears bluish because of the presence of the hot O and B main sequence stars, which indicates that there is active star formation in the

Bertil Lindblad (1895–1965) was a Swedish astronomer who was president of the Royal Swedish Academy of Sciences for 22 years and was chair of the Nobel Foundation when he died in 1965.

Harlow Shapley (1885–1972) quit school after the fifth grade, returned to school at age 16, and earned a Ph.D. in astronomy from Princeton at age 27.

disk (of a galaxy) The flat, dense portion of a spiral galaxy that rotates in a plane around the nucleus.

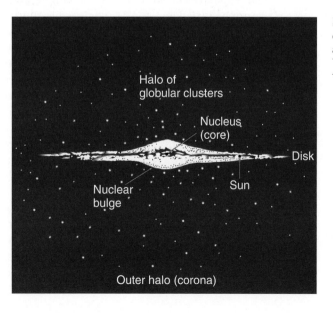

FIGURE 16-11 The Galaxy consists of a nuclear bulge, a rotating disk, and a halo. The halo probably extends farther than indicated here.

HISTORICAL NOTE

The Shapley-Curtis Debate

In the late 1910s, there was considerable controversy about the size of the universe and the nature of the spiral nebulae. Several scientific papers were published on these topics, and the controversy attracted national attention. Harlow Shapley of Mount Wilson Observatories and Heber Curtis of Lick Observatory became the primary spokesmen for two opposing views. In April 1920, the National Academy of Sciences invited these two astronomers to present papers before a meeting of that group in Washington, D.C. and to engage in a public debate concerning their research and conclusions.

At that time, Shapley had calculated the diameter of the Galaxy to be about 300,000 light-years. The reason that his result was too large is that he did not know about the interstellar medium. Another error led him to calculate the distance to the Magellanic Clouds to be only 75,000 light-years—less than his calculated diameter of the Galaxy. He was using Cepheid variables to determine the distance to the clouds, but he did not know that there are two types of Cepheids and that he was seeing the unknown type in the Magellanic Clouds.

Adriaan van Maanen, a Dutch astronomer who was a friend of Shapley, had published results showing that he could observe rotation of the Andromeda nebula. In seeing the rotation, he was observing proper motion of parts of the nebula. Because we cannot observe proper motion within an object unless it is relatively close to us, this indicated that the nebula is nearby. (Van Maanen's observation was simply in error.) Because of his own observations and those of van Maanen, Shapley concluded that not only are the Magellanic Clouds

and the Andromeda nebula part of our own system of stars, but also that other spiral nebulae are also part of that system.

Curtis held the opposite view. He believed that the group of stars that includes the Sun is much smaller than Shapley claimed and that the Andromeda nebula and other spiral nebulae are outside that group and are other "island universes." One bit of evidence he used to back his view was that Vesto Slipher of Lowell Observatory had investigated 15 spiral nebulae and found that 11 of them have significant redshifts. The redshifts indicated that they are moving away from us at great speeds, and this would make it unlikely that they are nearby. (These redshifts will be of major importance in Chapter 17.)

It is interesting that although Shapley thought that spiral nebulae are part of our Galaxy, he held that the universe is larger than Curtis envisioned. In his recollections of the debate, Shapley says that the subject assigned to the debaters was the size of the universe, but that Curtis turned the topic to the nature of the spiral nebulae.

Most of the 200 or 300 people present at the debate were members of the National Academy of Sciences, but the debate made headlines in the New York Times and excited public interest. When, just a few years later, Hubble discovered Cepheids in the Andromeda nebula and showed conclusively that it was indeed another "island universe" (as galaxies were called), it looked as though Shapley had been entirely wrong in the debate. In his own mind, however, Shapley felt that the score was even, for he had been more correct in predicting the overall size of the universe.

---

disk. The edges of the disk are not well defined and, therefore, its width and thickness cannot be stated exactly. As we noted earlier, its diameter is about 50,000 parsecs. Stars are most crowded near the plane of the Galaxy and become less crowded as we move from that plane. The disk is generally considered to be about 1000 parsecs thick, which makes its thickness about 2% of its diameter. It thus has the shape of about two compact disks stacked on one another.

Because star formation is more likely to occur where the interstellar material is most dense, stars are born with greatest frequency near the Galactic plane. Gradually, their motions result in them wandering away from the plane. Very massive stars have short lifetimes and, thus, we would expect to find few of them far from where they were formed. This is indeed the case, for almost all of the O-type stars lie within about 100 parsecs of the Galactic plane. Only less massive stars live long enough to move very far from the plane.

It is difficult to determine the structure of the Galactic disk from our position inside it because interstellar dust obstructs our view of distant locations. This dust is what limited early investigators' attempts to determine the Sun's position in the Galaxy. In Section 16-3 we describe how radio astronomy has provided evidence for the spiral arms, and in an Advancing the Model box, we outline new evidence that the Galaxy may not be a standard spiral galaxy after all.

The *nuclear bulge* of the Galaxy is about 2000 parsecs (6500 light-years) in diameter. If you imagine the disk to be a stack of two compact disks, you thus must add a peanut at their center to represent the bulge. This is necessary to represent the Galaxy correctly because the bulge is wider along the disk than it is perpendicular to the disk. Stars, dust, and gas are much more densely packed inside the nuclear bulge than they are anywhere else in the Galaxy. The bulge contains both young and old stars and appears reddish because of the presence of many red giants and supergiants. We discuss the bulge and its mysterious core in Section 16-5.

Figure 16-11 shows the Galactic *halo*, which contains the globular clusters that caused Harlow Shapley to conclude that the Sun is not at the center of the Galaxy. A person viewing the globular clusters from the Sun's position would see many more in some directions than in others; and their distribution is centered at the nuclear bulge. Besides the globular clusters, which mostly consist of old, metal-poor Population II stars, the halo contains small amounts of gas and dust.

Globular clusters feel a gravitational force toward the center of the Galaxy and thus they orbit the Galactic nucleus, passing through the disk twice during each orbit. FIGURE 16-12 indicates their motion. About 150 globular clusters are known to be associated with the Milky Way, and the orbits of 75 of them have been calculated. Of these, 63 are in the halo, and 12 are confined to the disk.

In the next chapter we discuss an unresolved problem in galactic astronomy: the motions of galaxies indicate that much more gravitational force is exerted on galaxies than can be explained by the amount of mass that we can see. Astronomers conclude that there is more invisible mass in the universe than visible mass—by a factor of about 6. It is now thought that much of this mass exists in *extended halos* that surround galaxies. Therefore, an outer halo (or corona) was included in Figure 16-11. It is thought that the corona extends to perhaps two or three times the radius of the disk and halo. We do not know what the corona consists of—small black holes, cool dwarf stars, great numbers of neutrinos, or other exotic particles have been hypothesized. The first direct detection of such cool dwarf stars was announced in 2001. Thirty-eight previously unseen cool white dwarfs were observed within about 450 light-years of Earth. Observations suggest that, at most, 35% of the unseen matter in our Galaxy is made of normal matter, composed of neutrons, protons, and electrons.

TABLE 16-1 shows today's values for various Milky Way properties. Keep in mind, however, that the numbers are approximate, not only because every measurement is uncertain to some extent, but also because there are no specific boundaries for the various parts of the Galaxy. The total diameter of the halo is little more than a guess.

Galactic Motions

One method for determining the motion of the Sun around the Galactic center relies on the observation that the orbits of globular clusters seem to be randomly distributed around the center of the Galaxy. If we assume that the average velocity of all of the clusters, relative to the nucleus, is zero, we can measure their velocities relative to the Sun and attribute the average motion that is observed to the motion of the Sun. An analogy would be a person in a boat drifting on the ocean far from land and trying to measure the boat's speed. Suppose that the person is able to measure the speed of planes flying overhead and that equal numbers of planes are seen flying eastward and westward. Assuming that planes fly eastward at the same speed as planes fly westward, the person would know that the average speed of all the planes is zero. But if the measurements show that relative to the person the planes' average speed is 10 miles per hour toward the east, the person could conclude that the boat is moving 10 miles per hour toward the west.

nuclear bulge The central region of a spiral galaxy.

halo (around a galaxy) The outermost part of a spiral galaxy; fairly spherical in shape, it lies beyond the spiral component.

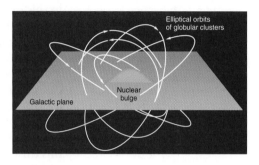

FIGURE 16-12 Globular clusters orbit the Galactic nucleus, passing through the disk twice during each orbit.

TABLE 16-1	
Galactic Data	
Radius of disk	80,000 light-years
Radius of nuclear bulge	3000 light-years
Total radius of halo	200,000 light-years
Sun's distance from center	26,000 light-years
Sun's orbital period	250,000,000 years
Thickness of disk	3000 light-years
Number of stars	200 to 400 billion

The Doppler effect is used to measure the radial speeds of the globular clusters.

In practice, wind speed would affect the speed of the planes, and the person's assumption would not be valid.

Scientists always look for different methods to measure a quantity, and in this case, another method exists. Oort and Lindblad studied the motions of great numbers of stars relative to the Sun and applied statistical methods to determine the motion of stars in the Sun's neighborhood. Similar studies yield a value for the speed of the Sun. The best measurements show that the Sun is traveling in a nearly circular path around the Galactic center at a speed of about 220 kilometers/second (135 miles/second). It is now moving toward the constellation Cygnus. Knowing that the radius of the Sun's orbit is 8000 parsecs (26,000 light-years), we can calculate the circumference of the Sun's path and find that it takes about 230 million years to complete one revolution. Although this seems a tremendously long time, the Sun has completed some 20 orbits during its 5-billion-year lifetime.

If the Galaxy had almost all of its mass concentrated at its center, its parts would rotate according to Kepler's third law, just as the planets do in the solar system. The dashed line in **FIGURE 16-13a** shows the *galactic rotation curve* that would be expected in this case. If the mass were distributed so that most of it is fairly near the center, but not in the center, the curve would look more like the solid blue line. In fact, the Galactic rotation curve is somewhat like the solid red line in this figure. This indicates that the Galactic mass is unevenly distributed, and the fact that the speeds of objects located farther from the center than the Sun are not lower than the Sun's speed indicates that large amounts of mass lie beyond the Sun's distance. The amount of matter we "see" in our Galaxy because of the light it emits amounts to about 10% of the total mass necessary to give the observed rotation curve. The remaining 90% of the mass of the Galaxy is not visible. In the next chapter we discuss the possible nature of this unseen matter.

As we might expect, however, the orbits of stars are not perfectly circular. Along with the general orbiting motion of each portion of the Galaxy, each star has its own peculiar motion—perhaps having a component of motion from one side of the disk to the other or perhaps moving closer to the Galactic center at one time and moving farther away at another time as it follows an elliptical path. In general, however, the paths of stars seem to be nearly circular.

The Mass of the Galaxy

In Section 3-5 we showed that Isaac Newton revised Kepler's third law to read

$$\frac{a^3}{P^2} = (m_1 + m_2)/m_{Sun},$$

where a is the average radius of the orbit (in AU), P is the orbital period (in years), and m_1 and m_2 are the masses of the two objects. In Section 7-2 we showed that the

> Saying that the Sun is moving toward Cygnus does not mean that it is getting closer to Cygnus; it means that the Sun is moving in that direction through space. Besides, each star of the constellation Cygnus has its own motion: some toward us and some away from us.

galactic rotation curve
A graph of the orbital speed of objects in the galactic disk as a function of their distance from the center.

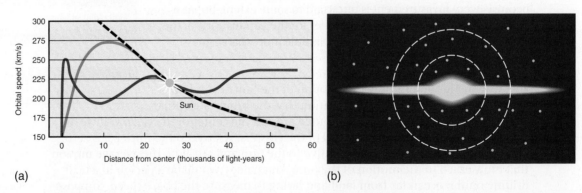

(a)

(b)

FIGURE 16-13 (a) If all the mass of the Galaxy were concentrated at its center, the Galactic rotation curve would be shown by the black dashed line in this graph. If the mass were distributed so that most of it is somewhat near, but not at the center, a rotation curve such as the one shown by the solid blue line would result. The observed rotation curve of the Galaxy is shown by the solid red line. It indicates that large amounts of mass orbit the center far beyond the Sun's orbit. The almost vertical segment of the rotation curve close to the center implies that matter in that region revolves as if it were a rigid body. (b) The mass enclosed in a sphere centered on the Galactic center increases with distance. As a result, the rotational speed stays approximately constant.

relationship holds not only for objects in orbit about the Sun, but also for satellites in orbit around the planets. Finally, in Section 12-6 we showed the law's application to binary star systems.

The discovery (in 1927) by Oort and Lindblad that the Galaxy in the Sun's neighborhood undergoes differential rotation meant that Kepler's third law might be applied to calculate the masses involved. In the case of a star revolving around the Galaxy's center, one of the masses in the equation is the mass of the star. The other mass is the mass of the *entire inner Galaxy*, including all objects in the Galaxy that are closer to the center than that star is. This may seem strange, as the inner portion of the Galaxy is made up of many objects rather than one, but it can be shown that this is the correct application of the equation.

The first step in using Kepler's third law to calculate the mass of the part of the Galaxy inside the Sun's orbit is to express the Sun's distance from the center (26,000 light-years) in AUs. Because 1 light-year is about 63,200 AU, the Sun is

$$26,000 \text{ ly} \times \frac{63,200 \text{ AU}}{1 \text{ ly}} \approx 1.6 \times 10^9 \text{ AU}$$

away from the Galactic center. Using this value in Kepler's third law, along with the Sun's period of 230,000,000 years, we find

$$m_1 + m_{Sun} = \frac{a^3}{P^2} \times m_{Sun} \approx \frac{(1.6 \times 10^9 \text{ AU})^3}{(2.3 \times 10^8 \text{ yr})^2} \times m_{Sun} \approx 10^{11} m_{Sun}.$$

The value obtained is the total of the mass of the Sun and the mass of the inner Galaxy. The mass of the Sun, of course, is negligible compared with this value and, thus, this answer is simply the mass of the part of the Galaxy that lies within the Sun's orbit. This value, 100 billion solar masses, must be taken as approximate. Recent analysis of the pattern of rotation in the outer parts of the Galaxy indicates that the total mass of the Galaxy is about 10^{12} (one trillion) solar masses, or about 10 times greater than we calculated above for the inner Galaxy. At present, the nature of this additional mass is unknown, for there are not enough stars to account for that much mass. We see later that a similar problem exists in the case of other galaxies.

Refer to the discussion of dark matter in Chapter 17 and again in Chapter 18.

16-3 The Spiral Arms

The **spiral** nature of the Galaxy is not obvious from observations in visible light because we can see only limited distances along the plane of the Galaxy. In 1951, however, astronomers at Yerkes Observatory discovered that as one looks either toward or away from the Galactic center, the distribution of O- and B-type stars is not uniform. They seem to be clustered at certain distances. **FIGURE 16-14** illustrates the concentrations of this type. This was the first hint of the spiral nature of the Galaxy. Unfortunately, because such stars are best observed in visible light, interstellar extinction allows us to observe them only in the vicinity of the Sun. The same is true for the red emission nebulae found near many of these hot, luminous, blue main sequence stars; however, more evidence for the spiral nature of our Galaxy came with the discovery during that same year of a specific emission line of neutral hydrogen.

You might guess, correctly, that if we could map the distribution of hydrogen clouds in our Galaxy, we

spiral galaxy A disk-shaped galaxy with arms in a spiral pattern.

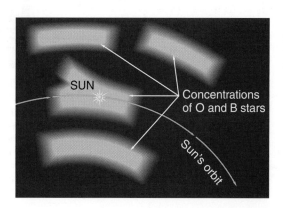

FIGURE 16-14 O- and B-type stars are found at greatest concentrations in the shaded areas.

should be able to infer its structure; however, most of the emission lines of hydrogen are at visible and ultraviolet wavelengths. Such radiation gets scattered or is absorbed easily by interstellar dust and this limits the range at which we could observe these clouds. Because the longer the wavelength, the easier it is for radiation to travel through interstellar dust without being scattered or absorbed, we need to look for radiation at infrared wavelengths. It turns out that cool hydrogen gas does emit such radiation in the radio portion of the spectrum (the ***21.1-centimeter radiation***).

The 21-cm emission photon results from a transition that a hydrogen atom makes from a higher energy level to a lower one (**FIGURE 16-15a**). To understand this transition, consider an approximate model in which the electron and proton of the hydrogen atom are small, charged spheres that spin around their axes. Because charges in motion create magnetic fields, we could visualize both the electron and proton as tiny magnets. When both magnets have their poles aligned in the same direction (as in the case of the electron and proton spinning in the same direction), they repel each other. When the two magnets are aligned in opposite directions (as in the case of the electron and proton spinning in opposite directions), they attract each other. The first configuration has a bit more energy than the second. When the transition occurs, a 21-cm photon thus is emitted. Of course, absorption can also occur that excites the hydrogen atom into aligning the spins of its electron and proton in the same direction.

Radio telescopes can use 21-cm radiation to detect high concentrations of cool hydrogen, such as exist in interstellar clouds. Because hydrogen is the main component of interstellar material—and the entire universe—this is an ideal method of detecting cool hydrogen clouds at great distances. Figure 16-15b is a false-color image of our Galaxy in 21-cm emission from interstellar hydrogen; this emission traces the "warm" interstellar medium, which on a large scale is organized into diffuse clouds of gas and dust that have sizes of up to hundreds of light-years.

As we have seen, new stars arise from gas and dust to which some of the material of a massive star returns at the end of the star's lifetime; therefore, hydrogen gas clouds detected by 21-cm radiation are located at the same places as newly forming stars. O and B stars are very massive and are therefore short lived. This means that they are found only where stars have recently formed and that we might expect to find hydrogen clouds at locations identified by astronomers as having high concentrations of O- and B-type stars.

It may seem that it would be impossible to determine the distance of a source of 21-cm radiation, for a radio telescope would seem to reveal only the direction of the source of the waves. In a sense this is true; however, the Doppler effect allows us to

21-centimeter radiation
Radiation from atomic hydrogen, with a wavelength of 21.1 centimeters.

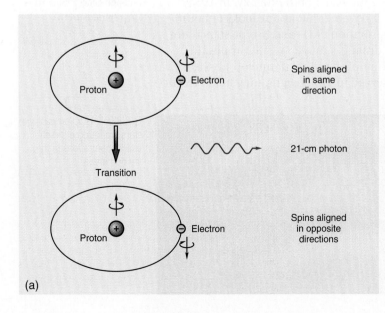

(a)

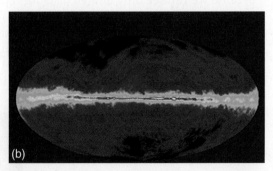

(b)

FIGURE 16-15 (a) A 21-cm wavelength photon is emitted when the spins of the electron and proton in a hydrogen atom make the transition from being aligned in the same direction to being aligned in opposite directions. (b) Our Galaxy in 21-cm emission of hydrogen. Hydrogen gas is concentrated along the plane of our Galaxy (which extends horizontally across the image). The range of hydrogen density is shown from black/blue (lowest) to red/white (highest).

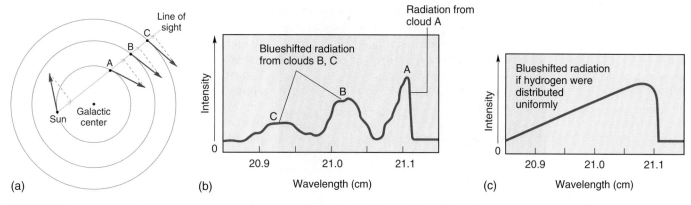

FIGURE 16-16 (a) The Sun and three hydrogen clouds (positions A, B, and C) are shown at different distances from the Galactic center. Assuming circular orbits and similar speeds, it seems that cloud A is stationary along the line of sight, while we seem to be approaching clouds B and C. (b) The 21-cm radiation observed along the line of sight shows peaks of blue-shifted radiation; the peaks occur because the hydrogen is concentrated in spiral arms. (c) A smooth blueshifted 21-cm line would be observed if hydrogen were distributed uniformly in the Galaxy.

determine the *radial* motion of the hydrogen with respect to us. For example, suppose that a radio telescope is pointed in a direction across the Galaxy, as shown in **FIGURE 16-16a**. Let us also make the simplifying assumption that our solar system and the clouds at positions A, B, and C move on approximately circular orbits around the Galactic center and have about the same speed. Their instantaneous velocities are shown in red and are tangent to their orbits; the components of these velocities along the line of sight are shown in yellow. Then, from our point of view, cloud A seems stationary along the line of sight, while we seem to be approaching clouds B and C. As a result, the 21-cm radiation from clouds B and C will be blueshifted, as shown in part (b). If hydrogen gas were distributed uniformly in the Galactic disk, instead of being mostly concentrated in the Galaxy's spiral arms, the 21-cm radiation would show the fairly regular blueshift of part (c); however, what is observed is a graph similar to that in Figure 16-16b, providing us further evidence for the spiral structure of our Galaxy.

If another cloud were between the Sun and cloud A in Figure 16-16b, we would observe a *redshifted* 12-cm line from that cloud.

Mapping the Galaxy in 21-cm radiation contains plenty of room for error, but there is no doubt that ours is a spiral galaxy. As we show in the next chapter, a spiral appearance is common for galaxies. **FIGURE 16-17** shows a spiral galaxy that may be similar to the Milky Way; recent work indicates that our Galaxy may have an elongated nucleus, as we discuss in the Advancing the Model box on page 467.

16-4 Spiral Arm Theories

It may seem at first glance that the spiral arms of a galaxy could be explained by the fact that stars near its center complete a circle in less time than those farther out. This would occur even if the stars do not orbit according to Kepler's third law, simply because stars near the center have less distance to cover to complete their orbits. Consider a group of stars that at a given time are in a straight line across the center of the galaxy, as shown in **FIGURE 16-18a**. As the stars move around the center (even if they all move at

FIGURE 16-17 If we could get outside our Galaxy and view it face on, it would probably look like this galaxy, M83, in the southern constellation Hydra. The bright stars in the image are part of our Galaxy. Because these stars are much closer to us than M83, we see them as we see flies on a car windshield when we look through it. (Compare this to Figure 16-2.)

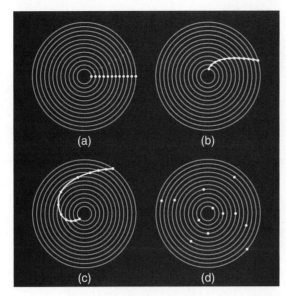

FIGURE 16-18 Stars that start out in a straight line from the galaxy's center (a) would begin to show a spiral pattern (b and c), but after a few revolutions, the pattern would be lost (d).

density wave theory A model for spiral galaxies that proposes that the arms are the result of density waves sweeping around the galaxy.

density wave A wave in which areas of high and low pressure move through the medium.

In a spiral galaxy, are all the stars located in the spiral arms?

the same speed), the line of stars will wind up into a spiral. The problem is that they would wind up so much that they would not be distinguishable as a line, as shown in Figure 16-18b, c, and d. The galaxy would appear as a fairly uniform disk. Yet we do perceive a spiral pattern in our Galaxy and in many other galaxies; therefore, this simple differential rotation hypothesis cannot be correct.

Currently, there are two competing theories to explain the spiral nature of galaxies: the density wave theory and the self-propagating star formation theory.

The Density Wave Theory

The **density wave theory** was first proposed in 1940 by Bertil Lindblad and holds that what we see in a spiral arm of a distant galaxy is not a simple fixed line of stars but a line formed by the brightest stars and the glowing nebulae surrounding them. The theory holds that stars revolve around the galaxy independent of the spiral arms and that the arms are simply areas where the density of gas is greater than at other places. According to this theory, there are almost as many stars per unit volume between the arms as in the arms, but the arms contain more of the brightest stars and a higher density of gas and dust. The areas of denser gas move around the galaxy in **density waves**, causing the formation of new stars and glowing emission nebulae. This is best explained by an analogy, illustrated in FIGURE 16-19.

Suppose that cars are traveling on a long highway at a speed exceeding the speed limit. Also traveling along the road, at less than the speed limit, is a traffic patrol car with its radar on. Observing from a helicopter high in the sky, we see the cars fairly evenly distributed along the highway except around the police car. For a short distance behind and in front of the police vehicle, the cars are bunched up. As a given car approaches the police cruiser from behind, the car slows down, slowly passing the feared patrol car. Then when the driver believes that his or her car is safely in front of the police cruiser, it again speeds up. As a result, there is constantly a high density of

FIGURE 16-19 An overhead view of a highway, seen as time passes. Car X is a police car, moving just slower than the speed limit. Bunched up around that car are other slow-moving cars that only gradually pass the police car. Thus, as time goes by, the cars near X are different, but a high density of cars remains there.

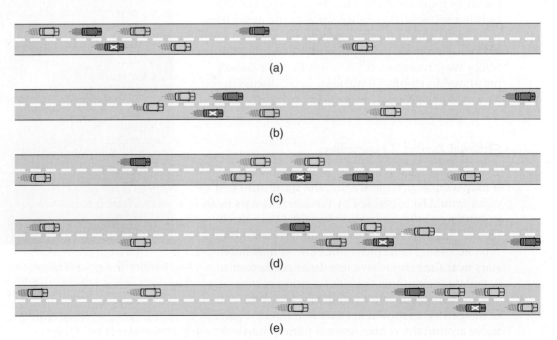

FIGURE 16-20 (a) The orbits of objects in the galaxy are very nearly circular. Because their orbital speed is greater than the speed of the density wave that forms the spiral pattern, they pass in and out of the spiral arms. (b) A cloud of dust and gas is about to enter a spiral arm. (c) The higher density of gas in the arm causes the cloud to compress, resulting in star formation. (d) As the cloud passes through the spiral arm, bright, massive stars make the arm itself appear brighter than the rest of the galaxy. (e) The bright stars burn out before they exit the spiral arm, as another cloud of dust and gas begins to enter the arm.

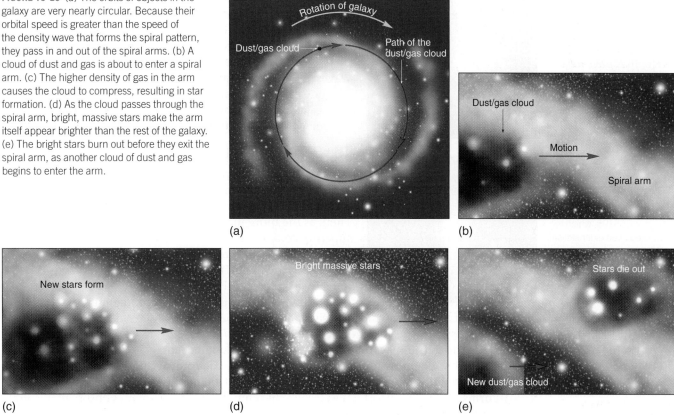

(a)

(b)

(c)

(d)

(e)

cars around the police car, even though the cars making up that group change all the time. The police car moves along, seeming to carry its high-density group along with it, in what we see from the helicopter to be a density wave.

Density waves are common here on Earth, for every sound wave is a density wave, with regions of high and low density making up the wave. In a galaxy, the density wave consists of a region of gas and dust that is denser than normal, perhaps by 10% to 20%. The major difference between sound waves and the density waves of a spiral galaxy is that in the atmosphere of Earth a sound wave travels faster than the particles of the gas itself; however, in the near-vacuum of a galaxy, the wave travels more slowly than the particles. The gas and dust particles—as well as stars—catch the wave from behind and pass through it in much the same way that cars in the analogy passed through the density pulse around the police car. FIGURE 16-20 illustrates the idea.

New stars are formed when an interstellar cloud of gas collapses. The density wave theory holds that the trigger for this collapse is the wave. As interstellar clouds approach the density wave from behind, they are compressed and stars are formed. Although stars of all masses are formed along the edge of the density wave, the brightest, most massive stars have ended their lives before they pass far from this edge. This means that when we look at a spiral galaxy, the spiral arms are obvious to us because they are the areas containing the bright stars.

One problem with the density wave theory is the question of how the density wave is sustained through the life of the galaxy. Computer simulations indicate that it would die out. In addition, infrared images of the center of the Whirlpool galaxy (FIGURE 16-21) indicate that spiral arms penetrate much farther into the nucleus of a galaxy than had previously been thought. The density wave theory had predicted that a galactic nucleus absorbs density waves, preventing the waves from extending into it. New information from the Whirlpool galaxy is causing a reexamination of the density wave theory.

It is possible that gravity is the driving mechanism behind the density wave. If the nucleus of a galaxy is not symmetrical (for example, if it is bar shaped, as it might be the case for our Galaxy), then the asymmetric gravitational field might

FIGURE 16-21 (a) A four-color composite in visible light of the Whirlpool galaxy (M51 or NGC5194); red corresponds to the H-alpha line. (b) This *Spitzer* image of M51 is a four-color composite; blue (3.6 μm), green (4.5 μm), orange (5.8 μm), and red (8.0 μm). The presence of the thin filaments between the arms is very puzzling. The star formation in M51 and its spiral structure are thought to be triggered by its on-going collision with its companion, NGC5195, which is composed of an older stellar population.

self-propagating star formation theory A model for spiral galaxies that explains the arms as resulting from a series of supernovae, each triggering the formation of new stars.

generate density waves by its interaction on stars and interstellar matter. Another possibility is that gravitational interactions between galaxies can generate and sustain density waves. Even if we resolve the issue of the driving mechanism, there is another problem. The spiral arms produced by density waves are well defined (for example, Figures 16-17 and 16-21); however, some galaxies have poorly defined arms. For such galaxies, another theory for the formation of spiral arms has been proposed: the self-propagating star formation theory.

The Self-Propagating Star Formation Theory

According to the ***self-propagating star formation theory*** of galactic spiral arms, the triggers that start the collapse of most interstellar clouds are nearby supernova explosions. Then, as the more massive stars finish their lives and become supernovae, they trigger more star formation, and so on. A simple analogy is a forest fire, with the flames jumping from one tree to the next. The formation of new stars thus is confined to areas where this process is taking place. Now differential rotation enters the picture. Computer analysis shows that at the rate at which massive stars would be formed, differential rotation would cause spiral arms to be formed and sustained.

The self-propagating star formation model is able to explain how a spiral arm would begin, as the process starts with a single supernova. Computer simulations based on this model can reproduce the spiral structure of galaxies with poorly defined arms, but the model is not successful in reproducing the structures observed in spiral galaxies with well-defined arms.

The two theories we presented for the formation of spiral arms are very different. In the density wave theory the spiral arms cause star formation, whereas the opposite is true in the self-propagating star formation theory. It is probable that we need a combination of these two theories to explain the observations.

16-5 The Galactic Nucleus

Observations by Shapley, Oort, and Lindblad in the early part of the 20th century revealed that the center of our Galaxy lies in the direction of Sagittarius. Because the presence of dust and gas in the Galactic plane dims visible light from the nucleus by about 28 magnitudes, astronomers had to await the development of nonoptical telescopes to learn more about that nucleus. To observe the Galactic nucleus, we use wavelengths in the infrared/radio part of the spectrum or in X-rays/gamma rays. Our observations helped us construct a picture of the nucleus of our Galaxy that includes a massive black hole and a history of violent events.

The *HST* image in **FIGURE 16-22**, taken in the infrared, shows a pair of the largest young star clusters, located less than 30 parsecs from the Galactic center. The ob-

FIGURE 16-22 These clusters are 10 times larger than typical young star clusters in our Galaxy; they will be destroyed in a few million years by the gravitational tidal forces in our Galaxy's core. The cluster on the left is so dense that 100,000 of its stars could fill the spherical volume between our Sun and its nearest neighbor star (about 4 light-years away). The cluster on the right is less dense; it has stars at the stage of becoming supernovae and is the home of the brightest star in our Galaxy (the Pistol star). The false colors in this *HST* image correspond to infrared wavelengths. Galactic center stars are white, red stars are located behind (or are surrounded by) dust, and blue stars are foreground stars (located between us and the Galactic center).

ADVANCING THE MODEL

The Milky Way: A Barred Warped Spiral Galaxy

In Chapter 17 we show that there are two types of spiral galaxies. About half are standard spiral galaxies, like the one shown in Figure 16-17. The other half have an elongated central bulge that appears as a bar across the center, like the one in FIGURE B16-1. Until recently, astronomers thought that the Milky Way is a standard spiral galaxy, but evidence is accumulating that our Galaxy is a barred spiral galaxy with an elongated nucleus that is probably shorter in proportion to the rest of the Galaxy than the one in Figure B16-1.

The new evidence comes primarily from data related to the 21-cm radiation from clouds of cool hydrogen gas near the nuclear bulge and observations using the *COBE* and *Spitzer* satellites. (The *Cosmic Background Explorer* was designed to measure short-wavelength radio waves from space, as we discuss in Chapter 18.)

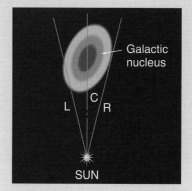

FIGURE B16-2 Opposite sides of an elongated nucleus—a bar—do not appear the same when viewed from the Sun, because lines that make equal angles with the central line pass through different volumes of the bar. Data indicate that the nucleus of the Milky Way is slightly elongated with its long axis at 45° to the line between the Sun and the Galactic center.

FIGURE B16-1 The Milky Way may be a barred spiral galaxy similar to this one.

Suppose the central bulge is elongated with its long axis oriented at some oblique angle to the line of sight, as in FIGURE B16-2. The Sun (and Earth) is located at the point indicated in the figure. Line C extends to the center of the Galaxy, and lines L and R are drawn at equal angles on each side of line C. The line of sight on the near side of the nucleus (the left side here) passes through part of the nucleus, but the equivalent line of sight on the other side does not. If the nucleus is shaped like this and has this orientation, astronomers should detect an asymmetry in the radiation

from near the nucleus. Data from *COBE*, as well as data from Japanese astronomers, correspond to what would be expected from such an elongated nucleus.

Further information comes from analyzing motions of the gas near the nucleus. The gas moves around the nucleus in a noncircular pattern, corresponding to what would be expected from a barred nucleus. Finally, using *Spitzer*, astronomers were able to see through interstellar dust clouds and survey more than 30 million stars around the Galactic center; the survey suggests the presence of a long bar of stars (about 27,000 light-years in length) at an angle of 45° to the line between the Sun and the Galactic center.

A map of Galactic atomic hydrogen gas emissions and the distribution of more than 100 million stars in the Galactic disk show that the disk is warped and that the warp extends across the entire diameter of the Galaxy. Computer simulations show that the warp is very dynamic and could be explained as the result of gravitational interactions between the Galaxy's dark matter and its prominent satellites, the Magellanic clouds.

served number density of stars is indeed increasing as we get closer to the Galactic center, down to about 2 parsecs from the center. For distances closer than 2 parsecs, observations of the velocities of stars suggest that there is a great deal of mass in a very small volume in the Galactic nucleus (less than 0.5 parsecs in radius). It seems that the Galactic nucleus harbors a massive black hole of mass about 3.7 million solar masses.

Radio observations show that a nuclear disk of neutral gas occupies a region between a few hundred parsecs to a thousand parsecs from the center. FIGURE 16-23a shows the region near the Galactic center, located at the edge of the bright object

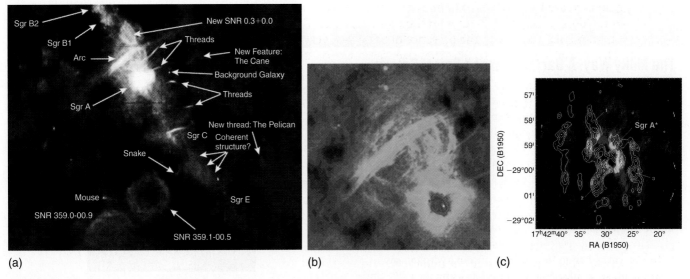

FIGURE 16-23 (a) This image was constructed from 1-meter radio data obtained by telescopes of the Very Large Array. It shows signs of a violent environment, including supernova remnants and filaments that may correspond to mass outflow from the Galactic center (located at the edge of the object Sgr A). (b) This image (in 20-cm) shows the inner region, about 60 parsecs across, of the Galactic nucleus. The nucleus is at the center of the strongest emission. The filament, running perpendicularly to the Galactic plane, may be associated with magnetic fields. (c) Ammonia emission (at 1.3 cm) in yellow contours in the region around the Galactic center; the resolution is as small as 0.13 light-years. The ring as seen at 3 mm is shown in the background. Streamers that possibly interact with the ring are numbered. The star labels the position of the massive black hole (Sgr A*).

labeled Sagittarius A (Sgr A). The Galactic plane runs diagonally across the image, which includes supernova remnants (SNRs), along with many elongated threads of ionized gas, almost perpendicular to the disk. One of the most intriguing features of this region is shown in Figure 16-23b. It is likely that this feature is the result of magnetic fields, with the radio emission being due to charged particles spiraling around the magnetic field lines. In the Sgr A region we find a large molecular ring, with an inner radius of 2 parsecs and outer radius of 8 parsecs, whose mass is at least 20,000 solar masses. The material in this ring is much warmer and denser than material in molecular clouds elsewhere in the Galaxy. The velocities of the ionized gas in this

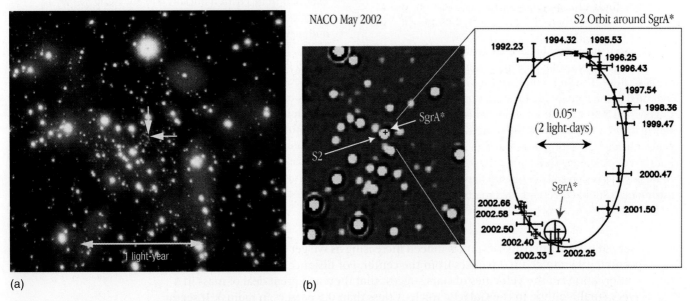

FIGURE 16-24 (a) The innermost area of our Galaxy, shown as a combination of three infrared images (wavelength between 1.6 and 3.5 μm). The compact objects are stars; blue denotes hot, and red denotes cool stars. There is also infrared emission from dust between the stars. The arrows mark the position of Sgr A*. (b) An infrared image of an area 60 × 60 light-days, centered on the position of Sgr A*. The inset shows the positions of a star, designated S2, during the period 1992–2002. The solid curve is the best-fitting elliptical orbit, with Sgr A* at one of the foci. As of 2009, astronomers have followed S2 making more than one complete orbit; using S2's orbital dynamics, they measured the distance between Earth and the Galactic center to be 7.94 ± 0.42 kpc, in good agreement with measurements by other methods.

region increase from about 100 km/s at the inner edge of the ring to 700 km/s at 0.1 parsecs from the Galactic center. These measurements, along with the presence of Sgr A*—a very small, unresolved (less than 20 AU in size) radio source inside the Sgr A region—suggest that at the center of our Galaxy we have a massive black hole of 3.7 million solar masses. This is consistent with the results obtained using stellar velocities in this region. Figure 16-23c shows the inner region of the Galactic nucleus. Narrow "streamers" can be seen connecting giant molecular clouds of gas and dust to the ring. The observed increase in velocity along a streamer suggests that gas/dust is being accreted from the molecular clouds and "fed" to the ring and then to the black hole as a result of its strong gravitational pull. Infrared observations of the innermost area of the Galactic nucleus, obtained in 2002 by ESO's VLT and shown in FIGURE 16-24a, show emission from interstellar dust between a great number of stars around the position of Sgr A*. In addition, years of observations allowed astronomers to trace the orbits of many stars around Sgr A*; one such orbit is shown in Figure 16-24b. The characteristics of these orbits leave little doubt that a supermassive black hole is present at the center of our Galaxy.

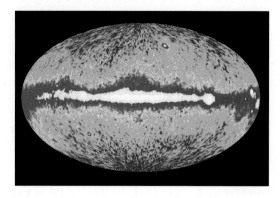

FIGURE 16-25 This image was taken by the *Energetic Gamma-Ray Experiment Telescope (EGRET)* on board the *Compton Gamma-Ray Observatory*. Our Galaxy is a diffuse source of gamma-ray light. Superimposed on this image are several gamma-ray pulsars and some active nuclei of other galaxies.

The case of a large black hole at the center of our Galaxy is also supported by observations of gamma rays emanating from a very small region in the Galactic nucleus. In addition, in late 2000, X-ray variability was observed from Sgr A* that lasted less than 10 minutes. As we discussed in Section 15-4, variability is a great indicator of the size of the radiating region because nothing can travel faster than light. This observation suggests that the hot, radiating gas could not occupy a region bigger than the distance between the Earth and the Sun. The most probable scenario is that the force that powers Sgr A* is a black hole, orbited nearby by a great number of stars. As matter falls into the black hole, collisions cause energy to be released. The release of energy here is similar to the energy released as matter falls into stellar black holes, which we discussed in Chapter 15. In the case of the Galactic nucleus, however, a *massive* black hole is at the center, and it is surrounded by a great amount of matter, causing enormous quantities of energy to be released. FIGURE 16-25 shows the intensity of high-energy gamma-ray emission in our Galaxy.

An X-ray look at our Galaxy's central black hole has revealed numerous outbursts and some large explosions. FIGURE 16-26a shows large lobes of hot gas (20 million K) in the innermost region around Sgr A*. These lobes show that many enormous explosions have occurred during the past 10,000 years. A possible jet of particles,

Black holes are common in the Galaxy. A few isolated stellar-mass black holes have been detected indirectly by the way their gravity bends the light of a more distant star behind them. We discuss this phenomenon in Chapter 17. Stellar-mass black hole candidates have been observed both in the Galactic disk and halo by the effect they have on the motion of their normal companion star.

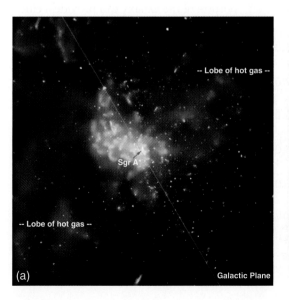

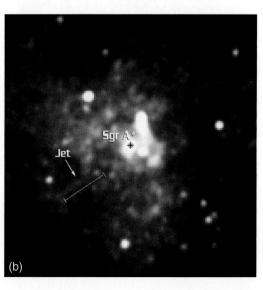

FIGURE 16-26 (a) A *Chandra* image that marks the position of the X-ray source associated with Sgr A*. Large lobes of gas (20 million K) are also shown. The image covers an area of 60 × 60 light-years. (b) A close-up Chandra image of the region around Sgr A*, showing a possible X-ray jet of length 1.5 light-years. The image covers an area of 9 × 9 light-years.

shown in Figure 16-26b, suggests that the observed X-ray flaring activity from the vicinity of the black hole has been occurring for many years.

Some other spiral galaxies also have tremendously powerful energy sources at their center, and we see signs of violent activity near those centers. The massive black hole hypothesis seems to be the best explanation for the energy source in our Galaxy and in similar galaxies. As we discuss in the next chapter, some galaxies produce even more energy in their nuclei. We are far from understanding the nucleus of our Galaxy, and further knowledge must come from observations of other galaxies as well as our own.

16-6   The Evolution of the Galaxy

The first step in developing a theory of the formation of the Galaxy is to consider the chemical content and the age of the Galaxy's components. After that is done, the pieces can be put together to give us a picture of how the Galaxy developed.

Age and Composition of the Galaxy

Astronomers say that a star with a large abundance of heavy elements is rich in metals, *or* metal rich.

In Section 15-9 we discussed the division of stars into Population I, Population II, and Population III according to their percentage of heavy elements. Although in reality there is a continuum from the low percentage of heavy elements in extreme Population II to the greatest percentage in extreme Population I, the grouping is still convenient. On the average, Population II stars contain only about 1% of the heavy elements of Population I stars. An extreme case is the star HE0107-5240, 36,000 light-years in the southern constellation Phoenix, which has the lowest abundance of heavier elements ever observed, 1/200,000 that of the Sun.

The abundance of heavy elements decreases by a factor of 0.8 for each thousand parsecs from the center of the disk.

Population I stars are stars that have formed from the remains of previous generations of stars, and most of them are found in the disk of the Galaxy. In fact, stars with the greatest percentage of heavy elements lie close to the nuclear bulge. The farther from the bulge, the lower the fraction of heavy elements. Most globular clusters, on the other hand, are made up of Population II stars. Of the known globular clusters in the Milky Way, about 70% have an average heavy-element content of about one twentieth of that of the Sun. The remaining 30% contain about one third the heavy-element content of the Sun.

In Section 14-2 we discussed how we determine the age of a star cluster.

Star clusters provide a convenient method of measuring the age of stars. All stars in a cluster were formed at about the same time. Knowing this, astronomers can determine a cluster's age by analyzing its H-R diagram. **FIGURE 16-27** is the H-R diagram of a cluster old enough that its heaviest stars have left the main sequence. The position of the main sequence turnoff point permits a calculation of the age of the cluster.

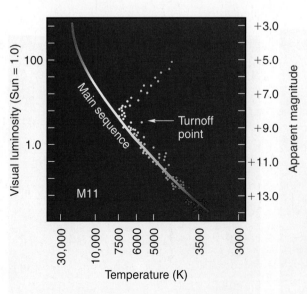

FIGURE 16-27 The H-R diagram of a galactic cluster shows a definite main sequence turn-off point. From the location of the turn-off point on the main sequence, astronomers can determine the age of the cluster.

The classification of stars in a specific portion of the Galaxy can provide a clue to the age of that portion. For example, there are no white dwarfs in the part of the Galactic disk near the Sun. Because white dwarfs do not form until after stars have gone through the main sequence and red giant stages, this sets an upper limit of about 10 billion years for the age of the Sun's portion of the disk.

Within the disk, O- and B-type stars are found primarily near the Galactic plane. Because these stars have short lifetimes, the fact that they are not found at great distances from the Galaxy's plane tells us that star formation does not occur to any great extent except along that plane.

Stars in the halo of the galaxy are older than those in the disk. While the oldest stars in the portion of the disk where the Sun is located are about 10 billion years old, the *youngest* stars in globular clusters are about 11 billion years old. Globular clusters thus formed before the disk portion of the Galaxy. The oldest globular clusters are about 13 billion years old, and because these are the oldest stars we find, we take this to be the minimum age of the Galaxy. Globular clusters were probably much more numerous in the early days of our Galaxy, but tidal interactions between a cluster and the Galaxy could tear the cluster apart. In 2002, astronomers announced the detection of a stream of stellar debris emanating from a globular cluster that is being torn apart by our Galaxy. Such streams provide a new way to determine how the mass of the dark matter halo of our Galaxy is distributed.

The age of stars in the nuclear bulge is difficult to determine because the bulge is obscured by dust. It can be studied only in the X-ray, infrared, and radio portions of the spectrum; however, observations through a few gaps in the dust have allowed astronomers to sample the outermost bulge population. We now know that the stars of the nuclear bulge are very rich in heavy elements, and they all appear to be red giants of types K and M. This tells us that the nuclear bulge must be very old, for two reasons: first, the stars themselves are old, for K- and M-type stars live long lives on the main sequence before becoming red giants. Second, they were formed from the remains of previous generations of stars.

The existence of a Galactic corona of hot gas has been confirmed based on data by NASA's *Far Ultraviolet Spectroscopic Explorer* (*FUSE*) satellite. Observations from *FUSE* show the presence in the corona of a specific species of highly ionized oxygen atom, which indicates the existence of coronal gas with temperatures close to a quarter million kelvin. The atoms observed by *FUSE* are believed to have been heated by supernova explosions to very high temperatures and pushed outward from the Galactic plane into the halo. The clouds of ionized oxygen appear in almost all directions and extend as much as 5 to 10 thousand light-years away from the Galactic plane.

FUSE operated between 1999 and 2007; it was designed to sample light (look at spectra) in the UV part of the spectrum. This light can only be observed above the Earth's atmosphere.

The Galaxy's History

Although astronomers disagree about the specifics, a general picture of the evolution of our Galaxy is emerging. The process is somewhat parallel to the way stars form. The Galaxy began as a tremendous cloud of gas and dust, bigger than the present Galactic halo (FIGURE 16-28). Mutual gravitation between the cloud's parts gradually pulled it together. The center portion was the first to become dense enough for stars to form, and it soon became a hotbed of star formation. Soon thereafter, stars began

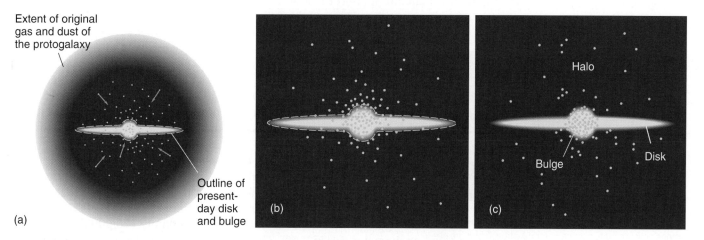

FIGURE 16-28 (a) According to a leading theory of galactic evolution, the Galaxy began as a sphere of gas and dust about 13 billion years ago. (b) As the sphere collapsed toward the center, stars and clusters formed so that the Galaxy appeared like this some 12 to 13 billion years ago. (c) The Galaxy reached its present state after the remaining gas and dust formed into a disk and spiral arms developed.

forming in dense pockets, hundreds of light-years across, which were in orbit around the center. They formed what are now the globular clusters.

The initial giant cloud had some rotation, and as it contracted, it spun faster. Just as occurs during a star's protostar stage, the rotating matter formed into a disk. In the case of a star, almost all the orbiting material becomes caught in the disk, but in the Galaxy, the density of matter in the disk is low enough that globular clusters pass through it without being captured. Finally, density waves formed in the Galaxy's disk, creating the spiral arms where star formation continues today.

Even though this scenario is very sketchy, not all astronomers agree with all of it. For example, in some models, several separate clouds of gas merge to form the Galaxy rather than one. High-velocity atomic hydrogen clouds have been observed since 1963. They have the mass of a small galaxy and are about 15,000 parsecs across. Several hundred of these clouds have been mapped, and their orbits do not follow the approximately circular orbits of most objects in the Galaxy. If such clouds are indeed the building blocks of our Galaxy, then they represent the earliest structures that formed in the universe. They are also one of the few places where the material has not been heavily contaminated by later star formation. The theory that such clouds are the seeds of the Milky Way makes several predictions. First, because these clouds formed long ago, the abundance of heavy elements should be much lower than in our Galaxy. Second, the clouds should be large and of low pressure. Third, if these clouds are a part of the local group of about 30 galaxies (of which Andromeda and our Galaxy are members), they should show little of the fluorescence seen in the relatively nearby atomic hydrogen clouds. Fourth, if these clouds collide with the galaxies in the local group, there should be evidence of a large cloud of hot, ionized gas around our Galaxy and Andromeda.

Recent observations seem to support this theory of gradual formation of a galaxy through accretion of large hydrogen clouds. FIGURE 16-29 is a composite image showing a high-velocity cloud "raining" material into the Galaxy, supplying it with material to make new stars. Such evidence suggests that these clouds play a crucial role in the chemical evolution of our Galaxy; they supply it with metal-poor gas that counteracts the buildup of heavy elements in stars and in the interstellar medium of our Galaxy. This explains why most stars in the Galactic disk have similar concentrations of heavy elements, independent of their age. It also explains how our Galaxy can form about a new star every year without running out of its supply of gas after a tenth of its lifetime.

It is also possible that our Galaxy, like others, grew from smaller galaxies by collisions. This "hierarchical" picture seems to fit better with current theories that try to explain the present structure of the universe as it evolved through time. Data from the *Hipparcos* satellite support the idea of a dwarf galaxy colliding with the Milky Way some 10 billion years ago. As a result of the collision, and over a long period, the dwarf galaxy broke up into streams of stars that keep a common memory of their origin. Evidence for two such streams of stars was found in the *Hipparcos* data; they are among the oldest stars in our vicinity and spend most of their time in the Galactic halo. A map of stars in our Galaxy constructed in 2006 with data from the Sloan Digital Sky Survey (SDSS) revealed a number of giant streams of stars, suggesting that multiple merging events are currently occurring. Also, studying 20,000 stars observed by the SDSS, astronomers discovered that the Galaxy's halo is made of two components; the inner halo follows the motion of the Galactic disk, albeit at a slower rate, while the outer halo rotates in the opposite direction and it has a different chemical composition. In early 2003, astronomers announced the discovery of a ring of stars around our Galaxy, approximately 120,000 light-years in diameter (FIGURE 16-30). The presence of the ring is an indication that at least part of our Galaxy was formed by collisions of many smaller galaxies billions of years ago. If indeed our Galaxy has accreted small nearby satellite galaxies, then many of its stars have originated in dif-

As discussed earlier, the density wave theory is not the only possible explanation for the spiral arms.

Studies of globular clusters in other galaxies suggest that some of these clusters may be the remnant nuclei of dwarf galaxies that were consumed by the host galaxy. Observations have also identified a large number of globular clusters wandering freely in the intergalactic space as a result of gravitational interactions between galaxies.

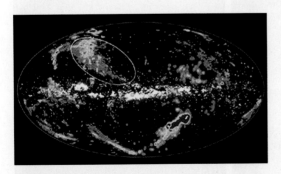

FIGURE 16-29 An all-sky composite image. The false-color orange-yellow clouds contain neutral hydrogen and are observed in 21-cm radiation, while the rendition of our Galaxy is in visible light. The edge-on Galactic plane appears as a white horizontal strip. *HST*'s spectrograph measured the composition and velocity of the encircled cloud.

ferent places. More studies are needed of the distribution of the halo stars (some of the oldest in the Galaxy), their chemical composition, and their motion so that we can better understand when and how the halo formed.

Our knowledge of the Galaxy has only recently progressed to the point where we can formulate meaningful models of its formation. Today's models will have to be developed, and perhaps combined, before we can be confident of their validity. In the next chapter we expand on our discussion of the formation of galaxies.

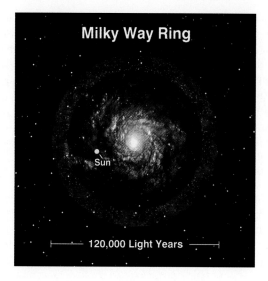

Milky Way Ring

Sun

120,000 Light Years

FIGURE 16-30 An artist's version of a ring of stars around our Galaxy was discovered by the Sloan Digital Sky Survey. The ring contains about half a billion stars, has a thickness of about 10,000 light-years, and encircles us like a giant doughnut.

Conclusion

Our Sun is one of some 200 billion stars in an enormous spiral galaxy that is itself just one of countless galaxies in the observable universe. The nature of our Galaxy was discovered only rather recently because our vision is obscured by gas and dust along the Galaxy's disk. Although measurements of the Galaxy's characteristics are necessarily imprecise, some of its properties can be measured using techniques dating from the time of Isaac Newton, such as the calculation of the Galaxy's mass using Kepler's third law. Detection and measurement of many other features of the Galaxy must rely on more modern methods and equipment, such as radio astronomy, which has opened many doors to understanding the Milky Way.

Analysis of 21-cm radiation leads us to conclude that our Galaxy is a spiral one, like many others we observe. Two principal theories have been proposed to explain the spiral arms of a galaxy: the density wave theory and the self-propagating star formation theory.

The nuclear bulge of our Galaxy contains a high density of stars. The tremendous energy that pours from the nucleus, as well as the rapid motions of objects near it, is consistent with the hypothesis that a massive black hole is at the center of the Galaxy.

Data that are accumulating about the Galaxy permit us to develop models of galactic formation, but these models, like many of astronomy's theories about the Milky Way, are very tentative and await further research by today's and tomorrow's scientists.

STUDY GUIDE

1. Our Sun is located
   A. at the very outer edge of the Galaxy.
   B. far from the center of the Galaxy, but not at the outer edge.
   C. near the center of the Galaxy, but not right at the center.
   D. at the center of the Galaxy.

2. Astronomers of the 18th and 19th centuries thought that the Sun was near the center of the Milky Way Galaxy because they counted the same number of stars in the disk of the Galaxy in every direction. The reason they were not correct is that the Galaxy
   A. is an irregular galaxy with a chaotic shape.
   B. contains dust that obscures its distant regions.
   C. has the shape of a tube with the Sun near one end.
   D. has two kinds of Cepheid variables, so that all distance measurements until recently were incorrect.
   E. has a giant black hole at its center.

3. The center of the Galaxy was first discovered by observing
   A. X-rays that it emits.
   B. radio waves that it emits.
   C. the intense light that comes from it.
   D. the bulge around it.
   E. its yellow glow.

4. When we look at the sky on a dark, clear night, sometimes we can see a faint band of light stretching across it. What we are observing is
   A. the plane of the Milky Way Galaxy.
   B. other galaxies.
   C. the stars of some other galaxy.
   D. the center of the Milky Way Galaxy.
   E. clusters of other galaxies that are drawn to the Milky Way.

5. The Sun is one of about how many stars in the Milky Way Galaxy?
   A. 100 to 200 million.
   B. One billion.
   C. 100 to 400 billion.
   D. One trillion.
   E. [We don't have any idea how many stars are in the Galaxy.]

6. Evidence for the spiral nature of the Galaxy comes primarily from
   A. Cepheid variable data.
   B. Doppler shift data of stars.
   C. 21-cm data.
   D. observations of globular clusters.
   E. [Both A and D above.]

7. We have learned that the Galaxy is a spiral galaxy by
   A. the reflection of light from nebulae.
   B. the Doppler shift of radio waves.
   C. observations of globular clusters.
   D. comparing ages of various clusters.
   E. plotting the stars of various clusters on an H-R diagram.

8. Why is observation of distant parts of the Galaxy limited in the visible region?
   A. Interstellar dust blocks visible light.
   B. The redshift of distant stars makes them difficult to detect.
   C. Opacity of our atmosphere blocks visible light.
   D. Relative motion of the spiral arms makes them difficult to detect.

9. Radio observations can provide information about the Galaxy that we cannot learn from observations in visible light because
   A. radio waves are able to pass through clouds of dust.
   B. stars emit a greater amount of energy in radio wavelengths than in visible wavelengths.
   C. the velocity of radio waves is greater than that of light.
   D. our Galaxy is spiral in nature.

10. William and Caroline Herschel counted stars in various directions in an attempt to determine the Sun's position in the Galaxy. Later, Jacobus Kapteyn tried to improve on their determination by taking into account
    A. the interstellar dust.

B. radio waves from stars.
C. bright nebulae between the stars.
D. redshift data on stars in various directions.
E. the density of stars at various distances.

11. Oort and Lindblad determined the motion of the Sun relative to other stars by analyzing
    A. the interstellar dust.
    B. radio waves from stars.
    C. bright nebulae between the stars.
    D. redshift data on stars in various directions.
    E. the density of stars at various distances.

12. A calculation of the mass of the inner Galaxy can be made based on
    A. Kepler's laws.
    B. observations of nebulae in spiral arms.
    C. star counts.
    D. distances to Cepheid variables.

13. According to the density wave theory of the spiral arms,
    A. the number of stars per volume is about the same between the spiral arms as in them.
    B. the number of stars per volume is much greater between the spiral arms than in them.
    C. the number of stars per volume is much less between the spiral arms than in them.
    D. [The theory predicts nothing about the number of stars per volume of space.]

14. How does the density wave theory of spiral arms explain why they are brighter than the regions between the arms?
    A. There are more stars per volume in the arms than between them.
    B. Brighter stars are in the spiral arms because they follow along with the arms as they move.
    C. Brighter stars are in the spiral arms because they are formed there and are short lived.

15. According to the self-propagating star formation theory of spiral galaxies' arms,
    A. stars revolve around the galactic center in perfect circles.
    B. stars obey Kepler's third law as they move around the galactic center.
    C. stars are about evenly spread around the galaxy.
    D. [Two of the above.]
    E. [None of the above.]

16. A black hole is thought to be at the center of the Galaxy because
    A. no light comes from that area of the Galaxy.
    B. large amounts of energy come from a small source there.
    C. motions of objects near the center indicate a large mass there.
    D. [Both A and C above.]
    E. [Both B and C above.]

17. Serendipity is the ability to
    A. detect small changes in data.
    B. make unexpected discoveries by accident.
    C. do mental calculations quickly and accurately.
    D. visualize situations in three dimensions.
    E. analyze data quickly and accurately.

18. The self-propagating star formation theory of spiral arms holds that
   A. nuclear reactions occur in cores of stars in chain reactions.
   B. the wave moves around the galaxy in chain reactions.
   C. new stars that form the arms are the result of shock waves from supernovae.

19. Use an analogy to explain the nature of a density wave.

20. Sketch the Galaxy both edge on and face on, showing its prominent features and the approximate location of the Sun.

21. Describe the Milky Way as seen from the Earth by the naked eye.

22. Distinguish between galactic clusters, globular clusters, and clusters of galaxies.

23. Name four principal components of the Galaxy and describe the location of each.

24. What two observations give evidence of spiral arms in the Galaxy?

25. What leads us to conclude that globular clusters are very old?

1. If stars revolve around the Galactic center independently of the spiral arms, it would seem that there should be as many stars per unit volume between the arms as inside them. Yet the text states that there is *almost* the same density of stars between the arms as inside them. Why would there be a difference at all?

2. Explain how Kepler's third law is used to calculate the mass of the Galaxy. Discuss the limitations of this method.

3. Explain why the hypothesized Galactic rotation curve of Figure 16-13a (solid blue line) extends down to very low speeds near the zero point on the bottom axis, whereas the dashed-line curve of Figure 16-13a does not.

4. If a radio telescope cannot reveal directly the distance to a source of 21-cm radiation, how can this radiation be used to reveal spiral arms in the Galaxy?

5. Write a report on various models of the formation of the Galaxy based on the article entitled, "How the Milky Way Formed," in *Scientific American*, January 1993, pp. 72–78.

6. What is the evidence that stars of the nuclear bulge formed before stars in other parts of the Galaxy?

7. The Milky Way looks more prominent in July than in December. Why?

8. How did William and Caroline Herschel seek to determine the Sun's position in the Galaxy? What measurements did Kapteyn make to answer the same question?

9. Explain why 21-cm radiation reveals the parts of the Galaxy in which O- and B-type stars are located.

10. We have never observed a globular cluster pass through the disk of the Galaxy. How do we know that they do? Approximately how long would it take for such a cluster to pass through the galactic disk?

11. Consider the H-R diagrams of two globular clusters (A and B), similar to the one shown in Figure 16-27. Assume that the turnoff point for cluster A is lower on the main sequence than that for cluster B. Which of the two clusters has a higher metallicity and why?

1. Show how you can calculate the period of the Sun's motion around the Galaxy, given that the Sun's distance to the center is 26,000 light-years and its speed is 220 kilometers/second.

2. Suppose that a portion of another galaxy is observed to move with a speed of 250 kilometers/second and to be at a distance of 40,000 light-years from the center of that galaxy. What is the period of revolution of that portion of the galaxy?

3. Use the data from Question 2 to calculate the mass of that galaxy that is inside the radius of the portion described.

4. Suppose we wish to construct a physical model of the Galaxy. If the Sun is represented by a small grain of sand 0.1 millimeters across, what will be the diameter of the model? (Hint: See the Activity below.)

The Scale of the Galaxy

Suppose that you wish to construct a scale drawing of our region of the universe. The Sun has an actual diameter of about 1,500,000 kilometers, and you represent it by a dot the size of a period on this page, about 0.5 millimeters across. The average distance between stars in our region of the Galaxy is about 5 light-years; therefore, you must calculate what the corresponding distance would be on your drawing, given that one light-year is equal to 9.5×10^{12} kilometers.

First, set up the ratio of the distance on your scale to the actual distance:

$$\frac{\text{distance on scale}}{\text{actual distance}} = \frac{0.5 \text{ mm}}{1.5 \times 10^6 \text{ km}}.$$

Next, calculate the distance between stars in kilometers:

$$5 \text{ ly} \times \frac{9.5 \times 10^{12} \text{ km}}{1 \text{ ly}} = 4.8 \times 10^{13} \text{ km}.$$

Now, use that value as the actual distance and solve for the distance in the drawing that represents the separation of stars:

$$\frac{\text{distance on scale}}{4.8 \times 10^{13} \text{ km}} = \frac{0.5 \text{ mm}}{1.5 \times 10^{6} \text{ km}};$$

$$\text{distance on scale} = 1.6 \times 10^{7} \text{ mm}.$$

Finally, change this to more appropriate units:

$$1.6 \times 10^{7} \text{ mm} \times \frac{1 \text{ m}}{10^{3} \text{ mm}} = 1.6 \times 10^{4} \text{ m}.$$

This is 16 kilometers! It means that on your drawing, stars (end-of-sentence periods) must be 16 kilometers, or 10 miles, apart.

The Galaxy is about 160,000 light-years in diameter, or 32,000 times the average distance between stars. This means that on your scale, the Galaxy would be 512,000 kilometers across, which is larger than the Earth–Moon distance! This is beyond imagining, and distances to other galaxies have not yet been calculated; therefore, you will have to try another scale.

Suppose you take the average distance between stars on your revised scale to be 5 centimeters. Although the stars would be too small to see on an actual drawing to this scale, you might cheat and make each star a very tiny speck.

On your drawing, 5 centimeters would correspond to 5 light-years. Now calculate the size of the Galaxy, the distance to the Andromeda galaxy (which is actually about 2 million light-years away) and the distance to the most distant objects we can see (which, as discussed in the next chapter, are about 13 billion light-years away).

1. "Meet the Milky Way," by V. Trimble and S. Parker, in *Sky & Telescope* (January, 1995).

2. "Searching for Dark Matter," by M. Mateo, in *Sky & Telescope* (January, 1994).

3. "Destination: Galactic Center," by R. Jayawardhana, in *Sky & Telescope* (June, 1995).

4. "How the Milky Way Formed," by S. van den Bergh and J. E. Hesser, in *Scientific American* (January, 1993).

5. "Our Growing, Breathing Galaxy," by B. P. Wakker and P. Richter, in *Scientific American* (January 2004).

6. "Redesigning the Milky Way," by W. H. Waller, in *Sky & Telescope* (September 2004).

7. "Tracing the Milky Way's History," by C. Chiappini, in *Sky & Telescope* (October 2004).

8. "Ripples in a Galactic Pond," by F. Combes, in *Scientific American* (October 2005).

9. "Messages from the Dark Side?" by G. Schilling, in *Sky & Telescope* (December 2006).

10. "The Ghosts of Galaxies Past," by R. Ibata and B. Gibson, in *Scientific American* (April 2007).

Quest Ahead to Starlinks
http://physicalscience.jbpub.com/starlinks

Starlinks is this book's online learning center. It features **eLearning**, which contains chapter quizzes and other tools designed to help you study for your class. You can also find **online exercises**, view numerous relevant **animations**, follow a guide to **useful astronomy sites** on the web, or even check the latest **astronomy news** updates.

A Diversity of Galaxies

SPIRAL NEBULAE WERE DISCOVERED IN THE LATE 1700s, but their nature remained a mystery for nearly 150 years. Early in the 20th century, Adriaan van Maanen, an astronomer at Mount Wilson Observatory, was convinced that they were clouds of gas rather than galaxies similar to the Milky Way. He determined to settle the question based on evidence, and between 1916 and 1923 he made careful measurements of pairs of photographs taken up to 12 years apart. He found that the nebulae rotated with periods of tens of thousands of years. Such a great rotation rate for a galaxy would mean that stars on its outer edge would have to be moving faster than the speed of light. His data thus proved that the nebulae were not galaxies.

Other astronomers obtained conflicting data, but van Maanen was using the largest telescope in existence, and he found a great rotation speed for each of the seven nebulae for which he had the necessary pairs of photographs.

In 1924, Edwin Hubble (also working at Mount Wilson Observatory) proved conclusively, by the use of Cepheid variables, that the nebulae were separate galaxies.

Why did van Maanen obtain incorrect results? In *Science and Objectivity* (Iowa State University Press, 1988), Norriss Hetherington argues that although van Maanen was attempting to be impartial, his preconceived ideas distorted his findings. Hetherington cites other examples of unintentional nonobjectivity in astronomy (including Lowell's

Several hundred galaxies are visible in this *Hubble Space Telescope (HST)* photo. The image covers a speck of sky 1/30th the diameter of the full Moon.

All cross references to chapters, sections, figures, and tables pertain to the main text, *In Quest of the Universe, Sixth Edition. In Quest of the Solar System* contains Chapters 1–11 and 19 of the main text. *In Quest of the Stars and Galaxies* contains Chapters 1–5 and 11–19 of the main text.

Martian canals) and warns that scientists must be constantly on guard against allowing their anticipated results to cloud the interpretation of their data.

Viewed with the naked eye in a good dark sky on a clear night, the Andromeda galaxy (see Figure 16-4) appears as a fuzzy little patch of light. After the invention of the telescope, many more such nebulous objects were found, but their nature remained a mystery. In 1924, Edwin Hubble found Cepheid variables in three of what had been called *spiral nebulae*, and he thereby showed that they were, in fact, spiral *galaxies*.

The importance of Hubble's discovery was tremendous, for it greatly expanded our appreciation of the size of the universe. Like the realization centuries earlier that our planet is just one of many planets, the realization that our Galaxy is just one of a multitude of galaxies was a giant leap in understanding our place in the universe. We know now that the Andromeda galaxy is a spiral galaxy much like our own and that it is about 2.5 million light-years (Mly) away. When you view the Andromeda galaxy, consider the fact that the light reaching your eye has been traveling through space for almost 3 million years.

In this chapter, we first describe how astronomers classify and measure properties of "normal" galaxies. Then we consider more unusual objects, including active galaxies and quasars.

17-1 The Hubble Classification

The serious study of galaxies began with their observation and classification. This is a common starting place in science. When we come on an assortment of objects about which we know little or nothing, we begin by grouping them into classes according to their observable properties.

Hubble provided the classification scheme that is the basis for the one still in use today. He divided galaxies into three types: ***spiral***, ***elliptical***, and ***irregular***, with subdivisions within each category. More recently, astronomers have discovered objects in intergalactic space that fit none of Hubble's categories, not even irregular. In this chapter we first discuss each category in turn and then look at how properties of galaxies are measured. Then we examine the objects that are too peculiar to be called "irregular."

Spiral Galaxies

Hubble divided spiral galaxies into two groups: ordinary spiral galaxies (**FIGURE 17-1**) and spiral galaxies with bars—called ***barred spirals*** (**FIGURE 17-2**). Ordinary spirals are designated by the capital letter S and barred spirals by SB. Each type is then further subdivided into categories a, b, and c, depending on how tightly the spiral arms are wound around the nucleus. Galaxies with the most tightly wound arms (type a) also have the most prominent nuclear bulges. When a spiral galaxy is seen edge on, it often displays the lane of dust and gas clouds that we see in our own Galaxy. Up to two thirds of all spiral galaxies contain bars. The bar systems provide an efficient mechanism for fueling star births at the centers of SB galaxies. A few galaxies seem to have the nuclear bulge and disk of a spiral galaxy, but no arms. Hubble called these S0 (or ***lenticular***) galaxies (**FIGURE 17-3**).

Differences between the three categories (a, b, and c) might be related to how much gas and dust they contain. Sc and SBc galaxies contain much more gas and dust than Sa and SBa galaxies, resulting in a larger proportion of their mass being involved in star formation. This may explain why Sc and SBc galaxies have larger disks and smaller bulges than Sa and SBa galaxies.

Most spiral galaxies are from 50,000 to 2,000,000 light-years across and contain from 10^9 to 10^{12} stars. (We see later how such measurements are made.) For comparison, recall that our Galaxy is about 160,000 light-years across and contains at least 2×10^{11} stars.

◆ Back through the ages the Andromeda *nebula* must have been a great source of wonder to curious people.

elliptical galaxy One of a class of galaxies that have smooth spheroidal shapes.

irregular galaxy A galaxy of irregular shape that cannot be classified as spiral or elliptical.

barred spiral galaxy A spiral galaxy in which the spiral arms come from the ends of a bar through the nucleus, rather than from the nucleus itself.

◆ Galaxies (and other objects) are named by their listings in the Messier catalog (prefix M) or in the New General Catalog (prefix NGC).

lenticular galaxy An S0 galaxy, having a flat disk like a spiral galaxy (with little spiral structure) and a large bulge like an elliptical galaxy.

FIGURE 17-1 (a) The Sombrero galaxy (M104, in Virgo) is a type Sa galaxy, so classified because of its large nuclear bulge. (b) M77 is a type Sb spiral galaxy in the constellation Cetus. (c) M51, the Whirlpool "nebula," comprises the large spiral galaxy NGC5194 (a type Sc galaxy) and its smaller, barred, and more amorphous companion NGC5195. The companion takes on a reddish tinge because it is behind the dust-filled arm connecting it to NGC5194.

Elliptical Galaxies

FIGURE 17-4 shows some elliptical galaxies and makes clear why they were given that name. Various elliptical galaxies show different eccentricities in their elliptical shape, depending in part on their orientation to Earth. (For example, a football appears round if viewed end on.) The actual eccentricity of an elliptical galaxy is difficult to determine because its orientation is unknown. Therefore, elliptical galaxies are classified based on how they appear, from round (E0) to very elongated (E7).

FIGURE 17-2 (a) The SBa barred spiral galaxy NGC4314 is in the constellation Coma Berenices. (b) M95 is an SBb barred spiral galaxy, possibly ringed, part of a cluster of galaxies at a distance of about 38 million light-years. (c) NGC1365 is a SBc barred spiral galaxy, part of a cluster of galaxies about 60 million light-years from our Galaxy.

FIGURE 17-3 The S0 galaxy NGC5866, 13.5 Mpc away in the constellation Draco, is seen near edge on in this *HST* image. A warped dust lane, a reddish bulge surrounding the nucleus, a blue disk of stars along the dust lane, a transparent outer halo, and many globular clusters are shown.

Most of the galaxies in existence are ellipticals, but most galaxies listed in catalogs are spirals. The reason for this is that although a few giant elliptical galaxies are larger than any spiral galaxy (having 100 times more stars than our Galaxy), most are small and dim (with one millionth as many stars as our Galaxy). Dwarf elliptical galaxies are visible from Earth only if they are relatively nearby.

Irregular Galaxies

Hubble found that some galaxies did not fit into either of the above categories, nor did they exhibit other common characteristics. **FIGURE 17-5** indicates why these were called "irregular" galaxies. Fewer than 20% of all galaxies fall in the category of irregulars, and they are all small, normally having fewer than 25% of the number of stars in our Galaxy.

The Magellanic Clouds are usually classified as irregular galaxies, although some astronomers think that the Large Magellanic Cloud is a barred spiral that has been disrupted by its proximity to the Milky Way and perhaps by a past collision with the Small Magellanic Cloud. Collisions between galaxies should not be unusual because on the average they are separated by distances only about 20 times their diameter. Stars within a galaxy, on the other hand, are separated by millions of times their diameter and therefore collide infrequently; however, as a result of their great distances from Earth, galaxies exhibit no proper motion (motion across our line of sight), and thus, we have to deduce past collisions from their present appearance. **FIGURE 17-6a** shows two galaxies, called "the Antennae," that are in the

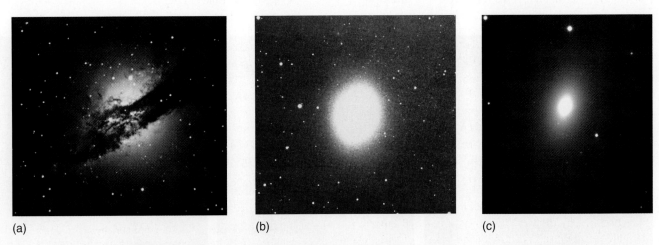

(a)   (b)   (c)

FIGURE 17-4 (a) NGC5128 (or Centaurus A) is a type E0 peculiar elliptical galaxy. It is one of the most luminous and massive galaxies known. The dark band across the galactic disk is debris from a smaller dusty galaxy that is being absorbed by Centaurus A. (b) M32 is a type E2 dwarf elliptical galaxy and is one of the companions to the giant Andromeda Galaxy M31. (c) M59 is a type E5 elliptical galaxy in the constellation Virgo, one of the many members of the Virgo Cluster.

Edwin Hubble

Edwin Hubble (**FIGURE B17-1**) was born on November 20, 1889, in Marshfield, Missouri. In high school, he showed promise in both academics and athletics, and when he was 16, he entered the University of Chicago. After graduation he received a Rhodes scholarship and enrolled at the University of Oxford, where he studied law. By 1914, Hubble decided instead to do graduate work in astronomy at Yerkes Observatory in Wisconsin.

FIGURE B17-1 Edwin Hubble (1889–1953).

Early in his graduate career, Hubble attended a presentation by Vesto M. Slipher. A hot topic in astronomy at the time was whether spiral nebulae are part of our Galaxy or are galaxies in their own right. Slipher's presentation included observational data that supported the latter hypothesis, but not everyone was convinced. Possibly inspired by that talk, Hubble began photographing nebulae using the Yerkes' 24-inch reflecting telescope. His Ph.D. thesis, entitled "Photographic Investigations of Faint Nebulae," grew out of this earlier work. Although this work was not scientifically rigorous, it led Hubble closer to the conclusion that the nebulae were extragalactic objects.

Before Hubble could reach any final conclusions, the United States entered World War I. Three days after receiving his Ph.D., Hubble enlisted, but the Armistice was declared by the time his division reached Europe. Hubble returned to his study of the nature of "nongalactic nebulae," and began observing NGC6822 using the 100-inch reflector at Mount Wilson Observatory. By 1923, he had found 12 variable stars within it, as well as several smaller nebulae. In 1924, he began observing M31 (the Andromeda "nebula"), in which he distinctly resolved six variable stars. In the course of this work, Hubble observed several Cepheid variables in M31 and M33. By comparing each star's luminosity with his observations of the star's apparent brightness, Hubble was able to deduce that M31 and M33 are about 930,000 light-years distant. (Hubble's measurement was off by more than a factor of two because he did not know that there are two types of Cepheids.) This distance placed them far outside the known boundaries of the Milky Way and, thus, Hubble concluded that they are galaxies outside our own and that other spiral nebulae may be separate, distinct galaxies. Hubble also attempted to classify the galaxies that he observed, which led to his so-called *tuning fork diagram*, which is still widely used today.

In the mid-1920s, Hubble started investigating the expanding universe hypothesis. As he observed galaxies such as M31 and M33, his colleague Milton Humason measured their radial velocities. By combining these radial velocities with the measured distances to the observed galaxies, they deduced what is called the Hubble law of redshifts: $v = H_0 d$. The law was published in a 1929 paper on the expansion of the universe. It sent shock waves through the astronomical community. The Hubble law indicates that the universe is expanding because the velocities of galaxies increase at increasing distances from any chosen point. From 1931 to 1936, Hubble concentrated on extending this law to increasingly greater distances.

Hubble next began researching how the density of galaxies changes with distance, and writing books on astronomical subjects for the general public. In the last few years of his life, he pushed for the construction of a 200-inch reflecting telescope at Mount Palomar in California. During World War II, he joined the staff of the U.S. Army's Ballistics Research Laboratory, where he used his early astronomical training to lead a group calculating artillery-shell trajectories. After the war, he continued his work at the now-completed Palomar Observatory until he died in 1953. His memory lives on with the many researchers who have built on his earlier work, as well as the work of the space telescope named for him.

* Much of the material for this box comes from Osterbrock, Gwinn, and Brashear, "Edwin Hubble and the Expanding Universe," *Scientific American* (July, 1993).

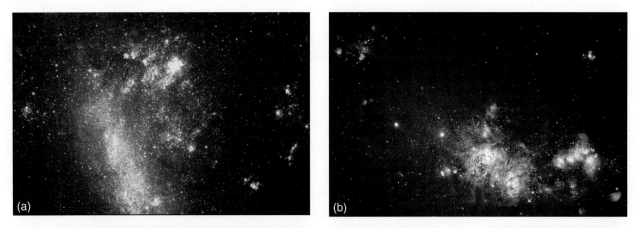

FIGURE 17-5 (a) The Large Magellanic Cloud, the nearest galaxy to the Milky Way at only 160,000 light-years away, is an irregular galaxy. (b) This is another irregular galaxy, NGC1313, which is visible from the Southern Hemisphere.

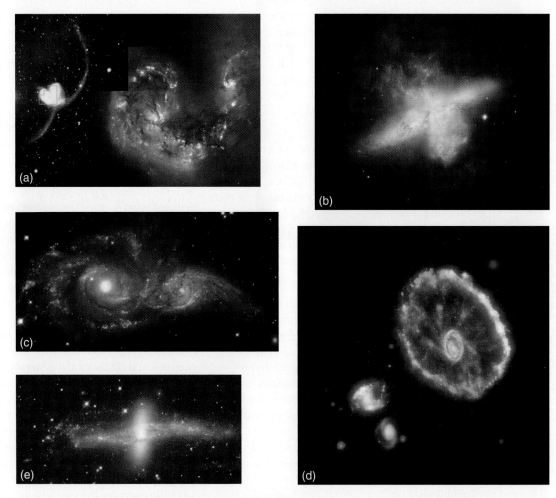

FIGURE 17-6 (a) NGC4038/4039 is known as "the Antennae." (Left) A ground-based view of the two galaxies. (Right) This *HST* image shows the orange cores of the galaxies, including filaments of dark dust. The collision has triggered star formation (bright blue star clusters along the spiral patterns). (b) M82 shows a "burst" of star formation (up to 50 times more than other galaxies). A composite image from *Chandra* (in blue), *Spitzer* (in red), and *HST* (hydrogen emission in orange, and the bluest visible light in yellow-green). The red dust clouds are believed to contain PAHs. (c) Tidal distortion as a result of a grazing encounter between two elliptical galaxies. This *HST* image (in the visible) reveals dust lanes in the spiral arms of NGC2207, silhouetted against IC2163. (d) A composite image of the Cartwheel galaxy from *GALEX* (far UV, in blue), *HST* (visible, in green), *Spitzer* (infrared, in red), and *Chandra* (X-ray, in purple). (e) The "polar-ring" galaxy (NGC4650A). The rotating inner disk of old red stars may be the remnant of one of the galaxies in the collision. The gas from the smaller galaxy probably got stripped off and captured by the larger galaxy, forming a new ring orbiting around the inner galaxy at right angles to the old disk.

process of colliding. Computer simulations show that gravitational forces between the galaxies produce such odd formations. When galaxies merge, especially in the case of a collision between a large and a small galaxy, the larger one swallows up the smaller, cannibalizing it. In the Antennae, observations by the *Infrared Space Observatory* have shown that shock waves generated by the collision excite the gas from which new stars will form. This process manifests itself in the infrared radiation emitted by molecular hydrogen. *Chandra* has also discovered rich deposits of neon, magnesium, and silicon, suggesting a high rate of supernova explosions caused by the collision.

Computer simulations also show that colliding galaxies actually pass through one another with few collisions between individual stars, as distances between stars are very large; however, the large sizes of the interstellar dust and gas clouds in the colliding galaxies make them more likely targets. The collisions between clouds of interstellar gas result in greater gas density and therefore increased star formation. Such "bursts" of star formation may even occur as a result of gravitational interactions among neighboring galaxies, as is the case for M82 (Figure 17-6b). In addition, gravitational forces drastically alter the shapes of the two colliding galaxies. Figure 17-6c shows a grazing encounter in which the strong tidal forces from the larger galaxy have distorted the shape of the smaller galaxy.

Figure 17-6d shows ripples of star formation that resulted when a smaller galaxy went through the heart of the Cartwheel galaxy about 100 million years ago. Figure 17-6e is an *HST* image of the "polar-ring" galaxy; the rings may be remnants of a collision between two galaxies at least a billion years ago.

Hubble's Tuning Fork Diagram

Edwin Hubble linked the three types of galaxies—ellipticals, ordinary spirals, and barred spirals—in what is called his ***tuning fork diagram*** (FIGURE 17-7). In his plan, S0 galaxies form the connecting link because they have characteristics of both elliptical galaxies and spiral galaxies. Astronomers once thought that the diagram represents an evolutionary sequence, with galaxies changing from one form to another; however, old stars are found in all three types, indicating that one type does not evolve to another, at least not as simply as indicated by the diagram. TABLE 17-1 summarizes the properties of the various galactic types.

tuning fork diagram A diagram developed by Edwin Hubble to relate the various types of galaxies.

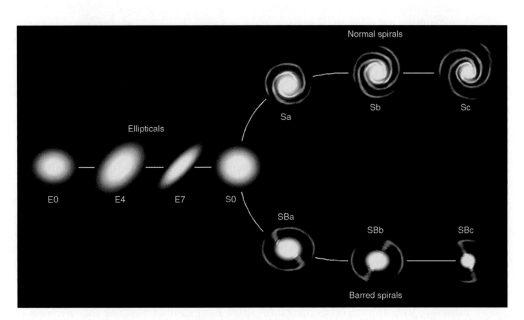

FIGURE 17-7 The tuning fork diagram developed by Edwin Hubble provides a convenient way to categorize galaxies; however, it does not represent an evolutionary sequence.

TABLE 17-1		
Types of Galaxies		
Type	Designation	Description
Elliptical	E0-E7	Galaxies that appear circular (E0) to very elongated (E7). Mostly Population II and old Population I stars.
Spiral	Sa, Sb, Sc	Disk-like galaxies with a nuclear bulge and spiral arms. Sa type have large nuclei and tightly wound arms. Sc type have small nuclei and open arms. Mostly young, Population I stars in the spiral arms, and a mix of Population II and old Population I stars in the disk and nucleus.
Barred Spiral	SBa, SBb, SBc	Spiral galaxies with elongated nuclei. Subtypes are determined as with spiral galaxies.
Lenticular	S0	Disk-like galaxies, but with no spiral structure.
Irregular	Irr	Galaxies that do not fit into any of the other types. Mostly Population I stars.

17-2 Measuring Galaxies

The most important properties of a galaxy we can measure are its distance, mass, and motion. As we have seen on several occasions, if we know the distance to an extended object, we can calculate its size and absolute luminosity from its angular size and apparent magnitude, respectively.

Distances Measured by Various Indicators

Distances to galaxies are measured using several different distance indicators. We have already discussed two of these, the Cepheid and RR Lyrae variables. The calculation of the distances to the Magellanic Clouds was the first use of Cepheid variables as distance indicators on an extragalactic (outside the Milky Way) object. Hubble also used two Cepheid variables in the Andromeda galaxy to show that the so-called spiral nebulae were indeed galaxies.

Although Cepheid variables are very bright stars, we can distinguish them only in relatively nearby galaxies, out to perhaps 200 million light-years. If we wish to measure the distance to a galaxy in which we are unable to see Cepheids, we must find other distance indicators. By analyzing giants, supergiants, and novae in our Galaxy and nearby galaxies, astronomers have learned that all have approximately the same range of luminosity. In addition, large globular clusters and supernovae are of consistent brightness from one galaxy to another. This allows us to use these bright stars and clusters as distance indicators in more distant galaxies.

The logic here is interesting. By determining the luminosities of nearby Cepheid variables, astronomers determined the period-luminosity relationship for Cepheids. Then they assumed that Cepheids in other galaxies are basically the same as the ones in our Galaxy and, thus, they used that same period-luminosity relationship in reverse to calculate the absolute luminosity and thereby the distances to Cepheids in the nearer galaxies. Knowing these distances, they learned that the brightest stars and globular clusters in galaxies of each type have approximately the same luminosity. That is, the brightest 50 stars in one spiral galaxy have approximately the same average luminosity as the brightest 50 in every other spiral galaxy. Likewise, this is also the case for globular clusters. Now astronomers can use these stars and clusters as distance indicators to learn the distances to galaxies in which they can see the individual objects. FIGURE 17-8 summarizes this chain, which allows us to determine distances as far as about 1000 million light-years.

At distances at which we can no longer see individual objects within a galaxy, the distance indicator becomes the galaxy itself. For example, suppose that we see a very distant spiral galaxy. Because we know the range of luminosities spiral galaxies have, we can make some judgment as to that galaxy's distance by assuming that it is an average spiral galaxy. If we assume that it is among the dimmest galaxies of its type, we find its

An object used to measure distance because its absolute magnitude is known is often called a standard candle.

Here again we assume—with successful results—that the same laws of science that apply here on Earth also apply in distant galaxies.

nearest probable distance. On the other hand, if we assume that it is among the brightest, we find its farthest probable distance. In this way, we find the range of distances within which we can be fairly certain the galaxy falls.

When applied to an individual galaxy, the whole-galaxy method of assessing distance is very imprecise. Fortunately, the method need not be restricted to individual galaxies. As we discuss later, galaxies exist in clusters, which contain from a few galaxies to thousands of galaxies per cluster. If we consider a cluster that contains a great number of spiral galaxies, we can logically assume that the brightest spiral galaxy in that cluster has about the same luminosity as the brightest spiral in another cluster. Thus, we use the brightest galaxies as a distance indicator to the cluster.

One measurement builds on another in a series of steps. As a result, if there is an error in a beginning step, the error will propagate up through the chain of steps. For example, if somehow there is an error in our understanding of the period-luminosity relationship for Cepheids, the stated distances to the farthest galaxies will have to be adjusted. This is the reason that astronomers constantly check their methods and assumptions as new data arrive; they also make cross-checks by comparing results obtained through different methods and tools and use statistical analysis wherever appropriate.

The previous analysis makes an important assumption: Galaxies in our neighborhood of the universe are basically the same as those farther away. This may seem reasonable, but remember that we are seeing distant galaxies not as they are today but as they were in the past. Light coming from a galaxy 100 million light-years away has been traveling 100 million years, and we cannot rule out the possibility that galaxies have changed in that time.

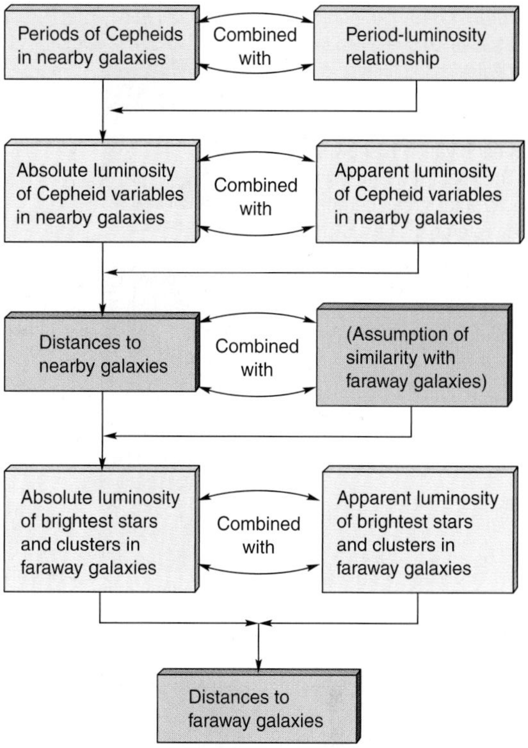

FIGURE 17-8 Starting with the period-luminosity relationship of Cepheids, astronomers are able to follow a chain of reasoning and observation that allows them to determine the distances to galaxies too far away for their Cepheids to be visible.

The Hubble Law

In 1912, Vesto M. Slipher, an astronomer working in Percival Lowell's observatory in Flagstaff, Arizona, was assigned the task of examining the spectrum of some of the spiral nebulae to learn about their chemical composition. A then-current hypothesis regarding the nature of these objects was that they were planetary systems in formation, and Lowell was seeking to discover life on other planets. Instead, Slipher found something else that had profound implications, namely a redshift in the spectra of most of the nebulae he examined. If the redshift were due to the Doppler effect, it meant that most of the other nebulae were moving away from us—as fast as 1800 km/s. There seemed to be no reasonable explanation for this strange finding.

William de Sitter, a Dutch astronomer, interpreted the redshifted spectra of the spiral "nebulae" as due to the Doppler effect and included this interpretation in his 1917 model of an expanding universe based on the general theory of relativity. In his 1920 debate with Shapley, Curtis used the relationship between redshifted spectra and spiral "nebulae" as evidence that they are extragalactic objects. In 1923, the German scientist Carl Wirtz suggested that there is evidence in Slipher's data that the recessional velocity of a spiral "nebula" is proportional to its distance.

In 1924, Edwin Hubble showed that the nebulae are, in fact, galaxies. Then he and Milton Humason used the 100-inch telescope on Mount Wilson to photograph the spectra of these objects. Their work not only confirmed Slipher's findings, but showed that there is a pattern in the speeds with which galaxies are receding from us. **FIGURE 17-9** is a diagram by Hubble and Humason on which they plot-

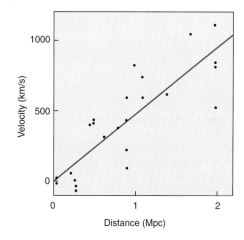

FIGURE 17-9 This graph, based on the data collected by Hubble and Humason in the 1920s, shows a relationship between the recessional velocities of galaxies and their distances.

FIGURE 17-10 The photos at the left in the figure are of galaxies in clusters at various distances from us. At the right, the spectrum of each galaxy is shown. (The spectrum is the white, horizontal streak between the reference lines.) Arrows indicate how far the two darkest absorption lines have shifted in each case. (λ = wavelength.)

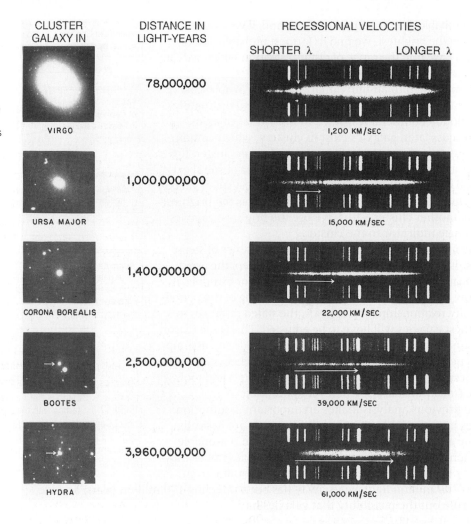

CLUSTER GALAXY IN	DISTANCE IN LIGHT-YEARS	RECESSIONAL VELOCITIES

VIRGO — 78,000,000 — 1,200 KM/SEC

URSA MAJOR — 1,000,000,000 — 15,000 KM/SEC

CORONA BOREALIS — 1,400,000,000 — 22,000 KM/SEC

BOOTES — 2,500,000,000 — 39,000 KM/SEC

HYDRA — 3,960,000,000 — 61,000 KM/SEC

Lowell's telescope on which Slipher worked had only a 24-inch aperture. The 100-inch telescope used by Hubble and Humason therefore had about $4^2 = 16$ times more light gathering power.

ted the distances to a number of galaxies against the galaxies' recessional velocities. It indicates that the more distant a galaxy is, the faster it is moving away from us. An example of this relationship is shown in **FIGURE 17-10**, which includes photographs taken years after Hubble's work, of five galaxies and their spectra. The distances to the galaxies were determined by methods described earlier in this section. Compare the location of the two prominent absorption lines in each of the spectra, noting that the farther a galaxy is, the farther to the right (toward longer wavelengths) is the pair of lines; therefore, assuming that the redshift is due to the Doppler effect, the data show that the farther a galaxy is, the higher its redshift and thus the faster it moves away from us.

Hubble's data, shown in Figure 17-9, were taken in the 1920s and were based on distance indicators that relied on incorrect Cepheid variable values, so the data had to be adjusted when more was learned about Cepheids; however, as **FIGURE 17-11** indicates, a better understanding of standard candles dramatically confirmed the relationship.

The implications of Hubble's findings for our understanding of the past and future of the universe are very important, for they imply that the universe is expanding. In fact, Hubble's work has become the foundation for today's theories of *cosmology*, which we study in Chapter 18. Even though the redshift that Humason observed is not really due to the Doppler effect, the difference is not important here; what is important is

FIGURE 17-11 The data on Type Ia supernovae by Riess, Press, and Kirshner (1996) provide a dramatic confirmation of the relationship found by Hubble and Humason. Reevaluations of distances have caused the slope of the line to change.

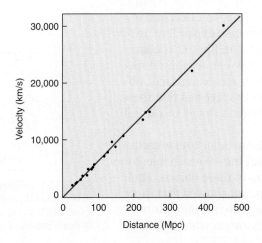

cosmology The study of the nature and evolution of the universe as a whole.

HISTORICAL NOTE

Milton Humason, Mule Driver/Astronomer

Milton Humason (1891–1972) was born in Minnesota. He grew up in southern California, however, where he liked the mountain on which he lived, Mount Wilson, more than he liked school. When he finished the eighth grade, he dropped out of high school to wander the mountain. Later, when George Hale chose that mountain as the site of a large observatory with a 60-inch telescope, Humason was hired as a mule driver to transport equipment up the steep slopes.

Humason learned about the observatory by talking to the construction workers, and when it was completed, he was hired as its first janitor. Young and very curious, he constantly talked to the astronomers about their work, and eventually he was allowed to help them develop photographic plates in the darkroom, and even take photographs with the large telescope.

Humason was aware of the inadequacy of his education and persuaded Harlow Shapley and Seth Nicholson, astronomers at the observatory, to teach him math, including calculus. In a few years he was hired as a full-time night assistant and gave up his janitorial duties. Finally, in 1930, when he was 39 years old, he was appointed to the professional staff of

the observatory, which by then had built a 100-inch telescope. Humason worked closely with Edwin Hubble, using the 100-inch telescope on Mount Wilson, but because Hubble was the leader in their research, some astronomy books fail to even mention the former mule driver who worked with him.

When the 200-inch Hale telescope was completed in 1948 on nearby Mount Palomar, Humason used that instrument in his research. His contributions were so significant that he was awarded an honorary doctor's degree from the University of Lund, Sweden.

It is interesting that Milton Humason almost discovered Pluto. At the same time that Percival Lowell was attempting to find a planet beyond Neptune, W. H. Pickering was doing similar calculations in an attempt to predict its position. He asked Humason to take photographs of certain areas of the sky where he predicted the planet to be, but Humason was unable to locate the planet. Later examination of his plates showed that he had photographed Pluto twice, but in one case it was too close to a star to be visible, and in the other it happened to coincide with a flaw on the photographic plate.

---

that astronomers observe a regular relationship between redshift and distance, a pattern that is valuable to them in determining distances to faraway galaxies, as we will show. Because it is convenient (and common) to think of the redshift as being due to the Doppler effect, that practice will be followed here.

Whenever data can be represented by a straight line on a graph, there is a direct proportionality between the quantities on the two axes. Thus, the relationship between the two quantities, radial velocity, v, and distance, d, can be written as

$$v = H_0 d,$$

where H_0 is the constant of proportionality, called the ***Hubble constant*** in honor of the person who discovered this ***Hubble law***. The value of the constant is simply the slope of the line on the graph, which is defined as the ratio of the change in the quantity plotted on the vertical (y) axis to the change in the quantity plotted on the horizontal (x) axis. In the following example we calculate the Hubble constant by determining the slope of a line that relates the recessional velocity and distance of galaxies.

> **Hubble constant** The proportionality constant in the Hubble law; the ratio of recessional velocities of galaxies to their distances.

> **Hubble law** The relationship that states that a galaxy's recessional velocity is directly proportional to its distance.

EXAMPLE

Calculate the Hubble constant based on the data for the five galaxies in Figure 17-10.

SOLUTION   FIGURE 17-12 shows that the recessional velocities of the five galaxies are linearly related to their distances from us. Calculating the slope of the line requires a comparison of the changes in the two quantities represented on the graph. We thus must choose two points on this line. Observe what happens to cross the point where the velocity is 30,000 km/s and the distance is 2000 million light-years. This will

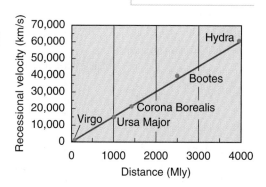

FIGURE 17-12

be one of the chosen points. The line also passes through the point where both quantities are zero; this will be the second point. Now calculate the slope as

$$\text{slope} = \frac{30{,}000 \text{ km/s} - 0}{2000 \text{ Mly} - 0} = 15 \, (\text{km/s})/\text{Mly}.$$

TRY ONE YOURSELF
Show that the slope of the line in the 1931 Hubble-Humason graph (Figure 17-9) corresponds to a value of the Hubble constant equal to 140 (km/s)/Mly, about seven times the modern value. This was due to their misunderstanding of the luminosity of Cepheids.

Astronomers usually use Mpc, megaparsecs (million parsecs), as their distance unit. One parsec is about 3.26 light-years; therefore, 15 (km/s)/Mly is about 50 (km/s)/Mpc.

According to the data represented in Figure 17-12, the Hubble constant thus is about 15 (km/s)/Mly (read as "kilometers per second per million light-years"). It is important to see what Hubble's constant means and what its units represent. This value means that for each million light-years a galaxy is farther from us, its speed is greater by 15 km/s. Refer again to the graph to check that a galaxy at a distance of 1000 million light-years has a speed of 15,000 km/s. In like manner, a galaxy 3000 million light-years away has a speed of 45,000 km/s.

The Hubble constant has been difficult to determine with accuracy. We can calculate the distance to galaxies in nearby clusters fairly accurately using established distance indicators, but the motion of a galaxy within a nearby cluster is significant compared with the motion of the cluster (so that the galaxy may even be approaching us). The velocity of a nearby galaxy thus is due in large part to random motions within its cluster. It therefore does not fit the Hubble law. A distant galaxy, on the other hand, is in a cluster whose motion away from us is much greater than the galaxy's individual random motion, so its redshift follows the Hubble law; however, for these galaxies, determination of the distance to the cluster is much less precise. In other words, in cases where we can measure the distance accurately, the value obtained for velocity is less meaningful, and where we can measure velocity accurately, the distance measurement is imprecise.

Through the years, the value of the Hubble constant obtained from observational data remained between 15 and 30 (km/s)/Mly, but this range has been decreasing as better data became increasingly available. The differences in the values were due to different methods and distance candles used and of course due to errors in the measurements. As we see in Chapter 18, we have entered the era of precision cosmology; the latest observations of the radiation left over from the hot big bang indicate that the value of the Hubble constant is about 21.5 ± 0.4 (km/s)/Mly or 70.1 ± 1.3 (km/s)/Mpc, in agreement with data from Type Ia supernovae, such as the ones shown in Figure 17-11. Note, however, that the Hubble constant does not remain constant as time goes on. As we see in Chapter 18, H_0 is the slope of the line in the graph of recessional velocity versus distance as measured during *this* period of time in the life of the universe. It is important to know the value of the Hubble constant because knowledge of the expansion rate of the universe is fundamental to our understanding of the universe as a whole. We discuss this idea again in the next chapter.

The Hubble Law Used to Measure Distance

As we've discussed, distances to nearby galaxies can be measured using Cepheid variables as distance indicators; however, as we progress to more distant galaxies, Cepheids can no longer be seen, and we must use globular clusters and the brightest stars as distance indicators. At even greater distances, these cannot be seen, and the brightest galaxies serve as distance indicators to clusters of galaxies. Finally, at the farthest distances, even this method cannot be used, and we turn to the Hubble law to indicate a galaxy's distance. We illustrate this idea with an example.

ADVANCING THE MODEL

Observations, Assumptions, and Conclusions

It is always important in science to separate observations from the conclusions that are based on the observations. Figure 17-10 serves as a good example of this rule. The figure shows simply that the dimmer a galaxy is, the greater the redshift of its spectrum. Although the figure only indicates this relationship for five galaxies, the relationship has been found to be true in innumerable cases and has been confirmed by a number of researchers. It is based directly on many *observations* and is considered indisputable.

The middle column of the figure shows the distance of each galaxy from Earth and allows us to conclude that the farther away a galaxy is, the greater its redshift. This conclusion, however, is *not* based directly on observations but on a conclusion about the distances to galaxies. Its validity depends on the accuracy of the distance determinations. One would suppose that the decreasing brightnesses of the galaxies in the figure are due to greater distance, but that is not *proven* by the figure. Actually, each of the galaxies in the figure is part of a cluster of galaxies, and the distance to the cluster has been measured using a variety of distance indicators. We are confident that the distances given are fairly reliable, although no one would pretend that they are exact.

Even though the distances to the galaxies are known with some accuracy, it is important to remember that the relationship between redshift and distance is a *conclusion*, not a direct observation. Now we take the next step: we assume that the redshift is caused by the Doppler effect and, therefore, that it is caused by the galaxies' motions relative to the Earth. The Hubble relationship between recessional velocity and distance depends on this *assumption*. Is there any other possible explanation for the redshift? In Chapter 18 we show that there is, although the explanation is similar to that of the Doppler effect and does not negate the Hubble law.

Scientists must be on guard to remember the difference between what is observed—the evidence—and the conclusions that they reach based on those observations. Between an observation and a conclusion lies one or more assumptions, and the validity of the conclusion is based not only on the accuracy of the observations, but on the validity of the assumptions.

The Precision of Science

It is sometimes said that science is an exact study. If those who make such statements mean that the measurements of science are exact, they are wrong. It is obvious that measurements of distances to galaxies are not exact. What is not so obvious is that *every* measurement in science is to some degree an approximation. No measurement is absolutely exact. What scientists attempt to do is to be aware of how *inexact* their measurements are. For example, in calculating the distance to a galaxy, parallel calculations are made to determine the probable error in the measurement. Measurements may show that a galaxy is 200 million light-years away, with a likely error of 50 million light-years. This means that the galaxy is measured to be between 150 and 250 million light-years away. Calculating the likely error is a common practice in all natural sciences. Scientists thus attempt to be specific about their inexactness.

In addition, scientists must take into account assumptions that are consciously or unconsciously included in the measurements. For example, in the case of determining distances to faraway galaxies, astronomers assume that there is negligible intergalactic material that might diminish the light from those galaxies. Early measurements of the size of the Galaxy were in error because they did not take into account the interstellar dust and gas. A scientist tries to be aware of the assumptions involved in each measurement. There is no claim that a measurement is exact. In fact, there is little room in science for the word "exact."

EXAMPLE

Suppose a very faint source is observed to have a redshifted spectrum that indicates a recessional speed of 120,000 km/s. Assuming that the Hubble law applies to this object, calculate its distance.

SOLUTION   The most recent data suggest that the value of the Hubble constant is 21.5 ± 0.4 (km/s)/Mly or between 21.1 and 21.9 (km/s)/Mly. This gives us a range of distances within which we can be fairly confident the object lies. Using 21.1 (km/s)/Mly:

$$v = H_0 d,$$

$$120{,}000 \text{ km/s} = 21.1 \text{ (km/s)/Mly} \times d,$$

$$d = 5687.2 \text{ Mly}.$$

Now, using 21.9 (km/s)/Mly:

$$120,000 \text{ km/s} = 21.9 \text{ (km/s)/Mly} \times d,$$

$$d = 5479.5 \text{ Mly}.$$

The object thus is between about 5480 and 5690 million light-years away. This is 5.48 to 5.69 billion light-years!

You may be uncomfortable with the uncertainty of the answer, but remember that without the Hubble law and this calculation we would have no idea whatsoever of the distance to the object.

FIGURE 17-13 A more massive galaxy would be expected to be bright and to have a wide 21-cm spectral line. Similarly, a galaxy of low mass is dim and has a narrow line.

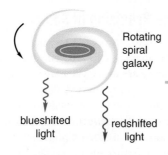
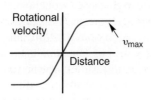
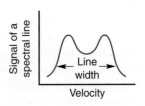
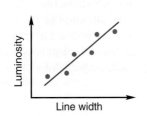
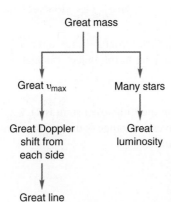

TRY ONE YOURSELF

Suppose an object is observed with a redshift that indicates a speed of 90,000 km/s. What is the range of distances within which we can feel reasonably sure that the object lies?

Other Relations

In 1977, astronomers Brent Tully of the University of Hawaii and J. Richard Fisher of the National Radio Astronomy Observatory discovered that *spiral galaxies* with wider 21-cm lines have greater absolute luminosities. The reason for the relationship is that a galaxy's spectral line is wide if the galaxy is rotating fast, because the Doppler effect causes a redshift from one side of the galaxy and a blueshift from the other side. The faster the rotation is, the wider the line. In addition, more massive galaxies would be expected to rotate faster and to be brighter. **FIGURE 17-13** shows the logic that indicates that a wide spectral line would be associated with a bright galaxy.

Using the ***Tully-Fisher relation***, astronomers can determine the absolute magnitude of a galaxy and use it as a distance indicator. This relationship is being used to check other methods of distance measurement and to improve our knowledge of distances to galaxies.

The ***Faber-Jackson relation***, noted in 1976 by Sandra Faber and Robert Jackson of the University of California at Santa Cruz, is a power-law relation between the absolute magnitude (or luminosity) of an *elliptical galaxy* and the central stellar velocity dispersion. Using spectroscopy, we can measure the Doppler shift of the light emitted by the stars and therefore the spread of their velocities. The relation then allows us to estimate the absolute magnitude of the elliptical galaxy; using the distance modulus and the observed apparent magnitude of the galaxy, we can estimate its distance.

Both the Tully-Fisher and Faber-Jackson relations relate the mass of a (spiral or elliptical) galaxy and the (rotational or random) speeds of its stars. In 2007 astronomers discovered a new relation between a galaxy's mass and a speed indicator that combines both rotational (ordered) and random (disordered) motion. As such, this relation applies to all galaxies, including irregulars. This new relation resulted from a study of about 500 galaxies at distances between 2 and 8 billion light-years. The study suggests that some galactic properties have changed very little over the last 8 billion years.

Tully-Fisher relation A relation that holds that the wider the 21-cm spectral line, the greater the absolute luminosity of a *spiral galaxy*.

Faber-Jackson relation A relation between the luminosity and central stellar velocity dispersion of an *elliptical galaxy*.

FIGURE 17-14 reviews the various methods used to measure distances in astronomy, from radar within the solar system to the Hubble law for the most distant objects.

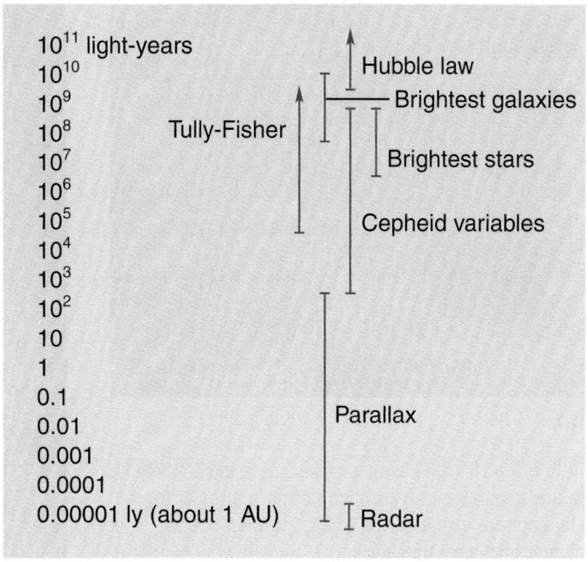

17-3 The Masses of Galaxies

A galaxy's mass can be determined in several ways, all of them limited in precision, primarily because of the tremendous distances involved. One method of measuring the mass of a galaxy is by observing the rotation periods of some parts of it. Naturally, we cannot wait for part of a galaxy to complete a revolution, for this takes millions of years. Instead, we use Doppler shift data to measure the velocity of part of a galaxy. Then, knowing the distance of that part from the center, we can determine the period of revolution of the stars located there. As in the case of our own Galaxy, Kepler's third law then allows us to calculate the mass of the material within the orbit of the part of the galaxy being studied.

Another method of measuring galactic masses is similar to that used to measure stellar masses. There are many cases of a pair of galaxies revolving around one another. Again, we cannot wait the long times necessary to measure the period of revolution, but we can use the Doppler shift to gauge the speeds of the galaxies. The problem with this method is that it is difficult to determine the angle of the plane of revolution to our line of sight (FIGURE 17-15), and knowledge of this angle is necessary for an accurate measurement; however, making measurements for great numbers of such binary galaxies allows us to determine an average mass for a given type of galaxy.

Clusters of Galaxies; Missing Mass

Most galaxies are part of clusters rather than lonely nomads drifting through space. FIGURE 17-16a shows a typical cluster of galaxies. Thousands of such clusters are known, and the largest clusters may contain as many as 10,000 galaxies. Figure 17-16b shows one of the most distant clusters known, about 8 billion light-years away. Thirteen of the galaxies in this cluster are remnants of recent collisions or pairs of colliding galaxies, the largest number of colliding galaxies found in a cluster so far. To determine whether a particular galaxy is part of a given cluster, astronomers must determine the galaxy's distance because a cluster of galaxies is not just a group of galaxies that are near each other on the celestial sphere (like a constellation), but a gravitationally linked assemblage.

FIGURE 17-14 A review of the various methods of measuring distance in astronomy. Approximate limits are indicated for each method.

Be careful with the terminology here. A *galactic cluster* is an open cluster of stars. It is not a *cluster of galaxies*.

All galaxies of a cluster have nearly the same redshift.

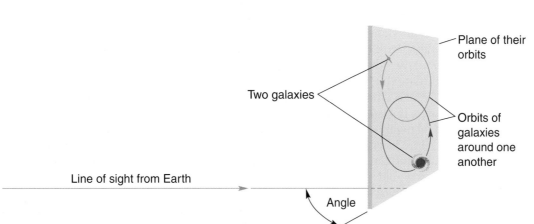

FIGURE 17-15 To use the Doppler shift to measure the speeds of two galaxies that orbit one another, we would have to know the tilt of their plane of revolution to our line of sight.

Line of sight from Earth

Two galaxies

Plane of their orbits

Orbits of galaxies around one another

Angle

(a)

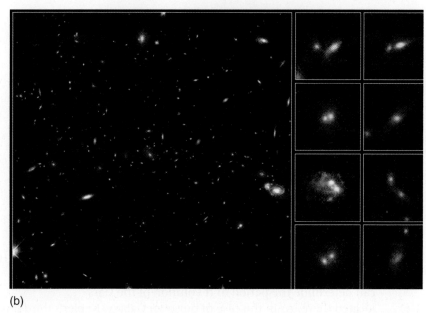

(b)

FIGURE 17-16 (a) Several galaxies are visible in this cluster located in the constellation Fornax in the Southern Hemisphere. (b) This *HST* image shows MS1054-03, a cluster of at least 81 galaxies. As a result of tidal interactions, streams of stars can be seen being pulled out of the galaxies (insets).

Recent (2007) measurements of the three-dimensional velocities of the Magellanic Clouds show that they are greater than previously thought. This suggests that either our Galaxy's gravity is stronger or the Clouds are not gravitationally bound to our Galaxy.

local group (of galaxies)
The cluster of over 30 or so galaxies that includes the Milky Way Galaxy.

supercluster A group of clusters of galaxies.

Our Galaxy is part of a group of over 30 galaxies, including the Magellanic Clouds and the Andromeda galaxy, that form a cluster called the **local group**. The Andromeda galaxy and the Milky Way are by far the largest members of this cluster, which appears to contain only one other spiral galaxy. Our Galaxy is currently on a collision course with the Andromeda galaxy, closing the gap between the two at about 500,000 kilometers an hour. Within 5 billion years, the two galaxies will merge.

If stars are grouped into galaxies and galaxies are grouped into clusters, do clusters group into something bigger? Yes, we call them **superclusters**. Our local supercluster includes the local group and the Virgo cluster as well as other clusters. It has a diameter of perhaps 100 million light-years and contains some 10^{15} (a thousand trillion) solar masses. Between superclusters are great voids with no galaxies. Patterns of clusters, superclusters, and voids are clear in **FIGURE 17-17**, which shows the sky distribution of about 230,000 galaxies, with the Milky Way at the center. It seems that matter in the universe forms a cosmic web in which galaxies are formed along filaments of matter, and clusters are formed at the intersections of these filaments. This structure results in smooth "flows" of galaxies caused by the gravitational pull of great concentrations of galaxies. Observations show that our Galaxy is in a filament at the edge of a huge void, and that all galaxies within 15 million light-years in this filament are moving together. However, the asymmetric distribution of matter around us has the net effect of a push away from the void at a speed of about 270 km/s (170 mi/s). In addition, we have known for decades that our Galaxy is moving toward the Virgo Cluster (55 Mly away) and two huge concentrations of galaxies, one behind the other, at distances of about 200 and 600 Mly.

A third method of measuring the masses of galaxies takes advantage of their clustering. There is only one force that could hold a galaxy within its cluster: the gravitational force. If we observe a galaxy near the outside of a large cluster and assume that it is in a relatively circular orbit, we can use the Doppler effect to measure its speed and therefore its period. This allows us to calculate the total mass of the galaxies within its orbit. Calculating the mass of the cluster based on measurements for only one galaxy would not be meaningful because there would be too many uncertainties, but the same measurement, repeated for several galaxies on the outer portion of the cluster, gives a useful value for the mass of the cluster and therefore an average value for the masses of the galaxies within the cluster.

When measurements are made by the various methods, we find that clusters have much more mass than is accounted for by the visible stars within their galaxies.

Based on Doppler shifts in their spectra, galaxies in large clusters are observed to be moving at speeds greater than the escape speeds from the clusters as calculated from the luminous matter in them. The fact that these galaxies are part of their respective clusters suggests there is enough nonluminous mass in the clusters that keeps them gravitationally bound; this ***missing mass*** can be calculated from the observed speeds and cluster sizes, and it is several times greater than the mass measured by other independent methods.

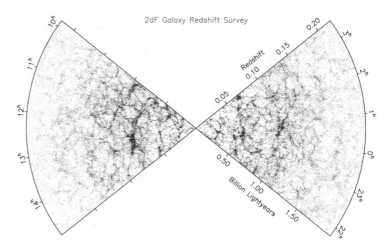

FIGURE 17-17 The cosmic web revealed by the Two Degree Field (2dF) Galaxy Redshift Survey. The universe contains long filamentary chains of galaxies, superclusters, and empty voids up to 200 million light-years across. The map contains about 230,000 galaxies, has a depth of 4 billion light-years, and covers 1/20th of the whole sky. The Milky Way is at the middle of this sky distribution.

Within our own Galaxy, the interstellar gas and dust account for at least 10% to 20% of the total mass, and the rotation curve of the Milky Way (Figure 16-13a) indicates that even considering the interstellar matter, we can account for as little as one tenth of the total mass of the Galaxy; recall that the rotation curve leads us to hypothesize a massive Galactic corona. The rotation curves of other spiral galaxies also suggest that there is a great amount of nonluminous mass at their outer reaches. Observations suggest that only 20% of the "ordinary" matter (such as electrons, protons, and neutrons) in the early universe converged to form the stars and galaxies we see today. There has been no accounting for the remaining 80% of this ordinary matter.

We know that intergalactic dust and gas exist between galaxies in a cluster, but this is also not nearly enough to account for the calculated mass. There are indications that the halo of our own Galaxy contains more material than previously thought, and perhaps galaxies' halos account for much or all of the matter that makes up the missing mass. For example, observations of 63 spiral and dwarf galaxies in the nearby Virgo cluster (chosen to be representative of the galactic population in the local universe) unveiled the presence of large amounts of very cold dust particles surrounding the galaxies. The data, taken by the *Infrared Space Observatory* and announced in 2002, also show that up to half the total energy output in normal galaxies has been converted from visible into infrared light. This implies that the total amount of emitted starlight is much greater than previously suspected. In addition, astronomers announced in 2003 that they discovered a warm fog in the local group of galaxies. The temperature of this diffuse fog is between 100,000 and 10 million K, causing it to shine faintly in X-rays. The data, taken by *Chandra* and the *FUSE* satellite, suggest that this warm fog contains as much as a trillion solar masses. This amount of matter is enough to gravitationally bind together the galaxies in the local group. The presence of hot gas around galaxies and in galactic clusters is shown in **FIGURE 17-18**. Although vast, this material is only a part of larger hot gas filaments that exist between all the galaxies in the universe; data taken by *Chandra* and *XMM-Newton* have provided evidence for such cosmic filaments.

Finally, when astronomers compare the total mass of a system with its total luminosity, the ratio is greater than 1. For stars in the solar neighborhood, the ratio is about 3. For clusters of galaxies, the ratio is about 300. For the universe, the ratio is about 1000 solar masses for every solar luminosity (assuming that the average mass and energy density of the universe is such that space has a flat geometry).

What does this nonluminous matter consist of? The search for the answer to this question is one of the most active areas of research in astronomy today. Several possibilities have been proposed, and the actual situation may be a combination of several kinds of matter.

missing mass The difference between the mass of clusters of galaxies as calculated from Keplerian motions and the amount of visible mass.

ESA's *Infrared Space Observatory* operated from 1995 until 1998.

Nature has played a trick on astronomers. We thought we were studying the universe; now we know we are studying only the small fraction that is luminous.
Astronomer Vera Rubin

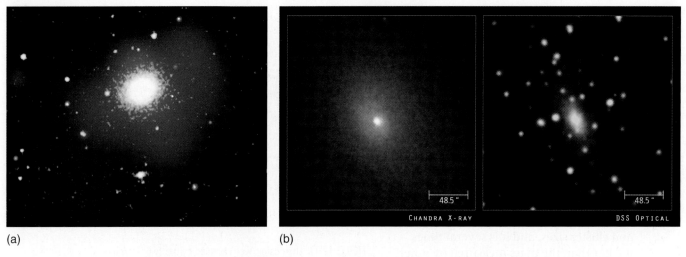

(a)

(b)

CHANDRA X-RAY   DSS OPTICAL

FIGURE 17-18 (a) A composite optical (in white/yellow) and X-ray (in blue) image of NGC4555, an isolated large elliptical galaxy embedded in a 10 million K hot gas. The gravity of the stars in the galaxy is too low to hold the cloud, requiring the presence of a massive halo of dark matter. The image's scale is 8 × 6.5 arcminutes. (b) A cloud of hot gas envelops the cluster Abell 2029, which is composed of thousands of galaxies. The presence of a massive (more than a hundred trillion Suns) dark matter halo is required to gravitationally hold the cloud.

> Because particle physicists often use the term *baryons* to refer to massive particles such as protons and neutrons, ordinary "dark" matter is sometimes called "baryonic dark matter."

1. **Ordinary nonluminous matter**. In previous chapters, we discussed objects such as white dwarfs, which no longer generate their own energy and are no longer easily observable, and brown dwarfs, which have never generated enough energy to become easily observable. These are common examples of ordinary nonluminous matter, as are planets, meteors and comets, and the vast dust and cold molecular hydrogen clouds of interstellar space. All of this matter consists of particles with which we are familiar, primarily atoms containing protons, neutrons, and electrons.

2. **Hot dark matter**. As we discussed in Chapter 11, experiments indicate that neutrinos have a small amount of mass. There are so many neutrinos zipping around the universe that even a tiny mass could contribute to the mass of the universe; however, observations suggest that this contribution is not significant (otherwise the Cosmic Microwave Background radiation, which we discuss in Chapter 18, could not look as it does). Astronomers refer to this mass as "hot" dark matter because the particles are moving at high speeds, comparable to the speed of light. Other particles have also been proposed for this category, including particles that have been hypothesized in several theories of particle physics but that have not been detected.

3. **Black holes**. Astronomers speculate that a great number of black holes of all sizes could contribute to the dark matter.

4. **Cold dark matter**. This is an exotic form of matter that is not made up of the protons, neutrons, and electrons of ordinary matter. Particles that might be moving through space at relatively slow speeds make up "cold" dark matter. Such particles have been given various names, such as photinos, axions, or neutralinos, but have not been detected. We surmise their existence because estimates of ordinary nonluminous matter and hot dark matter do not make up the amounts of mass we can detect by gravitational effects but cannot see. In addition, various theories of particle physics and of galaxy formation in the early universe indicate that this kind of matter exists.

> **dark matter** Matter that can be detected only by its gravitational interactions; it appears to be quite abundant throughout the universe.

Even though we do not yet know the composition of cold ***dark matter***, we do know that it can be detected by its gravitational interactions with ordinary matter, thus tracing its location through space. The evidence that dark matter is distributed in galaxies and clusters of galaxies in a way similar to visible matter comes from the

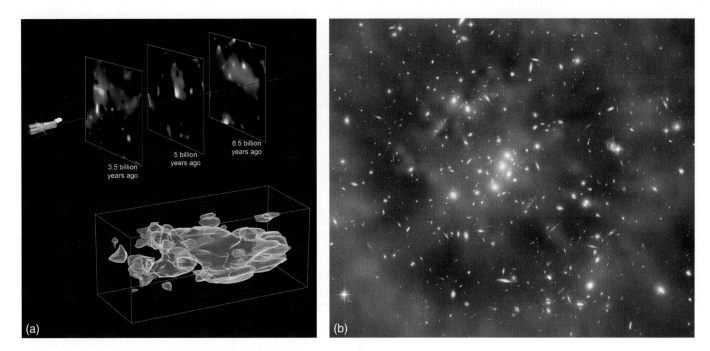

FIGURE 17-19 (a) A three-dimensional map of the dark matter distribution in the universe up to 6.5 billion light-years away. The *HST* survey covers an area of the sky nine times that of the Moon. The *HST* data were combined with multi-wavelength data from other telescopes such as the VLT, Subaru, VLA, and *XMM-Newton*. (b) An *HST* image of the cluster CL0024+17, 5 billion light-years away. The ring is 2.6 Mly across.

rotation curves of galaxies and from the way their gravity bends the light from an object farther away. **FIGURE 17-19a** is a three-dimensional map of the dark matter distribution in the universe. It is based on *HST* observations of the shapes of half a million galaxies at great distances from us, up to 6.5 billion light-years. The observed distortions of the shapes of the galaxies are the result of weak gravitational lensing due to the presence of dark matter between us and the galaxies. The map supports the idea that galaxies form along the densest concentrations of dark matter, in loose networks of filaments, which intersect at the locations of galactic clusters.

Several research teams are presently investigating the nature and amount of dark matter. One promising approach is to search the Galactic halo for stars that grow suddenly bright due to gravitational lensing as dark objects pass between them and the Earth. Such "microlensing" events have been observed, but the nature of the dark objects is not known for all events. Gravitational lensing has also been used to study the size and shape of dark matter halos surrounding galaxies; such halos appear flattened and extend more than five times further than the visible stars in a galaxy. The dark ring in the cluster CL0024+17, shown in Figure 17-19b, provides strong evidence that dark matter exists. Spectroscopic observations of the cluster's structure suggest that it is the result of a collision between two clusters about 1.5 billion light-years ago. Simulations on collisions between clusters support the idea that the ring is a ripple of dark matter that moved outward from the center of the combined cluster. This dense ripple, slowing down under the gravitational pull of the cluster's matter, causes changes in the observed shapes of the background galaxies, just as water ripples at the surface of a lake cause apparent changes in the shapes of stones at the bottom.

Other experiments assume that the particles of dark matter have a small probability of interacting with the ordinary matter of Earth, much like neutrinos do, and can be detected; however, the sensitivity required of the detectors is presently still beyond our capability. Modeling of faint shells that are observed around elliptical galaxies requires the presence of a massive halo composed of dark matter; these shells extend out a few times farther than most of the starlight in the galaxies. They appear to be remnants of mergers between the elliptical galaxies and smaller satellite galaxies, and their locations provide a measure of the gravitational field.

Some astronomers refer to "dark" matter in the Galactic halo (which might be brown dwarfs or stellar-mass black holes) as "massive compact halo objects," or MACHOs. Their name for the more exotic particles being looked for is "weakly interacting massive particles," or WIMPs. When you're looking for objects whose very existence has never been proven, it helps to have a sense of humor.

Chandra and *XMM-Newton* observations of hot gas clouds around galaxies and in clusters of galaxies (as shown in Figure 17-18) that cannot be held by the gravity of the stars in these systems suggest that massive halos of dark matter must be present. Advances over the next few years might enable us to begin setting limits as to the nature and amount of dark matter that interacts with the ordinary matter. A VLT study of tens of distant galaxies has found that over the past 6 billion years galaxies had the same amount of dark matter relative to that of their stars. This result, if confirmed by additional studies, suggests that there is a close interplay between normal and dark matter.

Different lines of evidence show that the universe is only about 4.6% ordinary matter, 23.3% dark matter, and 72.1% dark energy, which we discuss in Chapter 18; however, these numbers depend a bit on the specific cosmological model used. The quest for an understanding of the nature of dark matter and dark energy is a very active field in astronomy today because it is fundamentally important to cosmology.

17-4   The Origin of Galactic Types

◆ Evidence for the lack of interstellar material in ellipticals comes from the *Infrared Space Observatory,* which detected very little infrared radiation from ellipticals. Interstellar material radiates in the infrared region of the spectrum.

Astronomers once thought that Hubble's tuning fork diagram might represent the evolutionary sequence of galaxies. Perhaps spirals evolve into ellipticals or vice versa. This hypothesis can be dismissed quickly. First, we know that ellipticals do not evolve into spirals because elliptical galaxies contain very little gas and dust, whereas spirals contain significant amounts. Ellipticals have already used up their gas and dust forming stars. What about evolution the other way? This proposal presents mechanical problems. After a disk has formed in a galaxy, there is no way that the galaxy, on its own, could disperse its disk and acquire a symmetric elliptical shape.

Another discarded theory held that the difference is caused simply by rotation: spiral galaxies evolved from dust clouds that were spinning fast and ellipticals came from slowly spinning clouds; however, this theory cannot account for the fact that elliptical galaxies have completed their star formation and have little interstellar material.

Today, there are two leading theories that explain why galaxies exist in various types. We discuss each in turn.

◆ A puff of wind cannot pass through still air without carrying the still air along (and reducing the wind speed in the process).

1. **The Cloud Density Theory.** Early in the history of the universe, before galaxies, some gas/dust clouds must have been denser than others. One theory of galactic formation holds that elliptical galaxies are those that formed from the densest clouds. Because of the great density of these clouds, star formation would have proceeded quickly and would have used up all dust and gas before a disk had a chance to form. After stars have formed, they act as individual particles that orbit the center of the galaxy and have little interaction with one another. This is why galactic clusters are able to pass through the disk of our Galaxy without being captured. Thus, according to this theory, early star formation explains the lack of a disk in elliptical galaxies.

    Clouds that had a lower density of gas and dust would have formed stars less frequently when the cloud contracted, and the dust and gas would have collapsed into a disk before star formation used it all up. The reason that a rotating cloud of gas and dust forms into a disk is that gas acts like a fluid, where the particles interact a great deal. When two fluids collide, energy is dissipated and the fluids do not pass through one another. (In the case of star birth, a disk is a common occurrence.) Within a galactic disk, star formation proceeds slowly. Instabilities cause density waves that assist star formation, so that new stars are still being born in spiral galaxies today.

    Various hypotheses exist to explain why some spiral galaxies have bars and others do not. One idea is that faster spinning clouds form elongated nuclei, or bars. Another hypothesis notes that barred spiral galaxies generally do not have a halo of globular clusters. It proposes that the lack of a halo results in instabilities that cause a bar to form.

2. **The Merger Theory.** A more recent theory proposes that spiral galaxies formed before elliptical galaxies and that ellipticals result from mergers of spirals. The direction of spin of galaxies is random and, thus, merged galaxies would often have little or no overall rotation (which is what is observed for ellipticals). The small quantities of interstellar matter in ellipticals are explained by the fact that the merger would compress interstellar clouds and would cause strong density waves. During the merger, then, star formation would be frequent and interstellar matter would be used up.

    This theory is supported by computer simulations and by the observation that ellipticals make up a greater percentage of the galaxies in large clusters, where galaxies are packed close together. In these clusters, mergers would have been frequent and would have produced numerous elliptical galaxies. In loosely packed clusters of galaxies, on the other hand, ellipticals are fairly rare.

> The word *theory* is being used here for what we should probably call a *hypothesis*. There is no sharp dividing line between the terms.

    You may have noticed that nothing has been said about irregular galaxies. Neither theory explains these oddballs well. Astronomers know that some galaxies that were once classified as irregulars are actually pairs of galaxies that are in the process of collision. Others are spiral galaxies with large, dense dust clouds that hide their spiral patterns. Perhaps all irregular galaxies can be explained like this. In any case, irregulars make up a small fraction of all galaxies.

    It should be obvious that the question of why some galaxies are of one type and others are of another type is far from settled. This is just one of the unanswered questions that make extragalactic astronomy a lively source of research today.

Look-Back Time

As new observational techniques are developed, we are seeing objects farther and farther away. We have now detected objects that may be as far away as 13 billion light-years. If these objects are truly this far, the light we see left them 13 billion years ago. We are seeing far into the past. Astronomers speak of ***look-back time***, a term that emphasizes this idea. An object 13 billion light-years away has a look-back time of 13 billion years.

> **look-back time** The time light from a distant object has traveled to reach us.

    One of the problems with using large galaxies as distance indicators relates to the idea of look-back time. When we see these galaxies, we are not seeing them as they are today but as they were in the past. There is some evidence that within large clusters of galaxies, some galaxies combine to form supergalaxies. That is, some galaxies are "cannibalized." If this occurs, the largest galaxies in nearby clusters are not the same size as the largest galaxies in distant clusters because we are seeing nearby clusters at a later stage of their life than we see distant ones. In the distant clusters, not enough time has passed for galaxies to have gobbled up one another. This would invalidate the assumption that distant clusters are similar to nearby clusters and would call into question the practice of using galaxies as distance indicators. Although this is a possible problem, the Hubble law seems to apply equally as well to faraway clusters, where galaxies are used as distance indicators, as it does to nearby galaxies, where other methods are used for determining distance. This indicates that the dilemma posed by look-back time is not a major obstacle in this case.

> We are referring now to distant "objects" because, as we explain later, we cannot be sure that they are galaxies.

    Among the objects with the greatest look-back times, we find a class of objects not yet discussed—active galaxies. These objects play a major part in helping us to determine the scale of the universe.

17-5 Active Galaxies

Until now we centered our discussion in this chapter on normal galaxies, those that fit the Hubble classification. Now we turn to objects that do not fit into those categories—objects that were discovered since the late 1940s by radio astronomy. Before then, our understanding of the distant universe was based on observations in the visible part of the spectrum. Radio observations opened new windows on our understanding of the cosmos by introducing us to several exotic objects and phenomena.

ADVANCING THE MODEL

Galaxy Formation and Evolution

A typical spiral galaxy is thought to have started as an enormous, slowly rotating cloud that lost energy (as its atoms and molecules collided and radiated energy) and collapsed under gravity; additional mergers with smaller clouds of hot gas and dark matter finally resulted in a faster spinning disk. As part of its formation process, a large spiral should be immersed in a leftover halo of hot gas. Indeed, *Chandra* observations show the presence of such a halo around the massive spiral galaxy NGC5746 (**FIGURE B17-2**); the galaxy shows no other activity that could possibly create such a halo.

The presence of gas in spiral galaxies dampens out the disruptive effects of mergers, allowing spirals to survive a violent collision. Over its lifetime, a spiral can make many transitions between being barred or unbarred; computer simulations show that a bar survives at least 2 billion years, redistributing mass and angular momentum in the galaxy, before dissolving as a result of its own gravitational effects on the orbits of passing stars.

VLT observations show the presence of massive elliptical galaxies at redshifts between 1.6 and 1.9, suggesting that they must have formed when the universe was only 2 billion years old. This implies that the buildup of elliptical galaxies from mergers of smaller galaxies started earlier and was much faster in the early universe than currently thought. This is supported by a study of very young galaxies using the deepest images taken by the *HST*.

As a result of mergers between galaxies, matter gets spun off and forms small dwarf galaxies. *Spitzer* data for these tidal dwarfs show strong emission lines from organic compounds such as polycyclic aromatic hydrocarbons (PAHs) and warm molecular hydrogen. Tidal dwarfs, together with dwarf galaxies that may be primordial remnants of the formation of the universe, are far more common than the more massive spiral or starburst galaxies.

Large surveys of galaxies suggest that, in addition to mergers, the rate of star formation also depends on the galaxy's mass. *Spitzer* data allowed astronomers to see through the dust obscuring star forming regions in distant galaxies; the data show that early in the universe's life, massive galaxies formed stars early and rapidly, whereas less massive galaxies formed stars over longer periods of time.

Results from a number of multi-wavelength surveys of thousands of galaxies up to 9 billion light-years away consistently show that star-forming activity has gradually decreased with time; they also show that a galaxy's mass (and therefore the supply of raw material) is the dominant factor in star formation, while mergers account for only 20% of the activity. This is consistent with the current leading theory of how galaxies have evolved. According to the theory, galactic properties (mass, shape, and size) develop through mergers of less massive proto-galaxies. During the first few (3 to 4) billion years, the landscape of the universe was dominated by mergers; as time went on, the immediate environment of a galaxy and its supply of raw materials gradually started to dominate the rate of star formation. A byproduct of merging is that collisions destroy the disks of spiral galaxies, send stars into more chaotic orbits, and therefore produce football-shaped (elliptical) galaxies. This scenario is supported by computer simulations.

While this is the accepted theory, there are observations of a few galaxies very early in the life of the universe that challenge the theory; observations of massive *disk* galaxies show a furious rate of star formation even though they very likely formed from a massive cloud that collapsed rapidly.

Strong winds resulting from massive stars and supernova explosions early in a galaxy's life, or from the activity of a black hole at its center, can blast significant amounts of the galaxy's material into intergalactic space. Such winds have been observed. Their presence explains why key elements for planetary formation and life (including carbon, oxygen, and iron) are widely distributed throughout the universe from its early stages. They also limit the size of a galaxy by preventing further star formation.

FIGURE B17-2 A large halo of hot gas (blue) surrounds the disk (white) of the massive spiral galaxy NGC5746.

All galaxies emit some radio waves; our Galaxy radiates them from its nucleus (see Figure 16-23); however, the radio waves from a normal galaxy constitute only about 1% of that galaxy's total luminosity. In the late 1940s, astronomers began observing strong extragalactic radio sources, and in 1951, it was discovered that a radio

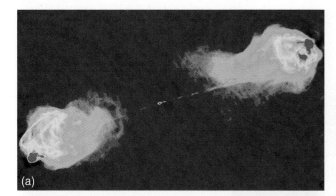

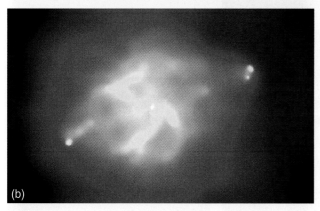

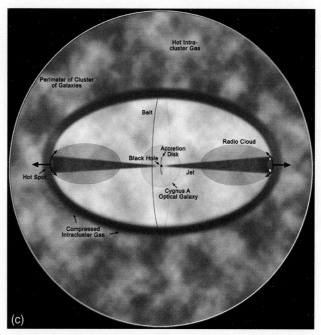

FIGURE 17-20 (a) This false-color radio image of Cygnus A (a peculiar elliptical galaxy about 700 million light-years away) shows a jet extending from the source (an AGN) toward one of the radio lobes. Regions of brightest (fainter) radio emission are shown in red (blue). (b) This *Chandra* image shows X-ray emitting hot gas (bright orange region) that surrounds Cygnus A and within it a giant cavity (the yellow and light orange inner region) created by the two jets emitted from the galaxy's central black hole region. The low-density hot gas (tens of millions K) fills the space between the galaxies in the cluster containing Cygnus A and provides enough resistance to slow down the jets. The jets terminate in radio and X-ray emitting "hot spots" some 300,000 light-years from the center of the galaxy. The pressure of the particles and magnetic fields in the jets inflates the cavity. (c) An illustration of the Cygnus A system.

source in the constellation Cygnus is actually a double source associated with a galaxy (**FIGURE 17-20**). Cygnus A, as the source is called, emits about a million times more energy in radio waves than does the Milky Way.

Since the discovery of Cygnus A, many more such ***radio galaxies*** have been discovered, typically emitting millions of times more energy in radio waves than does a normal galaxy. **FIGURE 17-21** shows Centaurus A. Observe that the visible galaxy in this case is an elliptical galaxy with a prominent lane of interstellar gas and dust, which is bisected at an angle by opposing jets of high-energy particles blasting away from the nucleus. The double-lobed radio image seen in Centaurus A is common in radio galaxies, and most of the galaxies associated with double-lobed radio sources are either giant ellipticals or spirals. The radio lobes are enormous, sometimes extending 15 million light-years from the visible galaxy. In general, radio lobes mark the positions where the outflows start interacting with the intergalactic medium.

Radio galaxies often appear unusual when viewed in visible light. **FIGURE 17-22** shows a jet of hot gas being emitted from Virgo A (M87). This feature is seen in several cases. More recently, galaxies have been observed that have properties like those of radio galaxies, but have their primary emission at wavelengths other than the radio region of the spectrum. Astronomers therefore classify radio galaxies as one type of a group of high-energy galaxies called ***active galaxies***. Because the energy of an active galaxy comes from its nucleus, astronomers often refer to them as ***active galactic nuclei*** or ***AGNs***.

Jets of material, such as those shown in this section and in Section 17-6, seem to be a universal phenomenon. They are the natural byproducts of accretion onto a compact object, emanating at right angles to the disk that surrounds this compact

radio galaxy A galaxy having its greatest luminosity at radio wavelengths.

The Milky Way is about 160,000 ly, or 0.16 Mly, in diameter. From end to end, large double lobes are 150 times this size!

active galaxy (active galactic nucleus, AGN) A galaxy with an unusually luminous nucleus.

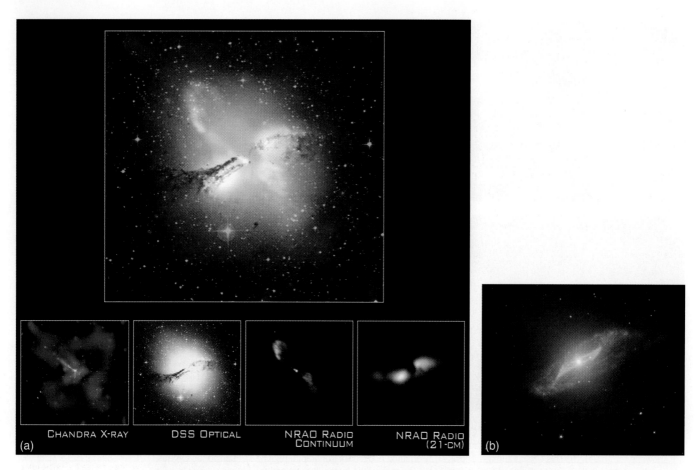

CHANDRA X-RAY   DSS OPTICAL   NRAO RADIO CONTINUUM   NRAO RADIO (21-CM)

(a)   (b)

FIGURE 17-21 (a) A composite image of Centaurus A (about 10 million light-years away) made with X-ray (blue), radio (pink and green), and optical (orange and yellow) data. The compact nucleus contains more than 200 million solar masses, suggesting the presence of a supermassive black hole. Two large arcs (part of a projected ring 25,000 light-years in diameter) of X-ray emitting, multimillion-degree hot gas exist in the outskirts of the galaxy on a plane perpendicular to the jets. The observations suggest that about 100 million years ago Centaurus A merged with a small spiral galaxy, shown as the parallelogram-shaped structure of dust in the *Spitzer* image in (b), which led to a burst of star formation in the galaxy's nucleus. About 10 million years ago a huge explosion occurred that produced the jets and a galaxy-sized shock wave that moved outward at about 450 km/s. Radio observations show that parts of the jet have speeds of about half the speed of light.

object. They occur not only in radio galaxies, but also in the entire range of objects from young stellar objects to extremely powerful galaxies. These "outflows" are mostly very well collimated (that is, they form straight beams, with very little spreading over their length) and can extend from a few to hundreds of thousands of light-years. They transfer energy, matter, momentum, and magnetic fields from the central region to the surrounding environment. The speeds of the observed outflows seem to remain constant or, in some cases, even increase along most of the length of the jets. Some outflows have speeds that are clearly close to the speed of light. Such speeds signify the presence of an acceleration mechanism that works along the entire length of the outflow. Magnetic fields in the disk and along the length of the jets, coupled with the rotation of the central object, offer a very likely mechanism for the formation, acceleration, and collimation of these outflows.

During the first decade of observing active galaxies, many types were discovered, and various hypotheses were put forward to explain them. Then, less than 10 years after the discovery of active galaxies, the mystery deepened with the discovery of what appeared to be an entirely new type of object.

Quasars

Before 1960, astronomers thought that intense radio waves like those from active galaxies necessarily come from large galaxy-sized objects. Individual stars are such weak sources of radio waves that before 1960 the Sun was the only star from which

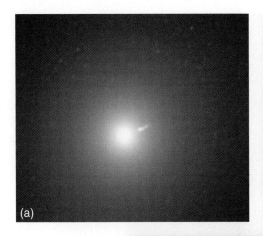
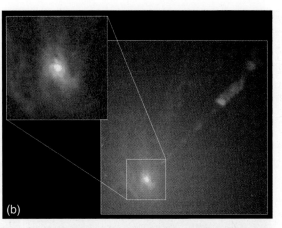
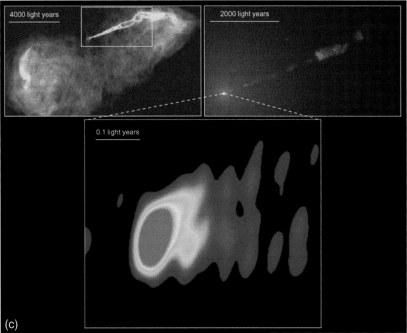

FIGURE 17-22 (a) This visible-light photo shows Virgo A (M87), a giant E0 elliptical radio galaxy, 50 Mly away in the constellation Virgo. (b) A short-exposure infrared image by the *HST* shows a jet of material coming from the nucleus of the galaxy. The nucleus is the bright white spot toward the left. *HST* observations show a spiral-shaped disk of hot gas in the nucleus. The rapid rotation of the disk suggests the presence of a black hole of about 3 billion solar masses, concentrated in a space no longer than our solar system. (c) Very Large Baseline Array (VLBA) measurements show that the jet is formed within a few tenths of a light-year from the core of the galaxy. The jet's opening angle is initially large (about 60°), but a few light-years away it is reduced to only 6°. Material in the jet is seen to be moving at apparent speeds greater than that of light; however, this superluminal motion is a geometric illusion, as we describe later in this chapter.

radio waves had been detected. In that year, however, two radio sources were found that were so small that they appeared to be stars. Recall from Chapter 7 that planets' sizes can be measured when they occult a star. The same phenomenon can be used in reverse to measure the size of a distant object. In the case of one of these radio sources, 3C 273, its visual and radio images were observed as it was occulted by the Moon. By observing 3C 273 as its visible light and radio waves were blocked out by the Moon and again as they reappeared on the other side, details of the source could be determined (FIGURE 17-23). The object appeared to be very small—more like a star than a galaxy. In addition, it had a small jet protruding from it, like the jets that have been observed from some radio galaxies. The radio waves from 3C 273 seemed to have two sources: the jet and the main body of the object. We have seen in this section that double-lobed radio sources are common for active galaxies.

The spectra of both 3C 273 and 3C 48 (the other source mentioned previously here) were found to be extremely unusual. Although *emission* spectral lines were prominent, the lines could not be identified with any known chemical element. Were we seeing objects that had entirely different chemical elements than are known to us? Because of their unusual nature (their star-like appearance and strong radio emission), the objects were called *quasi-stellar radio sources* or *quasars*.

In 1963, Maarten Schmidt of the California Institute of Technology found the solution to the puzzle of the unusual spectrum of 3C 273. He found that its prominent

3C 273 is so named because it is the 273rd object listed in the third Cambridge catalog.

quasar (quasi-stellar radio source) A small, intense celestial source of radiation with a very large redshift.

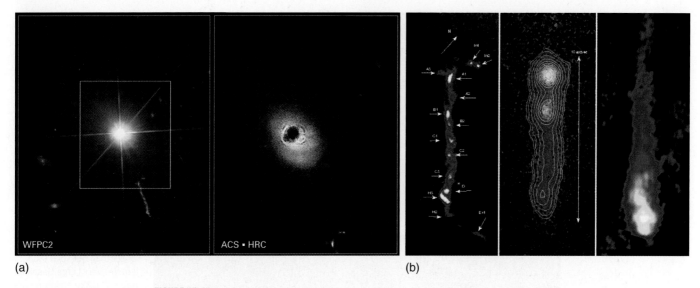

(a)   (b)

FIGURE 17-23 (a) The *HST* image at left shows the brilliant quasar 3C 273 (about 3 billion light-years in the direction of the constellation Virgo). The *HST* image at right is the clearest view yet in visible light of the quasar. After blocking the light from the central quasar, astronomers discovered the complexity of its host galaxy. A spiral plume wound around the quasar, a red dust lane, and a blue arc and clump in the jet's path are clearly seen. (b) The jet emanating from the core of quasar 3C 273, as seen, from left to right, in optical (*HST*), X-ray (*Chandra*), and radio (MERLIN). The scale is about 10′ from one end of the jet to the other. The source is beyond the top of the images.

The expression $z = \Delta\lambda/\lambda_0 = v/c$ relating redshift and speed should only be used for small speeds. After all, it implies that an object with redshift larger than 1 moves faster than the speed of light, which is not possible. The correct expression is

$$z = \frac{\Delta\lambda}{\lambda_0} = \sqrt{\frac{c+v}{c-v}} - 1,$$

which for speeds less than about $0.1c$ can be approximated by $z \approx v/c$.

spectral lines are simply hydrogen spectral lines that are very greatly redshifted. Each wavelength is redshifted to a value about 16% greater. (That is, the ratio of the change in wavelength to the nonshifted wavelength from a stationary source is 16%; $\Delta\lambda/\lambda_0 = 0.16$.) **FIGURE 17-24** illustrates the situation. Schmidt's colleague Jesse Greenstein examined the spectrum of 3C 48 and found that its spectral lines are shifted even farther—by about 37%.

If the redshifts of the two quasars are caused by the Doppler effect, then one quasar is moving at 15% of the speed of light and the other is moving at 30% of the speed of light. These were speeds far greater than any yet encountered for celestial objects (or for terrestrial objects, other than nuclear particles).

If these large redshifts follow the Hubble law, the closest of the two quasars must be as far away as the then-known farthest galaxies. Yet it is brighter than those galaxies, even though in size it is more like a star than a galaxy.

Since the early 1960s, more than about 23,000 quasars have been discovered. Unlike the first two, most (about 90%) are not sources of radio waves. Most quasars are bluish white objects and are X-ray emitters. In addition, almost all quasars show variability in their light output; some quasars quadruple in brightness in just a few hours, while the average is a change of 10% to 15% over a period of a year. This latter observation confirms their small size; they cannot be larger than a few light-months in diameter.

The first two quasars, seen with redshifts of 16% and 37%, have smaller redshifts than most. The smallest observed redshift is about 6%. The quasar SDSS J114815251, in the direction of the constellation Ursa Major, has the largest redshift observed to date ($z = 6.41$). It indicates that the quasar's speed is about 96% of the speed of light and that its distance, assuming that quasars' redshifts fit the Hubble law, is about 12.9 billion light-years away. We are seeing this quasar when the universe was very young, only about 800 million years after the big bang. The amount of dust observed in the quasar's region suggests that galaxies and stars formed earlier in the history of the universe and at a faster rate than previously thought. The black hole at this quasar's core has a mass of 3 billion solar masses (**FIGURE 17-25**).

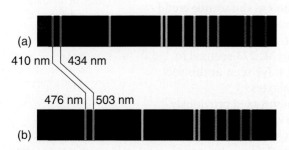

FIGURE 17-24 (a) An "at rest" spectrum. The wavelengths of two lines are indicated. (b) The same spectrum shifted 16%. Observe that the pattern of lines is the same. The *average* redshift of a quasar is determined from measurements of a number of spectral lines.

Now consider what the luminosity of such an object must be for us to be able to detect it at that distance. We can calculate the luminosity by using the apparent brightness of the quasar and the inverse-square law of radiation. We find that a bright quasar is 1000 times more luminous than a galaxy like ours. Finally, remember that we are speaking about a small object—far smaller than a galaxy.

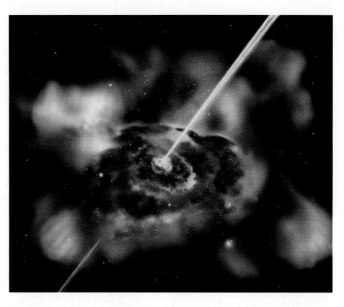

FIGURE 17-25 An artist's view of a quasar's core. The material in brown and yellow corresponds to the disk of dust and gas that hides the central black hole. The black hole's gravity pulls in this material, which swirls as it moves closer. The resulting friction heats the gas and makes it shine brightly.

Competing Theories for the Quasar Redshift

When astronomers were faced with the prospect of an object being so small and yet so luminous, they were forced to reexamine their assumptions. As was pointed out earlier, strictly speaking, the Hubble law is not due to the Doppler effect, although the basis for the law is mathematically equivalent to that effect. In Chapter 18 we discuss the theorized cosmological reason for the validity of the Hubble relationship. Nevertheless, the Hubble law indicates that quasars are at tremendous distances from us. Is there any other possible explanation for the redshift?

Einstein's general theory of relativity predicts that light leaving a massive object is redshifted because of the mass of the object. Calculations show that to produce redshifts such as those seen in quasars, the gravitational field near the object must be far greater than that near the most massive neutron star, and well-grounded nuclear theory tells us that there is a limit to the mass of a neutron star. After a certain mass is reached, a neutron star cannot exist, and a black hole is formed instead. The theory of relativity simply cannot be used to explain the redshift.

Perhaps quasars are nearby objects with enough local motion that they do not fit the Hubble law. (Galaxies within the local group do not fit the Hubble law either.) FIGURE 17-26 shows fast-moving objects ejected from our Galaxy; some are moving at nearly the speed of light. If this is what quasars are, then they do not have such tremendous luminosities after all.

The problem with this **local hypothesis** is not only that we can imagine no source for such objects, but that we must also ask why we do not see similar objects from other galaxies. We cannot expect our Galaxy to be unique in this regard. Yet if other galaxies emit such fast-moving objects, we would see some of them with tremendous *blueshifts* as they move toward us. In fact, however, such blueshifted objects are not seen, and the local hypothesis is not considered a likely explanation.

Some quasars appear to be linked to nearby galaxies of different redshifts. Even though such links seem to suggest that the Hubble law may not be valid, they do not prove that the objects are *physically* connected and they can easily be explained as chance alignments. A few scientists who support non-standard cosmologies have suggested that the observed redshifts are the result of other physical effects unrelated to the expansion of the universe; however, these effects cannot reproduce the Hubble law on the observed scales.

Perhaps there is some other explanation that might require drastically different laws of

Because $v = H_0 d$, we get $d = \dfrac{c}{H_0} \cdot \dfrac{v}{c}$, where the ratio v/c can be calculated from the redshift.

The largest redshift observed today for a galaxy is $z = 7$. It implies a distance of 13 billion light-years, based on recent research indicating that the universe's age is 13.73 billion years.

local hypothesis A proposal stating that quasars are much nearer than a cosmological interpretation of their redshifts would indicate.

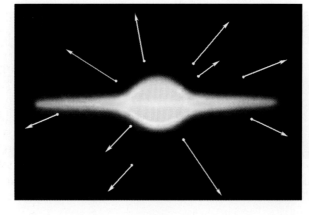

FIGURE 17-26 The local hypothesis holds that quasars are relatively nearby objects that have been ejected from the Galaxy at great speeds. If quasars are emitted by galaxies, however, some should be moving toward us from nearby galaxies, and these would show blueshifts. No observed quasar has a blueshift.

FIGURE 17-27 The core of the nearby active galaxy Circinus, which belongs in a class of mostly spiral galaxies called Seyferts. The galaxy is only 13 million light-years away, and most of the gas in its disk is concentrated in two rings (with diameters of 1300 and 260 light-years). At the center of the inner ring is a supermassive black hole that is accreting surrounding gas and dust.

Seyfert galaxy One of a class of spiral galaxies having active nuclei and spectra containing emission lines.

Seyferts and quasars are the same phenomenon, just with the rheostat turned up.
Astronomer Allan Sandage

gravitational lens The phenomenon in which the gravity caused by a massive body between a distant object and the viewer bends light from the distant object and causes it to be seen as two or more objects.

physics than those of today. We never can disregard such a possibility, but we must progress with what we know until it is definitely ruled out. The redshift *can* be explained by the Hubble law, even though that law leads us to conclude that quasars are at distances greater than we are accustomed to and are therefore much more energetic than we are accustomed to. Almost all astronomers now agree that quasars' redshifts do indeed fit the Hubble law.

Seyfert Galaxies

As we discussed previously in this section, one of the first two quasars seen (3C 48) has a double-lobed radio source and a jet from its center. In this regard, it resembles an active galaxy. As astronomers found more and more quasars, they discovered that most of the nearer ones are associated with clusters of galaxies and have essentially the same redshifts (and thus distances) as these galaxies. The quasars are more luminous than the galaxies, but the association causes us to look for more similarities.

One particular type of spiral galaxy, a **Seyfert galaxy** (FIGURE 17-27), has a very luminous nucleus that—although it is not as luminous as a quasar—in some cases varies in intensity in time periods even shorter than those of quasars. Similarities in the spectra of Seyfert galaxies and quasars point to a further link between galaxies and quasars, indicating that perhaps a quasar is a galactic nucleus.

Recent high-resolution photos of nearer quasars show a fuzz on the image near the quasar. Again, it appears that quasars are at the nuclei of some types of galaxies and that the fuzz is caused by the stars of the galaxies.

Quasars and Gravitational Lenses

Twin quasars were discovered in 1979 in Ursa Major. The two quasars are very close together, separated by only 6 arcseconds. They have the same luminosity, the same redshift ($z = \Delta\lambda/\lambda_0 = 1.4$), and identical spectra. The explanation for these identical twins lies with the theory of general relativity, which predicts the possibility of a large mass bending light from a more distant object so that two images of the distant object appear. The predicted phenomenon is called a **gravitational lens**, but until the discovery of the twin quasars, not much attention was paid to the prediction. Since the discovery of the twin quasars, the intervening galaxy has been found, and we are now confident that the twins are actually one quasar that has been made to appear double by the galaxy that lies between it and us (FIGURE 17-28). Many examples of gravitational lensing have been found since 1979. One interesting case is shown in FIGURE 17-29a. The two yellow-orange spots in the image are twin images of a single quasar behind a massive object that acts like a gravitational lens. The ring is caused by another object, behind whatever object causes the lensing effect, which is so well aligned that light passing on every side of the intervening massive object is bent toward Earth, making the distant quasar appear to be on all sides of it at once.

Arcs, such as in Figure 17-29a, are the most common type of lensing, although most cases are not as complete as that arc. Pairs of lensed objects are the next most common. Figure 17-29b shows a much rarer case, one with a quadruple pattern.

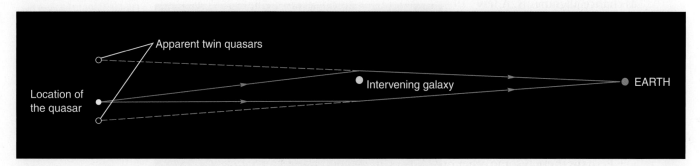

FIGURE 17-28 Light from the quasar at the left is bent as it passes by the intervening galaxy at the center. This causes the light to appear to come from twin quasars, one on either side of the galaxy.

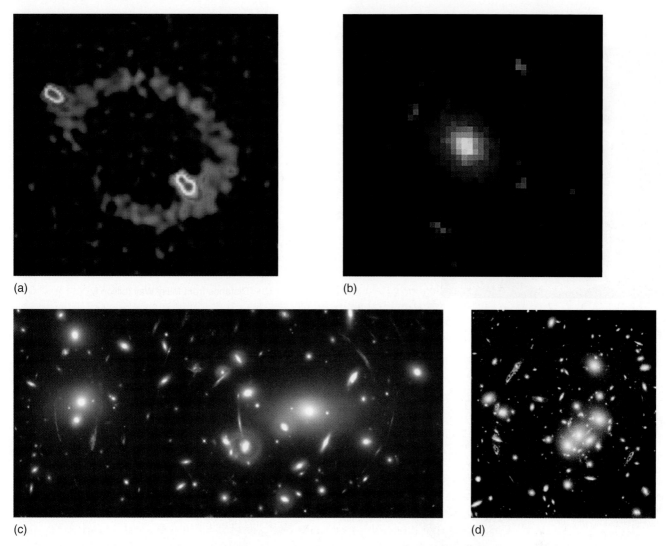

(a)

(b)

(c)

(d)

FIGURE 17-29 (a) This false-color image of object MG 1131 + 0456 was made by radio interferometry with the Very Large Array. The ring is called an "Einstein Ring." This 1987 image is the first such ring ever observed. (b) In 1995 the *HST* was used to discover this case of quadruple gravitational lensing. (c) Abell 2218 is a rich galaxy cluster. The *HST* captured this arc-like pattern spreading across the image like a spider web, a great example of gravitational lensing. Abell 2218 has a total of seven multiple systems. (d) This *HST* image shows five images of the same galaxy, distorted from a normal spiral to an arc-shaped object. The lensing cluster is 5 billion light-years away, and the blue galaxy is two times farther away.

Figure 17-29c reveals numerous arcs, which are the distorted images of a galaxy population located 5–10 times farther than the cluster of galaxies doing the lensing. Figure 17-29d shows several blue, loop-shaped objects, which are multiple images of the same galaxy seen through the "lens" of the cluster of yellow, elliptical, and spiral galaxies near the center of the image.

The importance of the discovery of gravitational lenses is that they provide another confirmation of the general theory of relativity and they indicate that quasars are indeed very distant—that their redshift fits the Hubble law.

Suppose that we plot the number of known quasars versus their distances from our Galaxy. The graph would not be very instructive because as the distance from our Galaxy (or from any point) increases, the volume of space in a shell increases. This happens because the area of a large sphere is greater than the area of a small sphere (**FIGURE 17-30a**); therefore, even if quasars are evenly distributed throughout the universe, we would expect to see more of them at greater distances.

To truly illustrate the distribution of quasars, the graph should show the *density* of quasars—the number of quasars per unit of volume of space—versus their distance

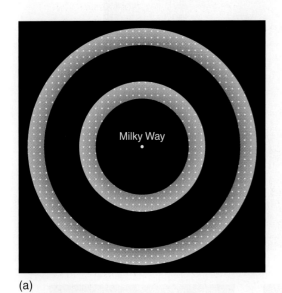

(a)

Density of Quasars at Various Distances

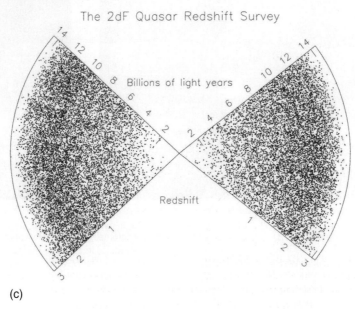

(b)

FIGURE 17-30 (a) If two equally thick shells are drawn at different distances from us, the more distant shell will contain more space. Even if quasars (dots in the figure) were equally spaced throughout the universe, more would be seen in the faraway shell. (b) The graph shows the relative abundance of quasars at various distances from the Milky Way. The distance scale is based on a Hubble constant of 23 (km/s)/Mly. (c) As seen in this two-degree field (2dF) survey of about 23,000 quasars, the quasars are clustered to the same extent as local, optically selected galaxies.

(c)

from us. Figure 17-30b is such a graph. The distance shown on the horizontal axis depends on the value chosen for the Hubble constant, and other choices are possible. Nevertheless, even if a different value were chosen for the Hubble constant, the shape of the graph would remain the same. The graph leaves no doubt that quasars appear at a fairly specific distance from the Milky Way. Because distance is proportional to time, this indicates that quasars existed during a relatively short period in the distant past.

Quasars, Blazars, and Superluminal Motion

In the early 1970s, astronomers discovered an important clue in their efforts to understand quasars. They found a new class of objects, called **blazars** or **BL Lac objects**. A blazar seems to be star-like and very luminous, just like a quasar. Its few faint emission lines show high redshifts, suggesting a very large distance, just like for a quasar. Blazars are unusual because they are highly variable—their luminosity can change by up to 30% in a day and by a factor of 100 in a few months. Radio observations showed that there is faint radio emission around a bright core, suggesting that a blazar is a double radio source oriented in such a way that one jet is coming straight (or

blazars (or BL Lac objects)
Especially luminous active galactic nuclei that vary in luminosity by a factor of up to 100 in just a few months.

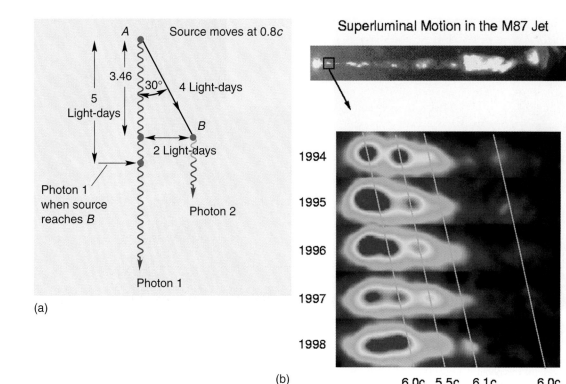

(a)

Superluminal Motion in the M87 Jet

(b)

6.0c 5.5c 6.1c 6.0c

FIGURE 17-31 (a) This diagram shows how an object moving at a speed slower than the speed of light can appear to be moving at a speed faster than the speed of light because of projection effects. (b) The top panel shows the jet emanating from the galaxy's nucleus (at far left; see Figure 17-22). The box indicates the location of the observed superluminal motion. The bottom panel includes the sequence of *HST* images taken between 1994 and 1998. The slanted lines track the observed moving features in the jet, and the apparent speeds of these features are given in units of the speed of light.

nearly so) at us. This was soon supported by observations of **_superluminal motion,_** which is also observed in some quasars.

FIGURE 17-31a shows a simple explanation for superluminal motion. Suppose that a source moves at 80% of the speed of light, along a line that makes an angle of 30° with the line of sight from Earth. At point *A*, it emits a photon toward us. After traveling for 5 days, it reaches point *B*, where it emits a second photon toward us. The real distance traveled by the source between points *A* and *B* is $5 \times 0.80 = 4$ light-days and, thus, the source has moved toward us along the line of sight by about 3.46 light-days. In the meantime, the first photon has traveled a distance of 5 light-days toward us; therefore, the first photon is ahead of the second by $5 - 3.46 = 1.54$ light-days. From our perspective, we see the source moving a horizontal distance on the sky (as a result of projection effects) equal to 2 light-days. Because this is the horizontal distance traveled by the source in the time period during which we receive the two photons, we conclude that the source is moving at a speed of $2/1.54 = 1.3$ times the speed of light! Figure 17-31b shows the observed superluminal motion in the M87 jet.

superluminal motion
Motion that appears to occur at speeds faster than the speed of light.

17-6 The Nature of Active Galactic Nuclei

The first question that has to be answered in trying to solve the puzzle of the nature of active galactic nuclei is this: "Why are these strange objects not found among nearby galaxies?" Here we use the term *active galactic nuclei* (AGNs) to mean the bright energy source at the centers of radio galaxies, Seyfert galaxies, quasars, and blazars. To answer this question, we invoke the idea of look-back time: AGNs are far away because they existed only in the distant past and are not part of today's universe; therefore, any theory about the nature of these mysterious objects must explain why they existed long ago but not today.

According to today's prevalent theories, the tremendous amount of energy that comes from an AGN is caused by an immense black hole at the nucleus of the galaxy. The black hole is surrounded by an accretion disk that is extremely hot because material falls into it with a very great speed. The temperature of the accretion disk explains the enormous energies emitted by the objects. What remains to be explained

Be aware of the tremendous scale of extragalactic astronomy; a few million light-years is a very small distance.

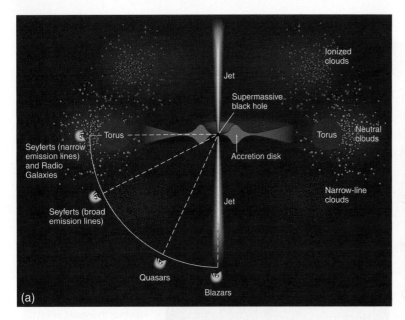

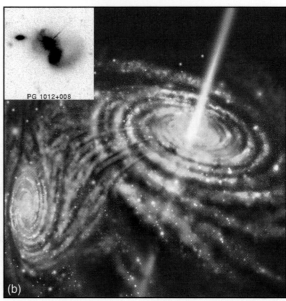

FIGURE 17-32 (a) Active galactic nuclei appear to be different because they are oriented at different angles to our line of sight. The leading model for AGNs holds that they are basically the same type of object, often involving a supermassive black hole. About 10% of quasars are BALs; they exhibit *b*road *a*bsorption *l*ines and are likely surrounded by a thick cocoon of gas. (b) This photo of quasar PG 1012+008 (inset) could be explained by a supermassive black hole in one galaxy attracting matter from a nearby companion galaxy.

That period [1960–1966] was incredibly electric. Every time you went to the telescope and came down you had a major new discovery.
Astronomer Allan Sandage

Recall from Chapter 13 (and from Figure 13-15) that jets of material have been observed coming from the poles of protostars. The galactic case is likely a similar phenomenon on a much larger scale.

is why some AGNs are so much more luminous than others, why some emit intense radio waves and others do not, why the luminosity of some of them changes so rapidly, and in general, why there appear to be so many different types of these peculiar objects.

Various theories have attempted to explain the many observations about AGNs, but none is completely satisfactory. The leading theory provides a link between the different types of objects. This "unification theory" holds that they are all basically the same and that they appear different depending on their orientation with respect to us. **FIGURE 17-32a** illustrates the hypothesized object, which is an active galactic nucleus. At its center is a supermassive black hole of many million (or even a few billion) solar masses. It is surrounded by an accretion disk, perhaps a few light-months across.

A much larger ring, shaped like a fat doughnut (a *torus*), surrounds the accretion disk in the same plane. The torus consists of relatively cool gas and dust. It may be several light-years across and is dense enough that light from the accretion disk cannot penetrate it. Light from the disk does reach the outside world, however, for it shines out of the "doughnut hole" in two directions, forming two cones of light. Radiation in all other portions of the electromagnetic spectrum, including high-energy wavelengths, pours out along with the light. A jet of material is ejected by some unexplained process along the axis of each of the cones. Finally, irregular clouds of gas and dust move in orbit around the center.

Now suppose that the active nucleus in the figure is oriented so that we see it edge on, so that the torus hides the accretion disk from our view. Telescopes would reveal infrared energy coming from the clouds, but would not show the high-energy radiation from the accretion disk. In some cases the jets cause emission of visible light, but in other cases they do not. If the active nucleus has energetic jets, astronomers would observe large radio-emitting areas at a great distance on either side of the galaxy. **FIGURE 17-33** (and 17-20a) shows the radio lobes and jets of some active galaxies.

When the torus is seen obliquely, radiation from the accretion disk reaches us, and the object appears as a very energetic source that emits radiation across the spectrum, including X-rays. These are the objects we call quasars. On the other hand,

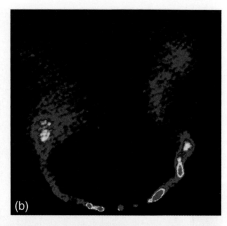

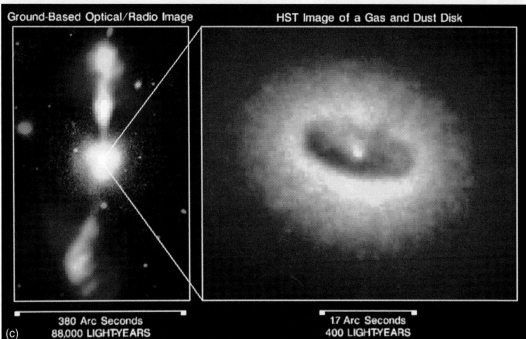

Ground-Based Optical/Radio Image   HST Image of a Gas and Dust Disk

380 Arc Seconds
88,000 LIGHT-YEARS

17 Arc Seconds
400 LIGHT-YEARS

(c)

FIGURE 17-33 (a) A false-color image of the radio emission (red) is superimposed on a visible-light image (blue-white) of the radio galaxy NGC1316 (Fornax A). Each radio lobe is about 600,000 light-years across. The two faint pink extensions just visible in the center show the paths of the outflows (jets). (b) A false-color image of the jets in the radio galaxy NGC1265 (3C 83.1). The red circle denotes the galaxy's center. Red (blue) shows regions of intense (fainter) radio emission. The galaxy moves through the intergalactic medium at ~ 2000 km/s, resulting in the "U" shape of the jets. (c) (Left panel) A composite image of the giant elliptical galaxy NGC4261 in the Virgo Cluster. In visible light (white), the galaxy appears as a fuzzy disk of hundreds of billions of stars; the radio emission (orange) shows two jets emanating from the nucleus. (Right panel) A giant disk of gas and dust fuels a possible black hole at the core in this IR image.

suppose the jet is aimed directly toward us. Again it would appear as a small, energetic source, but in this case, it might vary quickly in intensity as clouds of dust move across our line of sight. These are the objects we call blazars.

How could we test this "unification theory"? The best test is by detecting radiation from AGNs that is not blocked by the dust in the torus and is not affected by how the torus is seen by the observer. These requirements are fulfilled by far-infrared radiation. Astronomers announced in early 2001 that observations made by the *Infrared Space Observatory* showed that very hot and luminous quasar cores are found even in weak radio galaxies at large distances. These and additional observations support the conclusion that all AGNs are similar objects that appear different depending on how they are oriented with respect to us.

How does this model explain the observation that active galaxies are not found in our neighborhood? There is a limited amount of material near the black hole in the galactic center, and after that material spirals into the black hole, the galaxy calms down to become a more standard galaxy. In today's universe, no galaxy remains with a nucleus that is still in the quasar stage. This means that previous quasars and active galaxies are the ancestors of today's galaxies.

FIGURE 17-34a provides further evidence that galaxies have progressed from having quasars or blazars at their centers, to Seyferts or radio galaxies, to normal spiral

A study by the Sloan Digital Sky Survey shows that clouds of gas and dust surround the central black hole of many of the most energetic quasars. More than half of the quasars in the relatively recent universe are hidden quasars. X-ray observations by *Swift* and *Suzaku* confirm this result.

One of the most powerful object in the local universe is quasar PDS-456. With a redshift of $z = 0.184$, it is only about 2.2 billion light-years away. It is powered by a black hole of about a billion solar masses.

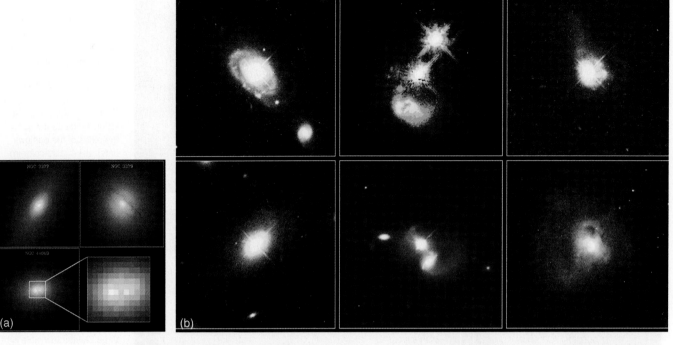

FIGURE 17-34 (a) These three active galaxies are believed to contain supermassive black holes in their nuclei. One has a double nucleus. (b) Examples of different home sites of quasars. The images in the left column represent normal galaxies, in the center, colliding galaxies, and in the right, peculiar galaxies.

or elliptical galaxies. The three galaxies in the figure are classified as normal galaxies, but they all show characteristics of having black holes at their centers. A census of 27 nearby galaxies carried out by the *HST* and ground-based telescopes in Hawaii suggests that nearly all galaxies may harbor supermassive black holes that once powered quasars. It seems that supermassive black holes are so common that nearly every large galaxy has one. The *HST* images shown in Figure 17-34b give examples of quasars found in normal, colliding, and peculiar galaxies.

In addition, the black hole's mass is proportional to the mass of the host galaxy. This suggests that the growth of the black hole is linked to the formation of the galaxy in which it is located.

Finally, the number and masses of the black holes found are consistent with what would have been required to power the quasars. Observations in X-rays (using *Chandra*) show that black holes at the centers of AGNs grew more rapidly in distant (and therefore younger) galaxy clusters and were much more active in the past than at present, whereas infrared observations (using *ISO*) show hundreds of young active galaxies with high rates of star formation. The ancestors of today's galaxies were much more active than previously thought. By combining observations in ultraviolet and X-rays, astronomers can study the link between violent stellar activity and accretion by a supermassive black hole at the nucleus of an active galaxy (**FIGURE 17-35**).

Observations suggest that quasar variability is related to the mass of the central black hole and the efficiency of the quasar in converting gravitational potential energy into light energy; less efficient quasars are more massive and variable. A study by the Sloan Digital Sky Survey of more than 4000 bright quasars in the distant universe shows that they are clustered, with the clusters separated by huge voids. This is consistent with our current understanding of the evolution of our universe; quasars are powered by supermassive black holes at the center of massive halos within massive concentrations of dark matter. Perhaps a giant step has been taken in our understanding of the evolution of galaxies. Time will tell.

Once a supermassive black hole reaches a certain size, its activity (through radiation and outflows) affects the evolution of its host galaxy by dispersing raw material (gas and dust) for star formation from the area around the galactic core. A theoretical upper limit for the mass of a black hole is about 10 billion solar masses.

FIGURE 17-35 (a) The spiral galaxy M81 in UV (obtained by *XMM-Newton* in 2001). Strong ultraviolet emission is a feature of star formation, supernova explosions, and accretion by black holes. The emission from the galactic nucleus (the bright, point-like, white region) could support the idea of a miniquasar or intense and violent stellar activity at the center. Blue denotes regions where young, hot stars are formed. Red denotes cool regions (corresponding to older, less massive stars). The bright, red, point-like objects are foreground stars in our Galaxy. (b) An artist's impression of the central region of the spiral galaxy MCG-6-30-15. Its spectrum (obtained by *XMM-Newton* in 2000) shows that the number of measured photons and their energies far exceed the predictions of standard models for accretion disks around supermassive black holes. It is possible that strong magnetic fields exert a breaking effect on the black hole, extracting some of its rotational energy and heating the surrounding medium in the process.

Conclusion

In this chapter, we presented knowledge about which astronomers are fairly confident, as well as knowledge that is very tentative. Galaxies can be classified with confidence, and Hubble's classification is based simply on appearance. Although the tuning fork diagram may appear to indicate an evolutionary sequence, it does not. At least two competing theories are available to explain why galaxies exhibit such different appearances.

Not much can be known about a galaxy unless we first know its distance; thus, the study of galactic distances is important, and various distance indicators are in use today. To measure distances to the farthest galaxies, new tools, such as the Tully-Fisher and the Faber-Jackson relations, have been added to the Hubble law.

Astronomers can calculate the mass of a galaxy by applying gravitational theory to the measured motions of parts of the galaxy. They measure the mass of a cluster of galaxies in a similar manner. Both calculations show that we can account for only a small fraction of the mass that must exist, and the mystery of the missing mass has not been solved.

A busy area of research today is the study of active galactic nuclei. These distant, energetic objects that have been discovered by the methods of radio astronomy have brought about major changes in extragalactic astronomy. They seem to produce new questions almost as prodigiously as they produce energy.

With increasing distance our knowledge fades, and fades rapidly, until at the last dim horizon we search among ghostly errors of observations for landmarks that are scarcely more substantial. The search will continue. The urge is older than history. It is not satisfied and it will not be suppressed.
Edwin Hubble

STUDY GUIDE

1. Cepheid variables are important in calculating
   A. the distances to galaxies.
   B. the composition of stars.
   C. the composition of the interstellar medium.
   D. the ages of galaxies.
   E. the temperatures of stars.

2. About how many galaxies are within range of the largest telescopes?
   A. 20.
   B. 103.
   C. 1000 to 2000.
   D. 1 to 4 million.
   E. More than 100 billion.

3. The Magellanic Clouds are
   A. galaxies.
   B. extremely high atmospheric clouds.
   C. globular clusters.
   D. supernovae.
   E. of unknown nature.

4. The local group consists of
   A. about 100 nearby stars.
   B. about 30 nearby stars.
   C. about 100 nearby galaxies.
   D. about 30 nearby galaxies.
   E. the closest planetary nebulae.

5. Which of the following objects are used as a distance indicator to galaxies?
   A. Cepheid variables.
   B. Globular clusters.
   C. The brightest supernovae.
   D. [All of the above.]
   E. [None of the above.]

6. The masses of galaxies are determined by using the Doppler effect to measure the speeds
   A. of parts of individual galaxies.
   B. of individual galaxies in a cluster.
   C. of galaxies that are part of a binary pair.
   D. [All of the above.]
   E. [None of the above.]

7. Hubble found that the objects that were first called spiral *nebulae* are instead spiral *galaxies* by observing
   A. supernovae in them.
   B. Cepheid variables in them.
   C. radiation characteristic of black holes coming from them.
   D. black holes in their centers.
   E. that their spectra are characteristic of stellar spectra.

8. If galaxy X is four times more distant than galaxy Y, then according to the Hubble law, galaxy X is receding
   A. 16 times faster.
   B. 4 times faster.
   C. 2 times faster.
   D. 1.6 times faster.
   E. [No general statement can be made.]

9. Measurements of the total mass in galactic clusters indicate that they contain far more mass than we see. The "missing mass" is now known to be

A. tiny black holes.
B. cool gas and dust between the galaxies.
C. black dwarfs within the galaxies.
D. [None of the above; we don't know what it is.]

10. Which of the following objects has the greatest diameter?
    A. A galactic cluster.
    B. A cluster of galaxies.
    C. The Milky Way.
    D. The Large Magellanic Cloud.
    E. [Both A and B above; they are the same.]

11. The Tully-Fisher method of measuring distances to galaxies depends on the relationship between a galaxy's absolute magnitude and
    A. the galaxy's size.
    B. the galaxy's redshift.
    C. the galaxy's apparent color.
    D. the width of a line in the galaxy's spectrum.
    E. [Both A and C above.]

12. The Hubble law relates
    A. absolute magnitude and temperature.
    B. apparent magnitude and temperature.
    C. proper motion and radial velocity.
    D. proper motion and distance.
    E. distance and radial velocity.

13. The Hubble constant refers to
    A. the fact that the speed of light never changes.
    B. a number in a formula that will never change its value.
    C. the rate at which the universe is expanding now.
    D. the amount of mass in the universe.
    E. the amount of mass in galactic cores.

14. Suppose that we know that a galaxy 5 million light-years away is receding at 20 miles per second. What would be the Hubble constant based on this single galaxy?
    A. 4 (mi/sec)/Mly.
    B. 15 (mi/sec)/Mly.
    C. 25 (mi/sec)/Mly.
    D. 100 (mi/sec)/Mly.
    E. [None of the above.]

15. The look-back time of an object is
    A. how long light from the object takes to reach Earth.
    B. numerically equal to the object's distance in light-years.
    C. greater for more distant objects.
    D. [All of the above.]
    E. [None of the above.]

16. The Hubble law is based primarily on
    A. calculations of galactic masses.
    B. calculations of galactic rotational speeds.
    C. Doppler shift data.
    D. knowledge of galactic evolution.
    E. data concerning the number of stars in galaxies.

17. Which of the following statements is true about quasars?
    A. They seem to be extremely large compared to most galaxies.

B. They seem to be very small for the energy released.
C. If they are as far away as they seem, they are very energetic.
D. [Both A and C above.]
E. [Both B and C above.]

18. The evidence for small size of quasars comes from
    A. the amount of energy they release.
    B. their distance from us.
    C. the rapidity of their luminosity changes.
    D. comparison with Cepheid variables.
    E. the magnitude of their blueshift.

19. We cannot judge quasars' distances by using their absolute luminosity as we do in the case of galaxies because
    A. they have no absolute luminosity.
    B. there are no nearby quasars with which to compare distant ones.
    C. their luminosity is too great.
    D. their recessional speeds are unknown.

20. The radio waves from a radio galaxy can come from an area many times bigger than the visible object.
    A. True.
    B. False.
    C. No general statement can be made concerning this.

21. Which of the following statements is an observation rather than the result of theory?
    A. Light from most galaxies is redshifted.
    B. Most galaxies are receding from us.
    C. The rate of expansion of the universe is slowing.
    D. Galaxies farther away are moving away faster.
    E. Quasars emit more energy than most galaxies.

22. If the local hypothesis for quasars were true, we would expect
    A. that quasars would be less luminous.
    B. that quasars would be more luminous.
    C. to observe quasars with blueshifted spectra.

D. quasars to be emitted from other galaxies in addition to ours.
E. [Both C and D above.]

23. What is meant by a distance indicator and what serves as one for nearby galaxies? For more distant galaxies?

24. Describe three methods for measuring the mass of a galaxy other than the Milky Way.

25. What is the local group?

26. What is meant by *dark matter* in galactic astronomy?

27. The chapter stated that we know that elliptical galaxies do not evolve into spiral galaxies. The reason given was that we observe that spirals contain gas and dust while ellipticals do not. Explain why the observation proves the statement.

28. State the Hubble law. What is the currently accepted range of the Hubble constant? Include units in your answer.

29. Describe some difficulties encountered in determining the Hubble constant with accuracy.

30. Explain how the Hubble law can be used to determine the distance to a galaxy. Discuss the limitations of this method of determining distance.

31. What is meant by *look-back time*?

32. In what way are active galaxies "active"?

33. After the spectra of quasars were understood, why were the objects still so difficult to explain?

34. What do we mean when we say that a quasar has a redshift of 25%? Which quasar is moving away faster: one with a redshift of 25% or one with a redshift of 30%?

35. What is the present theory of the origin of the energy of quasars?

36. Why are there no nearby quasars?

1. Explain how we determine the distances to galaxies that are too far away to allow us to see Cepheids in them.

2. The difficulty in determining the orientation of the plane of revolution of binary galaxies limits the usefulness of Doppler shift measurements of velocity. Why does the same problem not exist in using Doppler shift data to determine the speed of part of an individual spiral galaxy?

3. Discuss the limitations on the accuracy of measurements of distances to galaxies. Doesn't the lack of precision in these measurements make galactic astronomy less a science than other sciences?

4. Compare the use of bright galaxies as distance indicators with grading a class "on the curve."

5. If the slope of the Hubble graph were greater (with distance plotted on the x axis), would the Hubble constant be larger or smaller? Explain.

6. What is a gravitational lens and what does the fact that they have been observed tell us about quasars?

7. Why is a spectral line of a galaxy broadened by the galaxy's rotation?

8. In describing what was called the *merger theory* of galaxy formation, no explanation was given for why some spiral galaxies have bars and others do not. Propose an explanation.

9. List similarities and differences between the collapse of an interstellar cloud to form a star and the formation of a galaxy.

1. Determine the slope of the graph shown in **FIGURE 17-36**. (The graph might represent the changing speed of a race car.) Be sure to include units in your answer.

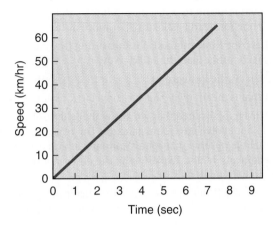

FIGURE 17-36 Graph for Calculation 1.

2. If the data for the Hubble expansion were as shown in Figure 17-9 or Figure 17-11, what would be the value of the Hubble constant for each graph?

3. Suppose that an object is seen with a redshift that indicates a speed of 100,000 km/s away from us. If this redshift fits the Hubble law, how far away is the object? (Assume some reasonable value for the Hubble constant.)

4. Suppose that Alex is moving away from you at a constant speed of 15 miles/hour and is 60 miles away at this moment. If he has maintained this speed since leaving you, how long ago were you together?

5. For the motion in question 4, calculate the "Alex constant" (analogous to the Hubble constant).

6. Figure 17-5a shows the Large Magellanic Cloud. It is 50 kpc from us, and has an angular size of 650′ by 550′.

Use the small-angle formula (in Chapter 6) to calculate the dimensions of the Large Magellanic Cloud in kpc.

7. **FIGURE 17-37** shows NGC4535. Its angular size is 6.8′ by 5′. Assuming that its longest dimension is about 100,000 light-years, calculate its distance from us.

FIGURE 17-37 Galaxy NGC4535, Calculation 7.

8. The existence of supermassive black holes is actually less exotic than you might think. The average density of an object is given by the ratio of its mass to its volume. For a spherical object of radius R, density = $M/(4\pi R^3/3)$. Writing the radius in km and the mass in solar masses show that density = $4.8 \times 10^{20} \times (M/R^3)$ kg/m³. The radius of a black hole (in km) is related to its mass (in solar masses) by the Schwarzschild expression: $R = 3M$. Now show that the density necessary to make a black hole of mass M (in solar masses) is $1.8 \times 10^{19}/M^2$ kg/m³. What density is needed to make a black hole of a billion solar masses? A hundred million solar masses?

1. "Understanding the Hubble Sequence," by G. Lake, in *Sky & Telescope* (May, 1992).

2. "When Galaxies Collide," by J. Roth, in *Sky & Telescope* (March, 1998).

3. "Active Galactic Nuclei: Sorting Out the Mess," by A. Finkbeiner, in *Sky & Telescope* (August, 1992).

4. "Massive Black Holes in the Hearts of Galaxies," by H. Ford and Z. I. Tsvetanov, in *Sky & Telescope* (June, 1996).

5. "The Most Distant Radio Galaxies," by G. K. Miley and K. C. Chambers, in *Scientific American* (June, 1993).

6. "A New Look at Quasars," by M. Disney, in *Scientific American* (June, 1998).

7. "The Evolution of Galaxy Clusters," by J. P. Henry, U. G. Briel, and H. Böhringer, in *Scientific American* (December, 1998).

8. "Unmasking Black Holes," by J.-P. Lasota, in *Scientific American* (May, 1999).

9. "Another Look at Cosmic Distances," by T. A. Weil, in *Sky & Telescope* (August 2001).

10. "Quasars Explained," by W. Keel, in *Astronomy* (February 2003).

11. "The Midlife Crisis of the Cosmos," by A. J. Berger, in *Scientific American* (January 2005).

12. "A Quasar in Every Galaxy?" by R. Irion, in *Sky & Telescope* (July 2006).

13. "The Galactic Odd Couple," by K. Weaver, in *Scientific American* Special Edition (January 2007).

14. "Colossal Galactic Explosions," by S. Veilleux, G. Cecil, and J. Bland-Hawthorn, in *Scientific American* Special Edition (January 2007).

Cosmology: The Nature of the Universe

18

THROUGH THE CENTURIES, ASTRONOMERS HAVE STUDIED THE EARTH–MOON SYSTEM, the solar system, the stars, and the galaxies, recognizing the need for a new scale of size with each step. What is left to study? The universe itself is the subject of cosmology, and now we must come to terms with its nature. Yet even here, our understanding is based on observation and deduction, in the spirit of science and not of guesswork.

The scientific study of the universe is a fairly recent field; only since the 1940s have astronomers realized that we can make quantitative predictions about the nature and behavior of the universe and verify these predictions through observations. One of the first physicists to point out the connections between particle physics (whose subject is subatomic particles) and cosmology was Stephen Weinberg, who shared a Nobel Prize in 1979 for his work on the forces between elementary particles. He applied the theories of high-energy physics to the conditions of the entire universe at the time of its beginnings, describing the sequence of events in the "First Three Minutes" of the universe. He concludes his book as follows:

> The universe will certainly go on expanding for a while. As to its fate after that, the standard model gives an equivocal prophecy: It all depends on whether the cosmic density is

The cosmic microwave background (CMB) sky over Mt. Erebus in Antarctica. In this fanciful picture, the CMB sky, as seen by the *BOOMERANG* project, is shown behind the prelaunch preparations for the balloon that carried the equipment. The images of the early universe have been overlaid onto the sky to indicate what size the fluctuations would appear if a standard 35-mm camera were sensitive to microwave light. The color map has been changed to match the rest of the picture.

All cross references to chapters, sections, figures, and tables pertain to the main text, *In Quest of the Universe, Sixth Edition. In Quest of the Solar System* contains Chapters 1–11 and 19 of the main text. *In Quest of the Stars and Galaxies* contains Chapters 1–5 and 11–19 of the main text.

less or greater than a certain critical value. . . . [W]hichever cosmological model proves correct, there is not much of comfort in any of this. It is almost irresistible for humans to believe that we have some special relation to the universe, that human life is not just a more-or-less farcical outcome of a chain of accidents reaching back to the first three minutes, but that we were somehow built in from the beginning. . . . It is very hard to realize that [our world] is just a tiny part of an overwhelmingly hostile universe. It is even harder to realize that this present universe has evolved from an unspeakably unfamiliar early condition, and faces a future extinction of endless cold or intolerable heat. The more the universe seems comprehensible, the more it also seems pointless.

But if there is no solace in the fruits of our research, there is at least some consolation in the research itself. Men and women are not content to comfort themselves with tales of gods and giants, or to confine their thoughts to the daily affairs of life; they also build telescopes and satellites and accelerators, and sit at their desks for endless hours working out the meaning of the data they gather. The effort to understand the universe is one of the very few things that lifts human life a little above the level of farce, and gives it some of the grace of tragedy.*

*Steven Weinberg, *The First Three Minutes*, New York: Basic Book, 1988.

Weinberg has since expressed some regret over the somewhat pessimistic tone of these words, but he is certainly not alone in feeling the enormity of the universe around us. It is easy to feel lost in the vastness of space; however, an increasing understanding of how the universe works can give us a more uplifting sense of how we fit into existence. We can also turn to religion as a way of dealing with our place in the universe; we discuss this briefly in an Advancing the Model box on Science, Cosmology, and Faith. However you feel about the size of the universe, the very contemplation of it seems to be a uniquely human endeavor.

In this chapter, we ask three simple questions: What is the nature of the universe? What was its past? What will be its future? These questions are the subject of cosmology, which studies the universe as a whole rather than its individual parts. Many of the ideas from Chapter 17 will be useful for answering cosmology's questions, but the questions are not easy to answer, and today's answers must be regarded as tentative.

Nature has provided us with some clues in our quest to understand the universe. Analyzing these clues, astronomers have been able to reach very meaningful answers to our three questions—questions that must have been among the first asked by intelligent humans.

18-1 The Search for Centers and Edges

Throughout the first 17 chapters of this book, a story has unfolded about changes in our basic ideas of the nature and extent of the universe. These changes have produced different answers to the question of whether we are located at the center of the universe. FIGURE 18-1 shows the life spans of the scientists whose cosmological contributions we describe in this section. Ptolemy's geocentric model (Chapter 1) placed the Earth at the center, and the stars were believed to be on a sphere surrounding the Earth.

Copernicus (Chapter 2) moved the Earth from the center of the universe and replaced it with the Sun, but the stars remained on a sphere surrounding the solar system. This sphere was generally considered to lie just beyond the most distant planet, and this was the entire physical universe; however, some important observations did not fit Copernicus' heliocentric model. It could not explain why the spinning Earth did not leave behind objects thrown upward, nor could it explain the lack of stellar parallax as the Earth moved around the Sun. A half century later, Galileo (Chapter 3) introduced the concept of inertia to explain motion, and he argued that stellar parallax was not observed because the stars are simply too far away. For this to be true, the universe had to be much larger than people had suspected.

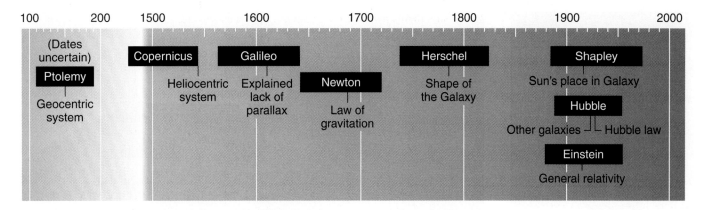

FIGURE 18-1 The life spans of the scientists whose cosmological contributions are described in this section.

Isaac Newton (Chapter 3) proposed that the universe is infinite in extent. He argued that if it were not, the force of gravity would cause the universe to collapse to its center. An infinite universe would not be in danger of collapse because each object in it would be pulled equally hard in all directions. Newton's idea did not catch on, however. Others claimed that the gravitational force has limited range and therefore cannot cause the universe to collapse. An infinite universe simply conflicted too strongly with "common sense."

During the 18th century, William Herschel's discoveries (Chapter 16) led to a better understanding of the Milky Way, again enlarging the boundaries of the known universe. Herschel's data, however, indicated that our solar system is at the center of the Galaxy, thereby returning the human race to center. Herschel proposed that the many fuzzy nebulae visible in telescopes are not part of the Milky Way, but are other galaxies like ours. Like Newton's idea of an infinite universe, Herschel's proposal was not accepted, and by the end of the 19th century, most astronomers were convinced that what we call the Milky Way Galaxy comprised the entire universe.

At the time, galaxies were called *isolated universes*, a name suggested by German philosopher Immanuel Kant.

In 1917, Harlow Shapley (Chapter 16) used globular clusters to conclude that the Sun is not at the center of the Galaxy, and in the next decade other astronomers calculated the Sun's motion. This showed finally that the Sun is not "special," either in nature or in position.

In 1923, Edwin Hubble (Chapter 17) discovered Cepheid variables in the Andromeda "nebula." This pushed back the limits of the universe tremendously, for astronomers recognized that our Galaxy of hundreds of billions of stars is just one of numerous galaxies. People perceived the size of the universe to be on a vaster scale than ever before imagined, and humans were far removed from any central position.

Einstein's Universe

While Hubble and others were using observations to advance our knowledge of the universe, several theoreticians were approaching the problem from a different perspective. Einstein's general theory of relativity (1915) substituted curved space for gravitational force. The theory holds that space itself is curved near a massive object and that the curvature causes the acceleration that previously had been attributed to the force of gravity. In Chapters 3 and 15 we described several observations that support Einstein's theory and show that it explains many phenomena better than does Newton's law of universal gravitation.

Newton knew that gravitational forces would cause a finite universe to collapse. The same situation appears in general relativity, but it is couched in different language. In the absence of any exotic forms of energy, if there is enough matter in the universe to eventually cause its collapse (because of the attractive nature of the gravitational force), space is curved enough that the universe is said to be **closed**. A closed universe has a positive curvature. Curvature of space is difficult to imagine because it involves a fourth dimension. The world of our direct experience is a three-dimensional one, and

closed universe The state of the universe if its total mass and energy density is greater than a specific value, called the critical density.

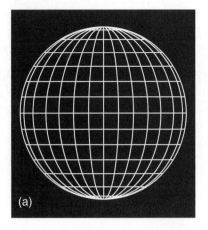

Our universe is curved in another direction. Um, that direction is—ahh, well, it is . . . just a different way.

Balloon surface

FIGURE 18-2 (a) A closed universe can be thought of as a two-dimensional plane curved into the third dimension so that it becomes a sphere. (b) Imagine that fleas on the surface of the sphere think that they live in only two dimensions. That is, the surface appears flat to them. Yet, if they travel far enough in one direction they will end up where they began.

Does a closed universe mean that nothing can get in our out?

open universe The state of the universe if its total mass and energy density is less than a specific value, called the critical density.

flat universe The state of the universe if its total mass and energy density is exactly equal to a specific value, called the critical density.

critical density The average mass and energy density of the universe at which space would be flat. It is equal to $3 \cdot H_0^2/(8 \cdot \pi \cdot G)$ and is equivalent to about 5.5 hydrogen atoms per cubic meter of space.

FIGURE 18-3 (a) An open universe can be represented as a plane curved into the shape of a saddle. The figure shows only a portion of the shape, for it extends forever in every direction. (b) A flat universe can be represented as a plane.

the best we can do to imagine a closed universe is to think of a two-dimensional plane being curved so that it forms a sphere (**FIGURE 18-2a**).

In Chapter 3 we compared our existence in a three-dimensional world that is curved into a fourth dimension with the situation of imaginary "flatfleas" on a two-dimensional surface that is curved (Figure 18-2b). If the creatures in the figure travel far enough in what they perceive as a straight line, they will return to their original positions. Likewise, in a closed universe, a beam of light sent in one direction will travel around the universe and return to its starting point (if the universe lasts long enough). Like the finite two-dimensional surface of a sphere, such a universe would have no boundaries, even though it would have a limited volume and a limited number of stars.

The alternative to a closed universe is an **open universe** that does not curve back onto itself. In the absence of any exotic forms of energy, the amount of matter in an open universe is not enough to slow down its expansion, and such a universe will continue expanding forever. An open universe has negative curvature; this can be visualized in a two-dimensional analogy as a saddle shape (**FIGURE 18-3a**). An open universe is infinite and continues on forever.

At the boundary between an open and a closed universe is the special case of a **flat universe**. In this case, the geometry that describes the universe is the well-known Euclidean geometry (Figure 18-3b). The universe is infinite and has just the "right" amount of matter and energy (the **critical density**) so that it shows no overall curvature; that is, it is flat.

Einstein believed that the universe is closed and finite. He would not accept the possibility that the universe was anything but static and unchanging (on a large scale). When he applied his equations of general relativity to the universe, he obtained a solution from which he concluded that such a universe would collapse on itself, just as Newton had concluded. (At the time, Einstein was unaware that his equations give rise to other solutions. One such solution, discovered by other scientists, corresponds to an expanding universe.) This seemed an unacceptable situation to Einstein, and he decided to adjust his theory to eliminate that conclusion. To do so, he inserted into the

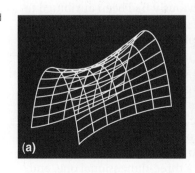

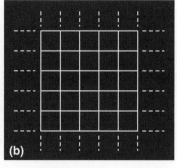

equations a *cosmological constant (Λ)*, a term that adds a cosmological repulsive force to support the universe against collapse. This force would tend to make the universe as a whole expand; however, this force would be very unusual, for instead of weakening with distance, it would become increasingly repulsive with distance. This explains why we do not experience the force at the small distances we normally deal with. The hypothetical cosmological repulsive force is significant only at large distances.

Einstein regretted having to adjust his theory in such an arbitrary, ad hoc manner. Adding the term violated the principle that a theory should be aesthetically simple. Nevertheless, he believed that the adjustment was necessary if his equations were to apply to a static universe. If he had rejected the cosmological constant, he probably would have concluded from his theoretical work that the universe is expanding. Instead, the expansion was discovered observationally.

Einstein abandoned the idea of a cosmological constant when evidence for the expansion of the universe and the big bang became almost completely accepted by the astronomical community; however, recent observations suggest the presence of an exotic form of energy, which results in a repulsive force helping the universe expand at an accelerating rate as time goes on. That is, the idea of a (positive) cosmological constant is supported by these observations. We do not really know yet the characteristics of this exotic form of energy.

> **cosmological constant**
> A term (denoted by the Greek capital letter "lambda," Λ) in the equations of general relativity that corresponds to a force throughout all space that helps the universe expand.

Something is said to be *ad hoc* if it is designed for one special purpose. An ad hoc committee, for example, is one that is created for a specific job.

18-2   The Expanding Universe

In 1929, Edwin Hubble announced that the pattern of redshifts of distant galaxies indicates that they are moving away from us and that the more distant the galaxy, the faster it is moving away. In Chapter 17, we explained how the resulting *Hubble law* has become a valuable tool to measure the distances to faraway galaxies. In addition, Hubble's law has tremendous implications concerning the nature of the universe.

First, if other galaxies are moving away from ours, does this not mean that our Galaxy must be at the center? Have we finally discovered that we are at a special location after all? Hubble published his findings in 1929, 12 years after Shapley's work with globular clusters had taken our Sun out of the central position within the Galaxy. By this time, it had become a working premise that our location within the cosmos is not central. To see why Hubble's discovery does not, in fact, conflict with this basic premise, consider the following analogy.

Imagine that you are a trainer in a flea circus and that you put a number of educated fleas on a balloon that your assistant is blowing up. You have instructed these fleas to hold their positions on the balloon. FIGURE 18-4 shows the balloon being blown up with the fleas in place.

Now imagine what a particularly intelligent flea sees when it looks out toward neighboring fleas. (Assume either that the flea can see around the curvature of the balloon or that the fleas are close enough together that the balloon seems flat to them.) The intelligent flea sees every other flea getting farther and farther away as the balloon is blown up. In addition, the more distant fleas are moving away at a greater speed than those nearby. The more distant a given flea is from the observer, the faster it is moving.

The important point is that the same result would be obtained no matter which flea is the observer. While the balloon is being blown up, every flea sees every other flea moving away with a velocity that depends on the flea's distance from the observer.

The analogy shows that if galaxies are part of an expanding universe, we would see exactly what Hubble and Humason observed: every galaxy is seen to be moving away from us at a speed that depends upon how far away

FIGURE 18-4 Fleas on a balloon get farther apart when the balloon is blown up.

it is. Many astronomers since Hubble have surveyed galaxies at greater and greater distances and have come to the same conclusion: the universe is expanding.

In Chapter 17 we included an Advancing the Model box concerning the differences between observations and conclusions. The distinction is important in this case. The pattern of redshifts is an *observation*. All spectral lines of distant (dim) galaxies show the redshift, not just the lines of the visible spectrum. The observations are 100% in agreement with what would be predicted for light from a receding object. We thus *conclude* that the galaxies are getting farther away from us. We do not, however, *observe* this.

What Is Expanding and What Is Not? The Cosmological Redshift

Saying that the universe is expanding certainly does not mean that the solar system is expanding, nor does it mean that stars within our Galaxy are getting farther apart. In fact, it does not even mean that all galaxies are getting farther apart.

The observed redshift leads us to conclude that other *clusters* of galaxies are moving away from ours. Some individual galaxies are actually moving toward us. This occurs in two ways. First, galaxies in our local group are moving randomly within the group as each responds to the overall gravitational force of the others. For example, the Andromeda galaxy and a half-dozen others in the local group are at this time moving toward our Galaxy.

Second, in nearby clusters, the same random motion of individual galaxies results in some of the galaxies moving toward us at the present time, even though the cluster in which each galaxy exists is moving away from the local group. A simple example can show how it is possible for an individual galaxy to show blueshifted lines even though it belongs to a cluster moving away from us. For a Hubble constant of, say, 70 (km/s)/Mpc, a cluster at a distance of 2 megaparsecs from us will be moving away from us at $70 \times 2 = 140$ km/s, due to the expansion of the universe; however, the random speeds of individual galaxies can easily exceed 140 km/s; therefore, if we were looking at an individual galaxy in this cluster with a random speed of 200 km/s, we could be seeing this galaxy moving at speeds between $140 + 200 = 340$ km/s and $140 - 200 = -60$ km/s. In the latter case, this galaxy will show blueshifted lines, even though its overall motion is away from us as a member of a cluster. No individual cluster—as far as we can tell—is expanding.

It is the clusters of galaxies that are moving farther apart. Although we often say that other galaxies are moving away from us, we thus should really say that other clusters are moving away from our cluster. Anything that is gravitationally bound is not expanding as a result of the expansion of the universe. This includes objects (like us) that are held together by forces stronger than the gravitational attraction between their individual components.

After the redshift of light from distant galaxies was discovered, astronomers assumed that the Doppler effect was the cause. Using the Doppler effect, they calculated the velocities of the galaxies. The Hubble law is a statement of the relationship between a galaxy's distance and its (Doppler effect) velocity, as we showed in Chapter 17; however, modern cosmology explains the redshift in an entirely different manner.

An expanding universe does not mean that clusters of galaxies are rushing *through* space. Instead, *space itself is expanding*. **FIGURE 18-5** illustrates the difference, using the balloon analogy again. This time, tiny seeds have been glued to the balloon, each seed representing a cluster of galaxies. Part (a) shows the Doppler interpretation of the expansion of the universe. The balloon is expanding against a background grid that represents space, indicating that the galaxies move through space. Part (b) shows the modern cosmological interpretation of the expanding universe. The grid that represents space is part of the surface of the balloon here. As the balloon expands, space itself expands.

Why don't individual galaxies expand along with space? General relativity shows that objects that are held together by their own gravity, such as the Earth, the solar system, the Galaxy, and the local group, do not expand as space expands. The force

The argument presented here also explains why we do not use the Hubble law to find distances to nearby galaxies.

The Doppler shift is still the correct explanation of other observations in astronomy.

ADVANCING THE MODEL

Cosmological Redshift

Objects have been discovered that have a redshift (z) greater than 5. A redshift of 5 means that the ratio of the shift in a particular wavelength ($\Delta\lambda$) to the unshifted wavelength (λ_0) is equal to 5. In the standard application of the Doppler effect, this means that the velocity of the object (v) is equal to five times the velocity of light (c), because

$$z = \frac{\Delta\lambda}{\lambda_0} = \frac{v}{c}.$$

This cannot happen! An object cannot travel at a speed greater than the speed of light. The resolution to this problem lies in the fact that the standard expression for redshift ($z = v/c$) is valid only in the approximation of speeds much less than the speed of light. For objects moving at large speeds (v larger than about $0.1c$), astronomers must take into account special relativity. When the rules of special relativity are used, we get the following general expression for redshift:

$$z = \frac{\Delta\lambda}{\lambda_0} = \sqrt{\frac{c+v}{c-v}} - 1.$$

This expression is valid for all speeds. For small speeds, it gives us our familiar expression $z = v/c$, whereas for large speeds (and thus large redshifts), the calculated speed of a receding cluster of galaxies is always less than the speed of light.

The fallacy here is that special relativity is not applicable in a universe where general relativity is the controlling rule. Special relativity can be applied only where space can be considered to be exactly flat. In the case of distant clusters, the curvature of space must be taken into account, and thus, special relativity cannot be used.

To review this relationship between laws, recall that Newton's law of gravitation should not be considered *wrong*. Rather, it is a law with limited applicability that must be replaced by general relativity in certain cases, but it is accurate enough for most everyday uses. Likewise, our standard concept of speed (where we do not worry about objects moving at speeds near the speed of light) works in everyday situations. But for accurate results, special relativity must be used for great speeds (especially for speeds greater than about $0.1c$). However, special relativity is also a law with limited applicability. It cannot be used when space curvature is a factor.

When we try to combine physical laws that are not consistent with one another, we violate the principle that a theory should be aesthetically pleasing. Consistency is important in applying scientific principles.

The Doppler effect thus cannot be used as the explanation for the cosmological redshift, as Doppler shifts are caused by an object's motion *through* space. The correct explanation lies in the idea of the expansion *of* space, as we discuss in the text.

of gravity (or, in terms of general relativity, the local curvature of space) holds them together just as each individual seed on the balloon holds itself together and does not expand with the balloon.

This interpretation of the expansion of the universe means that clusters of galaxies do not actually have a velocity through space and therefore that the Doppler effect does not explain the redshift in their spectra. A Doppler redshift is caused by the relative motion of an object that is emitting a wave. The *cosmological redshift* has a different cause. Long ago, when a distant galaxy emitted some electromagnetic waves, those waves were not redshifted, but as space expanded, the waves expanded—that is, they lengthened—along with space. The more time the radiation has traveled, the more its waves have stretched in wavelength. FIGURE 18-6 illustrates the difference between the two explanations.

When the Hubble law was written, astronomers assumed that the observed redshift was caused by the now-rejected Doppler explanation. The law is still valuable, however, if we think of the velocity in the Hubble equation as being caused by the expansion of space, rather than by velocity *through* space. Although astronomers sometimes use more modern ways of stating the redshift/expansion-rate relationship without using the concept of velocity, in this text, we continue to use the traditional Hubble relationship.

cosmological redshift
The shift toward longer wavelengths that is due to the expansion of the universe.

In equation form, the Hubble law is $v = H_0 d$, where v = the velocity of the object, d = the distance to the object, and H_0 is the Hubble constant.

Olbers' Paradox

Why is the sky dark at night? This appears to be a very simple question, but it actually has important cosmological implications. If stars (or galaxies or clusters of galaxies) are randomly distributed throughout space and if the universe were infinitely

FIGURE 18-5 Each tiny seed glued to the balloon represents a cluster of galaxies. (a) This is the old, incorrect interpretation of an expanding universe in which the universe expands through space. (b) This is the correct cosmological interpretation. Space expands, dragging the clusters of galaxies along with it.

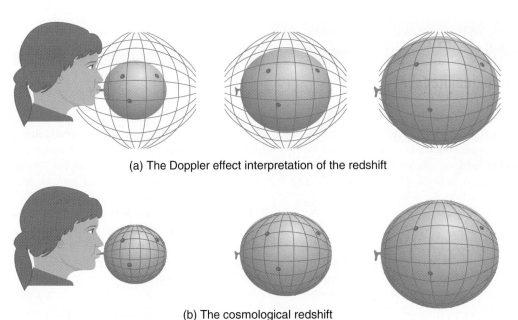

(a) The Doppler effect interpretation of the redshift

(b) The cosmological redshift

old, the sky should not be dark, because no matter in which direction you look, the line of sight should end on a star (**FIGURE 18-7a**). Light from any single star in the universe would have had infinite time to reach us. It is true that distant stars appear dimmer than nearby ones, but if we assume that stars are evenly distributed in our universe, then it is clear that there are more stars at great distances, and this eliminates the dimming effect. If you are deep within a forest, you see a tree no matter in which direction you look. Distant trees look smaller, but there are more of them (Figure 18-7b).

Perhaps the absorption of light by dust clouds prevents the sky from being bright, but dust clouds heat up as they absorb light. Because they absorb light from an infinite number of stars, they should get hot enough to glow. Dust clouds cannot be used to explain the dark sky.

FIGURE 18-6 (a) A Doppler redshift results from the *motion* of the source. The wave traveling toward the radio telescope is lengthened by the motion of the source. (b) When a wave is emitted by a *nonmoving* object, it has the same wavelength in all directions. As space expands, that wave lengthens. If the wave takes long enough to get from the source to the telescope, it will have lengthened significantly. This is the case for galaxies that exhibit the cosmological redshift.

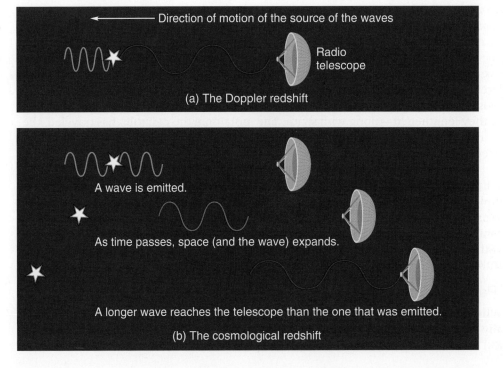

Direction of motion of the source of the waves

Radio telescope

(a) The Doppler redshift

A wave is emitted.

As time passes, space (and the wave) expands.

A longer wave reaches the telescope than the one that was emitted.

(b) The cosmological redshift

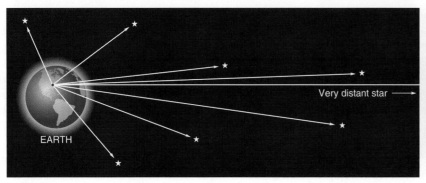 

FIGURE 18-7 (a) In an infinite, static universe, your line of sight would end at a star no matter in which direction you looked. (b) Deep in a forest, you see a tree in every direction.

The paradox between what is observed and the argument that the sky should not be dark was known long before it was popularized by Wilhelm Olbers in 1823. It is called *Olbers' paradox*.

The problem exists whether the universe is open or closed. An open universe extends without limit and, thus, the line of sight should extend through space until it hits a star. In a closed universe, the line of sight could possibly extend all the way around until the surface of a star is encountered.

The resolution of Olbers' paradox—the reason that the sky is dark—is based on two related ideas: one, that the universe has a finite age, and two, that it is expanding. First, a universe that has been expanding for billions of years, as suggested by observations, most likely had a beginning at one time in the past, emerging from a very dense state. This finite age for the universe (say about 14 billion years) implies that we cannot see any objects that are beyond a distance of 14 billion light-years from us. If we think of ourselves as being inside a sphere of radius 14 billion light-years, then our entire *observable* universe is located inside this sphere. Light from an object outside this sphere would have taken more than the age of the universe to reach us and thus we do not see it now. As a result, independently of whether the universe is finite or infinite, our line of sight does not always hit a star, which partially explains why the night sky is dark. Second, as a result of the expansion of the universe, light from distant galaxies is redshifted, and thus, it also loses energy. This occurs because the energy of a photon of light depends directly on the light's frequency, and thus, a photon of lower frequency and therefore greater wavelength (due to redshift) has less energy. This decrease in energy of the light we receive also helps explain (to a lesser degree) why the night sky is dark.

The resolution to the paradox depends on the fact that the universe is expanding and that it had a beginning. This latter point is at the heart of the major theory of modern cosmology, the *big bang theory*. Before discussing this theory, however, we need to describe the fundamental assumptions of modern cosmology.

> **Olbers' paradox** An argument showing that the sky in a static universe could not be dark.

> The resolution of Olbers' paradox due to the finite age of the universe was first pointed out in 1848 by the American writer Edgar Allan Poe.

> We discussed photons in Section 4-6. The frequency of a photon is inversely proportional to its wavelength.

18-3 Cosmological Assumptions

The assumptions of cosmology cannot be proved, but evidence indicates that they are reasonable. The first assumption is that the universe is *homogeneous*. This means that no matter where an observer is positioned in the universe, the universe looks essentially the same. This does not mean, of course, that there is another Earth from which an observer can look to see another Mars in the sky. It does not even mean that there is another local group of galaxies similar to our local group. Rather, it means that on the largest scale, the universe has about the same density and composition of matter at one location as it has at another. As we have described, humans have come to learn that they are not in a favored place in the universe. In cosmology, this becomes an underlying assumption.

> **homogeneous** Having uniform properties throughout.

FIGURE 18-8 Although the ocean surface is homogeneous, you see something different when you look in different directions because the surface is not isotropic.

isotropy (I-SOT-rah-pee) The property of being the same in all directions.

cosmological principle The basic assumption of cosmology that holds that on a large scale the universe is the same everywhere.

universality The property of obeying the same physical laws throughout the universe.

The second assumption is somewhat related to the first. The assumption of *isotropy* states that the universe looks the same in all directions. Although this sounds like homogeneity, it is actually a different quality. The ocean in **FIGURE 18-8** is homogeneous—for no matter where you are located on the surface of the ocean, the view is the same. You see, however, a different view when you look across the waves than when you look along the waves. This means that the ocean is not isotropic.

Like homogeneity, isotropy applies only on the largest scale. Recall from Chapter 16 that when we look out from Earth, we can see farther when we look out of the Galactic disk than if we look along the disk. The universe is not isotropic on this small scale. To apply the idea of isotropy, we must imagine being outside the Galaxy and even outside the local group of galaxies. In this case, the assumption is reasonable.

These two assumptions of cosmology are so much at the heart of the subject that they form what is called the *cosmological principle*. This principle formalizes the idea that we are not located at a special place in the universe. It brings to final completion the Copernican revolution that proposed that the Earth is not central. The cosmological principle not only states that the Earth is at no special place in the universe, but that there *is* no special place in the universe.

An additional cosmological assumption is *universality*. This means that the same physical laws apply everywhere in the universe. Although this may seem so obvious that it need not be stated, it was only fairly recently in human history that Newton found the first law of nature that could be shown to apply beyond the Earth. Now we have vastly more evidence for this assumption as we analyze radiation from the most distant objects, but we must remember that it is still an *assumption* that this principle holds throughout the universe.

These cosmological assumptions show our biases about the nature of the universe, but they are also necessary. Without them we would not be able to make any progress in understanding the universe. Unlike other disciplines, in cosmology we cannot control our subject, the universe, or make comparisons with another similar subject. There is only one observable universe. Based on these assumptions, our models have proven to be quite successful in describing the structure and evolution of the universe.

18-4 The Big Bang

big bang The theoretical initial explosion that began the expansion of the universe.

It's interesting that progress in understanding how the universe itself began depends on understanding the forces that act between the tiniest particles we know of and how these forces affect the behavior of these particles.

The present motion of galaxies leads us to believe that at one time in the past, the matter in the universe must have been much closer together. If we run the "movie" of the expansion of the universe backward, there must have been a time when the density of matter was extremely high and some sort of "explosion" initiated the expansion of the universe. This explosion is called the *big bang*, an irreverently trivial name for the event that began the universe. In its most basic sense, the standard big bang model is simply the idea that every bit of the matter and energy in the universe was once compressed to an unimaginable density. In the big bang, the material exploded apart.

We must be careful not to think of the material of the big bang as existing at a certain place, at one point in the universe. It *was* the universe—the entire universe. It did not exist as part of the universe.

Understanding the details of what happened in the big bang requires an understanding of particle physics, sometimes called "high-energy physics" because elementary particles are used in high-speed collisions to help us better understand what they consist of. Certainly the big bang represents the highest concentration of energy ever reached in the history of the universe. We describe some of the events that immediately followed the big bang in the Advancing the Model box "The Early Universe" on page 528.

After the initial explosion, the steady decrease in the temperature of the universe allowed the nuclear particles to form into atoms of low mass (hydrogen and helium). These atoms then clustered under the force of gravity into stars, clusters of stars, galaxies, and clusters of galaxies. The clusters of galaxies are still moving apart. Again, we must be careful not to think of the universe as expanding *into* empty space around it, for space does not exist apart from the universe. The universe is all there is, and its parts are moving away from one another.

A common *misconception* is to consider the big bang as being similar to the situation shown in **FIGURE 18-9**. Suppose a large firecracker is placed at the center of a small stack of sand. Soon after the explosion of the firecracker, grains of sand are at various distances from their original position, and—here is the important point—grains that are farther from their respective starting points are moving faster. That is how they got farther from one another. No matter which grain one considers, all others are moving farther from it. Although this example is useful to illustrate that the fastest-moving galaxies would naturally be the farthest away, the example is misleading if you do not remember that in the case of cosmological expansion, galaxies do not move *through* space, as the grains of sand do. Because we have no experience with the expansion of space in our everyday lives, Earthly examples must involve motions through space.

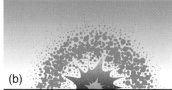

FIGURE 18-9 An illustration of the misconception that the big bang is an explosion. (a) A firecracker about to explode inside a small pile of sand. (b, c) The sand shown later. The sand grains that are farthest from the initial explosion are moving fastest.

Evidence: Background Radiation

The idea of the big bang arose from the evidence for the expansion of the universe. We must ask, however, if there is other evidence to support the theory. The first evidence was found quite accidentally when a prediction made by the big bang theory was found to be correct. As scientists learned more about nuclear processes in the 1940s, they were able to make predictions about the character of the early universe shortly after the big bang. The material of the big bang was originally extremely hot, but it cooled as it expanded outward, in much the same way that a gas cools when it expands. The hot matter of the very early universe was opaque to radiation, but when it cooled to about 3000 K, it became transparent. This happened when the universe was about 376,000 years old, at a redshift of about $z = 1100$. At this point, the gas that made up the universe was emitting radiation that had the characteristics of radiation from an object at 3000 K, and this radiation existed over the entire universe.

The universe has remained transparent to radiation since that time and, thus, it was predicted that this radiation should still be in existence. Remember that as we look at objects at greater and greater distance, we are looking back in time. If we were to detect the radiation left over from the big bang, it would be coming from far back in time and therefore from a great distance. According to the Hubble law, then, it would be greatly redshifted. The original 3000-K radiation peaked in the infrared region of the spectrum, but calculations showed that it would by now have been redshifted to the microwave region, with wavelengths of a millimeter or so. Such radiation would not be characteristic of radiation from a 3000-K object, but it would appear the same as radiation from a very cold object—one at about 3 K ($-270°C$). It was predicted that this ***cosmic microwave background radiation (CMB)*** should be striking Earth from all directions but that it should be very faint.

In 1948, when the prediction of the CMB radiation was made, there was no way to detect such weak waves. The prediction was laid aside and forgotten. Then in the mid-1960s, a group of physicists at Princeton University was studying the big bang and again predicted the existence of the CMB radiation, not realizing that the prediction had been made some 15 years before. They were confident that they could build a radio receiver that would be able to detect the radiation, and they set about building one on the roof of the Princeton biology building. They were too late.

cosmic microwave background radiation (or *CMB radiation*). Long-wavelength radiation observed from all directions; thought to be the remnant of radiation from the big bang.

ADVANCING THE MODEL

The Steady-State Theory

The cosmological principle holds that we are at no special place in space. It does, however, allow for us being at a special place in time. For at the present moment, the universe is less dense than it was in the past and more dense than it will be in the future because expansion reduces its density. In 1948, *a perfect cosmological principle* was proposed by astronomers Fred Hoyle, Hermann Bondi, and Thomas Gold. It states that our location in time is no more special than our location in space and that the universe of the past was the same as it is now and the same as it will be in the future. Because the overall universe does not change in this cosmology, it is called the *steady-state theory*.

Before looking at the implications of the perfect cosmological principle, consider why such an idea might be proposed. Remember that a theory should be aesthetically pleasing. The perfect cosmological principle is neater because it expands the nonspecialness to include time as well as space. The perfect cosmological principle carries the nonspecialness idea through both space and time.

It seems to be almost undeniable that the universe is expanding. How can this be squared with the steady-state theory? It seems that if clusters of galaxies are getting farther apart, the universe will necessarily be less dense in the future than it is now. Hoyle, Bondi, and Gold proposed that matter is created in the space between galaxies and that this new matter is sufficient for the universe to maintain its density (**FIGURE B18-1**). Such spontaneous creation of matter would violate a long-established principle of physics: the conservation of mass/energy, which holds that the total amount of mass and energy in the universe cannot change.

The principle of the conservation of mass/energy is dear to scientists, but it has changed in the past. Before this century, there were two such conservation principles: the conservation of mass and the conservation of energy. It was thought that neither matter nor energy could be created or destroyed. Then, in 1905, Einstein's theory proposed that mass could be changed into energy—a prediction that was dramatically demonstrated in the atomic bomb. Einstein combined the two principles into one, the conservation of mass/energy, which sees mass and energy as two forms of the same thing and holds that the total mass and energy in the universe remains the same.

The amount of mass that would have to be created to fill the gaps between galaxies as they move apart would be very small—only one new hydrogen atom in each cubic meter of space every 10 billion years or so. This means that if three new atoms were created every 10,000 years in the volume of Cowboys Stadium, the density of the universe would remain the same. The theory holds that as millions of years pass, newly formed atoms collect to form new stars and galaxies,

and that the expanded space is thereby refilled so that the universe is no different now than it was in the past and than it will be in the future.

The fact that the steady-state theory violates the principle of conservation of mass/energy should not be considered a strike against the theory. First, according to the predictions, the necessary creation events are so rare that they would be far below any possible detection limit and therefore far below the accuracy to which the principle has been verified. Second, the big bang theory may also involve a violation of the principle. In this case, it is a gigantic one-time violation, as all matter and energy are created in one big burst at the beginning.

One of the first predictions made by the steady-state theory involved the number of faint (very distant) radio sources. In this theory, the number density of such sources (that is, the number of radio sources per unit volume) remains constant, and so does the average energy radiated away by a radio source per unit time. This leads to a very specific relationship between the number density and the energy received per unit time per unit area; however, observations made in the 1950s implied a different relationship, suggesting that either the number density or the average energy increases with distance. This, of course, implies that the universe is evolving, in contradiction to the steady-state theory.

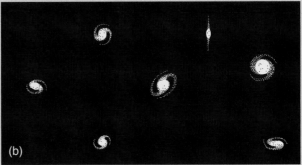

FIGURE B18-1 (a, b) If the universe does not decrease in density as galaxies move apart, matter must be created in the space between galaxies.

ADVANCING THE MODEL

The Steady-State Theory (Cont'd)

Another argument against this theory involves the amount of helium observed in the universe. This amount is much greater than what can be expected from stellar evolution; therefore, we will have to accept that either helium is also spontaneously created or that there was a special period in the universe's life during which a great amount of helium was created, along with other elements.

Finally, the discovery of the cosmic background radiation made it impossible to reconcile its existence with the steady state theory. Even if we assume an unknown mechanism that can produce such radiation in the steady-state theory, its temperature would have to be constant, independent of redshift; however, the measured temperature of this radiation depends linearly on redshift. That is, at redshift z, the measured temperature is greater by a factor $(1 + z)$. This implies that the universe is evolving.

Hoyle, Burbidge, and Narlikar proposed a quasi-steady-state theory in an effort to bring the theory closer to an agreement with the observations; however, none of the many versions of this theory has been successful so far. The historical development of both the big bang and steady-state theories provides another example in our discussion of how science is done. At the end, the theory that is accepted the most is the one that is simple and elegant, explains the current observations, makes predictions that can be tested, and allows itself to be shown wrong.

At the same time the predictions were being made at Princeton, only a few miles away two employees of Bell Telephone Laboratories, Arno Penzias and Robert Wilson (FIGURE 18-10), were doing applied research on microwave transmission in hopes of improving the transmission of messages. They were frustrated, however, by some low-intensity radio waves arriving at their receiver from all directions.

As you might suspect by now, the radiation they had found turned out to be the CMB radiation predicted by the physicists at Princeton and by astronomers years before—the prediction that was based on the big bang theory. Penzias and Wilson were awarded the Nobel Prize in 1978 for their discovery of a radiation for which they had not been searching and the cosmological implications of which they were unaware.

At the time of the big bang, the condensed material made up the entire universe. This is why the CMB radiation now comes to us from all directions. This radiation fills the universe just as it did at the beginning. Matter and energy comprise the entire universe, and no space is left over.

FIGURE 18-10 Arno Penzias (right) and Robert Wilson discovered the cosmic background radiation using Bell Laboratories' horn-shaped radio antenna, seen behind them.

Until the discovery of the CMB radiation, there was another popular cosmological theory (see the Advancing the Model box "The Steady-State Theory"), but this discovery established the big bang theory as the accepted explanation for the beginning of the universe.

In 1989, NASA launched the *Cosmic Background Explorer* (*COBE*, FIGURE 18-11). Its purpose was to measure the CMB radiation at various wavelengths and to map the sky in those wavelengths. *COBE*'s results provide an astounding corroboration of the accuracy of predictions concerning the CMB radiation. FIGURE 18-12 is an intensity-wavelength graph like those used to describe the temperature of stars in Chapters 12 through 15. The red line on the figure is the predicted intensity of radiation at each wavelength. The tiny black lines represent data points measured by *COBE*. Each data

FIGURE 18-11 The *Cosmic Background Explorer* (*COBE*), pronounced "KOH-bee," was launched in 1989 and for 4 years provided important data regarding the CMB radiation.

ADVANCING THE MODEL

The Early Universe

Stephen Weinberg's book, *The First Three Minutes*, was the first book written for the general public that describes the model cosmologists have developed to explain what happened at the beginning of the universe. Some of the events in this model have been refined over the past few decades as more data have accumulated from observations by satellites and balloon-borne instruments, but the sequence of events has become accepted enough to be called the "standard model of cosmology."

The model starts at about 10^{-43} seconds after the big bang itself. Why 10^{-43} seconds? Well, according to our present understanding of space and time, there is a limit to how closely you can look at tiny distances and still know what you're looking at. You know that if you look at a piece of paper, like a page of this book, through a microscope, it will no longer appear smooth, but will show a mixed-up jumble of wood fibers, along with filler that holds the fibers together. If you keep going in your imagination, you might think of the jumbled-up atoms that make up the page and the particles that make up the atoms, but if you keep going, eventually space itself becomes jumbled instead of smooth. Scientists have calculated the smallest distance that we can imagine before this jumbling occurs, coming up with a distance of about 4×10^{-35} meters. Light would take about 10^{-43} seconds to cross this distance, which makes this amount of time the smallest interval for which we can talk about the universe.

What was the universe like at this time? There were no atoms, and everything was scrunched up into a far smaller size than any particles we're familiar with. All we can say with any confidence is that the universe was hot, about 10^{32} K, and it was expanding. As it expanded, it cooled, very quickly. At about 10^{-35} seconds, a period of inflation may have occurred, lasting for only 10^{-32} seconds, but having tremendous implications for the history of the universe, as we discuss in Section 18-6.

By the time of one millionth of a second (10^{-6} s) after the big bang, the temperature had dropped to about 10^{13} K, which was cool enough (relatively speaking, of course!) for protons and neutrons to form. How do we know this? Because according to present theories, at temperatures higher than 10^{13} K, these particles break up into even smaller particles (called quarks; you may have heard about these littlest of all particles). After you get below this temperature, the quarks don't have enough energy to move away from each other and stay bound together in protons and neutrons.

To form nuclei more massive than hydrogen, the temperature of the background radiation must be low enough so that it does not break the nuclei apart. The first element more massive than hydrogen (whose nucleus is just a pro-

ton) is helium. Nuclei of helium consist of two protons and two neutrons or two protons and a single neutron. The easiest way of forming helium nuclei is by first forming deuterium nuclei, which consist of a single proton and a single neutron; however, for the first 10 seconds after the big bang, the temperature of the background radiation was too high (more than 5 billion K) for deuterium to form. As the universe cooled and deuterium nuclei formed, they led to the formation of helium nuclei. This continued for about 15 minutes, by which time the temperature had dropped to below about 1 billion K and no further nuclei were formed. Again, how do we know this? The evidence is in the proportions of hydrogen, helium, and deuterium we have observed in the universe. Other sequences of events in the early universe have been proposed, but they all predict formation of the light elements in different proportions than what we observe.

The formation of light elements (hydrogen, deuterium, helium, and lithium) in the early universe is called "Big Bang nucleosynthesis." Elements heavier than lithium were later formed in stars, as we described in an Advancing the Model box on "Nucleosynthesis" in Chapter 15.

FIGURE B18-2 shows how the predicted abundances of deuterium, helium, and lithium depend on the density of ordinary matter in the early universe. The circles correspond to the observed abundances. The predicted abundance for

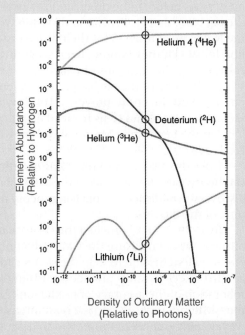

FIGURE B18-2 The curves show the dependence of the predicted abundances of helium, deuterium, and lithium on the density of ordinary matter in the early universe.

ADVANCING THE MODEL

The Early Universe *(Cont'd)*

helium-4 is about 24% of the ordinary matter in the universe, which agrees very well with the observations. The abundance of deuterium in the interstellar gas in our Galaxy was measured by astronomers at Haystack Radio Observatory; their results, along with observations of abundances of the other light elements, imply that only 4% of the total mass/energy density is in the form of ordinary matter. This is consistent with the "standard model" of a universe with a flat geometry, as we discuss in Section 18-7.

What happened next? Nothing much. The universe, dominated by radiation, continued to expand and to cool; however, as the universe expanded, matter became increasingly dominant. About 10,000 years after the big bang, when the temperature dropped to about 20,000 K (at redshift of about $z = 6700$), the universe became matter dominated and became increasingly so ever since. The expansion and cooling of the universe continued for the next few hundred thousand years. Only then, when the temperature dropped below a few thousand degrees, did electrons slow down enough to be captured by nuclei and form atoms. At this point, about 376,000 years after the big bang, at redshift of about 1100, an amazing

thing happened: the universe became transparent. Before this time, the background photons were energetic enough that they would not allow electrons and protons to get together and form atoms. Instead, the photons were continuously and vigorously colliding with the charged particles, resulting in a very opaque universe. As the temperature of the background radiation dropped below about 3000 K (376,000 years after the big bang), the photons did not have enough energy to keep electrons and protons from forming hydrogen atoms. After hydrogen atoms formed, the photons could now move freely through space, resulting in a transparent universe. It is these photons that make up the cosmic microwave background we observe today.

How reliable is this picture of the early universe? Some of it is conjecture, some of it is debatable; certainly none of it is proven. There is no photograph or eyewitness account of the big bang, but the description given here, although somewhat simplified, fits all the data we have so far acquired about the early universe. As more data become available, we will continue to revisit and refine our ideas about what happened in the beginning.

point is depicted as a line, which takes into account possible errors in the measurement. The longer lines at the left indicate that the measurements at shorter wavelengths are less accurate. The data fit is extraordinary.

The *COBE* data show that the temperature of the background radiation is 2.725 K (with an uncertainty of 0.002 K). This matches the theory to within 0.03%, which is 1000 times better than the best data before *COBE*.

Additional Evidence for the Big Bang

The big bang theory is the best current model for the universe. Observational data support the theory but do not prove it, because scientific theories are not proven. We have already mentioned some of these observations: the darkness of the night sky, the Hubble law, homogeneity, and isotropy at large scales.

Another argument in favor of the big bang involves the time dilation of distant supernova light curves. This dilation is the result of the standard interpretation of redshift. A supernova that takes 20 days to decay will appear to take $20 \times (1 + z)$ days to decay when observed at redshift z. Such dilation has been observed, providing a very strong confirmation of the cosmological nature of redshift.

Some additional observations show the evolution of the universe and are consistent

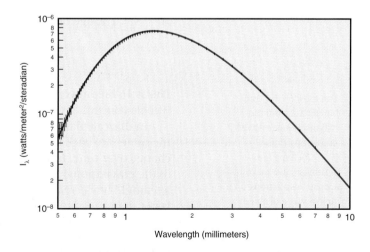

FIGURE 18-12 The theoretical intensity-wavelength graph for the cosmic microwave background radiation is shown in red. The tiny black lines represent data from *COBE*. The fit is remarkable!

with the big bang theory. For example, we mentioned in the Advancing the Model box on "The Steady-State Theory" that the observed relationship between the number density of distant radio sources and quasars and the energy received from them per unit time per unit area shows that the universe evolved. The existence of the CMB and the dependence of its temperature on redshift also show that the universe evolved from a hot and dense state.

Additional evidence that supports the big bang comes from the observed proportions of light chemical elements (such as deuterium, helium, and lithium). These proportions could not have formed in stars; however, they are consistent with a process of nuclear fusion in the first few minutes of a hot, young universe, as we described in the Advancing the Model box on "The Early Universe."

Finally, as we discuss later in this chapter, the observed anisotropies in the CMB radiation support a big bang model that includes an inflationary period and is dominated by dark matter and dark energy.

The Age of the Universe

If the big bang theory is correct, we can use the Hubble constant to determine how long ago the big bang occurred, for the constant tells us how fast galaxies are spreading apart. Here we make the simplifying assumption that the rate of expansion, given by the Hubble constant, has remained constant through time. Even though the assumption is not exactly correct, it allows us to find an approximate age for the universe. Instead of writing v for speed in the Hubble law, substitute it with the definition of average speed: the ratio of distance to time.

$$v = H_0 d, \quad \text{or} \quad \frac{d}{t} = H_0 d.$$

Now choose two widely separated galaxies and let d be the distance between them. Thus, t is the time taken to go that distance, or the age of the universe. Solving the equation for t gives

$$t = 1/H_0.$$

This indicates that the age of the universe is simply the reciprocal of the Hubble constant. This constant is normally expressed as (km/s)/Mly or (km/s)/Mpc. Before we can substitute numbers to calculate the age of the universe, both distances must be expressed in the same units, rather than one in kilometers and the other in millions of light-years. One million light-years is about 9.5×10^{18} kilometers, and thus, a Hubble constant of 20 (km/s)/Mly corresponds to 2.1×10^{-18} (km/s)/km or 2.1×10^{-18} s^{-1}. The age of the universe is the reciprocal of this, or

$$t = \frac{1}{2.1 \times 10^{-18}\,\text{s}^{-1}} = 4.8 \times 10^{17}\,\text{s} = 15 \times 10^9\,\text{yr}.$$

This is 15 billion years. On the other hand, using 25 (km/s)/Mly as the value of the Hubble constant gives an age for the universe of 12 billion years.

In this calculation we assumed that galaxies have continued to move apart at the same rate back through their history; that is, that the Hubble constant does not change over time. In fact, we would expect the speeds of galaxies to change. Because of the gravitational force they exert on each other, the recessional speeds of galaxies should be decreasing. This would mean that in the past, they were moving apart faster than they are now and that the age of the universe is therefore less than calculated above. For this reason and because the Hubble constant is not known precisely, 15 billion years is taken as the *upper limit* to the age of the universe. The universe is this age or younger.

The Hubble constant does *not* remain constant as time goes on. It is the slope of the line in the graph of recessional velocity versus distance (for example, see Figure 17-11) as measured during the *current* period of time in the life of the universe.

The most recent combined data from the *Wilkinson Microwave Anisotropy Probe* (*WMAP*), Type Ia supernovae, and baryon acoustic oscillations yield a value of about 70.1 ± 1.3 (km/s)/Mpc or about 21.5 ± 0.4 (km/s)/Mly for the Hubble constant. This puts the age of the universe at about 13.73 ± 0.12 billion years.

ADVANCING THE MODEL

Science, Cosmology, and Faith

Where did the matter and energy of the big bang originate? Science cannot answer that question at present. In fact, it seems possible—perhaps likely—that it will never be able to answer that question. For if matter existed in atomic form before the big bang, it did not carry information with it through the big bang. Matter itself thus can tell us nothing about its pre-big bang history. Actually, it is meaningless for us to talk about things "before" or "at the moment of" the big bang. Time did not exist until after the big bang. Because of the extreme conditions involved during the birth of our universe, the known laws of physics do not allow us to start describing its evolution until 10^{-43} seconds after the big bang. After this moment, we can consider the nature of space and time in the same way as we do today.

Does this then mean that science can say that there must be a creator and that science has, after all, found a need for God? An individual scientist may believe this, but science itself cannot use God as the explanation for the big bang or anything else. The reason that science avoids using God for explanations is basically that science has been successful in explaining the material world without reference to a God. Science, by intention, uses natural causes to explain natural effects.

We say that science is successful in its method because scientific explanations of the workings of the material world have led us to further understanding of that world. The fact that the success has come without reference to God indicates that the material universe seems to be describable by completely natural principles.

What about cases where science is unable to find an answer? If science, when it comes to something that it cannot explain at the time, were to explain it by reference to God, the search for an explanation would end. If Newton had used God as an explanation for why things fall to Earth, he would never have developed the theory of gravitation. If we use God as an explanation for the big bang, there would be no reason to look further for a natural explanation. Use of supernatural explanations would shut down science. History, however, tells us that it is profitable to look for natural explanations. Supernatural explanations cannot be used in natural science.

Testability

There is another reason that science cannot use God for an explanation, and this relates to the reason that traditional science does not accept creationism as a science: A theory of science must be able to be shown to be wrong. A theory must be testable. Every theory must be regarded as tentative, as being only the best theory we have at present. It must contain within itself its own possibility of destruction. The 1948 prediction concerning the cosmic microwave background radiation was such a case. If the background radiation had not been found, the big bang theory would have had to have been adjusted, or—if enough such contradictions appeared—the theory would have had to be dropped and replaced with another. This has happened many times in science. It happened to the steady-state theory and to other theories we presented in this text.

On the other hand, if science relied on a creator to explain the inexplicable, there would be nowhere to go, no way to prove that explanation wrong. The question would have already been settled. This is not to say that God might not be the explanation. Many people believe that a creator is the ultimate explanation of everything. Perhaps some questions, like the origin of the material for the big bang, simply cannot be answered without reference to a creator, but in that case, science simply cannot answer the question. The question is beyond the realm of science.

Science does not deny the existence of God. God is simply outside its realm.

As we noted in Chapter 17, the most distant quasar yet found has a redshift of 6.41. This indicates a look-back time of about 96% of the age of the universe. Although the age of the universe is not known exactly, Doppler shift data are accurate enough that we can be confident that the farthest quasar thus found has a look-back time of about 96% of whatever is the correct value for the age.

18-5 The Future: Will Expansion Stop?

What came before the big bang? As we discuss in the preceding Advancing the Model box, this question may not be a scientific one. However, to see how scientists might get a hint at what came before the big bang, we look into the *future*.

If galaxies are moving apart, they will obviously be farther apart in the future than they are now. Where does this stop? There *is* an agent to stop it: gravity.

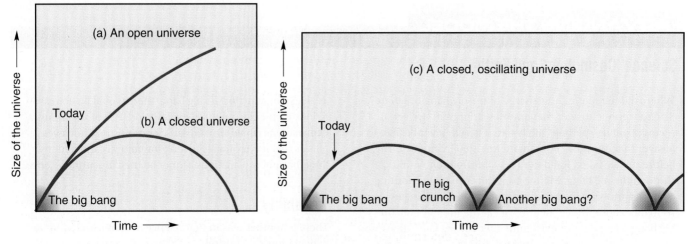

FIGURE 18-13 (a) An open universe would continue to grow in size indefinitely. (b) A closed universe would someday stop growing and begin to contract, assuming that there is no cosmological constant. (c) Could it be possible that a closed universe could oscillate, going through consecutive big bangs and big crunches?

◆ A closed universe will stop its expansion only in the absence of an exotic form of energy. An open universe won't stop at all.

oscillating universe model
A big bang theory that holds that the universe goes through repeating cycles of explosion, expansion, and contraction.

The force of gravity must be slowing the expansion of the universe. The question is, will gravity slow the expansion enough to stop it and bring the galaxies back together?

First, suppose the expansion does not stop. This means, simply, that the clusters continue to get farther apart. **FIGURE 18-13a** illustrates a universe that continues to grow. In this universe, the stars use up the hydrogen fuel that powers them, the glowing stars become fewer, the glow fades, and the universe fizzles out.

On the other hand, suppose the expansion stops, as shown in Figure 18-13b. Gravity begins to pull the galaxies back toward one another. Intelligent beings living on some planet will be able to use the Doppler effect to observe that the galaxies are then getting closer together. The infall will continue until all matter in the universe is condensed into a tremendously dense ball—the "big crunch." What might happen then? Perhaps another big bang, another universe. If this is the future, perhaps it was the past. Perhaps this is what preceded the big bang: a previous universe containing the same matter and energy as ours. If so, then today's universe may be just one phase of a number of oscillations, each having its own big bang. Such a scenario is called an *oscillating universe* (Figure 18-13c). It is a much more exciting prospect than a fizzling-out universe.

A model for an oscillating universe with no beginning or end was first proposed in the 1930s. This model was abandoned because it could not be reconciled with the second law of thermodynamics; this law states that the entropy of a closed system (a measure of its disorder) can never decrease. As a result, if the universe were oscillating, it would actually become larger in every cycle, with each oscillation becoming longer; therefore, looking back, it would also have to have a beginning, a big bang.

New models for an oscillating universe have been proposed. Indeed, computer simulations using quantum modifications to Einstein's general theory of relativity show that in place of a classical big bang there is actually a quantum bounce, leading to a new expansion; however, more work needs to be done to understand better the nature of quantum gravity. In another model, dark energy drives the expansion of the universe to a stage where all matter disintegrates and the universe fragments into patches that are so far apart that they cannot be reconnected. At this point, just before the end of time, *each* patch starts collapsing; it contracts individually, and then bounces out again creating a new universe. In this model, each "causal patch" becomes an independent universe, initially almost empty of matter and entropy; as a result, the model avoids the difficulties encountered during the contraction phase. Future observations could give us enough data to determine the equation of state for dark energy and thus test such models.

Evidence: Distant Galaxies and High-Redshift Type Ia Supernovae

As we described before, the borderline case between an open and a closed universe is called a flat universe. If the universe is flat and assuming no cosmological constant (that is, no influence that supports the universe against collapse), the universe will gradually slow its expansion, but will stop only after an infinite amount of time, never falling back inward. In order for the universe to be flat, however, it must meet an exact condition described by the value of the critical density. If the total mass and energy density of the universe is equal to the critical density, then the universe is flat. For simplicity, astronomers prefer to work with the ratio of the total mass and energy density of the universe to the critical density. This ratio is called the **density parameter** (Ω_0). If Ω_0 is equal to 1, the universe is flat and will expand forever, but only barely so. If it is greater than 1, the universe is closed, and it will eventually collapse. If it is less than 1, the universe is open and will expand forever.

In what we just described we assumed that there is no cosmological constant and that gravity is the main factor in determining the future of the universe. But do the observations support this assumption? To answer this question, astronomers study the relationship between distances and recessional velocities for distant galaxies and Type Ia supernovae. If the rate of expansion has decreased (because of gravity) or remained constant or increased (due to the effects of an unknown influence), any deviations from the Hubble law should become obvious at great distances.

FIGURE 18-14a shows the results of a project to measure H_0 based on a Cepheid calibration of a number of independent standard candles (such as Type Ia and Type II supernovae and the Tully-Fisher relation). The data show a linear relationship between recessional velocities and distances. This implies that the rate of expansion has not changed; however, the data go only as far as 400 Mpc, which corresponds to about 1.3 billion years into the past (only about 10% of the age of the universe). Even in the presence of gravity, the observations thus suggest that the rate of expansion has remained approximately constant over the past 1.3 billion light-years. The value of the Hubble constant obtained from the data is about 72 ± 8 (km/s)/Mpc. If we also take into account the current best estimates of the ages of globular clusters (about 12.5 billion years), these results favor a flat universe ($\Omega_0 = 1$) dominated by the existence of some form of exotic energy that supports the expansion of the universe. This

density parameter (Ω_0)
The ratio (denoted by the Greek letter "omega," Ω_0) of the total mass and energy density of the universe to the critical density.

The value of the Hubble constant changes with time. As a result, astronomers refer to times in the past in terms of redshift, not years. Since the time that light left an object at redshift z, the universe has expanded by a factor $1 + z$. At that time, distances between clusters of galaxies were $1/(1 + z)$ as great as they are now.

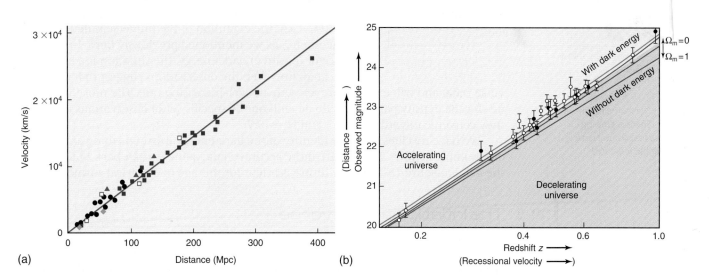

FIGURE 18-14 (a) The Hubble diagram for different standard candles calibrated by Cepheids. (The candles are shown by different symbols; filled squares correspond to Type Ia supernovae.) These are the final results of the HST Key Project (published in 2001). (b) The Hubble diagram for Type Ia supernovae. Supernovae at $z > 0.5$ appear dimmer (and thus farther away) than their redshifts would suggest if the universe were coasting or slowing down due to gravity. The data, published in 1998 by the Supernova Cosmology Project and the High-Z Supernova Search, suggest that the cosmic expansion has been accelerating since the universe was more than half its current age. The red curves correspond to models with $\Omega_\Lambda = 0$ (no dark energy) and range from an empty universe ($\Omega_m = 0$) to a universe of critical density ($\Omega_m = 1$). For a flat universe, the best fit is the blue line. It corresponds to a universe dominated by dark energy, with $\Omega_m = 0.3$ and $\Omega_\Lambda = 0.7$.

dark energy An exotic form of energy whose negative pressure currently accelerates the expansion of the universe.

energy, called **dark energy**, cannot be detected by its gravitational effects and emits no radiation. It provides a strong negative pressure that, just like the cosmological constant, is currently accelerating the expansion of the universe.

The data shown in Figure 18-14b go as far as 10 billion years (about 75% of the age of the universe). This graph shows the observed magnitude versus redshift of a number of nearby and distant Type Ia supernovae. These objects have very similar absolute magnitudes and thus are great standard candles. Their slow march to a sudden explosive end seems to erase most of the differences among the original stars. Recall from Chapter 12 that the apparent magnitude of an object is related to its distance through the distance modulus. Also, recall that redshift is related to recessional speed. These relationships are not linear; however, the greater the redshift of a supernova, the greater is its recessional velocity; the greater the apparent magnitude of a supernova, the greater is its distance. Figure 18-14b, therefore, is also a graph of distance versus recessional velocity. The low-redshift points in the diagram ($z < 0.1$) are not shown, but they do imply a linear Hubble law consistent with Figure 18-14a as the data go only as far as 2.5 billion years in the past; however, the high-redshift data (up to $z = 1$) go as far as 10 billion years in the past. It is in this region that we see deviations from the linear Hubble law. The different curves correspond to different cosmological predictions, depending on what we assume for the density parameters for matter and dark energy. For a flat universe (as indicated by the latest measurements of the CMB radiation), the best fit is the blue line. This implies that the universe is dominated by dark energy.

Contributions to the density parameter come from three sources: matter (Ω_m), radiation (Ω_{rad}), and any possible form of exotic energy (Ω_Λ); however, as we discussed in the Advancing the Model box on "The Early Universe," the contribution of radiation is negligible to that of matter. We can, therefore, write $\Omega_0 = \Omega_m + \Omega_\Lambda$.

The observational data shown in Figure 18-14 (and in Section 18-7) show that the cosmic expansion is now accelerating and that the universe is dominated by dark energy. **FIGURE 18-15** shows how the relative size of the universe (represented by the distance between two widely separated objects) evolves with time in different models. In the absence of dark energy ($\Omega_\Lambda = 0$), the evolution of the universe is determined by its geometry, as we mentioned previously here. In the presence of (positive) dark energy, the situation is a bit more complicated; however, the observations suggest that our universe is flat and will expand forever with an ever-increasing rate. The most likely values for the density parameters do depend on the specific set of observations used, but they converge toward $\Omega_m \approx 0.3$ and $\Omega_\Lambda \approx 0.7$.

We have no clues as to the identity of the dark energy. Distinguishing among the different possibilities will require more accurate data, going further back in time than the data in Figure 18-14b. The future is bright for this age of empirical cosmology.

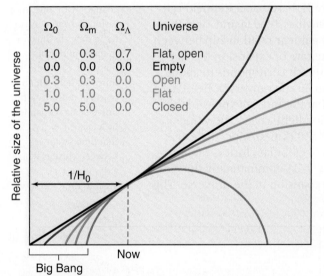

Ω_0	Ω_m	Ω_Λ	Universe
1.0	0.3	0.7	Flat, open
0.0	0.0	0.0	Empty
0.3	0.3	0.0	Open
1.0	1.0	0.0	Flat
5.0	5.0	0.0	Closed

Relative size of the universe

$1/H_0$

Now

Big Bang

FIGURE 18-15 The evolution of the universe in different models. There is growing evidence that our universe is following the red curve.

18-6 The Inflationary Universe

Scientists do not like coincidences, and the standard big bang model is unable to explain two cosmological observations that appear to be coincidences. Attempts to solve these two problems, as well as others, led astronomers to propose a modification of the standard model.

The Flatness Problem

The universe seems to be either flat or very nearly flat. This would be an extreme coincidence because a flat universe is a special case. Of all the possible densities that the universe might have, why would it have a density that puts it exactly on the border between open and closed? The universe could be a little more dense or billions of times more dense, and in either case, it would be closed. (If it were much more dense than it is, it would be short lived so that planets—and life—would not have had time to form.) On the other hand, it could be slightly less dense or billions of times less dense, in which case it would be open. (If it were much less dense than it is, gravity would not have gathered the diffuse matter into stars and planets, so again life would not exist.) The **flatness problem** asks, "Of all the possible conditions of the universe, why is it so nearly flat?"

The flatness problem is even worse than just described, however, because if the universe's density had not been exactly equal to the critical density in the beginning, the curvature of the universe would have increased as time passed, and the curvature would be even greater now. Consider the ratio of the actual density of the universe to the critical density. If, immediately after the big bang, that ratio was equal to 1, then the ratio would still be equal to 1 today. If, however, the ratio at the big bang was less than 1, calculations show that it would be *much* less than 1 today. If the ratio just after the big bang had a value greater than 1, the ratio would have become even greater today. For today's ratio to be somewhere between 0.1 and 2, as suggested by our best calculations and observations, it must have been extremely close to 1 at the beginning. Could this be coincidence?

The **inflationary universe model** has an explanation for the flatness of the universe. The model provides details of the first fractions of a second after the big bang, and according to its predictions, during the short time interval between 10^{-35} seconds and 10^{-32} seconds after the big bang, the universe suddenly expanded at an extremely fast rate (**FIGURE 18-16**). In fact, it grew by a factor of 10^{50} during that time. To understand how this solves the flatness problem, consider the simple analogy of a large balloon. If you balance yourself on the balloon, you will feel the curvature beneath your feet; however, if we expand the balloon to the size of the Earth, what you feel is nothing but flat space. If we now blow up the balloon to a cosmic scale, you can imagine how inflation could vastly flatten the observable universe. The inflationary universe model holds that an "inflation" of the universe took place so quickly that any portion of it that had a great curvature before the inflation became flat as a result of the inflation.

Like any analogy, the balloon story does not work perfectly, for even a large balloon has some curvature. The inflationary model predicts that that curvature was completely removed during the quick expansion; therefore, the universe's flatness is not a coincidence, but a necessary consequence of the brief period of inflation very early in history. Recent measurements of the polarization of the CMB radiation, which we discuss in the next section, confirm some of the predictions of cosmic inflation theories. It is unclear at this stage whether dark energy has any relationship to the period of cosmic inflation.

The Horizon Problem

The CMB radiation is remarkably uniform, no matter in what direction one looks. *COBE* data reveal that, after correcting for the motion of the Earth, the radiation varies in temperature by less than one part in 100,000. This would indicate that the temperature of the universe was extremely uniform at the time that the CMB radiation was emitted. This does not seem possible in the standard model of the big bang because in that model the various portions of the universe could not have been in "communication" with one another when the radiation

flatness problem The inability of the *standard* big bang model to account for the apparent flatness of the universe.

Some scientists support the idea that our universe appears fine tuned so that it allows life as we know it to exist. This *anthropic principle*, in its many variations, has created a controversy. References on this topic are included at the end of this chapter.

If the diameter of a helium nucleus expanded by a factor of 10^{50}, the nucleus would be 40 billion billion light-years across!

inflationary universe model A modification of the big bang model that holds that the early universe experienced a brief period of extremely fast expansion.

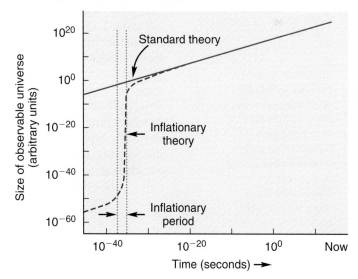

FIGURE 18-16 The solid line shows the size of the universe according to the *standard* big bang theory. The dashed line corresponds to the inflationary theory, according to which the universe's size was much smaller initially than in the standard theory. As a result, uniformity was established early before being stretched by inflation.

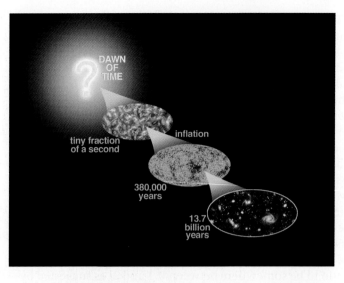

FIGURE 18-17 A summary of the big bang theory. The central image—the *WMAP* sky map—is a map of the CMB, the first light to break free in the infant universe. This light emerged about 380,000 years after the big bang. Its appearance is caused by minute temperature variations in the universe at the time the CMB was released. The observed patterns are the seeds for the development of the structures of galaxies we now see billions of years after the big bang.

horizon problem The inability of the *standard* big bang model to account for the directional uniformity of the background radiation.

The inflationary universe model was proposed in the early 1980s by Alan H. Guth of MIT and has been further developed by Guth and other cosmologists, including Stephen Hawking. Along with dark matter, it is now the best explanation we have for the development of the universe.

was emitted. To see why this is so, return to the balloon analogy one more time. First, imagine a balloon that is not expanding, and suppose that thermal energy is added to one portion of the balloon. That energy will spread around the balloon over a time period that depends on how fast the energy travels. Nevertheless, the energy will eventually distribute uniformly around a non-expanding balloon.

Now suppose that the balloon is expanding when energy is added to one portion of it. If the balloon is expanding rapidly enough, the energy will not be able to reach distant portions of the balloon because those portions are moving away too fast. There is a "horizon" beyond which the change in energy cannot be transmitted. In this rapid-expansion case, any portion of the balloon that experiences a change will not communicate this change to the rest of the balloon, and therefore, the properties of the balloon will not be uniform.

In the case of the real universe, a change in one portion can be communicated to another portion with a maximum speed equal to the speed of light. The *standard* big bang model predicts that at the time that the CMB radiation was emitted, the universe was expanding at such a rate that the "horizon" distance was much less than the distance across the universe; therefore, any little change in one part of the universe could not be communicated to faraway parts. This means that there would be no reason for the entire universe to be at the same temperature and astronomers should not observe the same temperature in the background radiation in various directions—hence, the ***horizon problem***.

According to the inflationary universe model, before inflation occurred, the universe was smaller than the horizon distance; therefore, any change in one portion of the universe was transmitted everywhere, and it was possible for the entire universe to be at a uniform temperature. Then inflation occurred. Two parts of the universe that were originally within each other's horizons were outside those horizons after inflation. They were still at the same temperature, however, as a result of their condition before inflation. This is why all parts of today's observable universe show a uniform background radiation.

FIGURE 18-17 is a NASA drawing that summarizes the inflationary big bang theory. At upper left is the universe at the big bang and then just after the big bang. At the latter time, the universe contained extremely small irregularities. The universe then expanded rapidly during the inflationary period. The central image is from the *WMAP*, and its production is explained in the next section. It pictures the universe about 380,000 years after the big bang. By that time, the universe had cooled down enough to become transparent to radiation. Galaxies and stars began to form. The bottom right part of the image shows today's universe, with each dot representing a cluster of galaxies.

18-7 The Grand Scale Structure of the Universe

The success of the inflationary theory at first caused a problem in explaining the formation of galaxies, for in a completely uniform universe, there would be no irregularities around which matter could group into galaxies and clusters of galaxies. Not only did galaxies form, however, but the discovery of extremely old quasars indicates that matter clumped together relatively soon after the big bang.

Discoveries made during the 1980s added to the problem. As we accumulated data on the distribution of clusters of galaxies (**FIGURE 18-18**), we were able to construct three-dimensional drawings to show their locations. The drawings showed that clusters of galaxies are not distributed uniformly at all, but are arranged as if to form bubbles, with a void between the walls of the bubbles. (See also Figure 17-17.) To visualize this, think of soap bubbles in a sink. Galaxies make up the surfaces of the bubbles, and where the bubbles join, we find the largest clusters of galaxies.

How is the problem of the grand scale structure of the universe resolved? The inflationary model includes a random distribution of density irregularities (perturbations) during the earliest moments of the universe. The amplitudes of these perturbations were smoothed out by the sudden and fast expansion of the universe during inflation. This model explains why the original perturbations would have left such small-scale anisotropies in the CMB radiation.

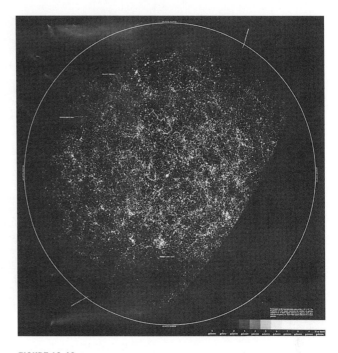

FIGURE 18-18 Each dot on the diagram is a cluster of galaxies. The grey tone in each picture element is a measure of the count of galaxies in this patch of sky. The figure shows that clusters are clumped together in an irregular way.

To understand the observed features in the CMB radiation, consider the universe during its infancy. For the first 376,000 years, the density and temperature of the universe were so great that it was opaque to electromagnetic radiation. The hydrogen and helium nuclei that were formed in the first few minutes of the universe's life, electrons, and radiation were all tightly coupled together through scattering and electromagnetic interactions. They behaved as a single fluid of **baryons** and photons with a single temperature. In this fluid, the baryons provided the inertia, the photons provided the pressure, and the density perturbations that remained after the big bang provided a mechanism for oscillations. The oscillations were caused by a competition between two factors: gravity and radiation pressure. As a region of excess mass would begin to collapse under its own gravity, becoming even denser, radiation pressure would become sufficient to stop the collapse and cause a rebound. As the universe aged, the horizon grew and oscillations of greater wavelength were included inside the horizon. These oscillations in the baryon-photon fluid sent out pressure waves that propagated at the local speed of sound.

baryons Subatomic particles, including the proton, the neutron, and a number of unstable, heavier particles. (The term *baryon* is derived from the Greek "barys," meaning "heavy.")

About 376,000 years after the big bang, the universe became cool enough (about 3000 K) that protons were able to capture electrons and form neutral hydrogen. It is at this stage, at redshift of about $z = 1100$, that matter and radiation decoupled from each other. It is also at this stage that the radiation last interacted strongly with matter and the universe became transparent. This radiation is what we see today as the CMB. After decoupling, the interaction between the radiation and the now neutral matter was not significant. As a result, the temperature of the radiation evolved differently from that of matter. The photons were free to propagate in the universe carrying on them the signature of the density perturbations just before the decoupling. Meanwhile, with the release of their pressure support, the baryons started collecting under the influence of the gravity of the dark matter. Dark matter does not interact with radiation, and therefore, any dark matter formed early on would have been free to gravitationally collapse much earlier. It is these regions, where baryons collect under the influence of dark matter, which will grow into the structures we see today.

The transition from an opaque universe to a transparent one was not instantaneous. It lasted for a time interval that corresponds to a change in redshift of about 100.

After decoupling, the wavelengths of the photons that last interacted strongly with matter have stretched so much due to the expansion of the universe that we see them as the CMB photons. These photons reach us from all directions and can tell us a great deal about the early universe. Photons that started their journey from very dense regions on this surface lost more energy escaping gravity than photons

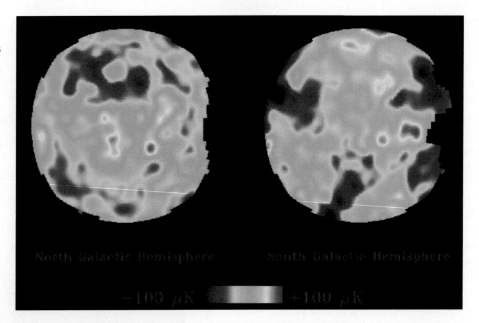

from less dense regions. Because energy is related to temperature, this implies that a certain density perturbation on this surface gives rise to a certain temperature perturbation. As a result, the CMB radiation carries with it information about the early universe in the form of temperature anisotropies.

In 1992, the *COBE* team reported finding tiny irregularities (variations) in the CMB radiation that amounted to a deviation of only 0.00003 K! To detect these variations, thought to be due to tiny differences in the temperature of the early universe, the *COBE* team first had to subtract irregularities caused by known sources of radiation. These include the effect of the Sun's radiation, the effect caused by known radiators within the Milky Way, and a slight Doppler shift in the radiation that is caused by our motion through space. **FIGURE 18-19** represents the *COBE* map of the early universe. The blue and red spots represent temperature variations and thus correspond to regions of greater or lesser density in the early universe; they give us a sense as to how matter and energy were distributed before matter gave rise to the structures we see today. These tiny irregularities are the roots of the structures we see in today's universe. Without the anisotropy, there would be no galaxies, no stars, no Earth, no us.

The CMB radiation gives us a view of the early universe, and we can use it as a tool to probe the conditions prevalent at the time. When we look at the CMB radiation, we are looking back to a time before the first stars and galaxies formed, when the universe was only about 376,000 years old. The spectrum of the CMB radiation and its almost perfect uniformity in all directions can only have one source: a hot, uniform, dense early universe. Five years after *COBE*, which had an angular resolution of about 7°, the *BOOMERANG* project gave us a map of the early universe that was about 40 times more detailed.

The *BOOMERANG* images, like the one shown in the chapter-opening figure, bring the CMB into sharp focus. This sharp snapshot of the early universe offers us a powerful tool in distinguishing among competing theoretical models on the evolution of the universe. As we discussed earlier, after decoupling and as the universe expanded, the photons that last interacted strongly with matter cooled down to become the CMB radiation, while carrying with them the pattern of the "sound waves" at the moment the universe became transparent. We now use these waves as a cosmic ultrasound to image the conditions in the early universe. The tiny temperature variations we observe correspond to tiny density fluctuations in the primordial soup of particles. These density fluctuations are thought to grow by gravitational attraction into the structures we see today. The visible patterns in the *BOOMERANG* images

BOOMERANG stands for *Balloon Observations of Millimetric Extragalactic Radiation and Geophysics*. The project involved cosmologists from Canada, Italy, the United Kingdom, and the United States.

The *BOOMERANG* telescope flew for 10.5 days (December 1998 to January 1999) around the Antarctic continent on a 800,000 m^3 balloon at an altitude of 36 km, above 99% of our atmosphere. It was launched in early January 2003 for about 10 days.

confirm predictions of the patterns that would result from sound waves racing through the early universe.

The sizes of the hot and cold spots in the *BOOMERANG* data reflect directly on the geometry of space (its curvature) and thus its fate. FIGURE 18-20 includes a *BOOMERANG* image (top) and three computer-simulated images, differing only in whether the universe is closed, flat, or open (bottom). Computer simulations suggest that in a universe whose geometry is flat, the images of the CMB radiation should be dominated by hot and cold spots of about 1° in size (bottom center). In a universe in which geometry is not flat, the curvature of space will distort the images. Specifically, in a universe that is closed, the images will be magnified by the curvature and structures will appear larger than 1° on the sky (bottom left). In a universe that is open, the images will appear smaller (bottom right). FIGURE 18-21 shows how the geometry of space affects the size of the spots in the CMB radiation. In a flat universe, lines that are initially parallel remain parallel when extended. In a closed universe, parallel lines converge, whereas in an open universe, parallel lines diverge. The *BOOMERANG* images indicate that our universe is very nearly flat.

The *BOOMERANG* results were not only confirmed, but also extended to finer resolution by *MAXIMA*. This was another balloon-borne project, which was more

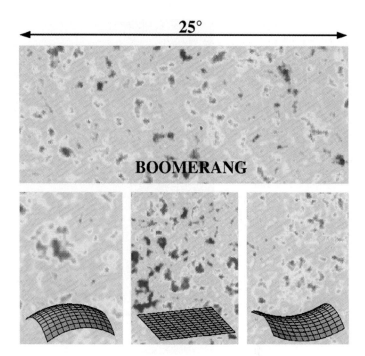

FIGURE 18-20 The *BOOMERANG* images allow us to determine the geometry of space. The image (top) is compared to images generated by computer simulations for a closed (bottom left), flat (bottom center) and open (bottom right) universe. The observations suggest that our universe is very nearly flat.

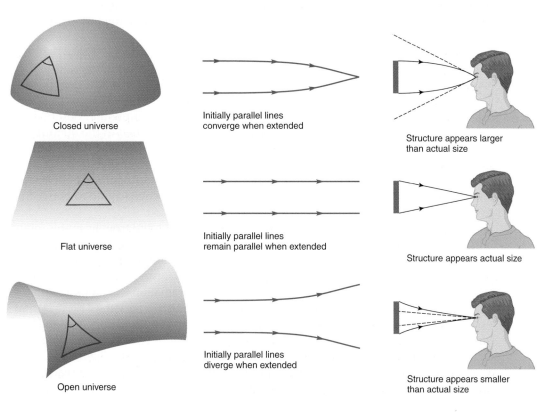

FIGURE 18-21 The geometry of space affects the angular size of an object because the curvature of space determines the path followed by a light beam. If the universe is closed, a structure in the CMB will appear larger than actual size. If the universe is flat, the structure will appear actual size. If the universe is open, the structure will appear smaller than actual size.

Closed universe

Flat universe

Open universe

Initially parallel lines converge when extended

Initially parallel lines remain parallel when extended

Initially parallel lines diverge when extended

Structure appears larger than actual size

Structure appears actual size

Structure appears smaller than actual size

FIGURE 18-22 *WMAP*'s image of the infant universe brings the *COBE* image into sharp focus. It shows temperature fluctuations, with red spots being the hottest and blue spots being coldest. Foreground radiation from our Galaxy, which *WMAP* distinguishes by observing at five microwave frequency bands, has been subtracted from the equatorial band. These fluctuations correspond to the seeds that grew to become galaxies. The new image provides strict limits to a number of cosmological parameters.

MAXIMA stands for Millimeter Anisotrophy Experiment Imaging Array. The project includes scientists from five countries. The balloon last flew in 1999. The project's sensitivity was from 4° to 10′.

WMAP observes the sky from an orbit about a semistable point on the line connecting the Earth and Sun (1.5 million km from Earth in the direction opposite the Sun). As the probe follows the Earth around the Sun, it observes the full sky every 6 months. *WMAP* is a partnership between Princeton and NASA.

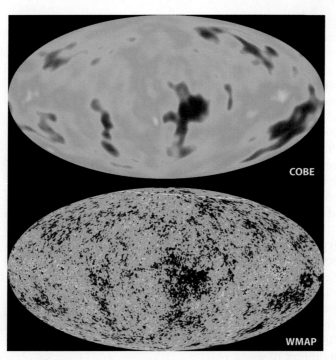

sensitive than *BOOMERANG* but looked at a smaller area of the sky. The *WMAP*, launched in June 2001, has made an accurate map of the relative CMB temperature over the full sky with angular resolution of about 0.2° and sensitivity of about 20 μK. The probe measures temperature differences between points about 140° apart on the sky using five different frequency bands. The *WMAP* image of the CMB is much sharper than the image obtained by *COBE*, as shown in **FIGURE 18-22**.

What do the tiny temperature fluctuations tell us about the universe? As we discussed earlier, the observed temperature fluctuations imprinted on the CMB radiation reflect the conditions set in motion just after the big bang. That is, these fluctuations are like a fingerprint of the infant universe that we could use to compare with fingerprints predicted by various cosmic theories. The results from the projects observing the CMB all point to the same match. Our universe has about 4.6% normal matter (which makes us and the stars), about 23.3% dark matter (a mysterious, nonluminous form of matter that is indirectly detected by its gravity), and the remaining 72.1% is dark energy. (We discuss this exotic form of energy responsible for accelerating the expansion of the universe in the Advancing the Model box on page 541.) Also, the *WMAP* results suggest that the big bang occurred 13.73 ± 0.12 billion years ago and that the CMB is a snapshot of the universe when it was only about 376 ± 3 thousand years old. The Hubble constant is found to be about 70.1 ± 1.3 (km/s)/Mpc, and a number of other cosmological parameters are given with high precision.

How do we compare the various CMB observations with theory? **FIGURE 18-23** shows such a comparison. It is a graph of the strength of the CMB temperature fluctuations (temperature-fluctuation power) versus their angular size. The data points (including error bars) correspond to observations from four different projects. The curve shows the best theoretical fit to these data plus a number of other observations. The angular size of a temperature fluctuation on the sky is related to the quantity "ℓ" on the horizontal axis approximately as $180°/\ell$. For example, because *WMAP*'s angular resolution is about 0.2°, the probe cannot follow the power spectrum beyond about $\ell = 180°/0.2° = 900$. In Figure 18-23, the temperature-fluctuation power for a given value of ℓ measures the average temperature difference (squared) between points on the sky separated by an angle of about $180°/\ell$.

The locations, heights, and shapes of the observed peaks and troughs in the angular power spectrum serve to constrain the cosmological parameters. As we discussed earlier, before decoupling, normal (baryonic) matter and photons behaved as a single fluid. In this fluid, density perturbations provided a mechanism for oscillations. The temperature fluctuations we observe today in the CMB radiation correspond to the density and velocity components of the oscillations during this period (**FIGURE 18-24**). An oscillation at its maximum density during the decoupling will appear brighter and hotter than average, whereas an oscillation at its minimum density will appear colder.

With this in mind, it is now possible to understand the peaks in Figure 18-23. The first peak corresponds to a perturbation that crossed the horizon and reached its maximum density just when the universe became transparent (stage "b" in Figure

ADVANCING THE MODEL

Dark Energy

Observations of Type Ia supernovae and the CMB radiation suggest the existence of dark energy and so does the observed large scale structure of the universe. An imprint of dark energy on the CMB radiation was found in 2003 by correlating millions of galaxies in the Sloan Digital Sky Survey and CMB temperature maps from *WMAP*. This physical evidence for the existence of dark energy can be understood as follows. As a CMB photon passes through a gravitational well of either ordinary or dark matter, it gains energy as it falls into the well and loses energy as it comes out of the well; the overall effect should be zero in the absence of dark energy; however, observations show that the CMB temperature is a bit higher in regions with more matter, suggesting the presence of a repulsive force.

Observations of Type Ia supernovae suggest that dark energy began boosting the universe's expansion when it was about 5 billion years old; it overtook gravity and started accelerating the expansion when the universe was about 8 billion years old.

What is the nature of dark energy? A number of ideas have been suggested, and the following are the leading two forms. The first form is Einstein's cosmological constant, a constant energy density that fills space in a homogeneous way. The second form is a dynamic field, called *quintessence*, whose energy density changes as a function of time and space. Precision observations of the acceleration of the universe's expansion are needed to distinguish between the two forms.

In the form of the cosmological constant, dark energy can be considered as the intrinsic energy of a volume of space, and as such, it is sometimes called *"vacuum energy."* It is estimated by cosmologists to be about $10^{-29}\,g/cm^3$.

(According to the special theory of relativity, energy is related to mass; as the universe expands, ordinary mass, and thus ordinary energy, gets diluted.) Even though most theories of particle physics predict that the quantum vacuum will have energy, its magnitude is 10^{120} times bigger than estimated by cosmologists!

In the form of quintessence, dark energy results in a slightly slower acceleration of the universe's expansion and it is very light so that it does not form structures like ordinary matter. Even though fields such as quintessence are predicted by the standard model of high-energy physics, such fields are supposed to have large masses. This is a similar problem to that of the cosmological constant. Support for a time-varying dark energy comes from observations of gamma-ray bursts. Even though their intrinsic energy varies greatly, their "wattage" can be determined fairly accurately; they are also brighter and more distant than Type Ia supernovae. A Hubble diagram similar to that of Figure 18-14b (covering GRBs up to a redshift of 6.30) shows a better fit for a time-varying dark energy.

The fate of the universe depends on the nature of dark energy. The expansion of the universe could continue forever, in which case we will slowly lose contact with the rest of the universe beyond our local supercluster. It is possible that the strength of dark energy decreases with time or that it even becomes attractive, in which case the influence of gravity could lead to a big crunch. If its strength increases with time, then it will overcome all other forces and lead to a big rip, tearing apart everything in the universe. It has also been suggested that the observed effects of dark energy are nothing but the effects of gravitational waves generated during the inflationary epoch of the universe.

18-24). The remaining odd peaks correspond to perturbations that underwent an additional integral number of oscillations by this time. The even peaks correspond to perturbations at their minimum density (for example, stage "d" in Figure 18-24); these perturbations give rise to power peaks of smaller heights because the rebound must fight against gravity. The troughs (for example, stage "c" in Figure 18-24) correspond to positions of maximum velocity. The amplitude of the fluctuations is reduced as ℓ increases (on the smallest scales) because the transition from an opaque universe to a transparent one was not instantaneous. The shape of the spectrum in Figure 18-23 depends then on the details of the oscillations, and it is a direct result of the conditions that existed at the time of decoupling; therefore, this spectrum serves to constrain the cosmological parameters. The curve in Figure 18-23 depends on the values of seven different cosmological parameters that were adjusted to fit the data.

The location of the first peak in Figure 18-23 is related to the geometry of the universe. That is so because the physical scale associated with the first peak is the horizon size at the time of decoupling. At this time, the universe became cool enough

FIGURE 18-23 The angular power spectrum of the CMB temperature fluctuations based on data obtained by four different projects. The curve shows the best theoretical fit to the observational data. The third overtone provides evidence for a neutrino background whose density is less than 0.65%. Courtesy of NASA and the WMAP Science Team. Modified from M.R. Nolta et. al., *ApJS* 180 (2009): 296–305. Reproduced by permission of the AAS.

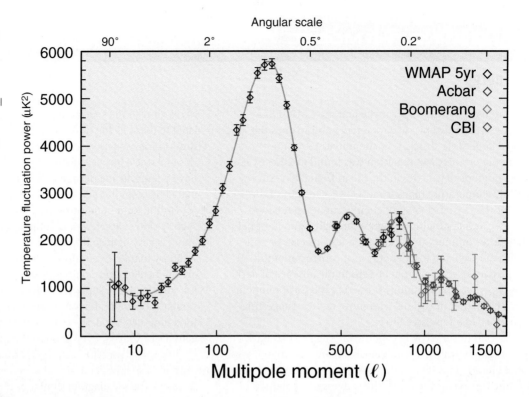

(about 3000 K) that protons were able to capture electrons and form neutral hydrogen. This stage clearly occurs at about $z = 1100$, and the angle subtended by the physical scale at this redshift depends on the geometry of the universe. The first peak occurs at $\ell \approx 200$, which corresponds to an angular size for the temperature fluctuations of about $180°/\ell \approx 1°$. This is consistent with a flat universe, as predicted by the inflationary universe theory.

Astronomers have hailed the *WMAP* results as the beginning of precision cosmology. Observing temperature fluctuations at the 10^{-6} K scale is not easy. Imagine that a band is playing random tunes at a location far away from us. Between the band and us there are many sources that generate noise. Our job is to understand the nature of these sources, remove their contributions to the overall sound we receive, and interpret the final signal we get from the band. *WMAP*'s precision strongly confirms the standard inflationary big bang scenario.

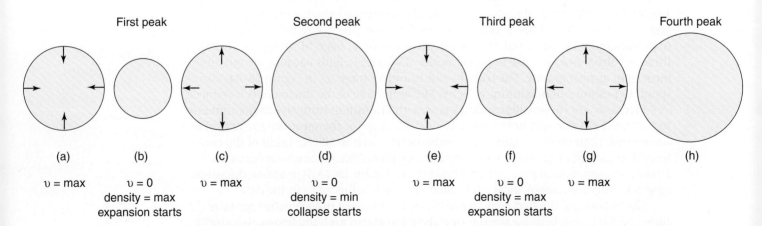

FIGURE 18-24 This illustration shows the relationship between the peaks and troughs in the angular power spectrum and the density and velocity components of the oscillations during the decoupling era.

In addition to the spectrum shown in Figure 18-23, *WMAP* mapped the intensity and direction of **linear polarization** everywhere on the CMB sky. Recall our discussion of light as an electromagnetic wave in Chapter 4. We say that light is linearly polarized when its electric field oscillates in a fixed plane, just like a length of rope that moves up and down in a vertical plane while the wave progresses horizontally. Natural light is a combination of random and rapidly changing fields that have a range of wavelengths. As a result, natural light is not polarized. A device that takes natural (unpolarized) light and transforms it into polarized light is a polarizer, as do certain sunglasses, for example; however, there are a number of natural processes that produce polarized light, one of which is scattering. The illustration in FIGURE 18-25 makes use of polarizing sunglasses to show how *WMAP* "sees" polarized light. The results of *WMAP*'s polarization measurements are shown in FIGURE 18-26, which correlates the temperature fluctuations and the polarization in the CMB radiation between points in the sky separated by an angle of about $180°/\ell$. The curve in Figure 18-26 is *not* a fit to the data; it is actually a *prediction* based on the observed temperature anisotropies in the CMB radiation. The overall shape of the cross-power spectrum contains a wealth of information about the conditions in the early universe. The rise in this curve at low ℓ values (large angles, about 90°) indicates that the first stars in the universe formed very quickly, about 430 million years after the era of decoupling.

Even though *WMAP* does not see the light of the first stars directly, the observed polarized light serves as the signature of the ultraviolet light released by the first stars. This ultraviolet light began the reionization of the neutral hydrogen that formed after decoupling. FIGURE 18-27 shows a sequence of events in the history of the universe, from the initial density perturbations to the formations of billions upon billions of stars and galaxies. FIGURE 18-28 is a more detailed illustration of important events in the early stages of the universe.

The *WMAP* results, along with results from other CMB projects and a range of different and independent observations (such as the galaxy and supernovae surveys), support the inflationary scenario. FIGURE 18-29 shows how the combination of data from different projects allows us to limit the range of values for the average density

Polarization: How It Works

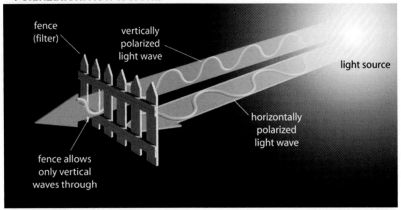

how we see it...

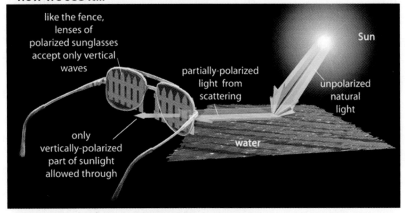

how WMAP sees it...

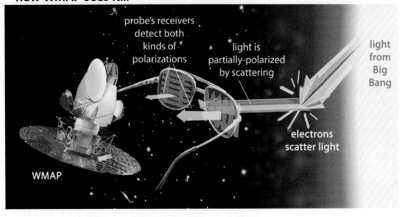

FIGURE 18-25 This illustration demonstrates how *WMAP* sees polarized light. The polarization of the CMB radiation was produced by the scattering of the cosmic light when it last strongly interacted with matter.

linear polarization The situation when the electric field in an electromagnetic wave oscillates in a fixed plane.

FIGURE 18-26 The cross-power spectrum that correlates the temperature fluctuations and the polarization in the CMB radiation between points in the sky separated by an angle of about 180°/ℓ. The curve is a *prediction* based on the observed temperature anisotropies in the CMB radiation. The rise in the curve at low ℓ values indicates that the first stars in the universe formed very quickly. Courtesy of NASA and the WMAP Science Team. Modified from M.R. Nolta et. al., *ApJS* 180 (2009): 296–305. Reproduced by permission of the AAS.

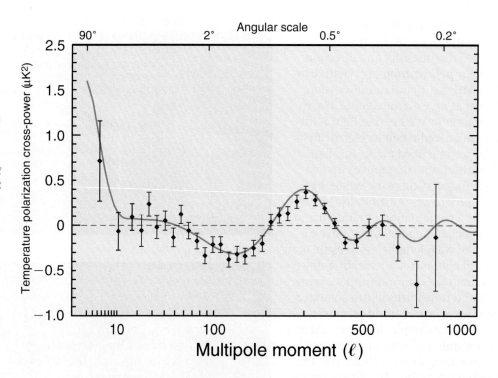

◆ Polarization data from the Cosmic Background Imager (CBI) provide strong support for the standard model of the universe. CBI is a microwave telescope array located at a high plateau (5 km above sea level) in Chile.

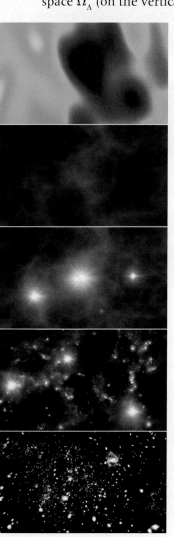

FIGURE 18-27 A sequence of events in the life of the universe. The first frame shows temperature fluctuations (as color differences) in the CMB radiation observed by *WMAP*. These temperature fluctuations correspond to slight density perturbations. The second frame shows matter condensing to regions of higher and higher density as a result of gravitational attraction. The third frame shows the period of formation of the first stars, about 350 million years after the big bang. The fourth frame shows a greater number of stars and galaxies forming along the filaments first seen in frame two. The last frame shows the modern era, billions upon billions of stars and galaxies that grew from the seeds planted in the early universe.

of matter in space Ω_m (on the horizontal axis) and the density of dark energy in space Ω_Λ (on the vertical axis). The results from CMB projects support cosmological models whose parameters lie in the blue region. This region is concentrated near the diagonal red line, which corresponds to a flat universe. The results from the supernovae studies support cosmological models whose parameters lie in the yellow region. The results from surveys of galaxy clusters support cosmological models whose parameters lie in the orange region. If all measurements are correct, then only models whose parameters lie in the green overlap area are allowed. This region indicates that our universe has a flat geometry, started with a big bang, and will continue expanding.

Cosmic Evolution

Our efforts to look back in time and understand the conditions of the early universe are continuing through a number of space missions and other projects. We are slowly putting together a photo album of the universe's life. The CMB projects are giving us a snapshot of the universe during its infancy while observations with space and ground telescopes are providing us with pictures of its early adolescent years.

As we discussed earlier, after decoupling, the first neutral atoms of hydrogen and helium started forming. This transition plunged the universe into darkness because stars had not formed yet. When stars started forming these *dark ages* slowly gave way to a *cosmic renaissance*; the first stars were large in size and massive. They formed at a very fast rate and emitted huge amounts of ultraviolet light that ionized atoms in their surrounding region, slowly lifting the fog of the dark ages, starting at about 430 million years after the big bang. This *reionization* and heating of the universe slowed down further galaxy formation, making it harder for gravity to form small galaxies.

Time		Temperature
0	Big Bang	∞
10^{-32} sec	End of Inflation	10^{19} K
100 sec	Formation of Deuterium and Helium	10^{9} K
1 month	CMB Spectrum Fixed	10^{7} K
10,000 yrs	Radiation Energy = Matter Energy	20,000 K
376,000 yrs	CMB Last Scattering (Decoupling)	2970 K
	Dark Ages	
430 million yrs	Reionization	32 K
13.7 billion yrs	Present Universe	2.7 K

FIGURE 18-28 Important events in the early stages of the universe, including the formation of deuterium and helium 100 seconds after the big bang. During the era of decoupling (about 376,000 years after the big bang), radiation last interacted strongly with matter and the universe became transparent. This "surface of last scatter" for the CMB radiation is analogous to the light coming through the clouds to our eyes on a cloudy day.

Indeed, *Chandra* observations of distant quasars show that the universe at an age of a billion years was dominated by fewer, larger galaxies instead of more numerous, smaller galaxies. Infrared observations using ESO's VLT (in 2001) reveal that many rather large galaxies (some with spiral structure similar to that seen in nearby galaxies) existed at a time when the universe was less than 1 billion years old. These galaxies appear to have already formed most of their stars, and they probably account for at least half the total luminous mass of the universe at that time. Also, *HST* observations of some of the most distant quasars show the presence of iron and other heavier elements in their spectra; because such elements form during a supernova explosion and the life cycles of the progenitor stars are 500-800 million years, these observations support the idea that the first quasar engines started slightly before the universe was 1 billion years old. Even though stars continue to be formed today, the rate of star formation has been declining since the early years. Surveys of galaxies, using space telescopes and the tool of gravitational lensing, show that galaxies existed very early in the universe's life, when it was about 500 million years old. The surveys also show that the number of bright galaxies increases between 700 and 900 million years probably due to merging; this supports the hierarchical theory of galaxy formation described in Chapter 17.

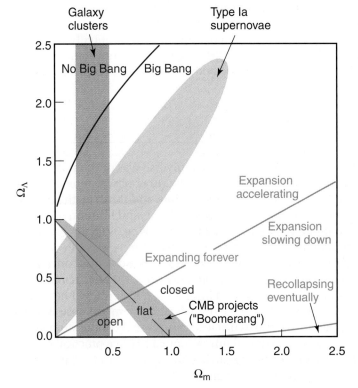

FIGURE 18-29 A comparison between results from CMB projects, surveys of galaxy clusters, and studies of Type Ia supernovae suggests that our universe is flat, started with a big bang, and will continue expanding. The horizontal axis shows Ω_m, the average density of matter in space compared with the critical density. Such matter (normal and dark) slows down the expansion of the universe. The vertical axis shows Ω_Λ, the density of the dark energy in the universe compared with the critical density. Such dark energy causes the expansion of the universe to accelerate. The latest results suggest that $\Omega_m = 0.279 \pm 0.015$, $\Omega_\Lambda = 0.721 \pm 0.015$, and $\Omega_0 = 1$ with an uncertainty of 1%. Courtesy of BOOMERANG Collaboration.

FIGURE 18-30 A computer model of the universe at an age of about 2 billion years. Gravity causes matter to arrange itself in thin filaments. Yellow, red, and blue correspond to high, medium, and low gas density, respectively. Stars will form in the yellow regions and slowly stream along the filaments. When the stars meet at the intersections of the filaments, they gradually form the galaxies we know today.

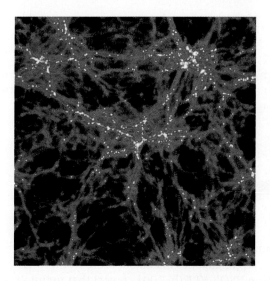

Composition of the universe	Universe at 376,000 years	Universe now
Atoms	12%	4.6%
Dark Matter	63%	23%
Dark Energy	< 1%	72%
Neutrinos	10%	< 1%
Photons	15%	< 1%

FIGURE 18-31 The make-up of the universe now and at its infancy.

Science progresses best when observations force us to alter our preconceptions.
Astronomer Vera Rubin.

Three-dimensional maps of the large-scale structure of the universe, including hundreds of thousands of galaxies, show variations in galactic distribution on a scale of 450 million light-years; this directly reflects the temperature variations observed in the CMB radiation. Such maps also support the case for dark matter and dark energy by showing that the observed density fluctuations in the universe agree with theoretical predictions. They also support the idea that the universe has a spongy structure (**FIGURE 18-30**), with galaxies forming in nests of dark matter, clustering close to each other along filaments and having their rotation axes oriented along the direction of the filaments.

Studies of the dark ages will provide even more data to test the inflationary universe model. Astronomers have started studying the shadows that the first atoms imprinted on the CMB radiation. Between 20 and 100 million years after the big bang, the temperature of the universe was such that hydrogen atoms could absorb a 21-cm CMB photon. To observe such absorption during the early stages of the universe's expansion and because of this expansion, astronomers must study the CMB radiation at longer wavelengths, between 6 and 21 meters.

All of the recent observations confirm the generally accepted cosmological model that astronomers constructed over the past decade. **FIGURE 18-31** compares our current understanding of the make-up of the universe now and at its infancy. The model is, however, problematic. For example, because it is probable that ordinary matter and dark matter originate through different mechanisms, why is the abundance of ordinary matter similar (within an order of magnitude) to that of dark matter? Also, because the total abundance of matter is rapidly changing compared with that of dark energy as the universe expands, why are the two so similar (within an order of magnitude)? Finally, what is the nature of dark energy? Even though there are still uncertainties, the model is not pure speculation, for each of its various modifications makes predictions about today's universe, and the predictions can be tested against observations.

As new observations bring us a flood of new data, undoubtedly we will have to revise the models. Some aspects of them will he discarded, but some will live on to be tested again and again until they join the body of firmly established science.

Conclusion

The big bang model states that in the past the universe was once much smaller, hotter, and denser. The model is supported mainly by three observations: first, the expansion of the universe; second, the CMB radiation; and third, by the observed proportions of light elements (such as deuterium, helium, and lithium). The latest results are consistent with a universe that started with a big bang, is about 13.7 billion years old, has flat geometry, and will continue expanding.

*New York: The Noonday Press, 1987.

In *The Nature of Reality*,* Richard Morris writes, "The scientific conception of the universe has changed so much that about the only thing that has remained the same is the word itself." In this chapter—and this book—we have shown how much our perception of the universe has changed. Nevertheless, we might pose the following question to a cosmologist: "What practical use can be made of the study of cosmology?" The answer is simply that not all pursuits need to have practical or useful consequences. We engage in some, like music, art, philosophy, and cosmology, for the sheer intellectual and artistic pleasure they provide. Curiosity is part of human nature, and questions like those raised in this chapter are among the most fundamental that can be asked—and perhaps answered—by science.

Our astronomical quest began 17 chapters ago in our home—the solar system. As you look back on the geocentric and heliocentric theories presented in the first few chapters, we hope that you see them as small, beginning steps in the development of today's view of the universe. After studying the solar system, we traveled outward to the stars and then much farther to galaxies. Finally, our study of cosmology considered the universe as a whole. We hope that this has given you a broader perspective on the natural world and a better appreciation of how the process of science helps us to understand the magnificent universe of which we are a part. We also hope that your quest for the universe will continue throughout your life.

Future data from space satellites should help astronomers choose among competing inflationary models. The European-based *Planck Surveyor*, launched in 2009, will map the CMB in unprecedented detail.

STUDY GUIDE

1. An open universe is one that
   A. will eventually collapse to form "a big crunch."
   B. will never stop expanding.
   C. [Either A or B above, depending on its density.]

2. Einstein's conviction that the universe is closed and finite
   A. resulted directly from his general theory of relativity.
   B. resulted directly from his special theory of relativity.
   C. was a belief that caused him to make a change in his general theory of relativity.
   D. was a belief that caused him to make a change in his special theory of relativity.

3. What *evidence* leads us to believe that the universe is expanding?
   A. The big bang theory.
   B. The inflationary universe theory.
   C. The expansion of the solar system.
   D. The redshift of most galaxies.
   E. [All of the above.]

4. The assumption of isotropy states that the universe looks the same
   A. at all times.
   B. in all locations.
   C. in all directions.
   D. [All of the above.]

5. The cosmological principle refers to
   A. a comparison of our location in the universe to other locations in the universe.
   B. the redshift used to determine distances.
   C. Cepheid variables used to determine magnitude and therefore distance.
   D. the Doppler effect used to determine velocities.
   E. [Both B and C above.]

6. The existence of the 3-K background radiation supports
   A. the big bang theory.
   B. the steady-state theory.
   C. the big bang theory and the steady-state theory about equally.

   D. neither the big bang theory nor the steady-state theory, for neither can explain it.

7. Big bang theories predict _____ speed for the most distant galaxies than do steady state theories.
   A. a lesser
   B. a greater
   C. the same

8. The Hubble law is a relationship between galaxies'
   A. redshifts and colors.
   B. distances and redshifts.
   C. redshifts and spectral types.
   D. colors and spectral types.

9. If present data are correct, the universe is
   A. closed, but not oscillating.
   B. closed and oscillating.
   C. flat.
   D. open.

10. A better knowledge of which quantity would most aid us in determining whether the universe is open or closed?
    A. The age of the Earth.
    B. The age of the solar system.
    C. The size of the Galaxy.
    D. The average density of the universe.
    E. [All of the above would help about equally.]

11. What *evidence* indicates that an explosion took place at the beginning of the universe?
    A. The density of matter in the universe.
    B. Low-intensity radiation from all directions.
    C. The big bang theory.
    D. The inflationary universe theory.
    E. The existence of supernovae.

12. If the average density of the universe is less than the critical density, the universe
    A. will expand forever.
    B. will eventually collapse.
    C. was produced in a big bang.
    D. is an inflationary universe.
    E. violates the laws of energy conservation.

13. Suppose that you read that astronomers calculate the maximum age of the universe to be 15 billion years. Their calculation is based on
    A. calculations of the density of matter in the universe.
    B. the Hubble constant.
    C. the age of rocks on Earth and Moon.
    D. the inflationary universe theory.
    E. [All of the above.]

14. The Hubble law is based on
    A. the big bang theory.
    B. the oscillating universe theory.
    C. redshift data.
    D. inflationary theories.
    E. [None of the above.]

15. Which of the following forces may halt the expansion of the universe?
    A. Nuclear forces.
    B. Electrical forces.
    C. Gravitational forces.

16. The cosmological redshift indicates that
    A. most stars in the Galaxy are moving away from the Sun.
    B. most planets are moving away from ours.
    C. stars at greater distances are hotter.
    D. space is expanding.
    E. [All of the above.]

17. The horizon problem and the flatness problem
    A. are solved by the standard big bang theory.
    B. are solved by the inflationary universe theory.
    C. have not yet been solved by modern cosmology.

18. Which of the following is/are *assumptions* of cosmology?
    A. The cosmological redshift.
    B. The homogeneity of the universe.
    C. Universality of physical laws.
    D. [Both B and C above.]
    E. [All of the above.]

19. The discovery of the cosmic background radiation was made by
    A. Jocelyn Bell, a British graduate student.
    B. cosmologists early in the twentieth century.
    C. two Bell Telephone employees.
    D. the *COBE* satellite.
    E. Edwin Hubble.

20. When we talk about the cosmological expansion of the universe, what is expanding?
    A. The planets are getting farther apart.
    B. The stars of the Galaxy are getting farther apart.
    C. All galaxies are getting farther apart from one another.
    D. Clusters of galaxies are getting farther apart from one another.
    E. [All of the above, for the entire universe is expanding.]

21. The *BOOMERANG* project
    A. had as its purpose a search for gravitational lensing.
    B. had as its purpose a search for the edge of the universe.
    C. examined the CBM before the *COBE* satellite was launched.
    D. examined the CBM after *COBE* data was all received.

22. It is said that the universe is expanding. Explain just what this means.

23. Explain how, if almost all galaxies are moving away from ours, we are not necessarily at or near the center.

24. Name and explain three basic assumptions of cosmology.

25. What is background radiation, and how does it enter the discussion of cosmological theories?

26. How is the age of the universe calculated?

27. Describe two methods of determining how quickly the expansion of the universe is slowing.

28. What is the oscillating universe?

---

1. Describe at least six steps in the historical progression of our ideas of the universe from Ptolemy's geocentric universe to Einstein's relativistic space curvature.

2. Quasars do not exist in our part of the universe. Why is this not a violation of the homogeneity principle?

3. Explain why a flat universe would be such a special case that it would seem to be an extremely unlikely state of the universe.

4. Why did Einstein add a term in his equation that predicted a repulsion force?

5. Popular science videos that illustrate the Big Bang commonly show an explosion as seen from outside. What would it mean to be "outside" the Big Bang? Comment on the noise that often accompanies such illustrations.

6. Discuss the observation of the CMB radiation as an example of the confirmation of a theory, comparing it with the prediction of stellar parallax in the 1600s.

7. The age of the universe calculated from the Hubble constant assumes a constant velocity of recession for galaxies. If, in fact, galaxies were moving faster at an earlier time, is the universe older or younger than calculated? Show your reasoning.

8. Where in the universe did the big bang occur? Explain.

9. It has been said—perhaps only half in jest—that when astronomers come on a problem they cannot solve, they use a black hole. Give two examples that may have led someone to say this.

10. Explain why the horizon problem exists in the standard big bang model.

11. The horizon problem results from difficulties explaining how the background radiation can be so uniform in all directions; however, Figure 18-19 shows that it is not perfectly uniform. Explain.

12. Explain how the *COBE* results of Figure 18-19 relate to the universe of today.

1. If the Hubble constant is 20 (km/s)/Mly, what is the maximum age of the universe? Repeat for a Hubble constant of 17 (km/s)/Mly.

2. Using Wien's law (Section 4-4), show that the blackbody curve for an object at temperature 3 K peaks in the microwave region of the electromagnetic spectrum.

At what part of the spectrum does the blackbody curve of a 3000-K object peak?

3. The observed redshift of a quasar is $z = 5$. Show that its speed due to the expansion of the universe is about 94.6% of the speed of light.

1. S. Weinberg, *The First Three Minutes* (New York: Basic Books, 1988).

2. S. Hawking, *A Brief History of Time* (New York: Bantam Books, 1988).

3. M. Rees, *Before the Beginning* (Reading, MA: Addison Wesley Longman, 1997).

4. M. Rowan-Robinson, *The Nine Numbers of the Cosmos* (New York: Oxford University Press, 1999).

5. D. Goldsmith, *The Runaway Universe* (New York: Perseus Books, 2000).

6. John D. Barrow and Frank J. Tipler, *The Anthropic Cosmological Principle* (New York: Oxford University Press, 1986).

7. Nick Bostrom's Web site devoted to the anthropic principle: http://www.anthropic-principle.com.

8. Gale E. Christianson, *Edwin Hubble, Mariner of the Nebulae* (Farrar, Straus, and Giroux, 1995).

9. "Inflation in a Low-Density Universe," by M. A. Buchner and D. N. Spergel, in *Scientific American* (January, 1999).

10. "Mapping the Universe," by S. D. Landy, in *Scientific American* (June, 1999).

11. "Cosmological Antigravity," by L. M. K. Krauss, in *Scientific American* Special Edition (December, 2002).

12. "Is Space Finite?" by J.-P. Luminet, G. D. Starkman, and J. R. Weeks, in *Scientific American* Special Edition (December, 2002).

13. "Surveying Spacetime with Supernovae," by C. J. Hogan, R. P. Kirshner, and N. B. Suntzeff, in *Scientific American* Special Edition (December, 2002).

14. "The Search for Dark Matter," by D. B. Cline, in *Scientific American* (March, 2003).

15. "Reading the Blueprints of Creation," by M. A. Strauss, in *Scientific American* (February, 2004).

16. "The Cosmic Symphony," by W. Hu and M. White, in *Scientific American* (February, 2004).

17. "From Slowdown to Speed-up," by A. G. Riess and M. S. Turner, in *Scientific American* (February, 2004).

18. "The Myth of the Beginning of Time," by G. Veneziano, in *Scientific American* (May, 2004).

19. "The First Stars in the Universe," by R. E. Larson and V. Bromm, in *Scientific American* Special Edition (September, 2004).

20. "The Midlife Crisis of the Cosmos," by A. J. Berger, in *Scientific American* (January, 2005).

21. "Does Dark Energy Really Exist?" by T. Clifton and P. G. Ferreira, in *Scientific American* (April, 2009).

Quest Ahead to Starlinks
http://physicalscience.jbpub.com/starlinks

Starlinks is this book's online learning center. It features **eLearning**, which contains chapter quizzes and other tools designed to help you study for your class. You can also find **online exercises**, view numerous relevant **animations**, follow a guide to **useful astronomy sites** on the web, or even check the latest **astronomy news** updates.

Life zone around galaxy

Life zone around star

The Quest for Extraterrestrial Intelligence

19

Life could exist only in certain regions of a galaxy, neither too close to the center nor too close to the edge. Similarly, planets that could support life could not be too close to a star nor too far away from it.

AS FAR AS THE GENERAL PUBLIC IS CONCERNED, probably the best-known astronomer in the last half of the 20th century was Carl Sagan (1934–1996). He appeared often on television, from late-night talk shows to educational programs about space flight, and hosted a 13-part series on astronomy, based in part on his own book, called *Cosmos*. His popularity was such that his reputation as a serious researcher suffered; critics assumed that anyone who spent so much time explaining fundamental astronomy to the public did not have much time for cutting-edge research. Yet Sagan published more than 600 scientific papers and nonscientific articles, including work on the environment of Venus, dust storms on Mars, the atmosphere of Jupiter, the origin of life on Earth, and the long-term environmental effects of nuclear war—the so-called "nuclear winter."

One of Sagan's strongest interests was in the *Search for Extraterrestrial Intelligence* (SETI). He played key roles in the *Viking* search for life on Mars and in designing the snapshots of life on Earth carried by the *Pioneer* and *Voyager* space probes. Here are some of his thoughts about the significance of SETI:

> Through all of our history we have pondered the stars and mused whether humanity is unique or if, somewhere else in the dark of the night sky, there are other beings who

All cross references to chapters, sections, figures, and tables pertain to the main text, *In Quest of the Universe, Sixth Edition. In Quest of the Solar System* contains Chapters 1–11 and 19 of the main text. *In Quest of the Stars and Galaxies* contains Chapters 1–5 and 11–19 of the main text.

contemplate and wonder as we do, fellow thinkers in the cosmos. Such beings might view themselves and the universe differently. Somewhere else there might be very exotic biologies and technologies and societies. In a cosmic setting vast and old beyond ordinary human understanding, we are a little lonely; and we ponder the ultimate significance, if any, of our tiny but exquisite blue planet. The search for extraterrestrial intelligence is the search for a generally acceptable cosmic context for the human species. In the deepest sense, the search for extraterrestrial intelligence is a search for ourselves.*

Carl Sagan, *Broca's Brain* (New York: Random House, 1979, p. 268).

Are we alone in the universe? Either answer to this question has immense consequences. If we are indeed alone, then our assumption that the universe is basically the same everywhere would be challenged. Earth once again will take a special place in the cosmos, a view that we have not subscribed to since the time of Copernicus. If other intelligent life forms exist, communicating with them will forever change the way we see ourselves as part of nature. Searching for extraterrestrial intelligence may not be a "scientific" activity. After all, the hypothesis that such intelligent species exist cannot be disproved; however, as Carl Sagan so eloquently put it, this is mostly "a search for ourselves."

19-1 Radio Searches and SETI

The presence of intelligent life on Earth, the huge number of stars in our Galaxy, and the enormous number of galaxies in the universe, coupled with the latest observations of planets around other stars, lead us to believe that, given sufficient time and the right conditions, life—maybe even intelligent life—can evolve elsewhere in the universe. An obvious way of searching for such life is by sending out unmanned spacecraft; however, with our current technology, even a trip to nearby stars would take thousands of years to complete. Instead, SETI is based on the detection of radio transmissions from other life forms. As we have emphasized in this text, radio waves can travel through gas and dust without significant degradation.

To search for signals from extraterrestrials, astronomers have had to decide not only where to look in the sky, but on what frequencies to expect the signals. In a sense, we must know what "channel" the extraterrestrials are broadcasting on. Taking into account the many natural sources of radio "noise," astronomers have selected certain ranges of frequencies as most likely. A primary consideration in making the decision is to select frequencies at which natural sources do not emit great amounts of energy. Strong natural signals would drown out intelligent signals over long distances. Early SETI receivers had to "listen" to one frequency at a time, but modern receivers are capable of tuning to tens of millions of different frequencies simultaneously. This makes the search much faster than was previously dreamed possible.

The best radio frequency range for communications is between 1000 and 10,000 MHz (megahertz). In this range, there is relatively little radio noise from other sources. At lower frequencies, we have strong emission from interstellar gas, whereas at higher frequencies, radio waves tend to be absorbed by our atmosphere. If an intelligent civilization chose to communicate with us, the frequency of choice could most likely be 1.4×10^3 MHz. This corresponds to a wavelength of $\lambda = c/f = (3 \times 10^{10}$ cm/s)/ $(1.4 \times 10^3$ MHz$) \approx 21$ cm, the same wavelength astronomers use to detect cool hydrogen clouds in our galaxy (as we discussed in Chapter 16). An intelligent civilization would very likely know that the 21-cm radiation is a useful probe for astronomers, and we could easily distinguish an ordered message from the random natural noise usually found at that wavelength.

Radio telescopes are designed and built to detect and measure radio waves coming from objects in space. If there are intelligent beings out there, these telescopes will also be useful both in detecting their presence and in communicating with them. Radio telescopes have been used on occasion to search the heavens for evidence of radio signals from intelligent beings. In 1960, American astronomer Frank Drake used a telescope of the National Radio Astronomy Observatory to search for signals from two nearby stars. The search, which Drake called Project Ozma, involved looking for unusual patterns in radio signals—patterns that were different from the signals emitted by inanimate objects such as stars and galaxies.

Since that time, several astronomers have conducted searches. For example, in an experiment conducted in the mid-1970s, more than 600 nearby stars were watched for about 30 minutes each. In 1971, a group of astronomers and engineers made plans for an elaborate array of radio telescopes to be devoted to a search. Their proposal, called Project Cyclops, would have cost billions of dollars to put into operation. Because of budget pressures, Congress canceled a search begun by NASA after $58 million had already been spent. This project was taken over by the SETI Institute, a private research group; Project Phoenix ran from 1995 to 2004, observed about 800 stars to a distance of about 240 light-years, and covered more than a billion frequency channels for each star. No extraterrestrial signals were detected. A number of new projects are underway, such as the Allen Telescope Array, a SETI-dedicated array of telescopes that will equal a 100-m radio telescope.

The Arecibo Radio Telescope in Puerto Rico receives more radio signals than we can search through for signals from extraterrestrial intelligence, even with the help of a large supercomputer. In May 1999, a project called SETI@home began using individual computers in people's homes and offices to examine the data. Over 5.5 million people worldwide have downloaded the necessary software and contributed computer time equal to 3 million years. The project is continuing using a different software platform that allows testing for more types of signals.

Plans have been made in case we detect a signal that we verify as coming from an extraterrestrial being. There is even a protocol for involving international bodies in decisions about possible replies.

You can find the latest on SETI at http://www.seti.org/ and the latest on SETI@home at http://setiathome.berkeley.edu/.

19-2 Communication with Extraterrestrial Intelligence

In Carl Sagan's science fiction novel *Contact* (later made into a movie), extraterrestrials first find out about civilization on Earth by detecting the TV broadcast of the 1936 Olympic games in Berlin, which was the first television transmission on Earth using significant power.

If our search for radio signals from a race of intelligent extraterrestrials is successful, it might succeed by detecting their stray, wasted radio signals. After all, we have been transmitting radio signals into space for decades now. If an extraterrestrial civilization exists nearby, say 20 light-years away, they have already received all the episodes of the "I Love Lucy" series, along with other radio and TV programming of the 50s and 60s. On the other hand, those beings may already be transmitting messages into space with the purpose of announcing their presence and telling others something about themselves. The same radio telescopes that are used to receive signals from space can be used to transmit radio signals. Considering the probable differences between beings in different parts of the Galaxy, what could one race of these beings communicate to another? The study of this question is called CETI, for *Communication with Extraterrestrial Intelligence*.

First, we must point out that if another intelligent race were found, the tremendous distances between stars would prohibit a dialogue. The nearest star is nearly 5 light-years away from us. It would require 5 years for our radio signals to reach a planet circling that star and another 5 years for the signal to return from beings on that planet, and the likelihood that life is so common that it exists around the nearest star is extremely remote. If extraterrestrial life exists, the closest life sites are likely to be much farther away, making it impossible to get a reply to our signal during the span of one generation on Earth.

At first glance, the language problems might seem insurmountable; however, we do have something in common with every other race of beings that might exist: the physical universe and its mathematical laws, which we believe to be the same everywhere. Every intelligent race knows that hydrogen is the most common element and that an atom of hydrogen is made up of one proton and one electron. The prime numbers—those numbers that cannot be divided evenly by any other numbers but one and themselves—are the same in every language. Those scientists studying the problem have concluded that an understandable message could indeed be sent if enough time were devoted to its transmission.

We Earthlings have already sent a very short message. In 1974, the reconditioned reflecting surface of the Arecibo telescope (**FIGURE 19-1**) was rededicated, and at the ceremony, the telescope was used to transmit a message toward a cluster of 300,000 stars in the constellation Hercules. This transmission lasted only about 10 minutes, and thus, the information that could be sent was very limited. The signal consisted of a series of Morse Code-type pulses containing 1679 data points. The number 1679 was chosen because, except for 1 and 1679, only one pair of numbers can be multiplied to obtain 1679: 23 and 73. If the data points are arranged into a rectangular array, there are only two ways to do it: either 23 across and 73 down or vice versa. One array produces no pattern, but the other array makes a pattern that includes crude pictures and numbers to tell the beings that intercept the signal a little about those who sent the message.

FIGURE 19-1 The Arecibo Radio Telescope has been used to send a message to anyone out there listening. It has also been used to search for messages from any other intelligent sources.

If extraterrestrial beings happen to detect our 10-minute message, will they be able to decipher it? Who knows? Although we cannot know how much of it they will be able to understand, we can be confident that they will know that it comes from an intelligent source rather than an inanimate object. If more time were available to transmit messages, a slower development of language could be used so that understanding would be much more likely, but with the time limitation that existed, scientists believed this message was the best that could be done.

When might we receive a reply? The cluster toward which the message was sent is 26,000 light-years away, and thus, we need only wait 52,000 years for an answer!

19-3 Letters to Extraterrestrials

In addition to our radio signals, we have sent "letters" into space in the form of a plaque and records.

The *Pioneer* Plaques

In late 1971, Carl Sagan learned that the trajectories of the *Pioneer 10* and *Pioneer 11* spacecraft would take them out of the solar system and that it might be possible to include on them a message to extraterrestrials. Sagan called NASA authorities and within 3 weeks got approval to put a plaque on both *Pioneer 10*, which was scheduled to be launched the following March, and *Pioneer 11*, which was to be launched a year later. Carl and his wife Linda, along with Frank Drake, designed the message, which was etched on a 6 × 9-inch gold-anodized aluminum plaque (**FIGURE 19-2**).

Some of the message on the plaque is easily recognizable (to humans, at least). The umbrella-shaped object behind the man and woman is an outline of the Pioneer spacecraft, drawn to scale with the people. The finder thus can determine the size of a human. (The man on this scale is 5 feet 9.5 inches tall.)

The two circles at the upper left represent hydrogen atoms emitting radiation of a particular wavelength. Binary numbers on the plaque (not visible in this small rendition) use this wavelength to show the size of the woman and of the solar system at the bottom, where the *Pioneer* spacecraft is shown leaving the third planet.

The spidery-looking feature at the left shows the directions from the solar system to pulsars, compact celestial objects that emit beams of radio waves that when

sweeping past the Earth are observed as radio pulses with regular periods (see Section 15-5). The binary numbers along the lines represent the periods of the pulses, expressed in terms of the period of the wave from hydrogen. By analyzing the directions and periods of the pulsars, the finders could tell where our solar system is located in the Milky Way. In addition, because pulsars slow down with time, they could tell when the craft was launched.

Could extraterrestrials interpret the message? Curiously enough, because observations of hydrogen atoms and pulsars are common to all creatures of the universe, it is thought that *they* would be more likely to interpret the pulsar sketch than the human figures, which *we* recognize immediately.

FIGURE 19-2 This plaque aboard *Pioneers 10* and *11* carried a message to extraterrestrials—and to Earthlings.

The *Voyager* Records

In 1977, *Voyager 1* and *Voyager 2* were launched into space to rendezvous with Jupiter and Saturn. Plans were later adjusted to allow *Voyager 2* to continue to Uranus and Neptune. Both *Voyager 1* and *2* are now headed for the outer boundary of the solar system, as we discussed in Chapter 7.

With more time available to plan a message, a group of people, including the three who prepared the *Pioneer* plaque, designed a much more complete message to go aboard *Voyager 1* and *2*. The messages on *Voyager* are contained on two copper phonograph records (**FIGURE 19-3**), designed to be played at 16-2/3 revolutions/minute. Instructions for playing the records are included in pictures on the cover. On the records are 90 minutes of music from around the world, 118 pictures, and greetings in nearly 60 languages (including one nonhuman message from a humpback whale).

The music on the *Voyager* includes parts of compositions from Bach (Brandenburg Concerto No. 2 and The Well-Tempered Clavier), Beethoven (Symphony No. 5, String Quartet No. 13 in B-flat), and Mozart (The Magic Flute). In addition, it includes Chuck Berry's "Johnny B. Goode," Australian Aborigine songs, a Peruvian wedding song, and numerous other selections from the cultures of our world.

The records contain pictures as well as sound. The pictures include many photographs of scenes of nature here on Earth, of the Earth from space, of the Earth as changed by humans, of people in various activities, and of the biology of humans.

FIGURE 19-3 The *Voyager* spacecraft carried these phonograph records beyond the solar system and into outer space.

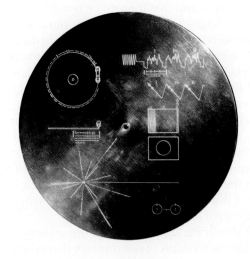

Will the Message Be Found?

The messages were not put aboard the *Pioneer* and *Voyager* craft with high hopes of them ever being found. They were sent in much the same manner as someone might put a message in a bottle and throw it into the sea. In fact, the space messages have even less chance than the bottle of being found. Even if the Galaxy abounds in intelligent life, the emptiness of space and the slow speed of the spacecraft make the chances slim indeed. The soonest any of the craft will pass within 2 light-years of a star is 40,000 years!

The messages are really a symbol of hope, a shout that "we are here." As stated in the book *Murmurs of Earth*—which tells the story of the *Voyager* records—after the Earth has been reduced to a charred cinder by an expanded and brighter Sun, the messages will continue to travel through space, "preserving a murmur of an ancient civilization that once flourished—perhaps before moving on to greater deeds and other worlds—on the distant planet Earth."

This quotation is from the book *Murmurs of Earth*, by Carl Sagan, F. Drake, A. Druyan, T. Ferris, J. Lomberg, and L. Sagan (New York: Ballantine Books, 1978).

19-4 The Origin of Life

The search for extraterrestrial intelligence is interdisciplinary by nature. It involves physics, astronomy, geology, chemistry, and biology. Biology is an integral component of our search, as it concerns itself with the question of how life began. The answer to this question is important not just for biology, but for astronomy as well, as it helps us determine the probability of the existence of extraterrestrial life.

The theory of evolution, under development since the last half of the 19th century, explains how higher forms of life evolve from more primitive forms, but this still did not answer the question of the beginnings of life. Then, in the 1920s and early 1930s, J. B. S. Haldane, a Scottish biochemist, and A. P. Oparin, a Russian biochemist, proposed that soon after the Earth's formation, the necessary chemical elements were present for complex molecules to form—molecules that are needed for life. At the time, it was thought that the seas of the early Earth were composed primarily of water, methane, ammonia, and hydrogen. The two scientists hypothesized that these molecules would spontaneously collect into more complex, organic (carbon-based) molecules.

The Haldane/Oparin hypothesis remained an interesting conjecture until 1953, when Harold Urey (the Nobel Prize winner who discovered deuterium) suggested that his graduate student, Stanley Miller, test the hypothesis by simulating the conditions of the early Earth. Miller put a mixture of water, hydrogen, methane, and ammonia into a sealed container (**FIGURE 19-4**). He heated the liquid—the young Earth was hot—and used an electrical source to produce sparks in the gas above the liquid. The sparks simulated lightning, which must have been common in the Earth's early atmosphere.

After Miller's apparatus had heated and sparked for a week, the mixture had turned dark brown. Analysis showed that it now contained large amounts of four different amino acids (organic molecules that form the basis for proteins). He also found fatty acids and urea (a molecule that is necessary in many life processes).

The Miller-Urey experiment showed that given the right chemicals and a source of energy,

See the Tools of Astronomy box in Chapter 8 for a discussion of the search for life on Mars conducted by the *Viking* landers.

In the beginning, this world was nothing at all, Heaven was not, nor earth, nor space. Because it was not, it bethought itself; I will be. It emitted heat.
Ancient Egyptian text.

In the beginning God created the heaven and the earth. And the earth was without form, and void; and darkness was upon the face of the deep. And the Spirit of God moved upon the face of the waters. And God said, Let there be light: and there was light.
Genesis 1:1–3.

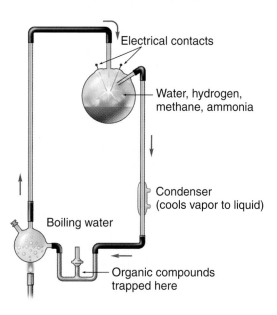

Electrical contacts

Water, hydrogen, methane, ammonia

Condenser (cools vapor to liquid)

Boiling water

Organic compounds trapped here

FIGURE 19-4 The Miller-Urey experiment showed that organic molecules necessary for life could form easily from the compounds and conditions of the young Earth.

chemical reactions could occur that produce the building blocks of life. Since then, researchers have learned that the early Earth's atmosphere was made up primarily of carbon dioxide, nitrogen, and water vapor, rather than the four compounds used in Miller's experiment. When these compounds are used in the sealed apparatus, even more complex organic molecules are produced. Furthermore, if instead of using electrical sparks for energy we use ultraviolet light (which strikes Earth from the Sun), we achieve the same results.

It must be emphasized that the Miller experiment did not produce life, only organic molecules. As yet, we have not been able to cross the gap between chemical evolution and biological evolution; however, the experiment confirmed the Haldane/Oparin hypothesis and showed that such molecules can form easily in a short time. The results of the experiment would indicate that the early Earth must have contained seas made up of organic goo.

As we have discussed in Sections 10-8 and 13-1, traces of amino acids have been found in some meteorites, and there is spectroscopic evidence that organic molecules exist in interstellar dust and gas clouds. The molecular building blocks of life thus seem to be common in the universe. How does life form from these blocks? We don't know, but it seems that nature's deck may be stacked in the direction of life.

For those who are studying aspects of the origin of life, the question no longer seems to be whether life could have originated by chemical processes involving nonbiological components but, rather, what pathway might have been followed.
National Academy of Sciences

19-5   The Drake Equation

So far we have not had any success in receiving a transmission from an extraterrestrial civilization. What are the chances that one day we may detect such transmissions? Astronomer Frank Drake proposed an equation for *estimating* the number (N) of technologically advanced civilizations in our galaxy whose signal we might be able to detect. The equation has been written in a variety of forms, including this one:

$$N = R_* \times f_p \times n_e \times f_l \times f_i \times f_c \times L,$$

where

R_* = the rate at which solar-type stars form in our Galaxy,
f_p = the fraction of these stars that have planetary systems,
n_e = the average number of planets per such system that are Earth-like enough to support life,
f_l = the fraction of those Earth-like planets on which life actually develops,
f_i = the fraction of those life-forms that evolve to intelligence,
f_c = the fraction of those intelligent species who are interested in interstellar communication and develop and use the necessary technology for it,
L = the average lifetime of such a technologically advanced civilization.

The factors in the equation, as you might surmise, are far from well known. Let's take a quick look at each factor.

Based on the average lifetime of stars and on the number of stars in the Galaxy, we have a pretty good idea of the rate of solar-type star formation; the first term on the right is known to be in the range of 1 to 10 solar-type stars per year. The emphasis on "solar-type" stars is necessary. We should probably exclude stars with masses larger than about 1.5 $M_\odot$ because their main-sequence lifetimes are shorter than the time it took for intelligent life to evolve on our planet. Stars less massive than the Sun have longer main-sequence lifetimes than our Sun, but they are also dimmer. This implies that only planets close enough to such stars would be suited for life, but short distances imply strong tidal interactions. It seems that the best candidates are stars with spectral types between F5 and M0. Other factors tend to limit the search for suitable stars to a range of distances from the center of a galaxy, sometimes called "Habitable Zones" or "Life Zones" (see the chapter-opening figure). Closer than the habitable zone, stars are too close together for stable planetary orbits and are bathed in X-rays and gamma rays from the Galactic center. Farther out than the habitable zone, stars don't have enough heavy elements to form solid Earth-like planets.

The hypothesis that life originated on Earth through moss-grown fragments from the ruins of another world may seem wild and visionary. All I maintain is that it is not unscientific.
Lord Kelvin

The more we learn about how planetary systems are formed, the better we can estimate what fraction of stars has planets orbiting around them. Better knowledge of the origin of our own system will help in this regard, and as we explained in Section 7-6, we are just beginning to understand the formation of our solar system. In addition, evidence is accumulating that planetary systems are common (see Section 7-7). It is then reasonable to assume that for solar-type stars, $f_p \approx 1$. We might suspect that if this is so, many systems probably have life-supporting planets. (After all, our solar system *almost* has three planets capable of supporting Earth-type life and a number of satellites where liquid water most likely exists.) A reasonable range for the value of n_e is, therefore, between 0.1 and 1. We see that we can make reasonably confident estimates of the values for the first three terms in the equation. Beyond this, things get fuzzier.

As we described earlier, the Miller-Urey experiment indicates that the most basic molecules from which life is made are formed easily and naturally under conditions expected to be prevalent on some young planets. Many biologists believe that, given the right conditions and enough time, life will develop from these components, but they have no way of testing this hypothesis. A reasonable estimate for f_l is about 1.

The last three terms are even less well known. Does life evolve easily to intelligence, and if so, is it common for intelligent species to be interested in communication with aliens? Finally, how long does the typical civilization last? How long will ours last? Even if we assume that evolution naturally leads to intelligence, which is far from certain, and that an intelligent species would naturally try to communicate with other intelligent species in our Galaxy, the last factor introduces the largest uncertainty in the estimated value of N. Since the invention of nuclear weapons, we have come very close to rendering ourselves extinct, and thus, a reasonable value for L might be only 100 years. On the other hand, it is certainly possible that L might have a value in the thousands or even millions of years.

Various astronomers have inserted their best guesses into the equation, and—depending on whether they are optimists or pessimists—have come up with answers anywhere from "only one" or "only a few" to "millions"; however, the value of the equation lies less in its mathematical answer than in showing us what we must learn to calculate an answer. It shows that we have a lot to learn.

Recall our discussion on the search for life on Mars and on the conditions on Jupiter's moons (especially Europa) and the similarities between Titan and young Earth.

Our obligation to survive is owed not just to ourselves but also to that Cosmos, ancient and vast, from which we spring.
Carl Sagan

19-6 Where Is Everybody?

The great size of our Galaxy and the tremendous number of stars in it suggest that the number of stars that could support a planet with intelligent life is very large. It is possible, therefore, that the value of N in the Drake equation is quite large.

There is an opposing point of view, however. The argument goes as follows: If the formation of life is common in the Milky Way, there should be a large number of civilizations like ours. We would expect many of the civilizations to be millions of years older than ours, because the Sun was formed fairly recently in the Galaxy's history. The older civilizations would be scientifically advanced enough to be able to travel to other stars. In fact, some would have been forced to leave their home planet when their star reached the end of its main sequence life and moved toward the red giant stage. These civilizations would have moved to hospitable planets near other stars and would have colonized space. (Aren't we likely to do so within the next million years, assuming our species lives that long?)

Therefore, if life begins and evolves as readily as some thinkers would have us believe, the Galaxy should be teeming with life. We should have encountered not only radio signals from extraterrestrial beings, but Earth should have been visited. The question is, "Where is everybody?" Although some people say that UFOs of extraterrestrial origin have visited Earth on many occasions, no report of UFOs controlled by extraterrestrial beings has been substantiated to the degree demanded for acceptance in science and, therefore, few astronomers consider the reports credible.

This question was first asked by the great physicist Enrico Fermi in 1950.

The argument presented here leads to either of two conclusions: (1) Earth has been (and continues to be) visited by extraterrestrials who for some reason do not want to reveal themselves, or (2) intelligent life is rare in the Galaxy. If we refuse to accept the first option, we are left with the second, and the money we spend searching is wasted. Or so the argument goes.

It is possible that conservative estimates for the parameters in the Drake equation are closer to reality; however, this would imply that we are somehow special, a view that would contradict the Copernican attitude that characterizes today's science. It is also possible that advanced civilizations outgrow technology and decide not to engage in interstellar communication. Maybe there is such a thing as "the prime directive," and other advanced civilizations are not interfering with our technological and biological evolution. The most likely answer to the question "Where is everybody?" is that the lifetime of a technological civilization is short. An order of magnitude calculation suggests that unless a technological civilization survives through some 10,000 to 10 million years of technology, we would not expect to find another such civilization still surviving in our Galaxy at the same time as we are.

There is another completely different argument against spending money on the SETI. A good theory in science must have within itself the seeds of its own destruction. The theory must be capable of being proved wrong. Some scientists point out that no matter how much we spend over the next decade in searching for extraterrestrial radio signals, it is very possible that we will find nothing. This would not prove that extraterrestrials are not broadcasting radio signals, however, but only that we have not found them. Will those in favor of the search continue to ask for more money? When do we stop? This line of thinking leads to the conclusion that SETI is bad science and is not worth the considerable amounts of money being proposed for it. What do you think?

Absence of evidence is not evidence of absence.
Frank Drake

Conclusion

Efforts to discover extrasolar planetary systems are intensifying. In the next few years we might be able to answer this question: "Are there Earth-like planets orbiting other stars?" If the results of these efforts turn out to be promising, searches for extraterrestrial intelligence will also intensify.

The potential benefits of knowledge and experience that would result from interstellar contact are immense. Beyond such benefits, SETI-type efforts will continue simply because we want to know what is out there, whether or not we are somehow unique. We want—and maybe need—to know where we came from and what it means to be human. Even if we never find the answers to these questions, it might still be worthwhile pursuing this quest. As in most cases, what is important at the end is the journey, not the destination.

STUDY GUIDE

RECALL QUESTIONS

1. Which of the following facts is the major factor that causes many astronomers to believe that extraterrestrial life is probable?
   A. UFOs that have been sighted on Earth.
   B. Laser signals that have been received from space.
   C. Radio signals that have been received from space.
   D. Signs of life that we have found on neighboring planets.
   E. The tremendous number of stars in the universe.

2. The most likely method of contact with extraterrestrial intelligence is likely to be
   A. radio waves.
   B. sound waves.
   C. light waves.
   D. actual visits.

3. The most likely wavelength at which to search for extraterrestrial intelligence is thought to be about
   A. 21 centimeters, the wavelength used to detect cool hydrogen clouds.
   B. 140 centimeters, the wavelength used to detect water.
   C. 1400 centimeters, the wavelength used to detect carbon.
   D. [Any of the above, for no wavelength is more likely than another.]

4. *SETI@home* is a project in which nonastronomers use
   A. dish antennae to search for extraterrestrial life.
   B. regular TV antennae to search for extraterrestrial life.
   C. radios to search for extraterrestrial life.
   D. antennae to send out signals to extraterrestrial life.
   E. computers to analyze data in a search for extraterrestrial life.

5. All of the following are projects that involve extraterrestrial life EXCEPT
   A. Project Phoenix.
   B. *COBE*.
   C. SETI.
   D. Project Ozma.
   E. *Voyager* records.

6. If intelligent life exists around a nearby star, it may be able to detect which of the following "signals" that we have sent?
   A. Sound waves emitted by loudspeakers at outdoor concerts.
   B. Radio and TV waves intended for communication on Earth.
   C. Radio waves sent from Arecibo intended to make contact.
   D. [All of the above.]
   E. [B and C above, but not A.]

7. Assuming that radio contact is established some day, which of the following do professionals in the field believe is true?
   A. Extraterrestrials will be able to figure out our message, but we will not be able to decipher theirs.
   B. We will be able to decipher their message, but extraterrestrials will not be able to understand ours.
   C. Neither we nor the extraterrestrials we contact will be able to understand the others' message.
   D. Communication will be monologues rather than dialogs because of the tremendous time lag.
   E. [Both C and D above.]

8. We have intentionally included messages to extraterrestrials on which of the following categories of spacecraft?
   A. The Space Shuttle.
   B. *Pioneer* and *Voyager*.
   C. *Magellan*.
   D. *Viking*.
   E. [All of the above.]

9. Assuming that extraterrestrial life is common, which of the following is true concerning the likelihood that one or more of our spacecraft will be picked up by extraterrestrials?
   A. It is so likely that it probably has already happened.
   B. It is likely to happen in the next few hundred years.
   C. It is likely to happen in the next few thousand years.
   D. It is unlikely to happen for many tens of thousands of years, if ever.

10. The *Voyager* record contains which of the following?
    A. Greetings in many languages.
    B. Music.
    C. Pictures of scenes on Earth.
    D. Pictures illustrating the biology of humans.
    E. [All of the above.]

11. The Miller-Urey experiment showed that if the chemical elements of the early Earth are put in a sealed container and energy is added,
    A. living matter in the form of viruses is formed.
    B. living matter in the form of bacteria is formed.
    C. some organic molecules are formed.
    D. [Two of the above.]
    E. [None of the above.]

12. The Drake equation shows us the factors that we need to know to calculate
    A. the rate at which stars form in the Galaxy.
    B. the number of planets in an average planetary system.
    C. the number of Earth-like planets in the Galaxy.
    D. the number of places in the Galaxy that contain the most basic life.
    E. the number of places in the Galaxy that contain technologically advanced civilizations sending out signals.

13. Which of the following factors in Drake's equation is *least* known?
    A. The rate at which solar-type stars form in our Galaxy.
    B. The fraction of solar-type stars that have planetary systems.
    C. The fraction of Earth-like planets on which life actually develops.
    D. The average lifetime of technologically advanced civilizations.
    E. [All of the above are about equally known.]

14. Most astronomers who hold the "Where is Everybody?" argument conclude that
    A. extraterrestrials must be hiding from us.
    B. extraterrestrial intelligence is rare or nonexistent.
    C. extraterrestrial communication is impossible because of technological constraints.
    D. there are not many Earth-like planets in the Galaxy.

15. What does "SETI" stand for? CETI?

16. Explain why the average lifetime of a technologically advanced civilization is a necessary factor in Drake's equation.

1. After years of listening at many frequencies, we have not picked up any artificial signals. Is this a proof that life has not evolved elsewhere? Is this a proof that technologically advanced life has not evolved elsewhere? Explain your answer.

2. What factors might limit the lifetime of a technologically advanced civilization? Explain your answer.

3. What would the consequences be if tomorrow we receive a radio signal from a civilization on a planet 10 light-years away, asking for a two-way communication? Explain your answer and consider both short- and long-term consequences.

4. Throughout our history, there have been many instances of devastation of certain cultures after contact with more "advanced" cultures. How does this historical fact affect your opinion on the searches for extraterrestrial intelligence?

1. Assume that there are 10,000 technologically advanced civilizations in a galaxy of 200 billion stars. How many stars would astronomers in this galaxy have to observe to have a reasonable chance of finding one such civilization?

2. Assume that in a galaxy that is about 10 billion years old, a thousand to a million technologically advanced civilizations have shown up at various times. On the average, for how long must a civilization last to communicate with another civilization?

3. Choose a reasonable value for each of the parameters in the Drake equation and estimate the value of N. Explain your reasoning behind your choices.

4. In the text, we mentioned that stars of mass greater than about 1.5 $M_\odot$ have main-sequence lifetimes that are shorter than the time it took for intelligent life to evolve on our planet. Show that this is true.

The Web sites for the SETI Institute (http://www.seti.org) and SETI at the University of California, Berkeley (http://seti.ssl.berkeley.edu) offer the latest information on the search for extraterrestrial intelligence.

1. "The Search for Extraterrestrial Intelligence," by C. Sagan and F. Drake, in *Scientific American* (May, 1975).

2. *Murmurs of Earth*, by Carl Sagan, F. Drake, A. Druyan, T. Ferris, J. Lomberg, and L. Sagan (Ballantine Books, 1978).

3. "No Greater Discovery," by F. Drake and D. Sobel, and "Surfing The Cosmos," by A. K. Dewdney, in *Taking Sides*, 4th ed. (McGraw-Hill, 2000).

4. "The Evolution of Life on the Earth," by S. J. Gould, and "The Origin of Life on the Earth," by L. E. Orgel, in *Scientific American* (October, 1994).

5. *Extraterrestrials—Where Are They?* by B. Zuckerman and M. Hart, eds., 2nd ed. (Cambridge University Press, 1995).

6. *Rare Earth: Why Complex Life is Uncommon in the Universe*, by Peter Douglas Ward and Donald Brownlee (Copernicus Books, 1999).

7. *Here Be Dragons: The Scientific Quest for Extraterrestrial Life*, by Simon Levay and David W. Koerner (Oxford University Press, 2000).

8. Searching for Extraterrestrials: "Where Are They?" by I. Crawford; "Where They Could Hide," by A. J. LePage; and "Intergalactically Speaking," by G. W. Swenson, Jr., in *Scientific American* (July, 2000).

9. "Refuges for Life in a Hostile Universe," by G. Gonzalez, D. Brownlee, and P. D. Ward, in *Scientific American* (October, 2001).

10. "Did Life Come from Another World?" by D. Warmflash and B. Weiss, in *Scientific American* (November, 2005).

11. "Origin of Life on Earth," by A. Ricardon, and J. W. Szostak, in *Scientific American* (September, 2009).

Quest Ahead to Starlinks
http://physicalscience.jbpub.com/starlinks

Starlinks is this book's online learning center. It features **eLearning**, which contains chapter quizzes and other tools designed to help you study for your class. You can also find **online exercises**, view numerous relevant **animations**, follow a guide to **useful astronomy sites** on the Internet, or even check the latest **astronomy news** updates.

Appendices

Appendix A

Units and Constants

Metric Prefixes	
nano (n)	$= 10^{-9}$
micro (µ)	$= 10^{-6}$
milli (m)	$= 10^{-3}$
centi (c)	$= 10^{-2}$
kilo (k)	$= 10^{3}$
mega (M)	$= 10^{6}$

Temperature	Kelvin (K)	Celsius (°C)	Fahrenheit (°F)
Absolute zero	0	-273.15	-459.7
Freezing point of water	273.15	0	32
Boiling point of water	373.15	100	212
		$T_K = T_C + 273.15$	$T_F = 1.8 \times T_C + 32$

Units of Length	
1 astronomical unit (AU)	$= 1.495979 \times 10^{11}$ m
	$= 92.96 \times 10^{6}$ miles
1 light-year (ly)	$= 6.324 \times 10^{4}$ AU
	$= 9.461 \times 10^{15}$ m
	$= 5.879 \times 10^{12}$ miles
1 parsec (pc)	$= 206{,}264.81$ AU
	$= 3.086 \times 10^{16}$ m
	$= 3.262$ ly
1 nanometer (nm)	$= 10$ Angstroms (Å)
	$= 10^{-9}$ m

Constants	
Speed of light	$= 2.99792458 \times 10^{8}$ m/sec
Electron mass	$= 9.1094 \times 10^{-31}$ kg
Proton mass	$= 1.6726 \times 10^{-27}$ kg
Planck's constant	$= 6.6261 \times 10^{-34}$ J·s
Stefan-Boltzmann constant	$= 5.6705 \times 10^{-8}$ W·m^{-2}·K^{-4}

Metric–English Conversion	
1 kilometer (km)	$= 0.6214$ miles
1 meter (m)	$= 39.37$ inches
2.54 centimeters (cm)	$= 1$ inch
1 kilogram (kg)	$= 1000$ g (weighs 2.2 pounds)
1 gram (g)	$= 0.001$ kg (weighs 0.035 oz)

Appendix B

Solar Data

	Value	Compared with Earth
Diameter	1,392,000 km	109.2
Mass	1.9891×10^{30} kg	332,980
Average density	1.41 g/cm^3	0.255
Surface gravity	274 m/s^2	28.0
Escape speed	618 km/s	55.2
Effective temperature	5778 K	
Luminosity	3.846×10^{26} watts	
Absolute magnitude	4.83	
Tilt of equator to ecliptic	7.25°	
Sidereal rotation period		
Equator	25.05 days	
40° latitude	27.40 days	
80° latitude	33.83 days	

Appendix C

Planetary Data

Physical Data

Planet	Equatorial Diameter* (km)	(Earth = 1)	Oblateness	Mass** (Earth=1)	Density (g/cm³)	Surface Gravity*** (Earth =1)	Escape Speed (km/s)
Mercury	4,879	0.383	0	0.055	5.43	0.38	4.30
Venus	12,104	0.949	0	0.815	5.24	0.91	10.36
Earth	12,756	1	0.00335	1	5.52	1	11.19
Mars	6,794	0.533	0.0065	0.107	3.93	0.38	5.03
Jupiter	142,984†	11.2	0.065	317.83	1.33	2.36†	59.5
Saturn	120,536†	9.4	0.098	95.2	0.69	0.92†	35.5
Uranus	51,118†	4.0	0.023	14.5	1.27	0.89†	21.3
Neptune	49,528†	3.9	0.017	17.1	1.64	1.12†	23.5
Pluto§	2,390	0.2	0	0.002	1.75	0.06	1.1

* Equatorial diameter of Earth = 12,756 km.
** Mass of Earth = 5.9736×10^{24} kg.
*** Surface gravity of Earth = 9.78 m/s².
† At 1 bar.
§ Dwarf planet

Orbital Data, Axis Tilt, and Rotation Period

Planet	Semimajor Axis (10⁶ km)	(AU)	Sidereal Orbital Period*	Mean Orbital Speed (km/s)	Orbital Eccen-tricity	Inclina-tion to Ecliptic (degrees)	Equatorial Tilt to Orbital Plane (degrees)	Sidereal Rotation Period**
Mercury	57.9	0.387	87.97 d	47.9	0.2056	7.0	0.01	58.65 d
Venus	108.2	0.723	224.7 d	35.0	0.0067	3.39	177.36	−243.02 d
Earth	149.6	1	365.256 d	29.8	0.0167	0	23.45	23.9345 h
Mars	227.9	1.524	1.881 y	24.1	0.0935	1.85	25.19	24.623 h
Jupiter	778.6	5.204	11.86 y	13.1	0.0489	1.3	3.13	9.925 h
Saturn	1433.5	9.582	29.46 y	9.7	0.0565	2.49	26.73	10.66 h
Uranus	2872.5	19.20	84.01 y	6.8	0.0457	0.77	97.77	−17.24 h
Neptune	4495.1	30.05	164.79 y	5.4	0.0113	1.77	28.3	16.11 h
Pluto§	5869.7	39.24	247.7 y	4.7	0.2444	17.2	122.5	−6.39 d

* The symbols "h," "d," and "y" imply "hours," "days," and "years," respectively. One day is 24 hours; 1 year is 365.256 days.
** The "−" signs indicate retrograde rotation.
§ Dwarf planet

The Sun

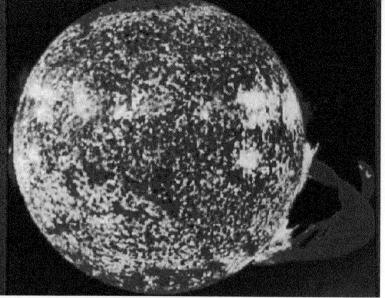

Solar wind shapes the magnetic fields of all planets in the solar system, as Earth's field shown.

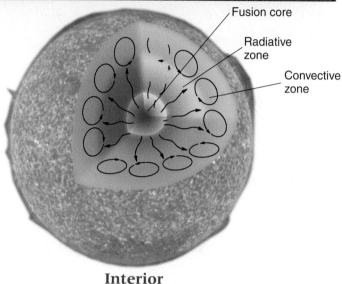

Interior

Sun Data

Sun	Value	Compared with Earth
Diameter	1,392,000 km	109.2
Mass	1.989×10^{30} kg	332,980
Density	1.41 g/cm^3	0.255
Surface Gravity	274 m/s^2	28.0
Escape speed	618 km/s	55.2
Effective temperature	5778 K	
Luminosity	3.85×10^{26} watts	
Tilt of equator to ecliptic	7.25º	
Sidereal rotation period		
Equator	25.05 days	
40º latitude	27.40 days	
80º latitude	33.83 days	

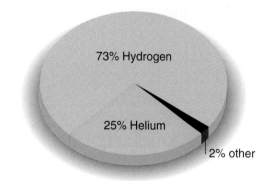

Overall Composition (by mass)

QUICK FACTS

- Typical star, consisting mostly (98% by mass) of hydrogen and helium. Contains 99.85% of all mass in solar system. Gravitational contraction causes nuclear fusion in the core, where four hydrogen nuclei fuse into a helium nucleus (proton–proton chain). Resulting energy transported to surface of Sun by radiation and convection. Complex magnetic field causes eruptions of gas as spicules, flares, and prominences. Solar wind of high-energy particles affects magnetic fields of all planets in the solar system. Sunspot activity follows 11-year cycle of activity.

Mercury

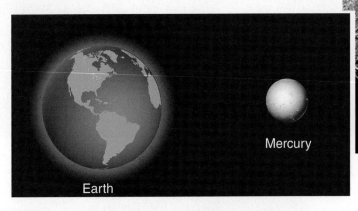

Earth

Mercury

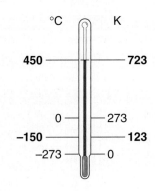

Mantle

Core
(nickel-iron)

Interior

Mercury Data

Mercury	Value	Compared with Earth
Equatorial diameter	4879 km	0.38
Oblateness	0	
Mass	3.302×10^{23} kg	0.055
Density	5.43 g/cm³	0.98
Surface gravity	3.7 m/s²	0.38
Escape speed	4.3 km/s	0.38
Sidereal rotation period	58.65 days	58.79
Solar day	176 days	176
Surface temperature	−150°C to +450°C	
Albedo	0.12	0.39
Tilt of equator to orbital plane	0.01°	
Orbit		
Semimajor axis	5.79×10^{7} km	0.387
Eccentricity	0.2056	12.3
Inclination to ecliptic	7.0°	
Sidereal period	87.97 days	0.24
Moons	0	

°C   K

450 — 723

0 — 273

−150 — 123

−273 — 0

Surface Temperature Extremes

QUICK FACTS

- Among the eight planets Mercury is the smallest; its diameter is one-third of Earth's, it is the closest planet to the Sun, and it has the most eccentric orbit. Greatest difference between maximum and minimum temperatures. Difficult to see from Earth. Almost no atmosphere. Surface cratered somewhat like Moon. Large iron core. Period of revolution 1.5 times period of rotation. Magnetic field not fully explained.

Venus

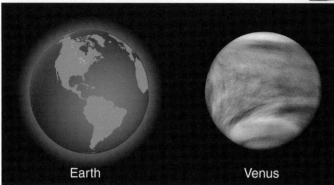

Earth   Venus

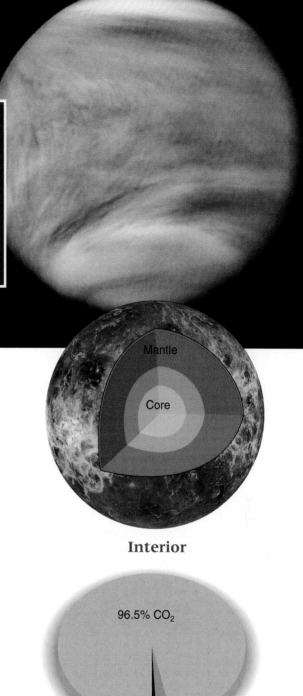

Venus Data

Venus	Value	Compared with Earth
Equatorial diameter	12,104 km	0.95
Oblateness	0	
Mass	4.869×10^{24} kg	0.82
Density	5.24 g/cm³	0.95
Surface gravity	8.87 m/s²	0.91
Escape speed	10.36 km/s	0.93
Sidereal rotation period	243 days	243.7
Solar day	117 days	117
Surface temperature	+464°C	
Albedo	0.75	2.5
Tilt of equator to orbital plane	177.4°	7.6
Orbit		
Semimajor axis	1.082×10^{8} km	0.723
Eccentricity	0.0067	0.40
Inclination to ecliptic	3.39°	
Sidereal period	224.7 days	0.62
Moons	0	

Interior

Mantle

Core

96.5% CO_2

Traces of SO_2, H_2O, others   3.5% N_2

Atmospheric Gases (by volume)

QUICK FACTS

- Visible as morning and evening "star"; Earth's twin in size, mass, and density. Surface is hidden by cloud layer, mapped by radar. Slow retrograde rotation. Dry, hot surface. No large tectonic plates. Surface primarily solidified lava. Surface older than Earth's, younger than Moon's. Dense carbon dioxide atmosphere. Greenhouse effect causes high atmospheric temperature.

The Earth

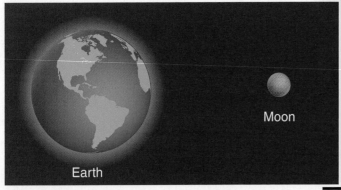

Earth

Moon

Earth Data

Earth	Value
Equatorial diameter	12,756 km
Oblateness*	0.00335
Mass	5.97×10^{24} kg
Density	5.52 g/cm³
Surface gravity	9.78 m/s²
Escape speed	11.19 km/s
Sidereal rotation period	23.9345 hours
Solar day	24.0 hours
Surface Temperature	10°C to 20°C
Albedo**	0.306
Tilt of equator to orbital plane	23.45°
Orbit	
Semimajor axis	1.496×10^{8} km
Eccentricity	0.0167
Sidereal period	365.26 days
Moons	1

*Recall that oblateness tells us how "flattened" an object is. The Earth's polar diameter is 12,714 km.

**The albedo of a solar system object is the fraction of incident sunlight that the object reflects without absorption.

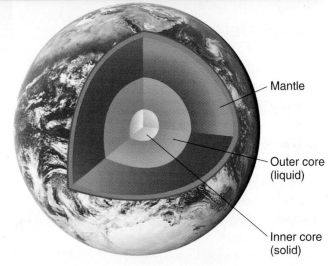

Mantle

Outer core (liquid)

Inner core (solid)

Interior

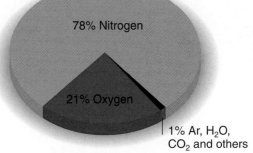

78% Nitrogen

21% Oxygen

1% Ar, H_2O, CO_2 and others

Atmospheric Gases (by volume)

QUICK FACTS

- Third planet from the Sun. Very circular orbit. Except for Pluto, has largest satellite compared with its size. Only planet with liquid water on surface. Active interior.

The Moon

Earth

Moon

Moon Data

Moon	Value	Compared with Earth
Equatorial diameter	3476 km	0.27
Oblateness	0.0012	0.36
Mass	7.35×10^{22} kg	0.0123
Density	3.35 g/cm³	0.61
Surface gravity	1.62 m/s²	0.166
Escape speed	2.4 km/s	0.21
Revolution period	27.322 days	
Sidereal rotation period	27.322 days	
Synodic period (phases)	29.531 days	
Surface temperature	−170° C to 130° C	
Albedo	0.11	0.36
Tilt of equator to orbital plane	6.68°	
Orbit		
Average distance from Earth	384,400 km (center-to-center)	
Closest distance	363,300 km	
Farthest distance	405,500 km	

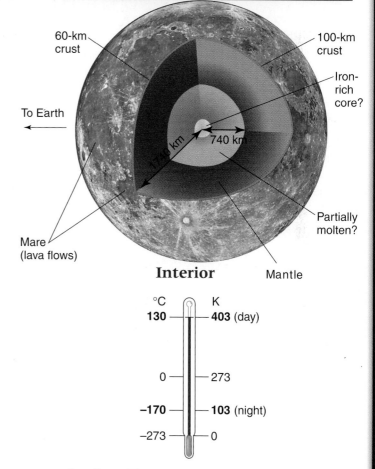

60-km crust

100-km crust

To Earth

Iron-rich core?

740 km

Mare (lava flows)

Partially molten?

Interior

Mantle

°C		K	
130		**403**	(day)
0		273	
−170		**103**	(night)
−273		0	

Surface Temperature Extremes

QUICK FACTS

- One fourth of Earth's diameter. Surface craters were caused by meteorite impacts. Maria were caused by lava flow. Extremely tenuous and transient atmosphere. Weak magnetic field. Likely to have formed as a result of a collision between Earth and a large object early in Earth's history.

Mars

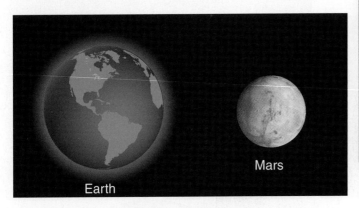

Earth

Mars

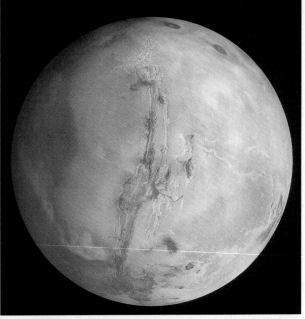

Mars Data

Mars	Value	Compared with Earth
Equatorial diameter	6794 km	0.53
Oblateness	0.0065	1.93
Mass	6.419×10^{23} kg	0.11
Density	3.93 g/cm^3	0.71
Surface gravity	3.69 m/s^2	0.377
Escape speed	5.03 km/s	0.45
Sidereal rotation period	24.623 hours	1.03
Solar day	24.66 hours	1.03
Surface temperature	$-140°C$ to $+20°C$	
Albedo	0.25	0.82
Tilt of equator to orbital plane	25.19°	1.07
Orbit		
Semimajor axis	2.28×10^8 km	1.524
Eccentricity	0.0935	5.6
Inclination to ecliptic	1.85°	
Sidereal period	1.881 years	1.881
Moons	2	

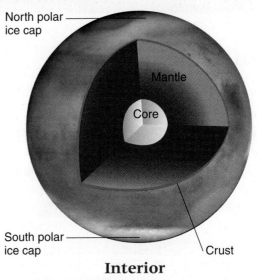

North polar ice cap

Mantle

Core

South polar ice cap

Crust

Interior

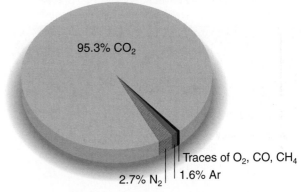

95.3% CO_2

Traces of O_2, CO, CH_4

1.6% Ar

2.7% N_2

Atmospheric Gases (by volume)

QUICK FACTS

- Known as the red planet. Half Earth's diameter. Least dense of terrestrial planets. Rotation rate and equatorial tilt very similar to Earth's. Has seasons. Water–carbon dioxide polar caps. Has largest mountain (Olympus Mons) and canyon (Valles Marineris) known. Low-pressure carbon dioxide atmosphere. Evidence of warmer, wetter past.

Jupiter

Earth   Jupiter

Jupiter Data

Jupiter	Value	Compared with Earth
Equatorial diameter*	142,984 km	11.2
Oblateness	0.065	19.36
Mass	1.899×10^{27} kg	317.8
Density	1.326 g/cm³	0.24
Surface gravity*	23.12 m/s²	2.36
Escape speed	59.5 km/s	5.32
Sidereal rotation period	9.925 hours	0.415
Albedo	0.34	1.12
Tilt of equator to orbital plane	3.13°	0.13
Orbit		
Semimajor axis	7.79×10^8 km	5.204
Eccentricity	0.0489	2.93
Inclination to ecliptic	1.3°	
Sidereal period	11.86 years	11.86
Moons	63	

* Measured where the pressure is one atmosphere, Earth's pressure at sea level.

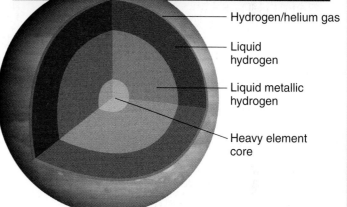

Interior

Hydrogen/helium gas

Liquid hydrogen

Liquid metallic hydrogen

Heavy element core

89.8% Hydrogen

10.2% Helium

H_2O, NH_3, and CH_4

Atmospheric Gases (by volume)

QUICK FACTS

- The largest planet. Contains more than twice the mass of all other planets combined. Fast rotation. Atmosphere shows bands and the Great Red Spot and is mostly hydrogen and helium. Giant magnetic field because of liquid metallic hydrogen that makes up much of the planet. Emits more energy than it receives. Has a large family of moons and a faint ring system.

Saturn

Earth   Saturn

Saturn Data

Saturn	Value	Compared with Earth
Equatorial diameter*	120,536 km	9.45
Oblateness	0.098	29.24
Mass	5.685×10^{26} kg	95.16
Density	0.69 g/cm^3	0.125
Surface gravity*	8.96 m/s^2	0.916
Escape speed	35.5 km/s	3.2
Sidereal rotation period	10.656 hours	0.445
Albedo	0.34	1.12
Tilt of equator to orbital plane	26.73°	1.14
Orbit		
Semimajor axis	1.434×10^9 km	9.58
Eccentricity	0.0565	3.38
Inclination to ecliptic	2.49°	
Sidereal period	29.46 years	29.46
Moons	61	

* Measured where the pressure is one atmosphere, Earth's pressure at sea level.

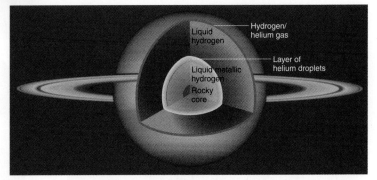

Interior

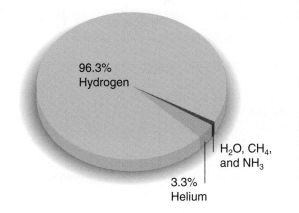

96.3% Hydrogen

H$_2$O, CH$_4$, and NH$_3$

3.3% Helium

Atmospheric Gases (by volume)

QUICK FACTS

- The ringed planet. Diameter is 84% of Jupiter's, but density is only half. Primarily hydrogen and helium, like Jupiter. Magnetic field 1/20 as strong as Jupiter's. Emits more energy than it receives—not fully explained. Large family of moons. Rings have detailed features and, seen from Earth, change greatly during its year.

Uranus

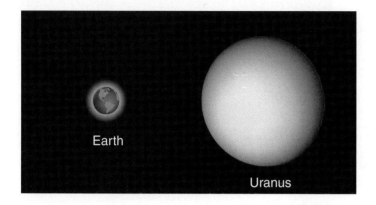

Earth

Uranus

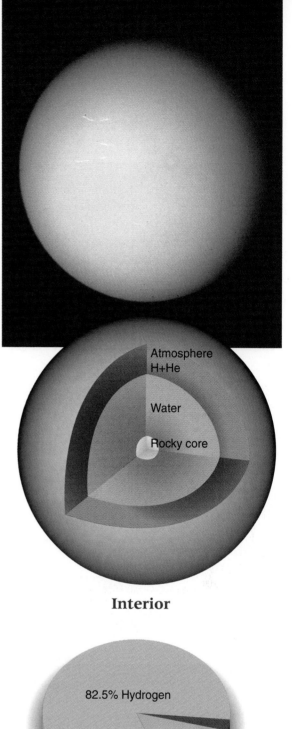

Atmosphere H+He

Water

Rocky core

Interior

Uranus Data

Uranus	Value	Compared with Earth
Equatorial diameter	51,118 km	4.0
Oblateness	0.023	6.84
Mass	8.683×10^{25} kg	14.5
Density	1.27 g/cm^3	0.23
Surface gravity*	8.69 m/s^2	0.89
Escape speed	21.3 km/s	1.9
Sidereal rotation period	−17.24 hours	0.72
Albedo	0.30	0.98
Tilt of equator to orbital plane	97.77°	4.17
Orbit		
Semimajor axis	2.872×10^9 km	19.20
Eccentricity	0.0457	2.7
Inclination to ecliptic	0.77°	
Sidereal period	84.01 years	84.01
Moons	27	

* Measured where the pressure is one atmosphere, Earth's pressure at sea level.

82.5% Hydrogen

15.2% Helium

H$_2$O, CH$_4$, and NH$_3$

Atmospheric Gases (by volume)

QUICK FACTS

- Four times Earth's diameter, but only one third of Jupiter's. Composition like other Jovian planets. Almost featureless atmosphere. Far greater tilt of axis than any other planet. Retrograde rotation. Uniform temperature. Many small moons, no large one.

Neptune

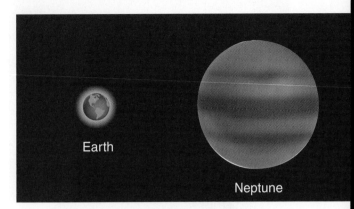

Earth

Neptune

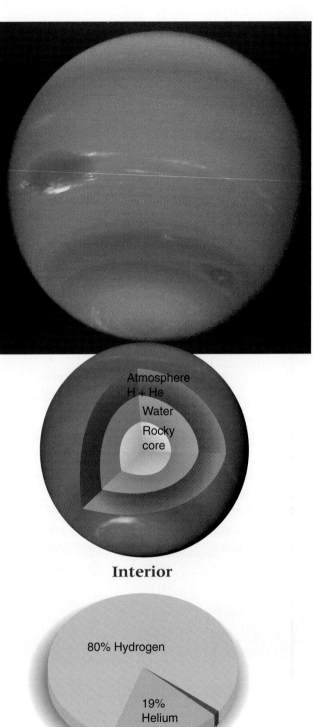

Neptune Data

Neptune	Value	Compared with Earth
Equatorial diameter	49,528 km	3.9
Oblateness	0.017	5.1
Mass	1.02×10^{26} kg	17.1
Density	1.64 g/cm³	0.30
Surface gravity*	11.0	1.12
Escape speed	23.5 km/s	2.1
Sidereal rotation period	16.11 hours	0.67
Albedo	0.29	0.95
Tilt of equator to orbital plane	28.3°	1.2
Orbit		
Semimajor axis	4.495×10^9 km	30.05
Eccentricity	0.0113	0.68
Inclination to ecliptic	1.77°	
Sidereal period	164.79 years	164.79
Moons	13	

* Measured where the pressure is one atmosphere, Earth's pressure at sea level.

Atmosphere
H + He

Water

Rocky core

Interior

80% Hydrogen

19% Helium

H_2O and CH_4

Atmospheric Gases (by volume)

QUICK FACTS

- Blue surface because of methane. Jupiter-like atmospheric features, including Great Dark Spot. Composition is thought to be similar to other Jovian planets. Unexplained internal heat source. Extreme differential rotation. Largest moon, Triton, has retrograde revolution. The other major moon, Nereid, has the most eccentric orbit of any moon in the solar system. Lumpy rings.

Pluto

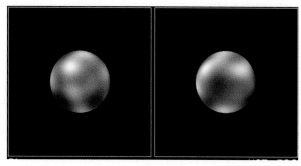

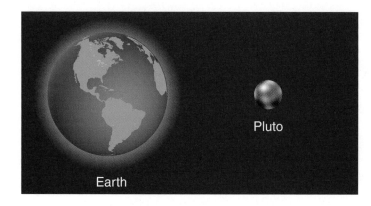

Pluto

Earth

Pluto Data

Pluto	Value	Compared with Earth
Equatorial diameter	2300–2340 km	0.2
Mass	1.3×10^{22} kg	0.002
Density	2.00 g/cm³	0.32
Surface gravity	0.58 m/s²	0.06
Escape speed	1.1 km/s	0.1
Sidereal rotation period	−6.39 days	6.4
Surface temperature	−223°C	
Albedo	0.15	0.47
Tilt of equator to orbital plane	122.5°	
Orbit		
Semimajor axis	5.870×10^{9} km	39.24
Eccentricity	0.244	14.6
Inclination to ecliptic	17.2°	
Sidereal period	247.7 years	247.7
Moons	3	

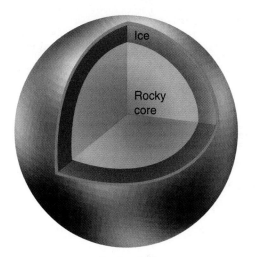

Ice

Rocky core

Interior

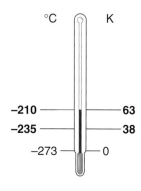

°C   K

−210 ——— 63
−235 ——— 38
−273 ——— 0

Surface Temperature Extremes

QUICK FACTS

- A dwarf planet; its eccentric orbit loops inside Neptune's orbit. Compared to the eight planets, Pluto's orbit has by far the greatest angle to the ecliptic. Receives 1/1600 the intensity of sunlight that Earth receives. Occultations by its moon Charon provided a method of determining its size and mass.

APPENDICES

Appendix D

Planetary Satellites

The Moon	Value	Compared With Earth
Equatorial diameter	3476 km	0.27
Mass	7.35×10^{22} kg	0.0123
Density	3.35 g/cm^3	0.61
Surface gravity	1.62 m/s^2	0.166
Escape speed	2.4 km/s	
Revolution period	27.322 days	0.213
Sidereal rotation period	27.322 days	
Synodic period (phases)	29.531 days	
Surface temperature	$-170°$C to 130°C	
Albedo	0.11	0.36
Tilt of equator to orbital plane	6.68°	
Orbit		
Average distance from Earth	384,400 km	
Closest distance	363,000 km	
Farthest distance	405,500 km	
Eccentricity	0.055	

Satellite Data				
Planet Satellite	Average Distance (1000 km)	Sidereal (Orbital) Period (days)	Diameter (km)	Mass (10^{18} kg)
Earth				
Moon	384.4	27.322	3476	73,490
Mars				
Phobos	9.38	0.319	$27 \times 22 \times 18$	0.011
Deimos	23.46	1.262	$15 \times 12 \times 10$	0.002
Jupiter				
Metis	128.1	0.295	40	0.095
Adrastea	128.9	0.298	$26 \times 20 \times 16$	0.019
Amalthea	181.4	0.498	$262 \times 146 \times 134$	7.5
Thebe	221.9	0.675	110×90	0.78
Io	421.6	1.769	3643	89,320
Europa	670.9	3.551	3122	48,000

Satellite Data *(Continued)*

Planet Satellite	Average Distance (1000 km)	Sidereal (Orbital) Period (days)	Diameter (km)	Mass (10^{18} kg)
Jupiter *(continued)*				
Ganymede	1,070	7.155	5262	148,190
Callisto	1,883	16.689	4821	107,590
Leda	11,170	240.9	10	0.0057
Himalia	11,460	250.6	170	9.5
Lysithea	11,720	259.2	24	0.078
Elara	11,740	259.7	80	0.78
Ananke	21,280	629.8 R*	20	0.038
Carme	23,400	734.2 R	30	0.095
Pasiphae	23,620	743.6 R	36	0.30
Sinope	23,940	758.9 R	28	0.078
Saturn				
Pan	133.58	0.575	20	0.003
Atlas	137.67	0.602	$37 \times 34 \times 27$	0.01
Prometheus	139.35	0.613	$148 \times 100 \times 68$	0.33
Pandora	141.70	0.629	$110 \times 88 \times 62$	0.20
Epimetheus	151.42	0.694	$138 \times 110 \times 110$	0.54
Janus	151.47	0.695	$194 \times 190 \times 154$	1.9
Mimas	185.52	0.942	$418 \times 392 \times 382$	37.5
Enceladus	238.02	1.370	$512 \times 494 \times 490$	65
Tethys	294.66	1.888	$1072 \times 1056 \times 1052$	627
Telesto	294.66	1.888	$30 \times 25 \times 15$	0.007
Calypso	294.66	1.888	$30 \times 16 \times 16$	0.004
Dione	377.40	2.737	1120	1100
Helene	377.40	2.737	$36 \times 32 \times 30$	0.03
Rhea	527.04	4.518	1528	2310
Titan	1,221.83	15.945	5150	134,550
Hyperion	1,481.1	21.277	$370 \times 280 \times 226$	20
Iapetus	3,561.3	79.330	1436	1590
Phoebe	12,952	550.5 R	$230 \times 220 \times 210$	7.2
Uranus				
Cordelia	49.8	0.34	40	
Ophelia	53.8	0.38	42	
Bianca	59.2	0.43	54	
Cressida	61.8	0.46	80	
Desdemona	62.7	0.47	64	
Juliet	64.4	0.49	94	
Portia	66.1	0.51	136	
Rosalind	69.9	0.56	72	
Belinda	75.3	0.62	80	
Puck	86.0	0.76	162	

(Continued)

Satellite Data *(Continued)*

Planet Satellite	Average Distance (1000 km)	Sidereal (Orbital) Period (days)	Diameter (km)	Mass (10^18 kg)
Uranus *(continued)*				
Miranda	129.4	1.413	480 × 468 × 466	66
Ariel	191.0	2.520	1162 × 1156 × 1155	1350
Umbriel	266.3	4.144	1169	1170
Titania	435.9	8.706	1578	3520
Oberon	583.5	13.463	1523	3010
Caliban	7,230	579.5 R	96	
Stephano	8,002	676.5 R	20	
Trinculo	8,571	758 R	20	
Sycorax	12,179	1283 R	190	
Prospero	16,418	1993 R	30	
Setebos	17,459	2202 R	30	
Neptune				
Naiad	48.2	0.29	66	0.2
Thalassa	50.1	0.31	82	0.4
Despina	52.5	0.33	150	2
Galatea	62.0	0.43	176	4
Larissa	73.5	0.55	208 × 178	5
Proteus	117.6	1.12	436 × 416 × 402	50
Triton	354.8	5.877 R	2707	21,400
Nereid	5,513	360.1	340	30
Pluto§				
Charon	19.6	6.387	1186	1900
Nix	50	26.0	140	
Hydra	65	38.6	125	

* The symbol "R" indicates retrograde orbit.
§ Dwarf planet

Appendix E

The Brightest Stars

Star	Popular Name	Apparent Magnitude	Apparent Brightness Compared*	Absolute Magnitude	Absolute Luminosity Compared**	Distance (ly)	Spectral Type
	Sun	−26.72	1.31×10^{10}	4.85	1	0.000015	G2
α CMa A	Sirius	−1.43	1	1.47	22.49	8.58	A1
α Car	Canopus	−0.62	0.474	−5.53	14,191	313	A9
α Boo	Arcturus	−0.05	0.281	−0.31	115.88	36.74	K2
α Cen A	Rigil Kentaurus	0.01	0.265	4.38	1.542	4.36	G2
α Lyr	Vega	0.03	0.261	0.58	51.05	25.32	A0
α Aur	Capella	0.08	0.249	−0.48	135.52	42.2	G5
β Ori A	Rigel	0.14	0.236	−6.62	38,726	733	B8
α CMi A	Procyon	0.38	0.189	2.66	7.516	11.42	F5
α Ori	Betelgeuse	0.45(v)†	0.177	−5.14 (v)	9,908	428	M2
α Eri	Achernar	0.45	0.177	−2.77	1,117	144	B3
β Cen	Hadar	0.61(v)	0.153	−5.42 (v)	12,823	525	B1
α Aql	Altair	0.77	0.132	2.21	11.38	16.79	A7
α Cru A	Acrux	0.77	0.132	−4.19	4,130	321	B1
α Tau A	Aldebaran	0.87(v)	0.120	−0.63 (v)	155.60	65.1	K5
α Vir	Spica	0.98(v)	0.109	−3.55 (v)	2,291	262	B1
α Sco A	Antares	1.06(v)	0.101	−5.28 (v)	11,272	604	M1
β Gem	Pollux	1.16	0.0920	1.09	31.92	33.74	K0
α Ps A	Fomalhaut	1.17	0.0912	1.74	17.54	25.09	A3
α Cyg	Deneb	1.23	0.0863	−7.23	67,920	1600	A2
β Cru	Becrux	1.25(v)	0.0847	−3.92 (v)	3,221	352	B0.5
α Cen B		1.34	0.0780	5.71	0.453	4.36	K0
α Leo A	Regulus	1.36	0.0765	−0.52	140.60	77.5	B7

* Compared with Sirius, the brightest star other than the Sun.
** Compared with the Sun.
† The symbol "(v)" indicates a variable star.

Appendix F

The Nearest Stars

Star	Apparent Magnitude	Apparent Brightness Compared*	Absolute Magnitude	Absolute Luminosity Compared**	Distance (ly)	Spectral Type
Sun	−26.72	1.31×10^{10}	4.85	1	0.000015	G2
Proxima Centauri	11.09	9.82×10^{-6}	15.53	5.3×10^{-5}	4.22	M5.5
α Centauri A	0.01	0.2655	4.38	1.542	4.36	G2
α Centauri B	1.34	0.0780	5.71	0.453	4.36	K0
Barnard's star	9.53	4.13×10^{-5}	13.22	4.5×10^{-4}	5.96	M4
Wolf 359	13.44	1.13×10^{-6}	16.55	2.1×10^{-5}	7.78	M6
Lalande 21185	7.47	2.75×10^{-4}	10.44	0.0058	8.29	M2
Sirius A	−1.43	1	1.47	22.49	8.58	A1
Sirius B	8.44	1.13×10^{-4}	11.34	0.00254	8.58	White dwarf
Gl 65 A***	12.54	2.58×10^{-6}	15.40	6.0×10^{-5}	8.72	M5.5
Gl 65 B (UV Ceti)	12.99	1.71×10^{-6}	15.85	4.0×10^{-5}	8.72	M6
Ross 154	10.43	1.80×10^{-5}	13.07	5.2×10^{-4}	9.68	M3.5
Ross 248	12.29	3.25×10^{-6}	14.79	1.1×10^{-4}	10.32	M5.5
ε Eridani	3.73	0.0086	6.19	0.291	10.52	K2
Lacaille 9352	7.34	3.10×10^{-4}	9.75	0.011	10.74	M1.5
Ross 128	11.13	9.46×10^{-6}	13.51	3.4×10^{-4}	10.91	M4
Gl 866 A (EZ Aquarii)	13.33	1.25×10^{-6}	15.64	4.8×10^{-5}	11.26	M5
Gl 866 B	13.27	1.32×10^{-6}	15.58	5.1×10^{-5}	11.26	
Gl 866 C	14.03	6.55×10^{-7}	16.34	2.5×10^{-5}	11.26	
Procyon A	0.38	0.1888	2.66	7.516	11.40	F5
Procyon B	10.70	1.41×10^{-5}	12.98	5.6×10^{-4}	11.40	White dwarf
61 Cygni A	5.21	0.0022	7.49	0.0879	11.40	K5
61 Cygni B	6.03	0.0010	8.31	0.0413	11.40	K7
Gl 725 A	8.90	7.4×10^{-5}	11.16	0.00299	11.52	M3
Gl 725 B	9.69	3.6×10^{-5}	11.95	0.00145	11.52	M3.5
Gl 15 A	8.08	1.6×10^{-4}	10.32	0.00649	11.62	M1.5
Gl 15 B	11.06	1.0×10^{-5}	13.30	4.2×10^{-4}	11.62	M3.5
ε Indi	4.69	0.0036	6.89	0.153	11.82	K5
DX Cancri	14.78	3.3×10^{-7}	16.98	1.4×10^{-5}	11.82	M6.5
τ Ceti	3.49	0.0108	5.68	0.466	11.88	G8

* Compared with Sirius, the brightest star other than the Sun.
** Compared with the Sun.
*** From Gliese and Jahreiss, *Catalog of Nearby Stars* (1991).

Appendix G

The Constellations

Name	Genitive	Abbreviation	Approximate Position		English Meaning
			Right Ascension (hours)	Declination (degrees)	
Andromeda	Andromedae	And	01	+40	Andromeda*
Antlia	Antliae	Ant	10	−35	Air pump
Apus	Apodis	Aps	16	−75	Bird of paradise
Aquarius	Aquarii	Aqr	23	−15	Water bearer
Aquila	Aquilae	Aql	20	+05	Eagle
Ara	Arae	Ara	17	−55	Altar
Aries	Arietis	Ari	03	+20	Ram
Auriga	Aurigae	Aur	06	+40	Charioteer
Bootes	Bootis	Boo	15	+30	Herdsman
Caelum	Caeli	Cae	05	−40	Chisel
Camelopardus	Camelopardis	Cam	06	−70	Giraffe
Cancer	Cancri	Cnc	09	+20	Crab
Canes Venatici	Canum Venaticorum	CVn	13	+40	Hunting dogs
Canis Major	Canis Majoris	Cma	07	−20	Big dog
Canis Minor	Canis Minoris	Cmi	08	+05	Little dog
Capricornus	Capricorni	Cap	21	−20	Sea goat
Carina	Carinae	Car	09	−60	Keel of ship
Cassiopeia	Cassiopeiae	Cas	01	+60	Cassiopeia* (Queen of Ethiopia)
Centaurus	Centauri	Cen	13	−50	Centaur*
Cepheus	Cephei	Cep	22	+70	Cepheus* (King of Ethiopia)
Cetus	Ceti	Cet	02	−10	Whale
Chamaeleon	Chamaeleonis	Cha	11	−80	Chameleon
Circinis	Cirini	Cir	15	−60	Compass
Columba	Columbae	Col	06	−35	Dove
Coma Berenices	Comae Berenices	Com	13	+20	Berenice's hair*
Corona Australis	Coronae Australis	CrA	19	−40	Southern crown
Corona Borealis	Coronae Borealis	CrB	16	+30	Northern crown
Corvus	Corvi	Crv	12	−20	Crow
Crater	Crateris	Crt	11	−15	Cup
Crux	Crucis	Cru	12	−60	Southern cross
Cygnus	Cygni	Cyg	21	+40	Swan (or Northern cross)

(Continued)

APPENDICES

| Name | Genitive | Abbreviation | Approximate Position | | English Meaning |
			Right Ascension (hours)	Declination (degrees)	
Delphinus	Delphini	Del	21	+10	Dolphin or Porpoise
Dorado	Doradus	Dor	05	−65	Swordfish
Draco	Draconis	Dra	17	+65	Dragon
Equuleus	Equulei	Equ	21	+10	Little horse
Eridanus	Eridani	Eri	03	−20	River Eridanus*
Fornax	Fornacis	For	03	−30	Furnace
Gemini	Geminorum	Gem	07	+20	Twins
Grus	Gruis	Gru	22	−45	Crane
Hercules	Herculis	Her	17	+30	Hercules*
Horologium	Horologii	Hor	03	−60	Clock
Hydra	Hydrae	Hya	10	−20	Hydra* (water monster)
Hydrus	Hydri	Hyi	02	−75	Sea serpent
Indus	Indi	Ind	21	−55	Indian
Lacerta	Lacertae	Lac	22	+45	Lizard
Leo	Leonis	Leo	11	+15	Lion
Leo Minor	Leonis Minoris	Lmi	10	+35	Little lion
Lepus	Leporis	Lep	06	−20	Hare
Libra	Librae	Lib	15	−15	Scales of justice
Lupus	Lupi	Lup	15	−45	Wolf
Lynx	Lincis	Lyn	08	+45	Lynx
Lyra	Lyrae	Lyr	19	+40	Harp or lyre
Mensa	Mensae	Men	05	−80	Table (or Mountain)
Microscopium	Microscopii	Mic	21	−35	Microscope
Monoceros	Monocerotis	Mon	07	−05	Unicorn
Musca	Muscae	Mus	12	−70	Fly
Norma	Normae	Nor	16	−50	Carpenter's level
Octans	Octantis	Oct	22	−85	Octant
Ophiuchus	Ophiuchi	Oph	17	00	Ophiuchus* (serpent bearer)
Orion	Orionis	Ori	05	+05	Orion* (hunter)
Pavo	Pavonis	Pav	20	−65	Peacock
Pegasus	Pegasi	Peg	22	+20	Pegasus* (winged horse)
Perseus	Persei	Per	03	+45	Perseus*
Phoenix	Phoenicis	Phe	01	−50	Phoenix
Pictor	Pictoris	Pic	06	−55	Easel

Name	Genitive	Abbreviation	Approximate Position		English Meaning
			Right Ascension (hours)	Declination (degrees)	
Pisces	Piscium	Psc	01	+15	Fishes
Piscis Austrinus	Piscis Austrini	PsA	22	−30	Southern fish
Puppis	Puppis	Pup	08	−40	Stern of ship
Pyxis	Pyxidis	Pyx	09	−30	Compass of ship
Reticulum	Reticuli	Ret	04	−60	Net
Sagitta	Sagittae	Sge	20	+10	Arrow
Sagittarius	Sagittarii	Sgr	19	−25	Archer
Scorpius	Scorpii	Sco	17	−40	Scorpion
Sculptor	Sculptoris	Scl	00	−30	Sculptor
Scutum	Scuti	Sct	19	−10	Shield
Serpens	Serpentis	Ser	17	00	Serpent
Sextans	Sextantis	Sex	10	00	Sextant
Taurus	Tauri	Tau	04	+15	Bull
Telescopium	Telescopii	Tel	19	−50	Telescope
Triangulum	Trianguli	Tri	02	+30	Triangle
Triangulum Australe	Trianguli Australi	TrA	16	−65	Southern triangle
Tucana	Tucanae	Tuc	00	−65	Toucan
Ursa Major	Ursae Majoris	Uma	11	+50	Big bear
Ursa Minor	Ursae Minoris	Umi	15	+70	Little bear
Vela	Velorum	Vel	09	−50	Sails of ship
Virgo	Virginis	Vir	13	00	Virgin
Volans	Volantis	Vol	08	−70	Flying fish
Vulpecula	Vulpeculae	Vul	20	+25	Fox

* These are proper names.

Appendix H

Answers to Selected Questions, Calculations, and Try One Yourself Exercises

Chapter 1

Try One Yourself (Measuring the Positions of Celestial Objects)

4.5 arcseconds × (1 arcminute/60 arcseconds) = 0.075 arcminutes;

0.075 arcminutes × (1 degree/60 arcminutes) = 0.00125 degrees.

Recall Questions

2. D   **4.** A   **6.** C   **8.** B   **10.** D   **12.** E   **14.** B
16. C   **18.** D   **20.** D

22. Meteor, Moon, Sun, comet, planet, Milky Way (although it is a galaxy), galaxy, nebula.

24. A group of gravitationally bound stars, from billions to perhaps a thousand billion.

26. Stars near Polaris appear to move in circles around the north celestial pole.

28. Summer solstice occurs around June 21, and winter solstice occurs around December 22.

30. Normally, planets move eastward among the stars.

32. Mercury and Venus.

34. The elongation of a celestial object is the angle, viewed from Earth, from the Sun to the object.

36. Umbra is the portion of a shadow that receives no direct light from a light source, whereas penumbra is the portion of a shadow that receives direct light from only part of the light source.

38. The Moon darkens only slightly during penumbral eclipses.

Calculations

2. 1 light-year = (3 × 10^8 meters/second) × (365 days/year) × (24 hours/day) × (3600 seconds/hour) = 9.46 × 10^{15} meters;

(9.46 × 10^{15} meters)/(1.5 × 10^{11} meters/AU) = 6.3 × 10^4 AU.

4. a) 109. b) No. The constellations are not actual objects. c) 733 years. We are seeing the star as it was 733 years ago, not as it is "now."

6. 15 arcseconds = 15 × (1/60) arcminutes = 0.25 arcminutes = 0.25 × (1/60) degrees = 0.0042 degrees.

8. 3 arcminutes = 3 × (1/60) degrees = 0.05 degrees.

10. The umbral shadow of the Moon barely reaches the Earth, and thus, the distance it reaches is about 380,000 km. The Earth's umbral shadow reaches a distance of about 1.4 million km. The ratio between the two is about 3.7, the same as the ratio between the diameters of Earth and Moon.

Chapter 2

Try One Yourself (Kepler's Third Law)

a^3/P^2 = (1 AU)^3/(1 year)^2;
a^3/(0.615 years)^2 = 1 AU^3/year^2;
a^3 = 0.615^2 AU^3;
a = 0.723 AU.

Recall Questions

2. A   **4.** B   **6.** C   **8.** C   **10.** B   **12.** E
14. C   **16.** A   **18.** A   **20.** B   **22.** B   **24.** B

26. The 2nd century AD.

28. An epicycle is the circular orbit of a planet in the geocentric (Ptolemaic) model, the center of which revolves around the Earth in another circle. It was used in the model to explain retrograde motion. Copernicus' assumption that planets move at constant speeds around the Sun forced him to also use epicycles.

30. (a) The theory must fit the data. (b) The theory must make falsifiable predictions. (c) The theory must be aesthetically pleasing.

32. By comparing the directions of the Sun with respect to the vertical at two different cities at the summer solstice, he found that they were off by 7°. By assuming that the Sun was far from Earth, he correctly attributed the difference to the Earth's curvature. Thus, the ratio of 7° to 360° (for a full circle) corresponds to the ratio of the distance between the two cities to the Earth's circumference.

34. The theory held that the Earth rotates on its axis, causing the entire sky to appear to rotate around the Earth.

36. He used epicycles to improve accuracy. They were necessary because planets' motions are not the perfect circles that he assumed.

38. The farther a planet is from the Sun, the slower it moves.

40. Put two tacks in a board and put a loop of string loosely around them. Stretch the string taut with a pencil and sweep the pencil around, always keeping the string taut. To make it less eccentric, put the tacks closer together or use a longer string. Do the opposite for greater eccentricity.

42. Speeds from fastest to slowest: Mercury, Venus, Earth, Mars, Jupiter, Saturn. Distances from nearest to farthest: (the same).

Calculations

2. 4.2 AU and 6.2 AU.

4. 0.387 AU.

6. 3600 quarters.

8. Using the answer to Recall Question 32 we have 10°/360° = 500 km/(2·π·R). Solving for R we find R = 36 · 500/(2·π) = 2865 km.

10. Dividing both sides of the expression by 360° we get: $P_{syn}/P_E = P_{syn}/P_{sid} + 1$. Solving for P_{sid} we find: $P_{sid} = (P_{syn} \cdot P_E)/(P_{syn} - P_E)$. Because $P_E = 1$ year $= 365.256$ days, we have $P_{sid} = P_{syn}/(P_{syn} - 1)$. For Jupiter, $P_{syn} = 398.9$ days $= 1.092$ years, and thus, $P_{sid} = 1.092/(1.092 - 1) = 11.86$ years. The angle θ corresponds to the ratio between Jupiter's synodic and sidereal periods: $\theta = (1.092/11.86) \cdot 360° = 33°$.

Chapter 3

Try One Yourself (The Law of Universal Gravitation)

6400 kilometers + 300 kilometers = 6700 (to get distance from Earth-center).

When the astronaut moves above Earth, only the distance changes. The force of gravity depends inversely on distance squared, so

$(6400/6700)^2 \times 160$ pounds $= 146$ pounds.

When in orbit, an astronaut is "falling" around the Earth along with the Space Shuttle. (The "Travel to the Moon" box and the discussion of Figure B3-4 should help you understand why we can say the astronaut is "falling.")

Try One Yourself (Newton's Laws and Kepler's Laws)

$a^3_{(AU)}/P^2_{(yrs)} = (m_{Moon} + m_{satellite})/m_{Sun}$;
$(4 \times 10^{-5})^3/(1.3 \times 10^{-3})^2 = m_{Moon}/2 \times 10^{30}$ because the mass of the satellite is negligible. Thus,

$m_{Moon} = 7.6 \times 10^{22}$ kilograms.

Recall Questions

2. D	**4.** D	**6.** C	**8.** C	**10.** D	**12.** B
14. A	**16.** A	**18.** B	**20.** A	**22.** A	

24. The phases of Venus.

26. The tendency of an object to retain its speed in a straight line. Countless examples are possible: a hammer resists stopping and thereby drives a nail; a car resists turning and slides sideways; a baseball is hit hard by a bat to change its motion.

28. See pages 75 and 77 for statements of the laws. Possible examples are:

Law 1) A ball rolling on a flat surface continues in a straight line unless a force changes its motion. 2) The more force that a bat exerts on a ball, the greater the acceleration of the ball. 3) An apple hanging from a tree exerts a downward force on the limb and the limb exerts an equal force upward on the apple.

30. The masses of the two objects and the distance between their centers of mass.

32. As a planet gets closer to the Sun, the distance between their centers of mass gets smaller, and thus (from Newton's law of gravity), the gravitational force from the Sun on the planet gets stronger. From Newton's law of motion, as the force acting on an object gets stronger, the object's acceleration increases.

34. The force of gravity from the Sun.

Calculations

2. 1/4 as much: about 63 pounds.

4. First: By Newton's revision of Kepler's third law: The star has two times the Sun's mass.

   Second: B is at 6 AU. Using the same law we get 10.4 years.

6. The Moon's speed on its orbit is: $v = (2 \cdot \pi \cdot r)/$period $= (2 \cdot \pi \cdot 3.84 \cdot 10^8 \text{ m})/(27.32 \cdot 24 \cdot 60 \cdot 60 \text{ seconds}) = 1020$ m/s. The Moon's centripetal acceleration is: $v^2/r = 1020^2/(3.84 \cdot 10^8) = 0.0027$ m/s$^2 = (9.8$ m/s$^2)/60^2$.

Chapter 4

Try One Yourself (Characteristics of Wave Motion)

$v = \lambda \times f$;
344 m/s $= \lambda \times 4000$ cycles/s;
$\lambda = 0.086$ m.

(The resulting units are actually meters/cycle. This is appropriate, however, for one wave is one cycle.)

Try One Yourself (The Doppler Effect as a Measurement Technique)

The difference between the two wavelengths is 0.048 nanometers. Substitute the values into the Doppler equation:

$v = c \cdot (\Delta\lambda/\lambda_0) = (3.0 \times 10^8 \text{ m/s}) \cdot (0.048/656.285) = 2.2 \times 10^4$ m/s.

Because the shifted wavelength is shorter than the unshifted one, the star is moving toward the Sun.

Recall Questions

2. B	**4.** A	**6.** B	**8.** A	**10.** A	**12.** C
14. B	**16.** E	**18.** B	**20.** D	**22.** D	**24.** D

26. The 400-nanometer light. The shorter the wavelength, the higher the frequency.

28. Radio, infrared, visible, ultraviolet, X-rays, gamma rays.

30. The electron is the negatively charged object that orbits the nucleus. The nucleus is the central, massive core of the atom. A photon is the least possible amount of electromagnetic energy of a given wavelength.

32. The amount of energy of a photon is directly proportional to the frequency of the light. In equation form, $E = h \cdot f$, where E is the energy of the photon, h is a constant, and f is the frequency of the light.

34. A continuous spectrum is produced by a hot solid (or a very dense hot gas). An emission spectrum is produced by a hot gas with a density less than that which produces a continuous spectrum. An absorption spectrum is produced when light having a continuous spectrum passes through a gas at a low enough temperature.

36. The Doppler effect is the phenomenon whereby the wavelength (and therefore the frequency) of a wave changes because of relative motion between the wave's source and the observer. The Doppler effect has nothing to do with the intensity of the wave.

Calculations

2. 1.34 meters.

4. Receding at 9140 m/s.

6. X releases 81 times as much energy per unit area as Y.

8. −40 degrees.

10. Its Kelvin temperature is one third as much, and thus, it releases $(1/3)^4$, or 1/81 as much energy per given area. Because it has five times the area, it releases 5/81, or 0.062 as much energy as the Sun.

12. The bulb 30 feet away provides one ninth as much illumination as the one 10 feet away.

Chapter 5

Try One Yourself (Angular Size and Magnifying Power)

1.5 meters is 1500 millimeters;
$M = f_{obj}/f_{eye} = 1500$ mm$/12$ mm $= 125$.

Try One Yourself (Light-Gathering Power)

Light-gathering power depends on the square of the diameter of the objective lens, but first we must express both diameters in the same units. Let's use millimeters, and 10 centimeters = 100 millimeters.

$100^2/5^2 = 10,000/25 = 4000$.

Thus, the small telescope collects 4000 times as much light as the eye.

Try One Yourself (Resolving Power)

We will use the equation for resolving power, but first we must change 600 nanometers to meters, as the telescope diameter is expressed in meters.

600 nm $= 600$ nm $\times (10^{-9}$ m/nm$) = 6.00 \times 10^{-7}$ m;
$\theta = 2.5 \times 10^5 \times \lambda/D = 2.5 \times 10^5 \times (6 \times 10^{-7}/2.4)$
$= 0.0625$ arcseconds.

Recall Questions

2. A   **4.** D   **6.** C   **8.** A   **10.** B   **12.** C
14. E   **16.** B   **18.** A   **20.** C   **22.** A

24. An achromatic lens is actually composed of two lenses. The second lens recombines the colors that have been separated by the first lens. The defect, called chromatic aberration, is the separation of colors that occurs when only a single lens is used.

26. Light-gathering power is the measure of how much light is collected by a lens or mirror. The area of the lens or mirror determines its light-gathering power.

28. Figure 5-12b shows the Newtonian focal arrangement. (The objective mirror is also called the primary mirror.) The image formed by the objective is located where the two rays cross in the drawing.

30. A mirror reflects all wavelengths in the same direction, and thus, chromatic aberration does not occur.

32. Telescopes are placed on mountains so that they are above most of the atmosphere (and NOT so that they are closer to celestial objects).

34. Blurring effects of the atmosphere are eliminated.

Calculations

2. Two eyes collect 0.005 (or 1/200) as much light as the 10-centimeter telescope.

4. Each must have a diameter of 8 meters.

6. Using the 1.3-centimeter observations and the greatest distance, the best resolution is achieved. It is 1.5×10^{-4} arcseconds.

Chapter 6

Try One Yourself (The Small Angle Formula)

$D \approx (A''/206,265) \times d = (0.52 \times 3600/206,265) \times 384,000$ km $= 3485$ km.

The slight difference from the accepted answer is because the angle had been rounded to 0.52 rather than 0.518.

Recall Questions

2. E   **4.** A   **6.** D   **8.** C   **10.** C   **12.** C
14. They are approximately the same, at about 0.5°.

16. The effect depends on the mass of the planet and its distance from Mercury. Jupiter's mass is about 320 times that of Earth's, while Jupiter, even at its closest to Mercury, is only about 8 times farther from Mercury than Earth is. The dependence of the effect on mass and distance is such that Jupiter's contribution is about 1.7 times that of Earth's and 0.54 times that of Venus'.

18. A magnetic field is a region where magnetic forces can be perceived. A magnetic field can be detected by placing a magnet in the suspected location and determining whether there is a magnetic force on the magnet. The magnet used is commonly a small one that is free to turn, that is, a compass.

20. (a) Continental drift has been measured. (b) Rift zones, from which spreading is occurring, have been found. (c) Plate motion explains a number of earthquake zones, mountain ranges, and ocean trenches.

22. Craters were caused by impacts of meteorites. Material thrown out by the impact caused the rays that radiate from some craters.

24. There is a high tide on the area closer to the Moon because that area experiences a stronger gravitational pull from the Moon than the Earth's center. There is a high tide on the area farther from the Moon because that area experiences a weaker gravitational pull from the Moon than the Earth's center; it is as if the Earth's center is "pulled from under it."

Calculations

2. Use the small-angle formula to obtain $d = 1.65 \times 10^8$ km.

4. The expression for the force of gravity between the Sun (m_{Sun}) and the small mass m at point A is $F_A = G \cdot m_{Sun} \cdot m/(d_{ES} - R)^2$. Similarly, $F_B = G \cdot m_{Sun} \cdot m/d_{ES}^2$ and $F_C = G \cdot m_{Sun} \cdot m/(d_{ES} + R)^2$. Therefore, $F_A/F_B = d_{ES}^2/(d_{ES} - R)^2 = (1 + R/(d_{ES} - R))^2 \approx (1 + R/d_{ES})^2 \approx 1 + 2R/d_{ES}$. Similarly, $F_B/F_C = (d_{ES} + R)^2/d_{ES}^2 = (1 + R/d_{ES})^2 \approx 1 + 2 \cdot R / d_{ES}$. Since $2 \cdot R = 12,756$ km and $d_{ES} = 1.5 \cdot 10^8$ km we get $2 \cdot R/d_{ES} \approx 0.0001$.

6. For the case of Io and Jupiter, the diameter of Io is $2 \cdot R_{Io} = 3643$ km, and its distance from Jupiter is $d_{JI} = 421,600$ km. Therefore, $2 \cdot R_{Io}/d_{JI} \approx 0.009$. That's about 100 times stronger than for the case between Earth and Sun.

Chapter 7

Try One Yourself (Measuring Distances in the Solar System)

One half of 4.7 minutes is 141 seconds.

$d = vt = (3 \times 10^5 \text{ km/s}) \times 141 \text{ s} = 4.23 \times 10^7 \text{ km}$;
$(4.23 \times 10^7 \text{ km}) \times (1 \text{ AU}/1.5 \times 10^8 \text{ km}) = 0.282 \text{ AU}$.

Venus is 0.72 AU from the Sun, and $1 - 0.72 = 0.28$, in agreement with our answer.

Try One Yourself (Measuring Mass and Average Density)

$(a^3/P^2) = (G/4\pi^2)m_{\text{Earth}}$;
$(3.844 \times 10^8 \text{ m})^3/(2.36 \times 10^6 \text{ sec})^2 = 1.69 \times 10^{-12} m_{\text{Earth}}$;
$m_{\text{Earth}} = 6.03 \times 10^{24}$ kilograms.

This is close to the value in the appendix, 5.97×10^{24} kilograms. The difference is caused by our assumption that the Moon's mass is negligible. In fact, the value that we calculated is the total mass of the Earth and Moon.

Try One Yourself (Calculating Average Density)

Refer to the solution to the "Try One Yourself" in Chapter 6, where we calculated the diameter of the Moon to be 3485 kilometers.

Radius = 1.74×10^6 meters.

Volume of Moon = $(4\pi/3) R^3 = (4\pi/3) (1.74 \times 10^6 \text{ m})^3 = 2.207 \times 10^{19} \text{ m}^3$.

Density = mass/volume = $(7.35 \times 10^{22} \text{ kg})/(2.207 \times 10^{19} \text{ m}^3)$

= 3330 kg/m³, in good agreement with the value 3350 kg/m³ in Appendix D.

Recall Questions

2. C	**4.** A	**6.** A	**8.** A	**10.** A	**12.** B
14. D	**16.** A	**18.** C	**20.** B	**22.** E	**24.** C

26. The total mass of the two objects.

28. Jupiter is the largest planet, with a diameter 11 times Earth's and a mass 318 times Earth's.

30. The masses of the planets that have moons were calculated using Kepler's third law. Mercury's and Venus' masses were calculated based on their gravitational effect on passing asteroids and comets, and later by their gravitational effects on spacecraft that passed near them.

32. All of the planets revolve around the Sun in the same direction. All except Venus, Uranus, and the dwarf planet Pluto rotate on their axes in the same direction as they revolve.

34. Compared with the Jovians, the terrestrials are nearer the Sun, smaller, less massive, more solid, slower rotating, more dense, and have thinner atmospheres and fewer moons. No terrestrial planet has a ring, whereas all Jovian planets have rings.

36. Earth is just slightly more dense than Mercury and Venus, somewhat more dense than Mars, and much more dense than the Jovian planets and Pluto.

38. The temperature of a gas is related to the speed of its molecules, and at the same temperature, the molecules of a more massive gas move more slowly.

40. Nonvolatile elements condensed in the inner solar system, but volatile elements were swept outward by the solar wind.

42. The asteroids are planetesimals that were prohibited from forming a planet by the effects of Jupiter's gravitational force. The Oort cloud is hypothesized to have resulted when small objects in the outer solar system were thrown outward by gravitational forces when Jupiter or Saturn passed near them.

Calculations

2. 77.2 AU.

4. 9.0×10^8 km, which is 6.0 AU.

6. Mass of Mars: 6.4×10^{23} kilograms. We assumed that the mass of Phobos is negligible compared with that of Mars. Density of Mars: 3.9×10^3 kg/m³, which is 3.9 g/cm³.

Chapter 8

Try One Yourself (Mercury via Mariner—Comparison with the Moon)

Time to cool is proportional to radius (and to diameter), and thus, because the diameter of Jupiter is 11.2 times that of Earth (from Appendix C), it takes Jupiter 11.2 times as long to cool as Earth does.

Recall Questions

2. B	**4.** B	**6.** D	**8.** C	**10.** E	**12.** D
14. C	**16.** B	**18.** D	**20.** B	**22.** D	**24.** A

26. Venus is most like Earth in size; Mars is most like Earth in rotation period.

28. Mercury's rotation period is exactly two thirds of its period of revolution. This has occurred because the mass of the planet is not evenly distributed through its volume. The planet has an eccentric orbit, and the gravitational forces toward the Sun have changed its rotation rate so that resonance occurs between its rotation and revolution.

30. There are two reasons for the extremes in temperature. First, day and night periods are very long (88 Earth days each), providing long periods of time for the surface to heat and cool. Second, there is no atmosphere to block sunlight during the day and to retain heat at night.

32. The greenhouse effect is the phenomenon of infrared radiation being prohibited from escaping through a planet's atmosphere, thereby causing a high temperature on the planet. In a florist's greenhouse an additional effect occurs: The hot air is trapped inside the greenhouse by the walls and roof.

34. The tilt of Mars' axis is very similar to that of Earth's. Mars' equator is tilted 25.2 degrees from its orbital plane, whereas Earth's is tilted 23.5 degrees. (Tables normally show equatorial tilt rather than axis tilt; they are numerically the same.)

36. The polar caps of Mars are made of frozen water and frozen carbon dioxide.

38. The temperature near the equator of Mars varies from −135°C to +30°C—a much greater difference than ever experienced on Earth. The reason for the difference is that Mars' atmosphere is very thin and therefore does not shield sunlight during the day or provide a greenhouse effect to blanket the planet at night.

Calculations

2. Jupiter's volume is 11^3, or 1331, times Earth's volume.

4. The smallest feature that can be observed is about 750 km across. Caloris Basin is 1400 km in diameter, about twice the smallest feature that can be seen.

6. $a^3/P^2 = 1$ when units of AU and years are used.

$1.524^3/a^2 = 1$;

$a = 1.881$ years.

Chapter 9

Recall Questions

2. B	**4.** C	**6.** A	**8.** C	**10.** D	**12.** D
14. B	**16.** B	**18.** C	**20.** B		

22. The band at Jupiter's equator has that rotation rate.

24. The outer atmosphere is primarily hydrogen and helium at a relatively low pressure. As we go deeper, pressure increases until the material would be judged a liquid. About 20,000 kilometers below the cloudtops, liquid metallic hydrogen is found, and it extends down to whatever heavy-element core exists.

26. Saturn's rings are aligned with its equator, which is tilted at about 27 degrees to the planet's orbital plane. Thus, as Saturn orbits the Sun, we on Earth (which is relatively near the Sun) sometimes see the southern side of Saturn and its rings and sometimes the northern side. Midway between these views, the rings are edge-on to our view and are nearly invisible.

28. 1) Knowing the distance to a planet and the planet's angular size, we can use the small-angle formula to calculate the diameter. 2) When a planet passes in front of a star, observation of the amount of time that the planet blocks the star's light, along with knowledge of the planet's speed, gives us another method to calculate the diameter.

30. Uranus' axis tilt (98 degrees) is such that each of its poles nearly points toward the Sun at one time during its revolution period.

32. When a moon changes its distance from its planet, the amount of tidal force exerted on the moon changes. This causes the moon to flex, and this produces heat. The more massive the planet, the greater the effect on its moon, and the closer the moon is to its planet, the greater the effect. Thus, the inner moons of Jupiter that have a noncircular orbit experience tidal heating the most. Io is the moon that shows the most prominent effect of tidal heating.

Calculations

2. Its density would be three fifths, or 0.6, of Earth's.

4. Period = 6 days and the circumference of the spot is about 60,000 km. Thus, the speed is 10,000 km/day, or 400 km/hour, or 0.1 km/s.

6. One percent of its mass is 8.9×10^{20} kilograms. At 1000 kilograms/second, it would take 8.9×10^{17} seconds, or about 28 billion years, much greater than the age of the solar system.

8. The energy per unit time emitted by the planet is $4\pi R^2 \cdot \sigma T^4$. The energy per unit time received by the planet (of cross-section πR^2) from the Sun is $4\pi R^2_{Sun} \cdot \sigma T^4_{Sun} \cdot (\pi R^2 / 4\pi d^2)$. The ratio then is $4 \cdot (T/T_{Sun})^4 \cdot (d/R_{Sun})^2$. For Jupiter this ratio is 1.3, but in reality the ratio is greater because Jupiter does not absorb all the Sun's light.

Chapter 10

Recall Questions

2. E	**4.** C	**6.** D	**8.** D	**10.** D	**12.** C
14. B	**16.** B	**18.** C	**20.** D	**22.** B	**24.** C

26. The most accurate values of Pluto's size come from observations of eclipses of its moon, Charon.

28. Before an asteroid is named, its orbit must have been determined accurately. The orbits of most observed asteroids have not been calculated.

30. Gravitational pull from Jupiter disrupts the orbits of asteroids so that the weaker gravitational forces between them cannot pull them together.

32. The nucleus of a comet is its solid core. Comet nuclei are irregular in shape and are typically a few kilometers across. The coma is made up of gas and dust surrounding the nucleus. It has a very low density and may be as large as 100,000 kilometers in diameter. A comet has two tails, one made up of ions and one of dust. Tails are typically from 10^7 to 10^8 kilometers long.

34. One tail is made up of ions swept away by the solar wind. It is straight because the ions move away from the coma at great speeds. The other tail consists of dust pushed away slowly by radiation pressure. The curvature results from the fact that the material moves away from the coma very slowly compared with the coma's speed.

36. First, the vast majority of long-period comets are in elongated elliptical orbits around the Sun. They must therefore spend most of their time far from the Sun. Second, comets are regularly being captured in the inner solar system. The existence of the cloud explains their origin.

38. A meteor shower results when the Earth encounters a swarm of meteoroids. As the Earth moves into the swarm, the resulting meteors seem to originate in the direction in which the Earth is moving.

Calculations

2. At 2×10^{-9} kilograms/meter^2 per day, it would take about 700,000 years.

4. The answer is simply the ratio of areas, 2.6×10^{-4}, which is about one chance in 4000.

6. Warning time = 10^6 seconds, which is less than 12 days. It would pass through the atmosphere in less than 10 seconds. (The asteroid will speed up as it approaches Earth.)

Chapter 11

Try One Yourself (Solar Energy)

Area of sphere = $4\pi R^2 = 4\pi (2.3 \times 10^{11}$ m$)^2 = 6.64 \times 10^{23}$ m^2.

Energy/area = 3.9×10^{26} watts/6.64×10^{23} m$^2 = 587$ watts/m^2.

Recall Questions

2. C   4. D   6. C   8. A   10. E   12. B
14. B   16. A   18. A   20. D   22. A   24. C

26. The most convincing evidence is that if a chemical process were the source of the Sun's power, the Sun would burn out over a few centuries, and we know that it has released energy fairly uniformly for (at least) many millions of years.

28. The number of protons in nuclei distinguish one element from another.

30. Most of the Sun is made up of gas.

32. The pressure at any depth within the Sun is caused by the weight (and therefore by the mass) of gas above that layer. To calculate the mass of gas above a given point within the Sun, one needs to know the density at levels above that point. This density is calculated from knowledge of pressures and temperatures. (Because the pressures and densities at each layer depend on pressures and densities of layers above, the calculations are best done by a computer.)

34. Photons from the Sun's core are continuously absorbed and reemitted by material within the Sun. Because a photon is absorbed after traveling only about 1 cm and because photons are reemitted in random directions, the radiation travels a very great distance before arriving at the photosphere.

36. First, most of their radiation is in the X-ray portion of the spectrum. Second, they have a very low density, and therefore, there is not much material in them to emit light.

Calculations

2. Volume of Sun = $(4/3) \pi R^3 = (4/3) \pi (6.95 \times 10^8 m)^3 = 1.4 \times 10^{27}$ m³.

   Mass/Volume = 1.99×10^{30} kg/1.4×10^{27} m³ = 1.4×10^3 kg/m³. The density of water is 1.0×10^3 kg/m³.

4. 500 nm. This is green (or yellowish green). It appears yellow because it has passed through the Earth's atmosphere, which scatters more light from the violet end of the spectrum than from the red end.

Chapter 12

Try One Yourself (Apparent Magnitude)

The difference in apparent magnitude between the two objects is 8. Referring to Table 12-1, we see that a difference of 5 corresponds to a ratio of 100, and a difference of 3 corresponds to a ratio of 16; therefore, we receive 1600 times more light from Mars than from Barnard's star.

Try One Yourself (Spectroscopic Parallax)

Using the H-R diagram in Figure 12-17 and the spectral type K1, we find that the absolute magnitude is between about +6.2 and +7.5. Because these numbers are greater than the star's apparent magnitude, we know that the star appears brighter than it actually is (because the greater the number, the dimmer the star). Thus, the star must be closer than 10 parsecs from us.

We can use the equation in the "Calculating Absolute Magnitude" box to get a better answer, as follows:

$m - M = 5 \times \log(d) - 5$ where m = apparent mag., M = absolute mag., and d = distance in parsecs.

Use 6.2 as the absolute magnitude:

$4.4 - 6.2 = 5 \log(d) - 5$;
$\log(d) = 0.64$;
$d = 4.4$ pc.

When we use the same method with the other limit of absolute magnitude, 7.5, we get

$d = 2.4$ pc.

Thus, we conclude that the star is between 2.4 parsecs and 4.4 parsecs from us.

(Actually, it is 3.2 parsecs away, close to the middle of our range.)

Try One Yourself (Luminosity and the Sizes of Stars)

Using solar units, luminosity = radius² × temperature⁴.

$10,000 = r^2 \times (2/3)^4$;
$r^2 = 50,625$;
$r = 225$.

Therefore, the radius of this star is 225 times that of the Sun. (If a star's radius is 225 times the Sun's, its diameter is also 225 times the Sun's.)

Recall Questions

2. C   4. D   6. B   8. D   10. D   12. E
14. C   16. B   18. A   20. A   22. D   24. C

26. We must know the object's distance as well as its angular velocity.

28. Apparent magnitude is a scale of the amount of light received from an object, whereas absolute magnitude is a scale of the amount of light emitted by an object. Luminosity is closely related to absolute magnitude, whereas brightness relates more closely to apparent magnitude.

30. The Doppler effect is used to measure a star's radial velocity.

32. Figure 12-17 is such a sketch.

34. If white dwarfs were as large as most stars, they would be very luminous. Their small size results in their being dim.

36. (1) A visual binary is visible as two stars (normally only with the use of a telescope).

    (2) A spectroscopic binary is a binary that can be detected by periodic Doppler shifts of the two stars. If the spectra of both are visible, each spectral line periodically splits into two lines.

    (3) An eclipsing binary system changes brightness as one star eclipses the other.

    (4) An astrometric binary system is one in which only one star can be seen, and that star is observed to move in a periodic motion that indicates the presence of an orbiting companion.

(5) A composite spectrum binary can be detected because of great differences in the spectra of the two stars.

38. If the plane of revolution of a binary is along a line from Earth, the maximum speed of each component that is observed by the Doppler effect is actually the speed of the star. If the plane of revolution is not along a line from Earth, however, that observed speed is less than the star's actual speed, and unless the angle between the plane of revolution and the line from Earth is known, the actual speed of the star cannot be calculated.

40. From the period of a Cepheid's light variation, its absolute magnitude can be determined. Then, knowing its absolute magnitude and its apparent magnitude, its distance can be calculated.

Calculations

2. They differ in magnitude by 2.5, and thus, we receive between 6.3 (which corresponds to a difference of 2) and 16 (which corresponds to 3) times as much light from Antares than from τ Ceti. (The actual ratio is about 9.9, which we calculate by raising the number 2.5 to the power of the difference in magnitudes—which happens in this case to also be 2.5.)

4. Moving α Centauri to a distance of 10 parsecs would entail moving it more than seven times farther away than it is now, thereby making it much dimmer; therefore, its absolute magnitude must be greater than zero. (Its actual absolute magnitude is 4.4.)

6. It is closer than 10 parsecs. (Actual distance: 5.1 parsecs.)

8. Using Figure 12-34, we can see that it is more than 100 times as luminous as the Sun. Using the equation relating the mass and luminosity of a star on the main sequence we find that $4^{3.5} = 128$, and thus, it is actually 128 times as luminous as the Sun.

10. The time period between points B and C on the diagram is about 0.4 days; therefore, the diameter of the smaller star is (0.4 days) $\times$ (6.9×10^7 km/day) = 27.6 million km.

Chapter 13

Recall Questions

2. C   **4.** A   **6.** D   **8.** D   **10.** B   **12.** C
14. A   **16.** D

18. A protostar is a "star" in the process of formation; it is heated by gravitational energy as its material falls inward.

20. Less massive stars take longer because the lesser mass means that there is less gravitational force pulling each particle toward the center.

22. There is a lower limit because if a star does not have some critical amount of mass there will not be enough pressure at its center for nuclear fusion to be sustained.

More massive stars release more radiation as they join the main sequence, and stars above about 150 solar masses probably emit radiation so intense that no more material can fall onto them.

Calculations

2. Luminosity = radius^2 × temperature^4 when solar values are used for each quantity. (We could also use the Sun's diameter, as diameter is twice the radius.)

$$1000 = r^2 \times (1/6)^4;$$
$$r = 1140.$$

Thus, the Sun at that time had a radius 1140 times greater than today's Sun.

This is 5.3 AU, which means the Sun would have extended to Jupiter's present position.

4. $\lambda = 2{,}900{,}000/T = 2{,}900{,}000/300 = 9700$ nm.

Visible light extends from about 400 to 700 nanometers, and 9700 nanometers is in the infrared region of the spectrum.

Chapter 14

Recall Questions

2. E   **4.** A   **6.** D   **8.** C   **10.** D   **12.** B
14. B   **16.** E   **18.** C   **20.** B   **22.** B   **24.** D
26. D

28. The star's mass.

30. After hydrogen fusion ceases, the core collapses and gravitational energy causes further heating.

32. The Earth will be absorbed in the outer part of the Sun. The Sun will expand so that its photosphere is somewhere between the present orbits of Earth and Mars.

34. (1) Pulsations in the core of a red giant increase in intensity until the outer layers of the star become unstable and are blown away.

(2) The stellar winds blow away the outer portions of the red giant, and two different stages of the wind overlap to cause the glowing shell.

36. Electron degeneracy.

38. Hydrogen is being deposited on a white dwarf from a binary companion. When enough pressure builds up, this hydrogen erupts in a fusion reaction. After it is fused to helium (and some of it is blown away) more hydrogen continues to build up again until fusion starts again.

Calculations

2. The small angle formula yields 1.05 light-years. At 20 km/s, this required 4.95×10^{11} s, or 15,700 years.

4. A magnitude change of 5 corresponds to a factor of 100 in brightness. Thus, a change of 10 corresponds to a factor of 100 × 100, or 10,000. A change of 20 corresponds to a factor of 100^4, or 100 million.

Chapter 15

Recall Questions

2. B   **4.** E   **6.** A   **8.** D   **10.** E   **12.** B
14. D   **16.** A   **18.** B   **20.** C   **22.** D   **24.** A

26. When mass is added to a main sequence star, the star becomes larger. White dwarfs and neutron stars become smaller when mass is added, however. If the size of a black hole is considered to extend to its event horizon, added mass makes a black hole larger.

28. The fact that the Chinese "guest star" was very bright and that today we see an expanding cloud of debris at its location leads us to conclude that it was a supernova.

30. Pulsars vibrate faster than the surface of a white dwarf could vibrate. On the other hand, they vibrate more slowly than neutron stars could vibrate.

32. A black hole is an object that is so dense that its gravitational force is great enough that light cannot escape. We might observe a black hole when material falls toward it, for that material would heat up as it falls and emit radiation. Another way to observe one is by observing the motion of another star (or stars) revolving around the black hole.

34. The escape velocity from inside the event horizon of a black hole is greater than the speed of light.

36. Radiation has been observed that has the characteristics predicted to occur when material falls toward the event horizon of a black hole. Other black holes have been detected by observing a star that is in orbit around an unseen companion, when calculations show that the companion is massive enough that if it were not a black hole it would be emitting light.

Calculations

2. The Schwarzschild radius is proportional to the mass. Thus, if the mass is doubled, the radius is doubled.

4. Section 15-6 indicates that the maximum mass of a neutron star is about 3 solar masses. Using 4 solar masses in Kepler's third law:

$a^3_{(AU)}/P^2_{(yrs)} = (m_1 + m_2)/m_{Sun} = 4\, m_{Sun}/m_{Sun} = 4$.
Using $P = 1$ second $= 3.2 \times 10^{-8}$ years, $a = 1.6 \times 10^{-5}$ AU $= 2400$ km.

6. $R_S = 2 \cdot G \cdot M/c^2$ and thus $R_S = 2 \cdot G \cdot (M/m_{Sun}) \cdot m_{Sun}/c^2 = 2 \cdot (6.67 \cdot 10^{-11}$ N $\cdot$ m^2/kg$^2) \cdot (2 \cdot 10^{30}$ kg$) \cdot (M/m_{Sun})/(3 \cdot 10^8$ m/s$)^2 \approx 3000 \cdot (M/m_{Sun})$ meters $= 3 \cdot (M/m_{Sun})$ kilometers. Thus, R_S (in km) $= 3 \cdot M$ (in solar masses).

Chapter 16

Recall Questions

2. B **4.** A **6.** C **8.** A **10.** E **12.** A
14. C **16.** E **18.** C

20. Figure 16-2 or Figure 1-2a,b is what your drawing should look like. The exact location of the spiral arms is not important, however.

22. Galactic clusters are groups of up to a few dozen stars that share a common origin and are located relatively near one another. Most are found within the Galactic disk. Globular clusters are groups of up to hundreds of thousands of stars. They are found primarily in the halo of the Galaxy. Clusters of galaxies are groups of individual galaxies that are held near one another by gravitational force.

24. The brightest stars, O- and B-type stars, are clustered at certain distances from the Sun. Better evidence is provided by the 21-cm radiation, which indicates that cool hydrogen gas is located in spiral arms.

Calculations

2. 40,000 light-years $= 3.8 \times 10^{17}$ km. The circumference of the path is then 2.4×10^{18} km. Dividing this by 250 km/s, we obtain 9.6×10^{15} seconds, or 3×10^8 years.

4. Diameter of Galaxy $= 160,000$ light-years $= 1.5 \times 10^{18}$ km and the Sun's diameter is 1.5×10^6 km. Setting up a ratio, we find that the scale model would be 100,000 km across!

Chapter 17

Try One Yourself (The Hubble Law)

Figure 17-9 indicates that a distance of 1 million parsecs corresponds to a speed of 470 km/s.

slope $= (470$ km/s $- 0)/1$ Mpc $= 470$ (km/s)/Mpc or about 144 (km/s)/Mly.

Try One Yourself (The Hubble Law Used to Measure Distance)

Using a Hubble constant of 21.1 (km/s)/Mly:
$v = H_0 d$;
90,000 km/s $= 21.1$ (km/s)/Mly $\times d$;
$d = 4265$ Mly.

Using a Hubble constant of 21.9 (km/s)/Mly, we similarly get $d = 4110$ Mly. Therefore, the object is between about 4110 and 4265 Mly away.

Recall Questions

2. E **4.** D **6.** D **8.** B **10.** B **12.** E
14. A **16.** C **18.** C **20.** A **22.** E

24. (a) Using Kepler's third law with measured values of rotation rates of various parts of a galaxy. (b) Application of Kepler's third law to binary pairs of galaxies. (c) Calculations of the mass necessary to prevent a galaxy from leaving the cluster of galaxies in which it is observed.

26. Not enough mass is directly observed in clusters of galaxies to hold the clusters together. Because they do hold together, more mass must exist than we can observe. Because it is not observed, it must be dark matter.

28. The Hubble law states that the velocity of recession of a distant galaxy is proportional to the galaxy's distance. The Hubble constant seems to be between 21.1 and 21.9 (km/s)/Mly, which is between 68.8 and 71.4 (km/s)/Mpc.

30. To use the Hubble law to determine the distance to a galaxy, we measure the recessional velocity of the galaxy and then calculate its distance based on a value of the Hubble constant. Although recessional velocity can be accurately measured, some of that velocity may not be due to Hubble expansion. In addition, the Hubble constant is not known accurately.

32. An active galaxy emits more than normal amounts of radiation from its nucleus.

34. For a quasar with a redshift of 25%, the ratio of the amount of shift in wavelengths of its spectral lines to the unshifted wavelenghts is 0.25. The quasar with a redshift of 30% is moving away faster.

36. If they were nearby and moving at the speeds indicated by their redshifts, they would have had to have been produced recently. It is thought, instead, that quasars do not exist in the universe of today, and we see them only because they are very distant and therefore have a great look-back time.

Calculations

2. The slope for the data in Figure 17-9 is about 470 (km/s)/Mpc. The slope for the data in Figure 17-11 is about 65 (km/s)/Mpc.

4. 4 hours.

6. 8.0 kiloparsecs $\times$ 9.5 kiloparsecs.

8. Density = d. Also, m is mass in solar masses and r is the radius in kilometers; as such, both m and r are dimensionless numbers.

$$d = M/(4\pi R^3/3) = (m \times m_{Sun})/(4\pi r^3 \cdot 1000^3/3)$$
$$= 4.8 \times 10^{20} \ (m/r^3) \ \text{kg/m}^3$$

For $r = 3 \cdot m$, we get $d = 1.8 \times 10^{19}/m^2$.
For $m = 10^9$, $d = 18$ kg/m³.
For $m = 10^8$, $d = 1800$ kg/m³. (The density of water is 1000 kg/m³.)

Chapter 18

Recall Questions

2. C **4.** C **6.** A **8.** B **10.** D **12.** A
14. C **16.** D **18.** E **20.** D

22. Clusters of galaxies are getting farther apart from one another. Theory tells us that space is expanding.

24. Homogeniety: The properties of the universe are the same throughout. On a sufficiently large scale, the universe looks the same everywhere. Isotropy: The universe is the same no matter in what direction we look.

Universality: The same physical laws apply throughout the universe.

26. The Hubble constant must first be determined. The maximum age of the universe is equal to the inverse of the Hubble constant, once million light-years (or million parsecs) have been converted to kilometers.

28. If the universe is an oscillating one, the present expansion will finally stop and compression begin. After compression leads to maximum density, the big bang will repeat, etc.

Calculations

2. $\lambda_{max} = 2,900,000/T$ where λ is in nm and T is in kelvin. Substituting 3 for the temperature, $\lambda_{max} = 9.7 \times 10^5$ nm. Figure 4-3 indicates that this is in the microwave region of the electromagnetic spectrum. 3000 K yields 970 nm, in the near infrared.

Chapter 19

Recall Questions

2. A **4.** E **6.** E **8.** B **10.** E **12.** E **14.** B

16. The longer the average civilization lasts, the more likely a civilization is to be in existence when we are observing. Analogy: Suppose that on the night of the Fourth of July, 1000 sparklers are lit on my street. If a sparkler stays lit only 5 seconds, a person in a low-flying plane may not have a chance to see one as he or she flies over, but if they stay lit 30 minutes, several will be lit at any particular time.

Calculations

2. If 1000 civilizations, $10^{10}/1000 = 10$ million years on the average. If there are a million advanced civilizations, this is 10,000 years.

4. From the Chapter 14 Tools of Astronomy Box "Lifetimes on the Main Sequence,"

$$t = t_{Sun}/M^{2.5} = 10^{10}/1.5^{2.5} = 3.6 \text{ billion years.}$$

Glossary

absolute magnitude   The apparent magnitude a star would have if it were at a distance of 10 parsecs.

absorption spectrum   A spectrum that is continuous except for certain discrete frequencies (or wavelengths).

accelerate   To change the speed and/or direction of motion of an object.

acceleration   A measure of how rapidly the speed and/or direction of motion of an object is changing.

accretion   The process by which an object gradually accumulates matter, usually due to the action of gravity.

accretion disk   A rotating disk of gas orbiting a star, formed by material falling toward the star.

achromatic lens (or achromat)   An optical element that has been corrected so that it is free of chromatic aberration.

active galaxy (active galactic nucleus, AGN)   A galaxy with an unusually luminous nucleus.

active optics   A technology that works by "actively" keeping a telescope's mirror at its optimal shape against environmental factors such as gravity and wind. It works on timescales of a second or more.

adaptive optics   A technique that improves image quality by reducing the effects of astronomical seeing. It relies on an active optics system and works on timescales of less than 0.01 seconds.

albedo   The fraction of incident sunlight that an object reflects.

altitude   The height of a celestial object measured as an angle above the horizon.

angular momentum   An intrinsic property of matter. A measure of the tendency of a rotating or revolving object to continue its motion.

angular separation   The angle between lines originating from the eye of the observer toward two objects.

angular size (of an object)   The angle between two lines drawn from the viewer to opposite sides of the object.

annular eclipse   An eclipse in which the Moon is too far from Earth for its disk to cover that of the Sun completely, and thus, the outer edge of the Sun is seen as a ring.

aphelion/perihelion   The point in its orbit around the Sun where a planet, or other object, is farthest from/closest to the Sun.

apogee/perigee   The point in the orbit of an Earth satellite where it is farthest from/closest to Earth.

Apollo asteroids   Asteroids that cross the Earth's orbit and have semimajor axes larger than Earth's.

apparent magnitude   A measure of the amount of light received from a celestial object.

asteroid   Any of the thousands of minor planets (small, mostly rocky objects) that orbit the Sun.

asteroid belt   The region between Mars and Jupiter where most asteroids orbit.

astrometric binary   An orbiting pair of stars in which the motion of one of the stars reveals the presence of the other.

astrometry   The branch of astronomy that deals with the measurement of the position and motion of celestial objects.

astronomical unit (AU)   A unit of distance equal to the average distance between the Earth and the Sun (about 150 million kilometers or 93 million miles).

astronomical seeing   The blurring and twinkling of the image of an astronomical light source caused by Earth's atmosphere.

astrophysics   Physics applied to extraterrestrial objects.

aurora   Light radiated in the upper atmosphere due to impacts from charged particles.

autumnal and vernal equinoxes   The points on the celestial sphere where the Sun crosses the celestial equator while moving south and north, respectively.

barred spiral galaxy   A spiral galaxy in which the spiral arms come from the ends of a bar through the nucleus, rather than from the nucleus itself.

baryons   Subatomic particles, including the proton, the neutron, and a number of unstable, heavier particles.

belts (and zones)   The dark-colored (and light-colored) bands that mark Jupiter's atmosphere.

big bang   The theoretical initial explosion that began the expansion of the universe.

binary star (system)   A pair of stars that are gravitationally bound so that they orbit one another.

binary system   A system of two objects orbiting each other due to their mutual gravitational attraction.

blackbody   A theoretical object that absorbs and emits all wavelengths of radiation, so that it is a perfect absorber and emitter of radiation. The radiation it emits is called **blackbody radiation.**

black dwarf   The theoretical final state of a star with a main sequence mass less than about 4 solar masses, in which all of its energy sources have been depleted so that it emits no radiation.

black hole   An object whose escape velocity exceeds the speed of light.

blazars (or BL Lac objects)   Especially luminous active galactic nuclei that vary in luminosity by a factor of up to 100 in just a few months.

blueshift   A change in wavelength toward shorter wavelengths.

Bohr atom   The model of the atom proposed by Niels Bohr; it describes electrons in orbit around a central nucleus and explains the absorption and emission of light.

brown dwarf   A star-like object that has insufficient mass to start nuclear reactions in its core and thus become self-luminous.

capture theory   A theory that holds that the Moon was originally solar system debris that was captured by Earth.

carbon (or CNO) cycle   A series of nuclear reactions that results in the fusion of hydrogen into helium, using carbon-12 in the process.

Cassegrain focus   The optical arrangement of a reflecting telescope in which a convex secondary mirror is mounted so as to intercept the light reflected from the objective mirror and reflect the light back through a hole in the center of the primary.

catastrophe theory   A theory of the formation of the solar system that involves an unusual incident, such as the collision of the Sun with another star.

celestial equator   A line on the celestial sphere directly above the Earth's equator.

celestial pole   The point on the celestial sphere directly above a geographic pole of the Earth.

celestial sphere   The imaginary sphere of heavenly objects that seems to center on the observer.

center of mass   The average location of the various masses in a system, weighted according to how far each is from that point.

centripetal force   The force directed toward the center of the curve along which the object is moving.

Cepheid variable   One of a particular class of pulsating stars.

Chandrasekhar limit   The limit to the mass of a white dwarf star, above which it cannot be supported by electron degeneracy and cannot exist as a white dwarf.

charge-coupled device (CCD)   A small semiconductor chip that serves as a light detector by emitting electrons when it is struck by light. A computer uses the pattern of electron emission to form images.

chemical differentiation   The sinking of denser materials toward the center of planets or other objects.

chromatic aberration   The defect of optical systems that results in light of different colors being focused at different places.

chromosphere   The region of the solar atmosphere between the photosphere and the corona.

closed universe   The state of the universe if its total mass and energy density is greater than a specific value, called the critical density.

cocoon nebula   The dust and gas that surround a protostar and block much of its radiation.

coma   The part of a comet's head made up of diffuse gas and dust.

comet   A small object, mostly ice and dust, in orbit around the Sun.

composite spectrum binary   A binary star system with stars having spectra different enough to distinguish them from one another.

conduction   The transfer of energy in a solid by collisions between atoms and/or molecules.

conservation of angular momentum   A law that states that the angular momentum of a system does not change unless there is a net external influence acting on the system, producing a twist around some axis.

constellation   An area of the sky containing a pattern of stars named for a particular object, animal, or person.

continental drift   The gradual motion of the continents relative to one another.

continuous spectrum   A spectrum containing an entire range of wavelengths, rather than separate, discrete wavelengths.

convection   The transfer of energy in a gas or liquid by means of the motion of the material.

core (of the Earth)   The central part of the Earth, consisting of a solid inner core surrounded by a liquid outer core.

corona   The outermost portion of the Sun's atmosphere.

coronal hole   A region in the Sun's corona that has very little luminous gas.

coronal mass ejection   An event in which hot coronal gas is suddenly ejected into space at speeds of hundreds of km/s.

correspondence principle   The idea that predictions of a new theory must agree with the theory it replaces in cases where the previous theory has been found to be correct.

cosmic microwave background radiation (or CMB radiation)   Long-wavelength radiation observed from all directions; thought to be the remnant of radiation from the big bang.

cosmic rays   Energetic charged particles that originate in outer space; they travel at nearly the speed of light and strike Earth from all directions.

cosmological constant   A term (denoted by the Greek capital letter "lambda," Λ) in the equations of general relativity that corresponds to a force throughout all space that helps the universe expand.

cosmological principle   The basic assumption of cosmology that holds that on a large scale the universe is the same everywhere.

cosmological redshift   The shift toward longer wavelengths that is due to the expansion of the universe.

cosmology   The study of the nature and evolution of the universe as a whole.

Coudé focus   The optical arrangement of a reflecting telescope in which two mirrors are used to reflect the light coming from the objective to a remote focal point.

crescent (phase)   The phase of a celestial object when less than half of its sunlit hemisphere is visible.

critical density   The average mass and energy density of the universe at which space would be flat. It is equal to $3H_0^2/(8\pi G)$ and is equivalent to about 5.5 hydrogen atoms per cubic meter of space.

crust (of the Earth)   The thin, outermost layer of the Earth.

dark energy   An exotic form of energy whose negative pressure currently accelerates the expansion of the universe.

dark matter   Matter that can be detected only by its gravitational interactions; it appears to be quite abundant throughout the universe.

dark nebula   An interstellar molecular cloud whose dust blocks light from stars on the other side of it.

declination   A star's angle north or south of the celestial equator.

density   The ratio of an object's mass to its volume.

density parameter   The ratio (denoted by the Greek letter "omega," Ω_0) of the total mass and energy density of the universe to the critical density.

density wave   A wave in which areas of high and low pressure move through the medium.

density wave theory   A model for spiral galaxies that proposes that the arms are the result of density waves sweeping around the galaxy.

deuterium   A hydrogen nucleus that contains one neutron and one proton.

differential rotation   Rotation of an object in which different parts have different periods of rotation.

diffraction   The spreading of light upon passing the edge of an object.

diffraction grating   A device that uses the wave properties of electromagnetic radiation to separate the radiation into its various wavelengths.

disk (of a galaxy)   The flat, dense portion of a spiral galaxy that rotates in a plane around the nucleus.

Doppler effect   The observed change in wavelength of waves from a source moving toward or away from an observer.

double planet (or **co-creation**) **theory**   A theory that holds that the Moon was formed at the same time as the Earth.

dynamo effect   The generation of magnetic fields due to circulating electric charges, such as in an electric generator.

dynamo model   The model that explains the Earth's (and other planets') magnetic fields as due to currents within a molten iron core.

eccentricity of an ellipse (e)   The result obtained by dividing the distance between the foci by the longest distance across an ellipse (the *major axis*).

eclipse season   A time of the year during which a solar or lunar eclipse is possible.

ecliptic   The apparent path of the Sun on the celestial sphere.

electromagnetic spectrum   The entire array of electromagnetic waves.

electron   A negatively charged particle that orbits the nucleus of an atom.

electron degeneracy   The state of a gas in which its electrons are packed as densely as nature permits.

ellipse   A geometrical shape of which every point is the same total distance from two fixed points (the foci).

elliptical galaxy   One of a class of galaxies that have smooth spheroidal shapes.

elongation   The angle in the sky from an object to the Sun.

emission nebula   Interstellar gas that fluoresces due to ultraviolet light from a star near or within the nebula.

emission spectrum   A spectrum made up of discrete frequencies (or wavelengths) rather than a continuous band of frequencies (or wavelengths).

epicycle   The circular orbit of a planet in the Ptolemaic model, the center of which revolves around the Earth in another circle.

escape velocity   The minimum velocity an object must have to escape the gravitational attraction of another object, such as a planet or star.

event horizon   A space-time boundary around a black hole from within which nothing can escape. Its radius is the Schwarzschild radius.

evolutionary track   The path on the H-R diagram taken by a star as its luminosity and color change.

eyepiece   The magnifying lens (or combination of lenses) used to view the image formed by the objective of a telescope.

Faber-Jackson relation   A relation between the luminosity and central stellar velocity dispersion of an *elliptical galaxy*.

fact   Generally a close agreement by competent observers of a series of observations of the same phenomenon.

field of view   The actual angular width of the scene viewed by an optical instrument.

fireball   An extremely bright meteor.

fission theory   A theory that holds that the Moon formed when material was spun off from the Earth.

flat universe   The state of the universe if its total mass and energy density is exactly equal to a specific value, called the critical density.

flatness problem   The inability of the *standard* big bang model to account for the apparent flatness of the universe.

fluorescence   The process of absorbing radiation of one frequency and re-emitting it at a lower frequency.

focal length   The distance from the center of a lens or a mirror to its focal point.

focal point (of a converging lens or mirror)   The point at which light from a very distant object converges after being refracted or reflected.

focus of an ellipse   One of the two fixed points that define an ellipse. (See the definition of *ellipse*.)

frequency   The number of repetitions per unit time.

full (phase)   The phase of a celestial object when the entire sunlit hemisphere is visible.

fusion (nuclear)   The combining of two nuclei to form a different nucleus.

galactic (or **open**) **cluster**   A group of stars that share a common origin and are located relatively close to one another.

galactic rotation curve   A graph of the orbital speed of objects in the galactic disk as a function of their distance from the center.

galaxy   A group of gravitationally bound stars, ranging from a billion to perhaps a thousand billion stars.

Galilean moons   The four natural satellites of Jupiter that were discovered by Galileo. Their names, starting closest to Jupiter, are Io, Europa, Callisto, and Ganymede.

general theory of relativity   A theory developed by Einstein that expands special relativity and presents an alternative way of explaining the phenomenon of gravitation.

geocentric model   A model of the universe with the Earth at its center.

giant star   A star of great luminosity and large size (10 to 100 times the Sun's diameter).

gibbous (phase)   The phase of a celestial object when between half and all of its sunlit hemisphere is visible.

globular cluster   A spherical group of up to hundreds of thousands of stars, found primarily in the halo of the galaxy.

granulation   Division of the Sun's surface into small convection cells.

gravitational lens   The phenomenon in which the gravity due to a massive body between a distant object and the viewer bends light from the distant object and causes it to be seen as two or more objects.

gravitational wave   Ripples in the curvature of space produced by changes in the distribution of matter.

greatest elongation   The configuration of an inferior planet at its greatest angular distance east or west of the Sun.

greenhouse effect   The effect by which infrared radiation is trapped within a planet's atmosphere through the action of particles, such as carbon dioxide molecules, within that atmosphere.

halo (around a galaxy)   The outermost part of a spiral galaxy; fairly spherical in shape, it lies beyond the spiral component.

heliocentric   Centered on the Sun.

helioseismology   The study of the propagation of pressure waves (similar to sound) in the Sun.

helium flash   The runaway helium fusion reactions that occur in the core during the evolution of a red giant.

hertz (abbreviated **Hz**)   The unit of frequency equal to one cycle per second.

Hertzsprung-Russell (H-R) diagram   A plot of absolute magnitude (or luminosity) versus temperature (or spectral type) for stars.

homogeneous   Having uniform properties throughout.

horizon problem   The inability of the *standard* big bang model to account for the directional uniformity of the background radiation.

Hubble constant   The proportionality constant in the Hubble law; the ratio of recessional velocities of galaxies to their distances.

Hubble law   The relationship that states that a galaxy's recessional velocity is directly proportional to its distance.

hydrostatic equilibrium   In a star or a planet, the balance between the downward pressure caused by the weight of material above a thin layer and the upward pressure exerted by material below.

hypothesis   An educated guess made in describing the results of an experiment or observation. A hypothesis can be wildly speculative but must be testable.

image   The visual counterpart of an object, formed by refraction or reflection of light from the object.

inclination (of a planet's orbit)   The angle between the plane of a planet's orbit and the ecliptic plane.

inertia   The tendency of an object to resist a change in its motion.

inferior planet   A planet that is closer to the Sun than the Earth is.

inflationary universe model   A modification of the big bang model that holds that the early universe experienced a brief period of extremely fast expansion.

instability strip   A region of the H-R diagram where pulsating stars are found.

interferometry   A procedure that allows several telescopes to be used as one by taking into account the time at which individual waves from an object strike each telescope.

interstellar cirrus   Faint, diffuse dust clouds found throughout interstellar space.

inverse square law   Any relationship in which some factor decreases as the square of the distance from its source.

ion   An electrically charged atom (or molecule), resulting from its loss or gain of at least one electron.

irregular galaxy   A galaxy of irregular shape that cannot be classified as spiral or elliptical.

isotope   One of two (or more) atoms whose nuclei have the same number of protons but different numbers of neutrons.

isotropy   The property of being the same in all directions.

Kelvin temperature scale   A temperature scale with its zero point at the lowest possible temperature ("absolute zero") and a degree that is the same size (same temperature difference) as the Celsius degree. ($T_K = T_C + 273$.)

kinetic energy   An object's energy due to its motion. For an object of mass m and speed v, its kinetic energy is equal to $(1/2)\ mv^2$.

Kuiper belt   A disk-shaped region beyond Neptune's orbit, 30–1000 AU from the Sun, closer to the solar system than the Oort cloud and presumed to be the source of short-period comets. The main Kuiper belt extends from 30 to 50 AU.

large impact theory   A theory that holds that the Moon formed as the result of an impact between a large object and the Earth.

law or **principle**   When a hypothesis has been repeatedly tested and has not been contradicted, it may become known as a law or principle.

lenticular galaxy   An S0 galaxy, having a flat disk like a spiral galaxy (with little spiral structure) and a large bulge like an elliptical galaxy.

light curve   A graph of the numerical measure of the light received from a star versus time.

light-gathering power   A measure of the amount of light collected by an optical instrument.

lighthouse model   The theory that explains pulsar behavior as being due to a spinning neutron star whose beam of radiation we see as it sweeps by.

light-year   The distance light travels in 1 year.

limb (of the Sun or Moon)   The apparent edge of the object as seen in the sky.

linear polarization   The situation when the electric field in an electromagnetic wave oscillates in a fixed plane.

local group (of galaxies)   The cluster of over 30 or so galaxies that includes the Milky Way Galaxy.

local hypothesis   A proposal stating that quasars are much nearer than a cosmological interpretation of their redshifts would indicate.

look-back time   The time light from a distant object has traveled to reach us.

luminosity   The rate at which electromagnetic energy is emitted.

luminosity class   One of several groups into which stars can be classified according to characteristics of their spectra.

lunar eclipse   An eclipse in which the Moon passes into the shadow of the Earth.

lunar month   The Moon's synodic period, or the time between successive similar phases.

lunar ray   A bright streak on the Moon caused by material ejected from a crater.

magnetic field   A magnetic field exists in a region of space if magnetic forces can be detected in that region.

magnetosphere   The volume of space in which the motion of charged particles is controlled by the magnetic field of the planet, rather than by the solar wind.

magnifying power, or magnification (of an instrument)   The ratio of the angular size of an object when it is seen through the instrument to its angular size when seen with the naked eye.

main sequence   The part of the H-R diagram containing the great majority of stars; it forms a diagonal line across the diagram.

mantle (of the Earth)   The thick, solid layer between the crust and the core of the Earth.

mare (plural maria)   Any of the lowlands of the Moon or Mars that resemble a sea when viewed from Earth.

mass   The quantifiable property of an object that is a measure of its inertia.

mass-luminosity diagram   A plot of the mass versus the luminosity of a number of stars.

meridian   An imaginary line that runs from north to south, passing through the observer's zenith.

meteor   The phenomenon of a streak in the sky caused not only by the burning of a rock or dust particle as it falls but also by the air molecules in the particle's path that, after being excited, then give off light as they de-excite.

meteor shower   The phenomenon of a large group of meteors seeming to come from a particular area of the celestial sphere.

meteorite   An interplanetary chunk of matter that has struck a planet or moon.

meteoroid   An interplanetary chunk of matter smaller than an asteroid.

microlensing   The effect of a temporary brightening of the light from a distant star orbited by an object such as a planet. Light rays from the star bend when they pass through the planet's gravity, which acts as a lens.

micrometeorite   A tiny meteorite.

Milky Way Galaxy   The galaxy of which the Sun is a part. From Earth, it appears as a band of light around the sky (**the Milky Way**).

minute of arc   One sixtieth of a degree of arc.

missing mass   The difference between the mass of clusters of galaxies as calculated from Keplerian motions and the amount of visible mass.

model (scientific)   An idea, a logical framework, that accounts for a set of observations and/or allows us to create explanations of how we think a part of nature works.

molecular cloud   A cold (10–50 K), dense (10^3–10^6 particles/cm^3), massive (from a few to a few million times the mass of the Sun) interstellar cloud; it consists mainly of molecular hydrogen, with traces of carbon monoxide and other molecules.

nanometer (abbreviated nm)   A unit of length equal to 10^{-9} meter.

neap tide   The least difference between high and low tide, occurring when the solar tide partly cancels the lunar tide.

nebula (plural nebulae)   An interstellar region of dust and/or gas.

neutrino   An elementary particle that has little rest mass and no charge but carries energy from a nuclear reaction.

neutron   The nuclear particle with no electric charge.

neutron star   A star that has collapsed to the point at which it is supported by neutron degeneracy.

Newtonian focus   The optical arrangement of a reflecting telescope in which a plane secondary mirror is mounted along the axis of the telescope in order to intercept the light reflected from the objective mirror and reflect it to the side.

nonvolatile element   An element that is gaseous only at a high temperature and condenses to liquid or solid when the temperature decreases.

nova (plural novae)   A star that suddenly and temporarily brightens, thought to be due to new material being deposited on the surface of a white dwarf.

nuclear bulge   The central region of a spiral galaxy.

nucleus (of atom)   The central, massive part of an atom.

nucleus (of comet)   The solid chunk of a comet, located in the head.

objective lens (or objective)   The main light-gathering element—lens or mirror—of a telescope. It is also called the primary lens.

oblate   Flattened at the poles.

oblateness   A measure of the "flatness" of a planet, calculated by dividing the difference between the largest (d_{large}) and smallest (d_{small}) diameter by the largest diameter: oblateness = $(d_{large} - d_{small})/d_{large}$.

Occam's razor   The principle that the best explanation is the one that requires the fewest unverifiable assumptions.

occultation   The passing of one astronomical object in front of another.

Olbers' paradox   An argument showing that the sky in a static universe could not be dark.

Oort cloud   The theorized spherical shell, lying between 10,000 and 100,000 AU from the Sun, containing billions of comet nuclei. The inner Oort cloud lies between 1000 and 10,000 AU.

open universe   The state of the universe if its total mass and energy density is less than a specific value, called the critical density.

opposition   The configuration of a planet when it is opposite the Sun in our sky. That is, the objects are aligned as Sun–Earth–planet.

optical double   Two stars that have small angular separation as seen from Earth but are not gravitationally bound.

oscillating universe model   A big bang theory that holds that the universe goes through repeating cycles of explosion, expansion, and contraction.

P waves   Seismic waves analogous to the waves produced by pushing a spring back and forth.

parallax   The apparent shifting of nearby objects with respect to distant ones as the position of the observer changes.

parallax angle   Half the maximum angle that a star appears to be displaced due to the Earth's motion around the Sun.

parsec   The distance from the Sun to an object that has a parallax angle of one arcsecond.

partial lunar eclipse   An eclipse of the Moon in which only part of the Moon passes through the umbra of the Earth's shadow.

partial solar eclipse   An eclipse in which only part of the Sun's disk is covered by the Moon.

particle density   The number of separate atomic and/or nuclear particles per unit of volume.

penumbra   The portion of a shadow that receives direct light from only part of the light source.

penumbral lunar eclipse   An eclipse of the Moon in which the Moon passes through the Earth's penumbra but not through its umbra.

perigee/apogee   The point in the orbit of an Earth satellite where it is closest to/farthest from Earth.

perihelion/aphelion   The point in its orbit around the Sun where a planet (or other object) is closest to/farthest from the Sun.

phases (of the Moon)   The changing appearance of the Moon during its revolution around the Earth, caused by the relative positions of the Earth, Moon, and Sun.

photometry   The measurement of light intensity from a source, either the total intensity or the intensity at each of various wavelengths.

photon   The smallest possible amount of electromagnetic energy of a particular wavelength.

photosphere   The visible "surface" of the Sun. The part of the solar atmosphere from which mostly visible radiation is emitted into space.

planet   Any of the eight large objects (Mercury to Neptune) that revolve around the Sun, or similar objects that revolve around other stars.

planetary nebula   The shell of gas that is expelled by a low mass red giant near the end of its life.

planetesimal   One of the small objects that formed from the original material of the solar system and from which a planet developed.

plate tectonics   The motion of sections of the Earth's crust (plates) across the underlying mantle.

positron   A positively charged electron emitted from the nucleus in some nuclear reactions.

power   The amount of energy exchanged per unit time.

precession   The conical shifting of the axis of a rotating object.

precession (of an elliptical orbit)   The change in orientation of the major axis of the elliptical path of an object.

pressure   The force per unit of area.

prime focus   The point in a telescope where the light from the objective is focused. This is the focal point of the objective.

principle or **law**   When a hypothesis has been repeatedly tested and has not been contradicted, it may become known as a law or principle.

principle of equivalence   The statement that effects of acceleration are indistinguishable from gravitational effects.

prominence   The eruption of solar material beyond the disk of the Sun.

proper motion   The angular velocity of a star as measured from the Sun.

proton   The positively charged particle in the nucleus of an atom.

proto-planetary nebula   An object in transition between the last stages of a red giant star's life and its planetary nebula phase.

proton–proton (p-p) chain   The series of nuclear reactions that begins with four protons and ends with a helium nucleus.

protostar   An object in the process of becoming a star, before it reaches the main sequence.

Ptolemaic model   The theory of the heavens devised by Claudius Ptolemy.

pulsar   A pulsating radio source with a regular period, between a millisecond and a few seconds, believed to be associated with a rapidly rotating neutron star.

quadrature   The position when a planet is 90° from the Sun in the sky as seen from Earth.

quarter (phase)   The phase of a celestial object when half of its sunlit hemisphere is visible.

quasar (quasi-stellar radio source)   A small, intense celestial source of radiation with a very large redshift.

radial velocity   Velocity along the line of sight, toward or away from the observer.

radiant (of a meteor shower)   The point in the sky from which the meteors of a shower appear to radiate.

radiation   The transfer of energy by electromagnetic waves.

radioactive dating   A procedure that examines the radioactivity of a substance to determine its age.

radio galaxy   A galaxy having its greatest luminosity at radio wavelengths.

redshift   A change in wavelength toward longer wavelengths.

reflection nebula   Interstellar dust that is visible due to reflected light from a nearby star.

refraction   The bending of light as it crosses the boundary between two materials in which it travels at different speeds.

resolving power (or **resolution**)   The smallest angular separation detectable with an instrument. It thus is a measure of an instrument's ability to see detail.

retrograde motion   The east-to-west motion of a planet against the background of stars.

revolution   The orbiting of one object around another.

rift zone   A place where tectonic plates are being pushed apart, normally by molten material being forced up out of the mantle.

right ascension   A star's angle around the celestial equator measuring eastward from the vernal equinox.

Roche limit   The minimum radius at which a satellite (held together by gravitational forces) may orbit without being broken apart by tidal forces.

rotation   The spinning of an object about an axis that passes through it.

RR Lyrae variables   Pulsating stars with periods shorter than a day.

S waves   Seismic waves analogous to the waves produced by shaking a rope, attached to a wall, up and down.

scarp   Transition zone from one series of sedimentary rocks to another of a different age and composition. They are found on Mercury, Earth, Mars, and the Moon.

Schwarzschild radius   The radius of the sphere around a black hole from within which no light can escape.

scientific method   A never-ending cycle of hypothesis, prediction, data gathering, and verification.

scientific model   An idea, a logical framework, that accounts for a set of observations and/or allows us to create explanations of how we think a part of nature works.

second of arc   One sixtieth of a minute of arc.

seeing   The best possible angular resolution that can be achieved.

self-propagating star formation theory   A model for spiral galaxies that explains the arms as resulting from a series of supernovae, each triggering the formation of new stars.

Seyfert galaxy   One of a class of spiral galaxies having active nuclei and spectra containing emission lines.

sidereal day   The amount of time that passes between successive passages of a given star across the meridian.

sidereal period   The amount of time required for one revolution (or rotation) of a celestial object with respect to the distant stars.

singularity   The center of a black hole, where density, gravity, and the curvature of space and time are infinite.

solar day   The amount of time that elapses between successive passages of the Sun across the meridian.

solar eclipse (or **eclipse of the Sun**)   An eclipse in which light from the Sun is blocked by the Moon.

solar flare   An explosion near or at the Sun's surface, seen as an increase in activity such as prominences.

solar system   The system that includes our Sun, planets and their satellites, asteroids, comets, and other objects that orbit our Sun.

solar wind   The flow of charged particles from the Sun.

space velocity   The velocity of a star relative to the Sun.

special theory of relativity   A theory developed by Einstein that predicts the observed behavior of matter due to its speed relative to the person who makes the observation.

spectrometer   An instrument that measures the wavelengths present in electromagnetic radiation. (A **spectrograph** is a spectrometer that produces a photograph of the spectrum.)

spectroscopic binary   An orbiting pair of stars that can be distinguished as two due to the changing Doppler shifts in their spectra.

spectroscopic parallax   The method of measuring the distance to a star by comparing its absolute magnitude with its apparent magnitude.

spectrum   The order of colors or wavelengths produced when light is dispersed.

spicule   A narrow jet of gas that is part of the chromosphere of the Sun and extends upward into the corona.

spiral galaxy   A disk-shaped galaxy with arms in a spiral pattern.

spring tide   The greatest difference between high and low tide, occurring about twice a month, when the lunar and solar tides correspond.

standard solar model   Today's generally accepted theory of solar energy production.

star   A self-luminous celestial object.

stellar parallax   The apparent annual shifting of nearby stars with respect to background stars, measured as the angle of shift.

stellar wind   The flow of particles from a star.

summer and winter solstices   The points on the celestial sphere where the Sun reaches its northernmost and southernmost positions, respectively.

sunspot   A region of the Sun's surface that is temporarily cool and dark compared with the surrounding regions.

supercluster   A group of clusters of galaxies.

supergiant   The evolutionary stage of a massive star after it leaves the main sequence; at this stage, the star has a very great luminosity and size (100 to more than 1000 times the Sun's diameter).

superior planet   A planet that is more distant from the Sun than the Earth is.

superluminal motion   Motion that appears to occur at speeds faster than the speed of light.

supernova   The catastrophic explosion of a star, during which the star becomes billions of times brighter.

synodic period   The time interval between two successive similar alignments between a celestial object and the Sun, as seen from Earth.

T Tauri stars   A class of stars that show rapid and erratic changes in brightness.

tail (of comet)   The gas and/or dust swept away from a comet's head.

tangential velocity   Velocity perpendicular to the line of sight.

theory   A synthesis of a large body of information that encompasses well-tested (by repeatable experiments) and verified hypotheses about certain aspects of the natural world. No theory can be proved to be true, but data can prove a theory to be false.

thermal energy   Energy that is due to the random motions of molecules and atoms of a substance.

tidal force   A gravitational force that varies in strength and/or direction over an object causing it to deform.

tidal friction   Friction forces that result from tides on a rotating object caused by its gravitational interactions with another object.

total lunar eclipse   An eclipse of the Moon in which the Moon is completely in the umbra of the Earth's shadow.

total solar eclipse   An eclipse in which light from the normally visible portion of the Sun (the photosphere) is completely blocked by the Moon.

transit   The passage of a planet in front of its star.

triangulation   The use of parallax to determine the distance to an object.

troposphere   The lowest level of the Earth's (and some other planets') atmosphere.

Tully-Fisher relation   A relation that holds that the wider the 21-cm spectral line, the greater the absolute luminosity of a *spiral galaxy*.

tuning fork diagram   A diagram developed by Edwin Hubble to relate the various types of galaxies.

turnoff point   The point on an H-R diagram of a cluster of stars where the stars are just leaving the main sequence.

twenty-one-centimeter radiation   Radiation from atomic hydrogen, with a wavelength of 21.1 cm.

umbra   The portion of a shadow that receives no direct light from the light source.

universality   The property of obeying the same physical laws throughout the universe.

vernal and autumnal equinoxes   The points on the celestial sphere where the Sun crosses the celestial equator while moving north and south, respectively.

visual binary   An orbiting pair of stars that can be resolved (normally with a telescope) as two separate stars.

volatile (element)   A chemical element that can be vaporized at a relatively low temperature.

watt   A unit of power. It corresponds to a specific amount of energy each second.

wavelength   The distance from a point on a wave, such as the crest, to the next corresponding point, such as the next crest.

weight   The gravitational force between an object and the planetary/stellar body where the object is located.

white dwarf   The burnt-out relic of a low mass star. (Typical diameter is 0.01 that of the Sun—about the size of Earth.)

winter and summer solstices   The points on the celestial sphere where the Sun reaches its southernmost and northernmost positions, respectively.

zenith   The point in the sky located directly overhead.

zero-age main sequence   The main sequence of newly formed stars that just started hydrogen fusion in their cores.

zodiac   The band that lies 9° on either side of the ecliptic on the celestial sphere.

zones (and belts)   The light-colored (and dark-colored) bands that mark Jupiter's atmosphere.

INDEX

Note: Page numbers in **bold** type refer to figures, tables, and activities.

A0620-00, 441
Abbott, Edwin A., 88
Abell 2029, **494**
Abell 2218, **505**
Absolute magnitude, 342–344, **347–348**, 490
Absorption spectra
  discovery of, 105
  Doppler effect and, **117**
  of galaxies, **486**
  interstellar gases and, 371–372
  of stars, 114, **117**, 349
  of Sun, **107**, 114, **115**
Acceleration, mass and, 75–77
Accretion, 381–382
Accretion disks, 410
Achromatic lenses (achromats), 128
Active galactic nuclei (AGNs), 499, 507–511
Active galaxies, 497–507
Active optics, 136–137
Adams, John C., 271
Adaptive optics, 137
Albedo, 215
Albireo, 354, **355**, **388**
Alcor, **10**, 131–**132**, 356, **387**
Aldebaran, **7**, 343
Aldrin, Edwin Jr. (Buzz), 69, 82
Algol, 356–**357**
ALH84001 (meteorite), **297**, 300
Alnitak, **369**
Alpha Centauri, **3**, 118
Alpha particles, 399
Alpha Tauri A and B, 343
Altitude, 15
Aluminum, radioactive, 446
Amino acids, meteorites and, 300
Andromeda galaxy
  images of, **141**, **454**
  local group and, 492
  locating, **388**, 453
  studies of, 458, 478, 481, 484, 517
Angular momentum, 194–195, 426
Angular power spectrum, 538–540
Angular separation, **9**
Angular size, 128, **129**, 154–155
Annular eclipses, **25**
Antennae galaxies, 481, **482**, 483
Anthropic principle, 535
Ant nebula, **407**
Aphelion, 59, 214, 279
Apogee, 156
Apollo asteroids, 284
Apollo missions, 82, 173, 174
Apparent magnitude, 339–341
Aquinas, Thomas, 48–49, 74
Archer cluster, 380
Arcminutes, 9
Arcs, 504–**505**
Arcseconds, 9
Arecibo radio telescope, 139, **140**
Aristarchus' heliocentric model, 45–48
Aristotle
  church support for theories, 48–49
  on earthly laws and heavenly objects, 79
  geocentric model of, 40–42
  ideas on light, 99, 112, 120
Armstrong, Neil A., 69, 82

Arrays, 143, 552, 553
Arroyos, on Mars, 229–230
Assumptions, scientific conclusions and, 489
Asteroid belt, 282, 283
Asteroids
  discovery of, 190
  naming, 285
  orbits of, 283–284
  origin of, 284–285
  as planetesimals, 197
  reclassification of, 180
  size of, 181
  as solar system debris, 281–283
  Titius-Bode law and, 184
Astrology, 46
Astrometric binaries, 357, 407
Astrometry, 144, 201, 342
Astronomical seeing, 133
Astronomical units (AU), 28, 183
Astronomy. *See also* Measurement(s)
  current developments, 29–**31**
  study of, 2–3
  units of distance, 27–28
Astrophysics, 29
Atlantis, **37**
Atmosphere (Earth)
  composition of, 164, 191–192
  effect on light waves, 102–103
  limits on telescopes, 132–133
  refraction in, 126–**127**
Atmospheres
  factors determining, 189–193
  of Jupiter, 247–249
  of Mars, 237–239
  Mercury's negligible, 212–213
  of Saturn, 257, 258–259
  of Sun, 322–327
  of Titan, 261–263
  of Uranus, 266
  of Venus, 219–223, **237**
Atom, Bohr's model of, 107–114
Auroras, **166**–167, 251, 258, **259**, 326, **333**
Autumnal equinox, 15

Bahcall, John, 319, 320
Balloon Observations of Millimetric Extragalactic Radiation and Geophysics (*BOOMERANG*), **515**, 538–540
Balmer, Johann Jacob, 111
Balmer series, 111, 349
Bands of Jupiter, 246–248
Barnard, Edward, 224, 372
Barnard 163, **369**
Barnard's star, 200, **344**
Barred spiral galaxies, 467, 478, **479**, **484**, 498
Barrel-shaped nebulae, **404**
Baryons, 494, 537
Becquerel, Henri, 307
Bell, Jocelyn, 427, 428
Belts of Jupiter, 247–248
Bessel, Friedrich, 407
Beta Centaurus, **3**
Beta Cygni, 105
Beta Pictoris, **199**–200
Beta Regio, **217**
Beta Tauri, 343
Betelgeuse, **7**, 8, **339**, 346, **420**
Bethe, Hans, 311
Biermann, Ludwig, 288

Big bang theory
  early universe description, 528–529
  features of, 524–525, 527, 529–530
  flatness problem, 534–535
  horizon problem, 535–536
  illustration of misconception about, **525**
  Olbers' paradox and, 523
  steady-state theory and, 526
  summary, **536**
  WMAP results on, 540, 542–544
Big Dipper
  ancient recognition of, **8**
  motions of, 6, **8**, 344–**345**
  resolving view of, 131–132
Binary star systems
  black holes and, **441**
  calculating orbits of, 118
  Doppler wobble and, 201–202
  measuring size and mass from, 357–358
  novae and, 410
  planetary systems around, 204
  pulsars in, 432–434, 436–437
  types of, 354–357
  visual wobble and, 200–201
Binary systems, 83–84
Bipolar planetary nebulae, 403, **404**
Blackbody, 105–106
Blackbody radiation, 105–106
Black dwarfs, 409
Black holes. *See also* Quasars
  in active galactic nuclei, 494, 507–511
  detecting, 440–441, 443–444
  formation of, 438–439
  in Milky Way nucleus, 455, 469–470
  properties of, 439–440
  in science, science fiction, and nonsense, 442–443
  in star life cycle, **447**
  very massive star death and, 437
Blazars, 506–507
Blink comparators, 277–**278**
BL Lac objects, 506–507
Blueberries, **232**
Blue light, **101**
Blueshifted light, 116, 375, **463**, **490**, **503**
Bode, Johann, 184, 190
Bohr, Niels
  biography, 108
  ideas on light, 111, 120
  model of atom, 107–114, **311**
Bohr atom, 108–109, **311**
Bondi, Hermann, 526
Boomerang nebula, **407**
BOOMERANG project, **515**, 538–540
Brahe, Tycho, 48, 54–55, 57, 410, 411, 424. *See also* Tycho's model
Bright line spectra, **107**. *See also* Emission spectra
Broad absorption lines (BALs), **508**
Brown dwarfs
  2M1207, **199**
  2M1207a, **391**
  definition of, 199
  GL229B, **391**
  life cycle of, **447**
  origin of, 390–392
  planetary systems around, 204
Bruno, Giordano, 73, 452

PHOTO CREDITS

Table of Contents

Page xi (top) X-ray courtesy of NASA/CXC/GSFC/M.Corcoran et al.; Optical courtesy of NASA/STScI; **page xi** (top middle) © Stocktrek Images, Inc./Alamy Images; **page xi** (bottom middle) © a. v. ley/ShutterStock, Inc.**; page xi** (bottom) Courtesy of G. Fritz Benedict, Andrew Howell, Inger Jorgensen, David Chapell (University of Texas), Jeffery Kenney (Yale University), and Beverly J. Smith (CASA, University of Colorado), and NASA; **page xii** (top) Courtesy of NASA, ESA, and the Hubble Heritage Team (STScI/AURA); **page xii** (bottom) Courtesy of NASA, ESA, M. J. Jee and H. Ford et al. (Johns Hopkins Univ.

Chapter 1

Chapter opener (top) Courtesy of Hubble Space Telescope Comet Team and NASA; **chapter opener (bottom)** Courtesy of STScI/NASA; **1-1** © Fred Espenak/Science Photo Library/Photo Researchers, Inc.; **1-2c** Photo by Dave Palmer; **1-3** Courtesy of T.A.Rector (NRAO/AUI/NSF and NOAO/AURA/NSF) and B.A.Wolpa (NOAO/AURA/NSF); **1-4** © Anglo-Australian Observatory/David Malin Images; **1-5** Courtesy of L.Vanzi, ESO; **1-6** Courtesy of AURA/NOAO/NSF; **1-8a** Courtesy of Akira Fujii/NASA; **1-9** Courtesy of the Adler Planetarium & Astronomy Museum, Chicago, Illinois. Hand-colored engraving by Johann Bayer (1572-1625); **1-15** Courtesy of Don Pettit, ISS Expedition 6 Science Officer/NASA; **1-24b** © J. Gatherum/ShutterStock, Inc.; **1-29** © Frank Zullo/Photo Researchers, Inc.; **1-32** Courtesy of Alex York; **1-33b** Courtesy of NASA/JPL-Caltech; **1-33c** Courtesy of Jacques Descloitres, MODIS Rapid Response Team at NASA GSFC; **1-35a** Courtesy of AURA/NOAO/NSF; **1-35b** Courtesy of Manfred Rudolf/EurAstro; **1-36a-b** Courtesy of NASA/JPL-Caltech; **1-39** Courtesy of Karl Kuhn; **1-41** Courtesy of NASA, ESA, Andrew Fruchter and the ERO team (STScI); **1-43** Courtesy of William P. Sterne, Jr.; **1-45** Courtesy of Karl Kuhn.

Chapter 2

Chapter opener Courtesy of STS-117 Crew/NASA; **2-4** © Photos.com; **2-10** © Photos.com; **2-11** © Photos.com; **2-16** Courtesy of NASA/JPL-Caltech; **B2-1** © Photos.com; **2-17** © Photos.com; **2-18** © Interfoto/age fotostock; **B2-2** Courtesy of National Library of Medicine; **B2-3** © Photos.com.

Chapter 3

Chapter opener Courtesy of JSC/NASA; **3-1** © Paschalis Bartzoudis/Dreamstime.com; **3-2** Courtesy of William P. Sterne, Jr.; **3-3c** Courtesy of NASA, Voyager 2 photo/JPL; **B3-1** © National Library of Medicine; **3-4a-d** Courtesy of Lowell Observatory Photograph; **B3-2** © Photos.com; **B3-5** © AIP Emilio Segrè Visual Archives.

Chapter 4

Chapter opener Courtesy of Astrophysics Data Facility at the NASA Goddard Space Flight Center; **4-1** © Comstock Images/Jupiterimages; **4-5c** Courtesy of Karl Kuhn; **4-6c** Courtesy of Karl Kuhn; **4-8** Courtesy of D. Golimowski, Johns Hopkins University/STScI/NASA; **4-10** © Deutsches Museum, Munich; **B4-4** © AIP Emilio Segré Visual Archive, Margrethe Bohr Collection; **4-16a** Courtesy of Karl Kuhn; **4-16b** Courtesy of Ryan Wilson.

Chapter 5

Chapter opener (center) Courtesy of NASA/CXC/SAO; **chapter opener (right)** Courtesy of UIT Team/NASA; **chapter opener (bottom left)** Courtesy of ESO; **chapter opener (top left)** Image courtesy of NRAO/AUI; **5-2a** Courtesy of Theo Koupelis; **5-4b** © Photodisc; **5-5a** Courtesy of Karl Kuhn; **5-8a-c** Courtesy of NASA; **5-9** Courtesy of Karl Kuhn; **5-10a** Courtesy of Don Pettit, ISS Expedition 6 Science Officer/NASA; **5-10b** Courtesy of George C. Atamian; **5-11** Courtesy of Gemini Observatory and Canada-France-Hawaii Telescope/Coelum/Jean-Charles Cuillandre; **5-13** Courtesy of Yerkes Observatory Photograph; **5-14b** Courtesy of Karl Kuhn; **5-15** Courtesy of California Institute of Technology; **5-17A** Courtesy of ESO - (The) European Organization for Astronomical Research in the Southern Hemisphere; **5-17b** Courtesy KI-Planet Quest/NASA/JPL-Caltech; **5-17c** Courtesy of G. Hüdepohl, ESO; **5-18a** Courtesy of AURA/NOAO/NSF; **5-18b** Courtesy of NASA/JPL-Caltech; **5-18c** Courtesy of Galileo Project/JPL/NASA; **B5-1** © Arizona Board of Regents, all rights reserved; **5-20** Courtesy David Parker, 1997, SPL and the NAIC-Arecibo Observatory; **5-21** Courtesy of NRAO/AUI; **5-23a** Courtesy of NOAO/AURA/NSF; **5-23b** Courtesy of JPL-Caltech/K. Gordon (University of Arizona)/NASA; **5-23c** Courtesy of Max-Planck-Institut fur Radioastronomie, Bonn (R. Beck, E.M. Berkhuijsen and P. Hoernes); **B5-2** Courtesy of NRAO/AUI; **5-26** Courtesy of NRAO/AUI; **5-27** Courtesy of CXC/S. Lee; **B5-3** Infrared, Courtesy of NASA/JPL-Caltech/L. Allen (Harvard-Smithsonian CfA); Visible, Courtesy of CIT, DSS; **B5-4** Courtesy of NASA/JPL-Caltech; **B5-5** Courtesy of ESA/MSSL; **B5-6** Courtesy of ESA, M. Revnivtsev (IKI/MPA); **B5-7** Courtesy of the STS-82 Crew/HST/NASA.

Chapter 11

Chapter Opener Courtesy of SOHO - EIT Consortium/ESA/NASA; **11-1** © AbleStock; **11-3a** Courtesy of William P. Sterne, Jr.; **11-3b** Courtesy of Louis Strous, SOHO - MDI Consortium, ESA, NASA; **B11-1a** Courtesy of Brookhaven National Laboratory; **B11-1b** Courtesy of Kamioka Observatory, ICRR (Institute for Cosmic Ray Research), The University of Tokyo; **B11-1c** Courtesy of the Sudbury Neutrino Observatory; **11-14A** Courtesy of AURA/NOAO/NSF; **11-14b** Courtesy of NASA/JPL-Caltech; **11-14c** Courtesy of NASA/ESA; **11-15** Courtesy Marshall Spaceflight Center/NASA; **11-16a** Courtesy of NASA/JPL-Caltech; **11-18a** Courtesy of Dr. Göran Scharmer, The Institute for Solar Physics, Stockholm University; **11-19a** Courtesy of UCAR/NCAR/High Altitude Observatory/NASA; **11-19b** Courtesy of NASA/JPL-Caltech; **11-20** This image, taken at the Swedish 1-meter Solar Telescope (SST) on the island of La Palma, Spain, is courtesy of Dr. Bart De Pontieu of the Lockheed Martin Solar & Astrophysics Laboratory; **11-21** Courtesy of NASA/JPL-Caltech; **11-22** 2009 Copyright, University Corporation for Atmospheric Research, High Altitude Observatory; **11-23a** Courtesy of SOHO (ESA & NASA); **11-23b** Courtesy of CFA/TRACE Team/NASA; **11-23c** 2009 Copyright, University Corporation for Atmospheric Research, High Altitude Observatory; **11-24a** Courtesy of NASA/JPL-Caltech; **11-24b** Courtesy of Marshall Space Flight Center/NASA; **11-25A** Courtesy of NASA/NSSTC/Hathaway 2009/04; **11-25b** Courtesy of NASA/NSSTC/Hathaway 2006/07; **11-26** Courtesy of NASA/NSSTC/Hathaway 2009/04; **11-28a-b** Courtesy of AURA/NOAO/NSF; **11-29** Courtesy of ESA/NASA/Office of Space Science/SOHO; **11-30a** Courtesy of Marshall Space Flight Center/NASA; **11-30b** Courtesy of SOHO/CDS, SOHO/EIT (ESA & NASA); TRACE (NASA); **11-31** Courtesy of SOHO - EIT Consortium/ESA/NASA; **11-32** The SOHO/LASCO data used here are produced by a consortium of the Naval Research Laboratory (USA), Max-Planck-Institut fuer Aeronomie (Germany), Laboratoire d'Astronomie (France), and the University of Birmingham (UK). SOHO is a project of international cooperation between ESA and NASA; **11-33a** Courtesy of Yohkoh/JAXA; **11-33b** Courtesy of SOHO (ESA & NASA); **11-34** Courtesy of SOHO/LASCO (ESA & NASA).

Chapter 12

Chapter Opener Courtesy of NOAO; **12-1** © Peresanz/Dreamstime.com; **12-8a-b** Courtesy of Yerkes Observatory Photograph; **12-12** Courtesy of Chad Trujillo; **12-13** Courtesy of Harvard College Observatory; **12-14** Courtesy of KPNO 0.9-m Telescope, AURA, NSF/NOAO; **12-24a** Courtesy of NASA/JPL-Caltech; **12-25** Courtesy of Sproul Observatory/Swarthmore College; **12-26** Image © UC Regents/Lick Observatory;

12-36 Courtesy of NOAO/AURA/NSF; **B12-2** Courtesy of Harvard College Observatory; **12-39** Courtesy of Dr. Wendy L. Freedman, Observatories of the Carnegie Institution of Washington and NASA.

Chapter 13

Chapter Opener Courtesy of Wolfgang Brandner (JPL/IPAC), Eva K. Grebel (Univ. Washington), You-Hua Chu (Univ. Illinois Urbana-Champaign), and NASA; **13-1a** Courtesy of T.A. Rector/University of Alaska Anchorage, H. Schweiker/WIYN and NOAO/AURA/NSF; **13-1b** Courtesy of ESO; **13-1c** Courtesy of NPS, Night Sky Program; **13-1d** Courtesy of ESO; **13-2a** Courtesy of Infrared Processing and Analysis Center, NASA/JPL-Caltech; **13-2b** Courtesy of NASA/JPL-Caltech; **13-5a** Courtesy of ESO; **13-5b** Courtesy of C.R. O'Dell, and S.K. Wong/Rice University/NASA; **13-5c** Courtesy of ESO; **13-7** Courtesy of NASA, ESA and AURA/Caltech; **13-8b** Courtesy of NASA/JPL-Caltech/J. Rho (SSC/Caltech); **13-9** Courtesy of Hubble Heritage Team (STScI) and NASA; **13-10a** Courtesy of Bill Shoening/AURA/NOAO/NSF; **13-10b** Courtesy of J.Hester and P. Scowen (Arizona State University), NASA; **13-10c** Courtesy of ESA & the ISOGAL team; **13-10d** Courtesy of ESO; **13-12a** Courtesy of ESO; **13-12b** Courtesy of NASA/JPL-Caltech/L. Allen (Harvard-Smithsonian CFA); **13-14a** Courtesy of NASA/JPL-Caltech/T. Megeath (University of Toledo) & M. Robberto (STScI); **13-14B** © Anglo-Australian Observatory/David Malin Images; **13-15a-c** Courtesy of J. Morse,AURA/STScI, NASA; **13-17a** Courtesy of Charles Robert O'Dell/Rice University; **13-17b** Courtesy of Charles Robert O'Dell/Rice University/NASA; **13-18a** Courtesy of NASA/JPL-Caltech/J. Bally (University of Colorado); **13-18b** Courtesy of NASA and the Hubble Heritage Team (STScI/AURA); **13-19** © Matthew Bate/Photo Researchers, Inc.; **13-20a** Courtesy of Don Figer/STScI/NASA; **13-20b** Courtesy of Andy Steere; **13-21** Courtesy of Hubble Heritage Team (AURA/STScI/NASA); **13-24a** Courtesy of NRAO/AUI; **13-24b** Courtesy of 2MASS/UMass/IPAC-Caltech/NASA/NSF; **13-24c** Courtesy of Palomar Observatory, Caltech/DSS; **13-24d** Courtesy of NASA/CXC/PSU/L.Townsley et al.

Chapter 14

Chapter Opener Courtesy of NASA, ESA, HEIC, and the Hubble Heritage Team (STScI/AURA); **14-1b** Courtesy of S. Kulkarni (Caltech), D. Golimowski (JHU), NASA; **14-1c** Courtesy of Paranal Observatory, ESO; **14-16** © Anglo-Australian Observatory/David Malin Images; **14-17a** Courtesy of Hubble Heritage Team (AURA/STScI/NASA); **14-17b** Courtesy of NASA/ESA/JPL-Caltech/J. Hora (CfA) & C.R. O'Dell (Vanderbilt); **DP14-1a** Courtesy of Robert Rubin (NASA Research Center), Reginald Dufour and Matt Browning (Rice University), Patrick Harrington (University of Maryland), and NASA; **DP14-1b** Courtesy of A. Hajian (USNO) et. al., Hubble Heritage Team (STScI/AURA), NASA; **DP14-3a** Courtesy of B. Balick (U. Washington) et al., WFPC2, NASA, HST, NASA; **DP14-4a** Courtesy of B. Balick (.U. Washington), et.al, WFPC2, HST, NASA; **14-19a** Courtesy of ESO; **14-19b** Courtesy of H. Bond/STScI/NASA; **14-20a** Courtesy B. Balick (U. Washington) et al./WFPC2/HST/NASA; **14-20b** Courtesy of R. Sahai and J. Trauger (JPL), the WFPC2 Science Team, and NASA; **14-21a** Courtesy of NASA, ESA, Hans Van Winckel (Catholic University of Leuven, Belgium), and Martin Cohen (University of California, Berkeley); **14-21b** Courtesy of ESA & Valentin Bujarrabal (Observatorio Astronomico Nacional); **14-22** Courtesy of NASA, ESA & A. Zijlstra (UMIST, Manchester, UK); **14-23a** Courtesy of ESA/NASA; **14-23b** Courtesy of NASA, ESA and the Hubble Heritage Team (STScI/AURA); **14-25a** Image © UC Regents/Lick Observatory; **14-25b** Kitt Peak National Observatory 0.9-meter telescope, National Optical Astronomy Observatories; Courtesy M. Bolte (University of California, Santa Cruz) and Harvey Richer (University of British Columbia, Vancouver, Canada) and NASA; **14-29a-b** Image © UC Regents/Lick Observatory; **14-31** X-ray: NASA/CXC/Rutgers/J.Hughes et al.; Optical: Rutgers Fabry-Perot; **14-32** © Royal Observatory, Edinburgh/AATB/SPL/Photo Researchers, Inc.

Chapter 15

Chapter Opener © Anglo-Australian Observatory/David Malin Images; **15-2** Courtesy of A. Dupree (CfA), R. Gilliland (STScI), NASA; **15-5a-b** © Jack Newton, www.jacknewton.com; **15-6** Courtesy of ESO; **15-7a** Courtesy of T. Rector/University of Alaska Anchorage and WIYN/NOAO/AURA/NSF; **15-7b** Courtesy of NASA/CXC/SAO; **15-8a** © Anglo-Australian Observatory/David Malin Images; **15-8b** Courtesy of J. Pun (NASA/GSFC), R. Krishner (CFA), and NASA; **15-8c** X-ray: Courtesy of NASA/CXC/PSU/S.Park & D.Burrows.; Optical: Courtesy of NASA/STScI/CfA/P.Challis; **15-9** X-ray: Courtesy of NASA/CXC/

GSFC/M.Corcoran et al.; Optical: Courtesy of NASA/STScI; **15-10a** Image courtesy of University of Cambridge, Mullard Radio Astronomy Observatory. With compliments of Professor Antony Hewish; **15-10b** Courtesy of ESO; **15-14a** Courtesy of NASA/CXC/SAO; **15-14b** Courtesy of NASA/CXC/HST/ASU/J. Hester et al); **15-14c** Courtesy of N.A. Sharp/NOAO/AURA/NSF; **15-15** Courtesy of ESA & Francesco Ferraro (Bologna Astronomical Observatory); **15-16a** Courtesy of NASA/PSU/G. Pavlov et al.; **15-16b** Courtesy of NASA/CXC/J. Hughes et al.; **15-21** © Stocktrek Images, Inc./Alamy Images; **15-23** Courtesy of NASA/CXC/M.Weiss; **B15-3** © ESA 2002.Illustration by Medialab; **B15-4** Courtesy of NASA/CXC/M.Weiss.

Chapter 16

Chapter Opener (right and left) Courtesy of NASA Goddard Space Flight Center; **16-1** © a. v. ley/ShutterStock, Inc.; **16-2** Courtesy of Robert Hurt (SSC/JPL/Caltech)/NASA; **16-3** Courtesy of Bill Keel, University of Alabama; **16-4** Courtesy of AURA/NOAO/NSF; **16-6** Courtesy of Yerkes Observatory Photograph; **16-7a** Courtesy of N.A.Sharp, REU program/NOAO/AURA/NSF; **16-7b** Courtesy of N.A.Sharp, Mark Hanna, REU program/NOAO/AURA/NSF; **16-8a** Courtesy of NASA and The Hubble Heritage Team (STScI/AURA); **16-8b** Courtesy of Rubina Kotak and Henri Boffin, ESO; **16-15b** Courtesy of NASA Goddard Space Flight Center; **16-17** Courtesy of FORS Team, 8.2-meter VLT, ESO; **16-21a** Courtesy of DSS/NASA/JPL-Caltech; **16-21b** Courtesy of NASA/JPL-Caltech/Robert C. Kennicutt, Jr. (Steward Observatory, Univ. of Arizona); **16-22** Courtesy of D. Figer (STScI) and NASA; **B16-1** Photo by Kim Zussman, Thousand Oaks, CA; **16-23a** Produced at the Naval Research Laboratory by Dr. N.E. Kassim and collaborators from data obtained with the National Radio Astronomy's Very Large Telescope, a facility of the National Science Foundation operated under cooperative agreement with Associated Universities, Inc. Basic research in radio astronomy at the Naval Research Laboratory is supported by the Office of Naval Research; **16-23b** Courtesy of NRAO; **16-23c** Reproduced by permission of the American Astronomical Society; 195th AAS Meeting, #05.01; *Bulletin of the American Astronomical Society*; McGary, R. S. and Ho, P. T. P.; Infalling Gas onto the Circumnuclear Disk; Vol. 31, p.1372. Photo courtesy of Doctor Paul. T. P. Ho; **16-24a-b** Courtesy of ESO; **16-25** EGRET All-Sky Gamma Ray Survey Avve 100 MeV; Courtesy of EGRET, NASA, GSFC; **16-26a** Courtesy of NASA/CXC/MIT/F.K. Baganoff et al.; **16-26b** Courtesy of CXC/MIT/F.K. Baganoff et al./NASA; **16-29** Image composite by Ingrid Kallick of Possible Designs, Madison Wisconsin. The background Milky Way image is a drawing made at Lund Observatory. High-velocity clouds are from the survey done at Dwingeloo Observatory (Hulsbosch & Wakker, 1988). Image courtesy of Dr. Bart Wakker; **16-30** Courtesy of Rensselaer Polytechnic Institute & the Sloan Digital Sky Survey (SDSS), www.sdss.org.

Chapter 17

Chapter Opener Courtesy of NASA, ESA, S. Beckwith (STScI) and the HUDF Team; **17-1a** Courtesy of ESO; **17-1b** Courtesy of NOAO/AURA/NSF; **17-1c** Courtesy of Todd Boroson/NOAO/AURA/NSF; **17-2a** Courtesy of G. Fritz Benedict, Andrew Howell, Inger Jorgensen, David Chapell (University of Texas), Jeffery Kenney (Yale University), and Beverly J. Smith (CASA, University of Colorado), and NASA; **17-2b** Courtesy of NOAO/AURA/NSF; **17-2c** Courtesy of ESO; **17-3** Courtesy of NASA, ESA, and The Hubble Heritage Team (STScI/AURA); **17-4a-c** Courtesy of NOAO/AURA/NSF; **B17-1** Courtesy of California Institute of Technology; **17-5a** Courtesy of AURA/NOAO/NSF; **17-5b** Courtesy of NASA and the Hubble Heritage Team (STScI); **17-6a** Courtesy of Brad Whitmore (STScI) and NASA; **17-6b** Courtesy of NASA/ESA/CXC/JPL-Caltech; **17-6c** Courtesy of NASA/ESA/and The Hubble Heritage Team (STScI); **17-6d** Courtesy of NASA/JPL-Caltech/P. N. Appleton (Spitzer Science Center/Caltech); **17-6e** Courtesy of Hubble Heritage Team (AURA/STScI/NASA); **17-10** Courtesy of California Institute of Technology; **17-16a** Courtesy of NASA, ESA, and the Hubble Heritage Team (STScI/AURA); **17-16b** Courtesy of P. van Dokkum (University of Groningen), ESA, and NASA; **17-17** Colless et al. (2001), *MNRAS*, 328, 1039. Image courtesy of the 2dF Galaxy Redshift Survey team and Professor Matthew Colless, Director of Anglo-Australian Observatory; **17-18a** Courtesy of NASA/CXC/E.O'Sullivan et al; Optical: Palomar DSS; **17-18b** X-ray courtesy of NASA/CXC/UCI/A.Lewis et al.; Optical courtesy of Pal.Obs. DSS; **17-19a** Courtesy of NASA, ESA, and R. Massey (California Institute of Technology); **17-19b** Courtesy of NASA, ESA, M. J. Jee and H. Ford et al. (Johns Hopkins Univ.); **B17-2** X-ray: Courtesy of NASA/CXC/U. Copenhagen/K.Pedersen et al; Optical: Courtesy of Palomar DSS; **17-20a** Courtesy of NRAO/AUI; **17-20b** Courtesy of

NASA/UMD/A. Wilson et al.; **17-20c** Courtesy of NASA/CXC; **17-21a** X-ray (NASA/CXC/M. Karovska et al.); Radio 21-cm image (NRAO/VLA/J. Condon et al.); Optical (Digitized Sky Survey U.K. Schmidt Image/STScI); **17-21b** Courtesy of NASA/JPL-Caltech/J. Keene (SSC/Caltech); **17-22a** Courtesy of NOAO/AURA/NSF; **17-22b** Courtesy of NASA/Wide-Field Planetary Camera 2/Hubble Space Telescope/STScI; **17-22c** Courtesy of NASA, NRAO, and J. Biretta (STScI); **17-23a** WFPC2 image: NASA and J. Bahcall (IAS); ACS image: NASA, A. Martel (JHU), H. Ford (JHU), M. Clampin (STScI), G. Hartig (STScI), G. Illingworth (UCO/Lick Observatory), the ACS Science Team and ESA; **17-23b** Optical courtesy of NASA/STScI; X-ray: NASA/CXC; Radio courtesy MERLIN; **17-25** Courtesy of NASA Education and Public Outreach at Sonoma State University/Aurore Simonnet; **17-27** Courtesy of NASA, Andrew S. Wilson (University of Maryland), Patrick L. Shopbell (Caltech), Chris Simpson (Subaru Telescope), Thaisa Storchi-Bergmann and F.K.B. Barbosa (UFRGS, Brazil), and Martin J. Ward (University of Leicester, UK); **17-29a** Courtesy of NRAO/AUI; **17-29b** Courtesy of K. Ratnatunga (JHU), NASA; **17-29c** Courtesy of W.Couch (University of New South Wales), R. Ellis (Cambridge University), and NASA; **17-29d** Courtesy of W.N. Colley (Princeton University), E. Turner (Princeton University), J.A. Tyson (AT&T Bell Labs), and NASA; **17-30c** Reproduced from Smith R.J., Croom S.M., et al., *MNRAS*, 359 (2005), 57-72. Image courtesy of Dr. Robert Smith, Astrophysics Research Institute; **17-31b** Courtesy of John Biretta, STScI; **17-32b** Inset photo: John Bahcall, Institute for Advanced Study and NASA; **17-33a** Courtesy of NRAO/AUI; **17-33b** Courtesy of NRAO/AUI and C. O'Dea & F. Owen; **17-33c** Courtesy of Walter Jaffe/Leiden Observatory, Holland Ford/JHU/STScI, and NASA; **17-34a** Courtesy of Karl Gebhardt (University of Michigan), Tod Lauer (NOAO), and NASA; **17-34b** Courtesy of John Bahcall (Institute for Advanced Study, Princeton), Mike Disney (University of Wales) and NASA; **17-35a** Courtesy of ESA; **17-35b** Courtesy of Dana Berry/NASA; **17-37** Courtesy of Achut Reddy/Adam Block/NOAO/AURA/NSF.

Chapter 18
Chapter Opener Courtesy of BOOMERANG Collaboration; **18-7b** © Valenta/ShutterStock, Inc.; **18-8** © AbleStock; **18-10** Courtesy of AT&T Archives and History Center; **18-11** Courtesy of NASA/JPL-Caltech; **B18-2** Courtesy of NASA/WMAP; **18-17** Courtesy of NASA/WMAP; **18-18** Courtesy of Seldner, M., Siebers, B., Groth, E.J. and Peebles, P.J.E, 1977; **18-19** Courtesy of NASA Goddard Space Flight Center and the COBE Science Working Group; **18-20** Courtesy of NASA/NSBF; **18-22** Courtesy of NASA/WMAP; **18-23** Courtesy of NASA and the WMAP Science Team. Modified from M.R. Nolta et al., *ApJS* 180 (2009): 296-305. Reproduced by permission of the AAS; **18-25** Courtesy of NASA/WMAP; **18-26** Courtesy of NASA and the WMAP Science Team. Modified from M.R. Nolta et al., *ApJS* 180 (2009): 296-305. Reproduced by permission of the AAS; **18-27** Courtesy of NASA/WMAP; **18-29** Courtesy of BOOMERANG Collaboration; **18-30** Courtesy of ESO.

Chapter 19
Chapter Opener Courtesy of NASA; **19-1** The Arecibo Observatory is part of the National Astronomy and Ionosphere Center, which is operated by Cornell University under a cooperative agreement with the National Science Foundation; **19-2** Courtesy of Pioneer Project, ARC, and NASA; **19-3** Courtesy of Pioneer Project, ARC, and NASA.

Appendix Data Pages
Sun Courtesy of Skylab/NASA; **Mercury** Courtesy of NASA/JPL-Caltech; **Venus** Courtesy of NASA/JPL-Caltech; **Earth** Courtesy of NSSDC/NASA; **Moon** Courtesy of NASA/JPL-Caltech; **Mars** Courtesy of NASA/JPL-Caltech; **Jupiter** Courtesy of NASA/JPL-Caltech; **Saturn** Courtesy of STScI; **Uranus** Courtesy of NASA/JPL-Caltech; **Neptune** Courtesy of NASA/JPL-Caltech; **Pluto** Courtesy of Hubble Space Telescope/NASA

Unless otherwise indicated, all photographs and illustrations are under copyright of Jones and Bartlett Publishers, LLC, or have been provided by the author(s).